Interactive Forecasting

UNIVARIATE and MULTIVARIATE METHODS
Second Edition

Spyros Makridakis
Professor of Management Science
INSEAD, Fontainebleau, France

Steven C. Wheelwright
Associate Professor of Business Administration
Harvard Business School, Boston, Mass.

HOLDEN-DAY, Inc. · San Francisco
Düsseldorf · Johannesburg · London
Panama · Singapore · Sidney

INTERACTIVE FORECASTING

Univariate and Multivariate Methods Second Edition.

Copyright © 1978 by Holden-Day, Inc.
500 Sansome Street, San Francisco, California 94111

Library of Congress Catalog Card Number: 76-27396

Printed in the United States of America

4567890

to Peniko and Matt

Foreword

If Makridakis and Wheelwright had not come on the scene when they did, they would have had to be invented. They have produced the first forecasting package that is truly interactive, comprehensive and, above all, practical.

My own work in developing forecasting techniques began in earnest in 1960, after the computer came into commercial use. Prior to that time, the arithmetic needed to implement sophisticated techniques was beyond our reach, because the size of the clerical staff required was simply not acceptable. The computer made such techniques feasible.

In the early days of the computer, we focused our efforts on adapting to company use the seasonal adjustment programs of the Census Bureau. In the corporate world, this went a long way toward improving our knowledge about the behavior of the time series of our individual product lines. Then an implementation explosion developed as literally dozens of techniques — some suggested decades ago — became practical, workable tools of the everyday forecaster. Indeed, whole eclectic systems came into being, including my own FORAN II (published in 1968 by the American Marketing Association). This system, which took six years to develop, included for the first time in a single computer program a whole host of ideas: naive models, moving averages, leading indicators, seasonal adjustments and trading day factors; exponential smoothing, paired indexes and pressures; correlation, econometrics and graphic schemes; raw forecasts, final forecasts and optimal forecasts.

Plans for a FORAN III began almost immediately after the publication of FORAN II. The time shared computer — with its new power of interaction — made a FORAN III inevitable. It was, like FORAN II, to include the "state of the art." But the project was massive and not easy to support within the confines of a single corporation. At about the time the project was floundering, I received a telephone call from Dr. Steven C. Wheelwright of the Harvard Business School. He wanted to know if he and Dr. Spyros Makridakis (European Institute of Business Administration, Fontainebleau, France) could discuss with me a major project on forecasting that they had underway. I, of course, jumped at the chance and it was on that day I realized that the big responsibility of FORAN III was settled. These two men in effect, were already building it. And that is what this book is all about.

In this Foreword, I come to praise. Makridakis and Wheelwright have given us the current "state of the forecasting art." Nearly every known device that could be useful in forecasting is included in their computer system. And it is at the call of anyone with a typewriter terminal. It is all there, with a minimum of input and an economy of printout — two considerations of extreme importance in the interactive world of time sharing. Here we see adaptive models, multiple regression, autoregressive schemes, ARMA, ARIMA, and MARMA models, the exponential smoothing models of Holt, Brown and Winters, the "BJ" models of Professors Box and Jenkins and their own contribution in the form of "generalized adaptive filtering," which will almost certainly become affectionately known as "MW" models after the names of these two authors.

ROBERT L. MC LAUGHLIN
SCOVILL, INC.

Preface

Over the past two decades forecasting has gained widespread acceptance as an integral part of business planning and decision making. A wide range of companies and government organizations use forecasting in such areas as sales planning, marketing research, pricing, production planning and scheduling, financial planning, program planning, and overall corporate strategy formulation. To meet the needs of these widely varied situations, a number of time series techniques and explanatory or causal approaches (such as regression) have been developed. As in many areas of management science, the development and application of these techniques have been somewhat spotty and fragmented. Thus it is today that, in spite of the number of techniques that are available and the almost universal agreement on the need for more and better forecasting in organizations, little work has been done to rationalize the many different approaches available and to provide users with a simplified and integrated approach to forecasting.

For several years now the authors have taught forecasting to graduate students of business and to practitioners in the United States and western Europe. They have become keenly aware of the need for more straightforward approaches in teaching forecasting techniques and of the desirability of developing materials that could be used effectively by people with limited training in mathematics and statistics. The set of materials in this book is the culmination of several years of experience in the forecasting area and is designed to be used by both the uninitiated and the expert.

The objectives of the material in this book are five-fold.

1. Minimize start-up costs for the user desiring to obtain a forecast.
2. Encourage and support self-directed learning about forecasting, especially in bridging the gap between theory and practice.
3. Provide a complete set of materials for the management-oriented forecaster who may have a limited background in mathematics and statistics but who clearly needs systematic forecasting tools.
4. Serve as a user's guide for the set of interactive forecasting programs known as SIBYL/RUNNER.
5. Provide a set of practice-oriented teaching materials for the faculty member teaching forecasting.

The first objective has been achieved by designing a set of interactive forecasting programs for use on a timesharing system. These programs enable the user to quickly apply one or more forecasting techniques in a particular situation. The system can be used equally well by both the novice and the forecasting expert. The second objective of self-directed learning has led the authors to include several options in the interactive programs themselves — as well as in the material in this book — that allow one to use a streamlined approach to most forecasting situations and, when more information is needed, to obtain that information either through the special HELP command or through the appendices and supplements to this book. (To keep the book a manageable length, much of the detailed theory is omitted and references are simply provided for those seeking such additional background.)

In order to accomplish the third objective of providing a complete set of materials, the authors have developed an overview of time series forecasting, guidelines for handling a range of situations, as well as information on how to use particular methods. Finally, a number of case examples that provide exposure to the practical considerations involved in management forecasting problems have been developed. The authors have sought to maintain simplicity for the management user who does not have extensive mathematical background in forecasting, while enabling the professional forecaster to move quickly from data entry to completion of one or more forecasts for a particular situation.

Most of the chapters describe a single forecasting technique and its application. In these chapters the background of the technique is presented, and references for further study are provided. For the non-technical user, these references include books like the authors' *Forecasting Methods for Management* (John Wiley and Sons, Second Edition, 1977). For the user seeking a more complete and perhaps more technical discussion, references like the authors' *Forecasting: Methods and Applications* (John Wiley and Sons, 1978) are included.

To accomplish objective number four, each technique chapter describes the inputs and outputs for the related program in SIBYL/RUNNER and includes sample results from that program. Finally, each of these chapters includes a solved exercise illustrating the results obtained from the appropriate computer program and the basic computational procedures used to obtain those. Objective number five is met through the integrated use of the text chapters, the case studies, and the SIBYL/RUNNER programs for interactive forecasting.

The material in this publication and complementary items can be grouped into five categories.

1. A set of computer programs that represents each of several different forecasting techniques and provides control programs for moving between the different techniques and performing preliminary analysis of the data. These programs are currently available from Dr. Steven C. Wheelwright, Applied Decision Systems, 15 Walnut Street, Wellesley Hills, Mass. 02181. While originally written in Hewlett/Packard BASIC, they are currently available in FORTRAN for a wide range of computer systems. The programs are also available on a number of national timesharing networks.

2. A set of twenty-seven chapters, each of which gives an overview of a particular technique and illustrates its application. Nineteen of these chapters deal mainly with time series methodologies developed for a single data series (univariate time series analysis). Eight chapters, most of them new in this second edition, deal with methodologies such as multiple regression, leading indicators, and transfer functions that can handle multiple data series (multivariate time series analysis).

3. A series of business cases that have been used in teaching forecasting in such areas as marketing, production, finance, and general management are included. These are grouped after major segments of text to serve as additional exercises. (However, many of these cases can be analyzed using the methodologies from several chapters in addition to those that immediately precede the case.) The purpose of these cases is to provide sufficient background information to complement the historical time series data so that selection of a forecasting technique can be based not only on accuracy but also on other factors such as ease of management understanding, the cost of the technique, and the time horizon for which a forecast is required. (While the computational mechanics of forecasting can be learned using simple exercise data, the authors believe that effective managerial use of forecasting requires knowledge of the forecasting situation.)

4. An extensive glossary of forecasting terms, arranged alphabetically, has been included. Those terms that commonly occur in forecasting are described in some detail so that the reader can build a more extensive base of knowledge as the need arises.

5. A teacher's manual has been prepared and is available to faculty members through Holden-Day. The manual suggests how a course in forecasting might be structured to utilize this publication and the related interactive forecasting programs, SIBYL/RUNNER. It also gives a description of the analysis that can be performed on each case study and the way in which individual cases might be used in teaching. (These teaching notes are based on the authors' use of the cases at INSEAD and at the Harvard Business School with both business students and managers.)

The authors' experiences have shown that these materials are suitable for both students and managers. The student can use them in a variety of situations ranging from the self-study needed by a doctoral student requiring specific forecasts for a research project, to use in a class dealing only with forecasting and planning, to meeting the requirements of the forecasting segment of a functionally oriented course. In this last case, a faculty member might choose to use some of the cases and only a subset of the forecasting techniques in order to teach students how to prepare forecasts that will then be used as part of management decision making in such areas as sales force management, production planning, financial management, capital budgeting, or corporate planning. The authors have found that particularly effective classroom use can be made of the complementary computer programs, by using an on-line overhead projection of the computer input and output. This enables the entire class to learn about individual techniques from the programs and to make subsequent better personal use of them.

A similar range of usage is also available to the practitioner. If that practitioner is a manager, the materials facilitate rapid preparation of a forecast and learning more about forecasting and the ease with which it can be applied in a range of situations. For the support staff professionally trained in the preparation of forecasts, this set of materials rapidly expands the range of applications for which forecasting can be provided and allows those professionals to become more productive in using their time for preparing and evaluating such forecasts.

As indicated in Robert McLaughlin's foreword, development of these materials and the associated computer programs has been a major task that has taken several years and required the efforts of many. At the risk of missing a few, we would like to thank INSEAD and Harvard for their support and use of their facilities; Anne Hodgsdon and Michele Hibon for programming assistance; and Pat Beh, Pat Murphy, Andrea Truax and others for typing and proofing numerous drafts. Finally, we'd like to thank our families for being so patient in letting us continue this project for so many years, and Professor Claude Faucheux for his support and encouragement, without which we'd never have had the energy needed to develop SIBYL/RUNNER.

SPYROS MAKRIDAKIS, FONTAINEBLEAU, FRANCE
STEVEN C. WHEELWRIGHT, BOSTON, MASS.
JANUARY 1977

Contents

Cases for Interactive Forecasting

The following cases were prepared as a basis for class discussion rather than to illustrate either effective or ineffective handling of an administrative situation.

Special Credits

The authors wish to give special thanks to the following faculty members for granting permission to include the indicated cases. (All other cases were prepared by the authors.)

Associate Professor Jay O. Light, Harvard Business School, Ferric Processing and Harmon Foods

Associate Professor Sherwood Frey, Harvard Business School, Shasta Timber

Assistant Professor Barbara Jackson, Harvard Business School, Tempo Inc.

Associate Professor Jeffrey Miller, Harvard Business School, Perkin Elmer Instruments Div.

OVERVIEW OF INDIVIDUAL CASES

A.   Alpha Concrete Products

The management of this company desires to predict sales and profits for each of the next several years in order to determine what price it should be willing to pay to acquire control of the company. Historical data for company sales and key profit and loss statement items are presented, as are data for related indicators that are thought to be closely related to company sales. The case describes a consultant's report that has applied multiple regression to the problem and allows that application to be evaluated. An alternative approach such as multivariate time series analysis can also be considered and tested.

B.   Arizona Security Bank

As part of the corporate planning process at a major bank, annual data have been collected on several different variables that might be used to provide background forecasts for those preparing the long-range plan. These data can be analyzed both with time series methods and regression techniques. The case presents several opportunities for taking different cuts at the forecasting problem.

C.   Atwood Appliance (A)

This case presents the results of a market research study that can be used either to simulate the impact of various marketing actions on company sales or to forecast sales with judgmentally based methods.

D.   Basic Metals Limited

This case presents data on weekly shipments of a commodity product manufactured by Basic Metals Limited. Accurate forecasts for this series are needed for effective production planning. Among the issues in this forecasting situation are the need for trading day adjustments and the cyclical impact of a commodity business.

E.   CFE Specialty Writing Papers

This case presents data similar to that used in Chapters 21 through 23 of the text. Monthly data for the sales of a particular product category are presented as well as monthly data for a related leading indicator. This case can be used for multivariate time series analysis as well as univariate time series analysis.

F.   Eagle Airlines

The financial planning office of a major airline must predict cash flows and profitability for each of the next several months. Monthly historical data are presented on profitability. These data are suitable for analysis with either time series methods or regression techniques.

G.   Exercises on Simple Exponential Smoothing

Exercises suitable for use with simple moving averages, single exponential smoothing, and higher order smoothing methods are presented.

H.   Ferric Processing

This case is suitable for either regression analysis or multivariate time series analysis. It presents a production situation where a company is trying to determine the relationship between the size of plants and their cost per ton.

I.   Forecasting IBM Stock Prices – Paris Exchange

This case presents a comparison of two multivariate forecasting methods applied to a single situation. The situation is the relationship between prices of IBM common stock on the Paris Exchange with previous prices on the New York Exchange. (This represents an autoregressive relationship.) The two methods examined are multiple regression and multivariate Box-Jenkins transfer functions.

J.   Gruner and Jahr

The production and operations management team of this magazine publishing firm is faced with the decision of how many copies of different magazines to print for each issue to maximize profit. The task involves predicting demand and its variability by region so that the cost of unsold copies can be balanced against lost profits from unfilled demand. While monthly and weekly data are presented, the focus is on structuring a forecasting and planning system that uses available techniques and best meets the needs of the decision situation.

K.   Harmon Foods

This case presents monthly data on shipments and various forms of promotion for a cereal firm. These can be used either through regression analysis or multivariate time series analysis to predict future sales. The impact of different types of promotion on the firm's profitability and sales can also be examined.

L.   INSEAD's Restaurant

The management issue in this case is forecasting the number of meals to be required on a daily basis in a captive educational institution restaurant. The forecast requires that certain adjustments be made in the historical data to reflect the effects of weekends, examination schedules, and holidays. The case is appropriate for use with several different time series methods including those that handle seasonal patterns.

M.   Jackson Hole Airport

This case presents an operations planning issue at a small regional airport that has witnessed a substantial increase in passenger boardings in recent years. The decision is whether or not the airport should be expanded. Annual data for passenger boardings are presented as well as data from two parallel series that may be correlated closely with airport activity. The issue is not only one of trend analysis but of those problems surrounding forecasting situations for which only a minimum amount of data is available.

N.   Kool Kamp Corporation

This case presents monthly sales data for a young, high growth company. The data demonstrate a strong seasonal pattern as well as growth. The company's need for effective forecasting to aid both marketing and production is clearly outlined. In addition, subjective estimates based on salesmen's judgments have recently begun to be collected.

O.   Lenex Corporation

This case describes a company that has recently found itself in financial difficulty, partly because of a lack of effective planning and forecasting. Several different time series are presented. These can be analyzed individually to discover the source of the company's problems, and they can also be correlated with one another to determine the degree to which the individual series move together. The data in this case are particularly appropriate if one is simply looking for exercise data that can be used for several different methods of forecasting.

P.   Lyon Plate Glass

The issue in this case is financial. An individual is evaluating two available alternatives for selling a major block of stock in a glass firm. Annual data are presented for the company as well as for two important customer categories that the company serves. Univariate or multivariate time series or regression analysis techniques can be considered as a basis for predicting future sales.

Q.   Perkin Elmer Instrument Division

This case describes the forecasting procedure adopted by a rapidly-growing division of the Perkin Elmer Company. Many of the strengths and weaknesses of the existing system are presented, and students have a chance to examine the interaction of forecasting with operations management, and the tasks associated with forecasting, other than simply selecting an appropriate methodology.

R.   Perrin Freres

This case presents the situation facing a company desiring to predict monthly industry sales of champagne as a basis for eventually preparing its own sales and cash flow projections. The case contains monthly champagne sales data for several years and provides background information on the conditions motivating the company to forecast industry sales. The data are seasonal.

S.   Phoenix City Council

This forecasting situation involves a local government and its desire to predict population for use in certain decision-making situations. Issues regarding bias in those forecasts and the need for providing more than a single point estimate are raised. Annual data for both the city and the county are provided.

T.   Shasta Timber

The issue in this case is estimating the production output from various acreages of government timber. The forecasts have historically been prepared by two different experts and questions of calibrating those experts and measuring their accuracy are raised.

U. Tempo Incorporated

A temporary help firm is faced with the problem of forecasting demand both for the short and intermediate term based on several different factors. The company has little historical data available for forecasting, and the focus is on what kind of data it should begin to collect and how it will eventually use the data once a history has been built up.

V. Utah Industrial Promotion Board

The situation in this case is that of a state government administrator faced with using annual historical data to predict nonagricultural employment. The case presents data for four sub-regions and allows comparison of direct forecasts of aggregate nonagricultural employment with forecasts made by sub-region and then aggregated.

W. Valley Litho

In this case management is faced with the task of pricing several printing contracts that follow a similar structure but have individual specifications that differ somewhat. The technique of regression analysis is applied to two different sets of data and varying results are obtained. These alternative models must be examined to understand their key differences and implications and to decide which is the most appropriate for this situation.

CHAPTER 1

Introduction: Interactive Forecasting System

Time-sharing computer configurations have introduced a new dimension in applying statistical and mathematical models to sequential decision problems. When the outcome of one step in the process influences subsequent decisions, then an interactive time sharing system is of great help. Since the forecasting function involves such a sequential process, it can be handled particularly well with an appropriate timeshared computer system. This book describes a system that allows the user to do preliminary data analysis, to recommend a set of appropriate forecasting techniques, and to apply those techniques in developing a forecast. This interactive forecasting system has met with excellent success both in teaching the fundamentals of forecasting for business decision making and in applying those techniques in a variety of practical situations.

The purpose of this chapter is to outline the philosophy used in the formulation and development of this interactive forecasting system. We hope that this chapter will increase understanding of what the system is doing (and why) during its actual use. In addition, the chapter outlines the overall structure of the system and the criteria it uses in recommending particular forecasting methods for a given data series.

*A portion of the material in this chapter is adapted from an article by the authors that appeared in *The American Statistician*, November, 1974. (Used by permission.)

The System

The past decade has seen the development of a number of forecasting methods that can be used for practical forecasting purposes. These advances in both theory and practice have been largely in response to requirements placed on individual firms by the increasing complexity and competitiveness of the business environment. Companies of all sizes now find it essential to make forecasts for a number of uncertain quantities which may affect their decisions or their performance.

Like the development of most management science techniques, the application of these methods has lagged behind their theoretical formulation and verification. Thus, while most managers are aware of the need for forecasting methods, few managers are familiar with the range of techniques that has been developed and the characteristics of these which must be known in order to select the most appropriate technique for a given situation.

Unfortunately, because of the lack of experience among managers in formalized forecasting procedures, there is very little reported work that deals directly with the management side of forecasting problems. Rather, the existing forecasting literature consists of a number of books and articles that deal with a particular forecasting technique or with a narrow class of techniques and associated technical characteristics. (For example, there are books like Box and Jenkin's on their approach to time series forecasting, books like Johnston's on regression techniques, and books like Brown's on smoothing approaches to time series forecasting.) However, as operations researchers have found in the past, in order to gain widespread management acceptance of forecasting procedures it is necessary not only to describe their theoretical and technical aspects but to translate those characteristics into practical management concerns. Of primary importance to managers is the practical experience that others have had in using the forecasting method, the cost of applying the method, the limitations of the forecasts developed by the method, and the accuracy of the resulting forecasts.

Those who have tried to apply forecasting methods in recent years have become aware of a number of requirements that are difficult to meet with existing techniques and systems. Three of the most important of these are discussed below.

1. Maintaining Flexibility in Approaching New Situations

Managers have found that it is extremely easy to develop a preference for one forecasting method and then to use it almost exclusively in any new situation. However, they generally recognize the need to consider a range of alternative techniques. One source of this difficulty is relying on a single technical person in the firm as the source of knowledge concerning forecasting methods. It is difficult to expect that person to feel unbiased towards each of several alternative forecasting methods simply because of the magnitude of the intellectual task involved. Thus, a system that explicitly supports the consideration of many different techniques for each new forecasting problem would be clearly attractive to managers.

2. Considering All Relevant Factors in Selecting a Forecasting Technique

Managers are well aware of the need to consider not only accuracy but a number of other factors in selecting a forecasting technique for a new situation. Since trade-offs inevitably occur among these various factors, the manager needs to be involved actively in selecting any forecasting system in order to apply his judgment to those considerations.

3. Rapidly Screening Alternative Techniques

A third point that has created problems for many managers in forecasting is that they have felt compelled to turn any new situation directly over to a technical person on their staff because they did not feel competent to do the preliminary analysis themselves. In many situations, however, if the manager could do the preliminary screening of alternative techniques considerable time and effort might be saved in the long run. It would also encourage the adoption of formal forecasting procedures in situations where it may not seem to be worthwhile to make the commitment required to obtain the involvement of a specialist.

In addition to the problems that have been recognized by managers trying to apply forecasting, those teaching in business management programs have also identified some major problems. One of these is a tendency to get bogged down in the classroom in technical details to the exclusion of more practical considerations. This is a natural tendency, given that most of the literature on forecasting is technical in its orientation and that there is little written about actual experiences in forecasting.

Another problem faced by those teaching forecasting is that it takes a lot of time to have students thoroughly apply any one forecasting technique to a single situation. Since most courses are relatively short, they do not allow the time necessary to apply a number of those techniques to each of several problem situations.

Finally, teachers have found it particularly difficult to teach about the assumptions inherent in each alternative forecasting method and the implications these have in practice. An obvious solution to this last problem would be to give the students some practical experience in applying the methods. But again, the time generally required to do this for a wide range of situations makes it impractical in all but the most specialized courses.

As a result of the recognition of the above problems, the authors have developed an interactive system for teaching forecasting methods and applying them in practice. This system has been installed on several timeshared computers and is used by many students and managers. The remainder of this chapter describes this interactive system, how it works, and the practical experiences of the authors in using it both in teaching situations and in identifying and applying forecasting techniques for particular management problems. To date, the results have been extremely encouraging and it seems to be meeting its objective of overcoming the specific problems outlined above.

DESCRIPTION OF THE COMPUTERIZED SYSTEM

The interactive forecasting system developed by the authors is divided into two sequential segments. The first segment, referred to as SIBYL, is aimed at allowing the user to perform a preliminary analysis of the data in order to identify forecasting techniques that may be appropriate for a given situation. As shown in Figure 1-1, the user begins by inputting the data to be examined and used as a basis for forecasting. This segment of the system queries the manager concerning judgments about the data and about the characteristics of the situation that are most important in selecting a forecasting method. Those factors that need to be considered include the following:

1. The time horizon for decision making: immediate term (less than 1 month), short term (1 to 3 months), medium term (3 months to 2 years), and long term (more than 2 years).
2. The pattern of data: seasonal, horizontal, trend (non-stationary), cyclical, or random.
3. The type of model desired: time series or causal.
4. The value of the forecast, and thus the amount of time and money that can be spent in obtaining it.
5. The accuracy that is required and justified.
6. The complexity that can be tolerated.
7. The availability of historical data.

Some of these factors can best be analyzed through statistical tests while others involve value judgments only the user can supply. Furthermore, a number require information about the forecasting techniques themselves, which can well be supplied by previous users. All of these factors are important and specific consideration of them must be made by the manager before the most appropriate forecasting method available can be selected.

For the user to decide upon the best method available, knowledge of all forecasting techniques must be possessed and the user must be able to evaluate all of the factors in the specific situation that will influence the

FIGURE 1-1 Flow-Chart of the Interactive Forecasting System (1973)

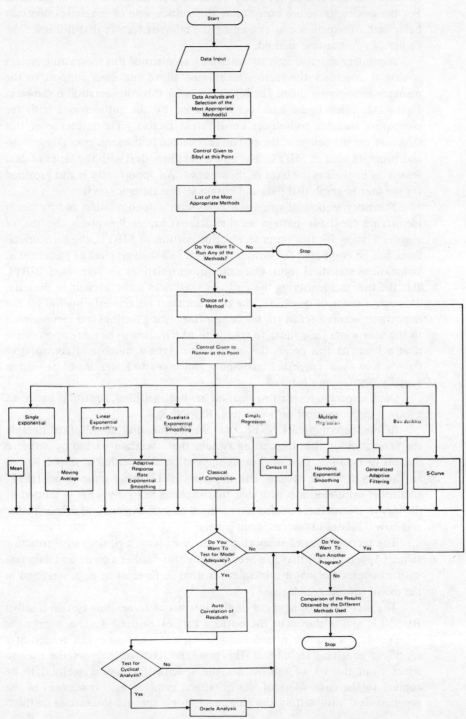

selection. Such a task is not easily handled, even for the expert in the field. For the novice, there are many more difficulties, one of the most important being lack of experience in evaluating the relevant factors that influence the choice of a forecasting method.

A primary characteristic of the SIBYL segment of this forecasting system is that it considers the factors mentioned above and gives support to the manager in applying them. The logical basis for this consideration is shown in Figure 1-2. (This figure has been developed by the authors as a basis for comparing available techniques on different factors.) The influence of this structure on the design of the system is reflected in the sequence of questions and prompts used in SIBYL. The initial questions deal with the series of data that is to be used as the basis of the forecast. An opportunity is also provided for the user to graph that data and obtain several statistics on it.

The next section of questions deals with autocorrelation and its use in identifying the basic pattern in the data. (Chapter 3 explains the use of autocorrelation for this purpose.) In this portion of SIBYL, the autocorrelations can be computed for various time-lags and then graphed or presented in summarized statistical form. Once the autocorrelations are computed, SIBYL aids the user in identifying the patterns that seem to be present in the data. Through a series of questions, the program then obtains information on the important factors needed for selecting a forecasting method and recommends to the user some techniques to start with which seem to be more appropriate than others. At this point, the user is also given a number of comparative statistics on those suggested techniques and asked to select one to be used in actually developing a forecast.

Once a specific forecasting method has been selected, control is passed to a second major segment of the system, RUNNER.

Before describing RUNNER, we should mention some other notable characteristics of this forecasting system that facilitate its use by either a student or manager. One of these is the option the user has of responding to any inquiry from the system with the word HELP. When a user types HELP, additional information is provided that explains not only what is wanted in the way of a response, but also gives a more general explanation of that factor and how it relates to forecasting in general.

The second form of support given to the user is a glossary of forecasting terms (Appendix 00) that has been prepared to further explain and illustrate with examples the major principles involved in forecasting in general and in the use of specific techniques.

The second major segment of this interactive forecasting system is called RUNNER and is shown in the bottom half of Figure 1-1. This segment is composed of a number of subroutines, each the computerized version of a specific forecasting technique. The program RUNNER allows the user to select from the set of subroutines the specific forecasting method to be applied to the data. Most of the questions involved in this segment of the program deal with setting the parameters for a specific forecasting method.

FIGURE 1-2 Factors to be Considered in Selection of a Forecasting Method

FACTORS		Naive	Mean	Simple Moving Average	Simple Exponential Smoothing	Linear Moving Average	Linear Exponential Smoothing	Classical Decomposition	Census	Foran System	Adaptive Filtering	Box Jenkins	Generalized Adaptive Filtering	Simple Regression	Multiple Regression	Econometric Models	Life Cycle Analysis	Surveys	Leading Indic or Diffusion Indexes	Input Output Analysis	Inventory Control	Mathematical Programming	Delphi	S-Curves	Historical Analogies	Morphological Research	Relevance Trees	System Analysis
		SMOOTHING						DECOMPOSITION			CONTROL			REGRESSION			OTHERS						EXPLORATORY (TECHNOLOGICAL)				NORMATIVE	
TIME HORIZON OF FORECASTING	Immediate — Less Than One Month	X	X	X	X	X	X														X	X						
	Short — One to Three Months	X	X	X	X	X	X	X	X	X	X	X	X	X	X	X					X	X						
	Medium — Less Than Two Years							X	X	X	X	X	X	X	X	X	X	X	X	X		X	X	X	X	X	X	X
	Long — Two Years and More													X	X	X	X		X	X		X	X	X	X	X	X	X
PATTERN (TYPE) OF DATA	Horizontal — Auto / Non Correlated	X	X	X	X			X	X	X	X	X	X	X	X	X												
	Trend — Auto / Non Correlated					X	X	X	X	X	X	X	X	X	X	X	X											
	Seasonal — Auto / Non Correlated							X	X	X	X	X	X			X												
	Cyclical — Auto / Non Correlated							X	X	X	X	X	X	X	X	X	X											
Minimum Data Req's		5	30	5–16	2	13–20	3	60	72	24	60	72	72	30	30	FEW 100's	15–30		0	FEW 1000's								
TYPE OF MODEL	Time Series	X	X	X	X	X	X	X	X	X	X	X	X				X											
	Causal													X	X	X		X	X	X	X	X						
	Statistical	X	X	X	X	X	X	X	X		X	X	X	X	X	X	X		X									
	Non-Statistical									X								X					X	X	X	X	X	X
	Mixed								X	X								X			X	X					X	X
COSTS (0 smallest, 1 highest)	Development	0	1	05	2	15	1	4	6	5	4	8	5	3	6	8	5	8	0	10	5	6	5	5	5	9	8	8
	Storage Req's — Program	NA	09	09	08	13	12	1E	6	5	28	9	35	17	34	48	34			NA	5	7						
	Storage Req's — Data	01	1	025	006	05	039	33	33	00	33	33	33	2	33	10	08			NA	01	10						
	Running	NA	03	006	007	037	005	2	65	2	3	88	70	1	21	7	2			NA	01	3						
ACCURACY (0 smallest, 1 highest)	Predicting Pattern	1	15	2	3E	2	25	5	7	7	7	9	85	5	8	10	5	8	5	6			5	5	5	5	5	5
	Predicting Turning Points	3	0	0	0	0	0	3	8	7	8	8	75	0	4	6	0	0	10	0			7	6	0	7	0	0
APPLICABILITY OR COMPLEXITY (0 smallest, 1 highest)	Time Required To Obtain Forecast	5	2	05	01	09	01	3	5	5	4	7	5	25	6	9	5	10	0	10	1	3	7	6	9	7	10	10
	Easiness To Understand And Interpret The Results	10	10	10	9	9	9	7	7	5	7	4	6	3	6	3	8	10	10	3	8	8	8	6	9	7	8	8

For example, if the forecasting method is simple exponential smoothing, it is required that the user define one parameter: the value for the smoothing constant, alpha (α)

Once the parameter value(s) are known, they can then be used to prepare a forecast and to compare that with actual values in order to determine the technique's accuracy. This is done in terms of mean percentage error, mean absolute percentage error, and mean squared error. Finally, RUNNER can be used to compare the performance of several different techniques on the same set of historical data.

After using the SIBYL-RUNNER combination, the user will have completed:

1. A general analysis of the data.
2. A screening of available forecasting techniques.
3. A detailed examination of a few of the most appropriate techniques.
4. Final selection of a technique for the situation.

EXPERIENCE IN USING THE INTERACTIVE FORECASTING SYSTEM

The use of this timeshared system has been well-received in teaching situations and in practical forecasting applications. Students report that they feel more motivated by this approach than they do by more traditional textbook approaches and that they gain a better understanding of the practical application of the alternative techniques (in addition to their theoretical basis). The authors and other colleagues have used this program in conjunction with a text on forecasting in a regular classroom course.

One of the attractive features of this program is the ease with which it can be used by someone familiar with it and with forecasting in general, since it allows suppression of much of the descriptive printout. Thus, students find it to be useful not only in their initial learning, but — once they understand a number of techniques — it facilitates their application of these to specific situations.

In general, experience with this system in teaching indicates that it overcomes many of the problems identified at the outset of this chapter. These include avoiding the expenditure of a disproportionate amount of class time on technical details, allowing students to gain experience rapidly in a number of situations in order to infer the important characteristics of each technique, and minimizing the amount of time required to reach a given level of competence.

The SIBYL-RUNNER system has also been thoroughly tested in actual business situations including its use in project consulting on forecasting, its use by corporate forecasting and planning staffs, and its use by individual line managers. It has proved to be successful in virtually all of the forecasting situations in which it has been applied. Some of the advantages of this system for management forecasting are that it allows the manager to obtain rapidly, and at a minimal cost, a forecast for a given situation. In addition, it overcomes many of the existing problems managers face by guiding their examination of a full range of alternatives for any new situation, helping them to consider those factors that are important in selecting the most appropriate technique for a given situation, and providing them with a wide range of techniques that can actually be applied.

Having such an interactive forecasting system available within a company can greatly encourage the use of formalized forecasting procedures, not only because of the range of techniques covered, but also because the techniques can be applied directly by the manager and his staff with support from the use of the HELP command, the glossary of forecasting terms, and the chapters in this book. It is practical for the manager to use the system at any time without remembering precisely the method that was used in the past. It is the authors' conclusion that this system meets many of the existing needs in the area of forecasting and through further development and application it should continue to find wide acceptance in management education and in business practice.

Of course there are limitations to this system. One is that both managers and students must be introduced to forecasting and the range of available techniques before they can begin to use the SIBYL-RUNNER system effectively. Another limitation is that the questions and prompts can become routine over time and thus fail to elicit the user's judgment. Also, the simplicity of the system and its ease of usage may lead to its application where a more sophisticated analysis under the direction of a forecasting specialist is warranted. This package is designed for expanding the application of systematic forecasting methods, not replacing existing systems that already meet the requirements of very specific situations.

Keeping these limitations in mind, the authors have found SIBYL-RUNNER to be a very useful tool in facilitating the appropriate use of existing forecasting techniques in management situations. Additional experience and improvements based on that experience have helped to enhance the system's effectiveness. The extent of these enhancements can be seen by comparing Figure 1-1 with Figure 1-3. The latter figure summarizes the major elements of this interactive forecasting system in 1977, three years after its initial development.

At present a series of operating computer programs incorporate the features of the SIBYL-RUNNER system illustrated in Figure 1-3. In addition,

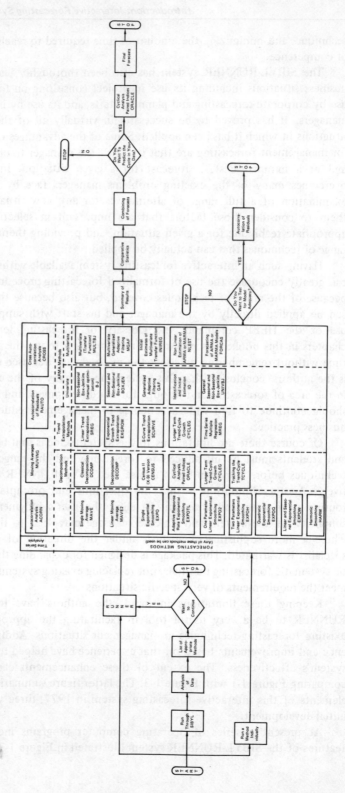

FIGURE 1-3 Flowchart of the Interactive Forecasting System (1976)

FIGURE 1-4 Programs in the Interactive Forecasting System (1977)

Chapter	Program Name	Program Function
3	ACORN	Autocorrelations
4	MUVING	Different Types of Moving Averages
5	EXPO	Simple Exponential Smoothing
6	EXPOTL	Adaptive Response Rate (Trigg and Leach's) Single Exponential Smoothing
7	EXPO2	Linear (Brown's) Exponential Smoothing.
8	EXPOH	Linear (Holt's) Exponential Smoothing
9	EXPOQ	Quadratic (Brown's) Exponential Smoothing
10	EXPOW	Linear and Seasonal (Winter's) Exponential Smoothing
11.A	MAVE	Single Moving Average
11.B	MAVE2	Linear (Brown's) Moving Average
12	SREG	Time-Series Simple Regression (Trend Fitting)
13.A	SCURVE	S-Curve (Life-Cycle Analysis) Fitting
13.B	EXGROW	Exponential Growth Trend Fitting
14	HARM	Harmonic (Harrison's) Smoothing
15	DCOMP/DECOMP	Classical Decomposition
16	CENSUS	Census II Decomposition Program (X-9 version)
17 (and 27)	AF	Univariate Adaptive Filtering
18	ID	Identification of ARMA Models
18	BOXJEN	Univariate Box-Jenkins
18	B-J	Non-Seasonal Univariate Box-Jenkins
19	SIBYL	Main Program
20	ORACLE	Main Program
21	CYCLEG	Long Term Trend-Cycle Growth
22	TCYCLE	Tracking the Cycle
23.A	MREG98	Multiple Regression
23.A	MREG99	Multiple Regression, Unlimited Observations
23.B	MREG	Time-Series Multiple Regression
25	CROSS	Cross Autocorrelations and Impulse Response Parameters

FIGURE 1-4   (continued)

Chapter	Program Name	Program Function
26	INIREG	Initial Regression Estimates for MARMA Models
26	MULTBJ	Main Program for Multivariate ARMA Models (Transfer Function and Adaptive Filtering)
26 (and 18)	UNIVBJ	Univariate Box-Jenkins
27	MGAF	Multivariate Adaptive Filtering
28	RUNNER	Main Program
	AUXILIARY PROGRAMS	
29.A	FINPUT	Enters data into a file
29.B	FEDIT	Edits a data file
29.C	UPDATE	Updates the contents of a data file
29.D	FLIST	Lists the contents of a data file
29.E	FPLAY	Performs most forms of data and/or file manipulations (e.g. adding and deleting elements, adding and sub-tracting data series, doing data transformations.)
	SIBYLL	Gives an alphabetic listing and short description of the programs in SIBYL/RUNNER

a computer program exists to represent each of the forecasting techniques included in this system. For example, there is one program for exponential smoothing, another for simple regression, and so on. The programs that represent each forecasting method are called and controlled directly by the SIBYL-RUNNER programs; however, they can be run individually too, using a simple control program called METHOD. The individual programs and their relationship to chapters in this book are summarized in Figure 1-4.

The SIBYL-RUNNER system has been adapted for use on a number of different computer systems in FORTRAN. It is also available in Hewlett/ Packard BASIC, although because of the much more general nature of FORTRAN it is the latter version that is more fully supported. In addition the programs have been adapted and are presently running on over 20 different computer systems. For those interested, information on how a copy of the programs can be procured, can be obtained by writing to Dr. Steven C. Wheelwright, ADS, 15 Walnut Street, Wellesley Hills, Mass. 02181.

CHAPTER 2

Introduction: Time Series Analysis and Forecasting

A basic assumption of the interactive forecasting system is that the forecaster/manager can best learn about alternative forecasting methods through an iterative process involving decisions based on the outcome of previous steps. Furthermore, it is useful to be able to apply several forecasting methods and observe their accuracy, effectiveness, and appropriateness in dealing with specific situations. This approach has proven successful in learning more about the characteristics of individual techniques as well as in learning how to select the most appropriate forecasting methods for a given situation. Since the techniques discussed in subsequent chapters are based mainly on the concepts of time series methodologies, this chapter introduces some of the common underlying principles of time series, illustrates their mathematical representation, and presents criteria for comparing the forecasting accuracy of alternative methodologies.

The central theme of quantitative techniques of forecasting is that the future can be predicted by discovering the patterns of events in the past. Such patterns are generally assumed to take one of two forms: first, is a pattern that is determined solely as a function of time. Such a pattern can be identified directly from historical data for the variable to be forecast and related variables. The time series forecasting techniques discussed up through Chapter 22 of this book assume that the pattern is of such a nature. For the univariate case, the assumption means that using historical data relating to the single items to be forecast is sufficient for detecting the basic time pattern and extrapolating it to predict the future. The alternative pattern that is often assumed to exist consists of a relationship between two or more variables. In this case, the historical data from a single variable is not assumed to contain all of the information about the underlying pattern. Rather, it is necessary to have data on several variables in order to identify the relationship between them. The multiple regression and multivariate time series methods discussed in Chapters 23 through 27 assume this type of pattern.

For purposes of forecasting, the underlying pattern is assumed to be constant over at least two subsequent sets of time periods, the first being the period in which data are collected and analyzed to identify the pattern and the second being that time in the future when the identified pattern will be used as the basis for forecasting. Thus, the notion of constancy of pattern, or relationship is fundamental to any quantitative forecasting method. Since different techniques vary in their ability to handle certain patterns of data, it is important to understand the most common patterns.

THE UNDERLYING PATTERN OF THE DATA

As noted above, all forecasting methods assume that some pattern or relationship exists that can be identified and used as the basis for preparing a forecast. In the case of quantitative methods of forecasting, each technique makes explicit assumptions about the underlying pattern. Because of these assumptions, the ability of a given technique to forecast effectively in a specific situation depends largely on accurately identifying the pattern and selecting a technique that can handle it. The four types of patterns that are most frequently encountered are horizontal, non-stationary, seasonal, and cyclical.

A horizontal pattern exists when there is no trend in the data, when the average or mean value does not change (i.e. it is independent) over time. When such a pattern exists, the series is often referred to as being stationary in the mean — there is no consistent tendency for the series either to increase or decrease in any systematic way. Thus it is equally likely that future values of the series will be above the stationary mean as below it. Figure 2-1 shows a typical horizontal pattern for a variable.*

*The term "stationarity" is a statistical one used to describe data series of horizontal pattern. In this book "stationary" pattern and "stationary" will be used simultaneously for horizontal pattern or horizontal data.

FIGURE 2-1 Horizontal Data Pattern

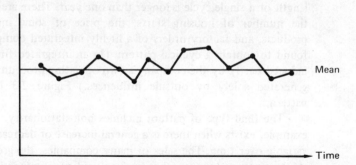

Typical situations that commonly contain a horizontal pattern include products with very stable sales, the number of defective items occurring in a production process under control, and the percentage of sales a company gets from each of several categories over a fairly short period of time. The element of time is important in identifying horizontal patterns because even patterns that over several years might exhibit a definite trend can often be assumed to have a horizontal pattern in the very short run.

A seasonal pattern exists when a series fluctuates according to some seasonal factor(s). In such a case, the seasons may be the months of the year or the quarters of the year, or they could be the days of the week or the days of the month. Seasonal patterns exist for a number of different reasons ranging from the way in which a firm has chosen to handle certain operations such as accounting and advertising practices (internally-caused seasons) to external factors such as the weather or consumer behavior.

Items that typically follow seasonal patterns include the sales of soft drinks, heating oil, and other items conditional on the weather; the receipt of revenues at a utility company, which may depend on the pattern used in sending out the bills as well as the pay period in the local community; and the number of new cars sold, which may depend on the timing of style changes and even tradition. Figure 2-2 illustrates a seasonal pattern where the seasons correspond to the four calendar quarters.

FIGURE 2-2 Seasonal Data Pattern

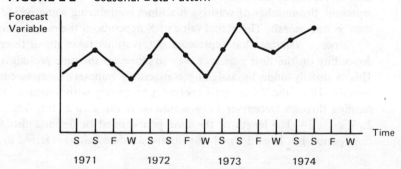

A cyclical pattern is similar to a seasonal pattern, but generally the length of a single cycle is longer than one year. There are many series, such as the number of housing starts, the price of some meat and agricultural products, and factory orders of a highly integrated company, that have been found to contain a cyclical pattern. (In an integrated firm, the cycle may be induced partly by the company's own operating procedures rather than being generated solely by outside influences.) Figure 2-3 illustrates a cyclical pattern.

The final type of pattern includes non-stationarity. A trend pattern, for example, exists when there is a general increase or decrease in the value of the variable over time. The sales of many companies, the gross national product, stock prices, and many other business and economic indicators exhibit trend patterns similar to those in Figure 2-4 in their movements over time.

While there are a number of other patterns that can be found in specific series of data, the four discussed above are the most important. These four patterns can often be found together as well as individually. In fact, some series are actually combinations of the above described patterns.

MATHEMATICAL NOTATION FOR QUANTITATIVE FORECASTING TECHNIQUES

In preparing a forecast using a quantitative forecasting technique, it is almost always the case that one has a number of historical *observations* or *observed values*. These observations may represent many things, from the actual number of units sold to the cost of producing each unit, to the number of people unemployed. Since these observed values vary over time, they are generally represented by a *variable*, such as X. (A variable is simply a symbol used to represent the value of some item.) For example, X could be used to represent the number of washing machines sold during some period of time such as one month. The actual value of X depends on the month in question.

Since a variable that represents observations takes on different values depending on the time *period*, a way to represent the time periods is needed. This is usually done by assigning consecutive numbers to consecutive time periods. Thus, the 24 months (periods) beginning with January, 1973 and running through December 1974 would be referred to as the time periods 1, 2, 3, . . . , 24. The length of the time period must be defined also. Once the time period has been defined, the observed values can be referred to with the

FIGURE 2-3 Cyclical Data Pattern

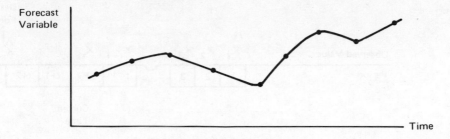

FIGURE 2-4 Trend Data Pattern

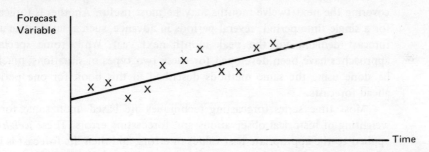

use of *subscripts*. For example, X_{10} would refer to the observed value in period 10 and X_{13} would refer to the observed value in period 13.

While X or some other symbol is generally used to represent the actual (historical) observed values of a variable, a different symbol is often used to represent the *forecasted value* of that variable. In this book, the symbol $\hat{X}_t$ or Γ_t will be used to represent the forecasted value for time period t. As a summary of the relationship between observed values over time and forecast values, one can consider Figure 2-5. It will be seen in this figure that the desired forecast is for two periods into the future rather than one period into the future. This is simply to illustrate the point that a forecast can be prepared for any number of periods in advance; however, in the discussions in this book forecasts will generally be made for one period(s) in advance. This does not exclude multiple period forecasts since the forecast for period t + 1 can be used as the actual value to forecast for period t + 2, etc. The authors have found this multiple period ahead forecasting procedure to be of greater value than forecasts involving a forecast lead time of one period. In budgeting, for example, what is desired is twelve monthly forecasts for each of the months of the next year and not a single forecast for twelve months ahead.

Two other time horizons for forecasting that are encountered in practice deserve mention at this point. One is the forecasting of cumulative values

FIGURE 2-5 Defining Observed Values for Individual Time Periods

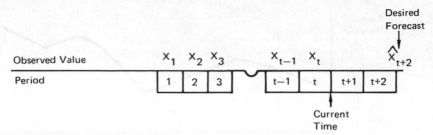

covering several time periods, an approach that is particularly useful in an area like aggregate production planning where a rolling cumulative forecast covering the next twelve months may be most useful. Another is forecasting for a single time period, several periods in advance, such as an electric utility forecast demand for the peak month next year. While some specialized approaches have been developed for these two types of situations, much can be done using the same methods described in this book for one period(s) ahead forecasts.

Most time series forecasting techniques are based upon some form of weighting of historical observations and forecasting errors. These *weights* are applied to the appropriate past values or errors, and then the forecast is based on this weighted sum. If the symbol w_i is used to represent the weight given to the ith observed value, and if one wanted to weight the last three observations, this would be written as:

$$\hat{X}_t = w_1 X_{t-1} + w_2 X_{t-2} + w_3 X_{t-3}$$

Obviously, any number of weights could be used as long as the observed values were available. When forecasting more than one period into the future, the one period forecast is often used as though it were an actual value. This approach is frequently referred to as *bootstrapping*. While it enables one to prepare forecasts further into the future, the fact that it uses forecasts in place of actual values can lead to a rapid compounding of forecasting errors.

Since many forecasting techniques require the use of weights and the summing of several values, it is convenient to represent such a sum using a *summation* sign Σ. Thus the sum of a set of weights can be written as:

$$w_1 + w_2 + \ldots + w_n = \sum_{i=1}^{n} w_i$$

This can be read as "the sum of the values of w_i taken from $i = 1$ to $i = n$ (inclusive)." Using this summation sign, a forecast based on a weighted sum of n historical values can be written as:

$$w_1 X_1 + w_2 X_2 + \ldots + w_n X_n = \sum_{i=1}^{n} w_i X_i$$

Any time a forecast is prepared, it is useful to be able to compare that forecast to the observed value for the same period, once the observed value is available. This can be done by computing the *error* in the forecast, e_i. This computation is simply:

$$e_i = X_i - \hat{X}_i = \text{Actual} - \text{Forecast}$$

FORECASTING ACCURACY AND ERROR MEASUREMENT

When forecasting is used over several periods, a series of error values is obtained, one for each time period. A summary of these errors provides a measure of accuracy, that is, of how well the forecasting method does in predicting the actual series. Obviously, one could simply sum these forecasting errors, divide by the number of error values, and obtain the average (mean) error. However, such an approach might be misleading because negative errors would partially cancel positive errors, and thus the average error might be close to 0, even though there had been a sizeable error for each individual forecast.

A better approach for examining the accuracy of a forecasting technique is to compute the *mean squared error* (MSE). This is done by first squaring the individual errors (i.e., e_i^2), which gives all positive values, and then taking the average (mean) value of these. This approach not only eliminates the canceling of positive and negative errors, but by the very nature of taking the square of a number, gives more importance to large errors in the mean squared error computation than small errors. The reader may want to prove this by working through a simple example.

An alternative measure of forecasting accuracy is the *mean absolute percent error* (MAPE). This involves finding the absolute percentage error for each forecast and computing the average of these values. The advantage of MAPE is that it allows comparisons among different series which are not possible with the MSE. If the absolute errors (deviations) are used in place of the absolute percent errors, the *mean absolute deviation* (MAD) is obtained. A final measure is the *mean percentage error* (MPE). The MPE shows the bias (the consistent over or under estimation) in forecasting.

Figure 2-6 illustrates the computations involved for each of these measures of forecasting accuracy.

FIGURE 2-6 Weekly Supermarket Sales

1	*2*	*3*	*4*	*5*	*6*	*7*	*8*
Week	*Sales*	*Forecast*	*Error*	*Absolute Error*	*Percent Absolute Error*	*Squared Error*	*Percent Error*
1	9	–	–	–	–	–	–
2	8	9	-1	1	13%	1	13%
3	9	8	1	1	11	1	11
4	12	9	3	3	25	9	25
5	9	12	-3	3	33	9	-33
6	12	9	3	3	25	9	25
7	11	12	-1	1	9	1	-9
8	7	11	-4	4	57	16	-57
9	13	7	6	6	46	36	46
10	9	13	-4	4	44	16	-44
11	11	9	2	2	18	4	18
12	10	11	-1	1	10	1	-10
		SUM	1	29	291	103	-41
		MEAN	.09	2.6*	26.4**	9.4***	-3.7****

*Mean Absolute Deviation (MAD)
**Mean Absolute Percent Error (MAPE)
***Mean Squared Error (MSE)
****Mean Percentage Error (MPE)

In comparing two forecasting methods applied to a specific situation, one of the questions that arises is the time span that should be used in computing the mean absolute percent error, or the mean squared error. This problem arises because forecasting methods require a set of data in order to determine the underlying pattern for a series. If the error is computed for the same set of data used for fitting the model, one would expect that the forecasting method would do quite well. However, the real test occurs when one takes a subsequent set of data (one not used in fitting the basic pattern) and tests the accuracy of the forecasting model on that new data.

Whenever a manager is evaluating the accuracy of alternative forecasting techniques, it is also necessary to look at resulting errors when the pattern changes. In that situation the common error calculations may not be appropriate criteria for evaluation. When the basic pattern changes, what the user is most interested in is whether, and how fast, the forecasting procedure can respond to that basic change. This means that the procedure must identify the change, and then alter its forecast accordingly. As will be seen in subsequent chapters, the ability of alternative forecasting methods to make such changes varies tremendously.

NAIVE METHODS OF FORECASTING

Naive "methods" are those approaches which provide forecasts without the utilization of a formal forecasting technique. For example, one can use the sales of March as a forecast for April. This figure is well known and easily obtained and can be used as a predictor of next month's sales with no cost and very little effort. This naive "method" will be referred to as Naive "Method" Number 1. It involves using the current period's value as a forecast for the next period, or

$$\hat{X}_{t+1} = X_t \qquad (2\text{-}1)$$

Naive "Method" Number 2 consists of using the current month's value as an estimate for next month, after adjusting for seasonality (i.e., the variation due to seasonal effects are eliminated). Thus,

$$\hat{X}_{t+1} = X_t' \qquad (2\text{-}2)$$

Naive "Methods" Numbers 1 and 2 are easily obtained, with little or no cost. They will be used as the basis for comparing the results from alternative forecasting methods in this book because the errors of a single forecasting method do not indicate much without a reference point. When a specific technique is compared with the above two naive methods, it is easy to see the improvement provided over naive "methods", and the cost associated with obtaining such an improvement. Thus, one can obtain a much better idea of the value and cost of each forecasting method. These concepts will be illustrated later on.

In summary, the naive forecasting methods to be used are as follows:

Naive Method Number 1 $\hat{X}_{t+1} = X_t$

Naive Method Number 2 $\hat{X}_{t+1} = X_t'$

where $\hat{X}_{t+1}$ is the variable to be forecast for the time period $t + 1$, X_t is the observed value in period t, and X_t' is that observed value adjusted for seasonality.

References for further study:

Makridakis, S. and S. Wheelwright (1978)
McLaughlin, R. and J. Boyle (1968)
Montgomery, D. and L. Johnson (1976)
Wheelwright, S. and S. Makridakis (1977)

EXERCISE 2.1

Given the sequence of 20 numbers shown below, forecast the next 8 values. [Hint: The series is noiseless and it took about 4 minutes of computer connect time (including printing) to perfectly identify the pattern and to prepare error-free forecasts. See Chapter 17 for the solution.]

Period	Actual Value	Period	Actual Value
1	19	11	320
2	24	12	383
3	32	13	451
4	63	14	504
5	99	15	560
6	120	16	639
7	144	17	723
8	191	18	792
9	243	19	864
10	280	20	959

CHAPTER 3

Autocorrelation Analysis

Autocorrelation coefficients provide important information about the pattern in time series data and its subcomponents (mainly trend, seasonality, and randomness). They also aid in determining whether or not one has been using the correct model or method. Autocorrelations are used by many different forecasting methods as the basis for time series analysis. Their main uses are for identifying the pattern in an original data series and in checking to be sure the residuals are random.

The concept of autocorrelation is similar to that of correlation but applies to values of the same variable at different time lags. It is perhaps easiest to understand autocorrelations by first considering regular correlations. A correlation coefficient refers to the degree of relationship between two variables. It is a relative measure of association and indicates what will happen to one variable (on average) if there is a change in the other. The correlation coefficient can vary from -1 (which indicates a perfectly linear NEGATIVE correction) to $+1$ (which indicates a perfectly linear POSITIVE correlation). When the coefficient is greater than 0 the two variables are positively correlated and when less than 0 they are negatively correlated.

For example, there is a POSITIVE CORRELATION between:

- A person's height and weight: taller people tend to be heavier, i.e., as height increases, so does weight.

- Income and spending: as people are paid more money, they spend more, i.e., as one increases, so does the other.

On the other hand, there is a NEGATIVE CORRELATION between:

- The price of steak and its demand: as steak becomes more and more expensive, demand for it becomes smaller, i.e., as one increases, the other decreases.

Autocorrelation is similar to correlation, but instead of indicating the relationship between two separate variables, such as income and spending, it deals with values of a single variable. Figure 3-1 shows how a single variable such as income (X), can be used to construct another variable (X1) whose only difference from the first is that its values are lagged by one time period. X and X1 can then be treated as two variables and their correlation found. Such a correlation is referred to as autocorrelation and shows how a variable relates to itself for a specified time lag. Similarly, one can construct X2 and find its correlation with X. This correlation will indicate how values of the same variable that are two periods apart (two time lags) relate to each other.

One could construct from one variable another time-lagged variable which is twelve periods removed. If the data consists of monthly figures, a twelve month time lag will show how values of the same month but of different years correlate with each other. If the autocorrelation coefficient is positive, it implies that there is a seasonal pattern of twelve months' duration. On the other hand, a near zero autocorrelation indicates the absence of a seasonal pattern. Similarly, if there is a trend in the data, values next to each other will relate, in the sense that if one increases the other will tend to increase, too, in order to maintain the trend. Finally, in case of completely random data, all autocorrelations will tend to be zero (or not significantly

FIGURE 3-1   Example of the Same Variable with Different Time Lags

Time	X Original Variable Income	X1 One Time Lag Variable Constructed From Income	X2 Two Time Lags Variable Constructed From Income
t = 1	13	8	15
t = 2	8	15	4
t = 3	15	4	4
t = 4	4	4	12
t = 5	4	12	11
t = 6	12	11	7
t = 7	11	7	14
t = 8	7	14	12
t = 9	14	12	
t = 10	12		

different from zero) since successive values of random data do not relate to each other. By definition, such a relationship is randomness.*

The application of autocorrelations is sometimes difficult to understand, and therefore several examples will be used to illustrate it. Figures 3-2 through 3-9 provide several illustrations of autocorrelations. A careful examination of these figures is the best way to understand the usefulness of the approach in allowing one to determine:

    a. whether the data are stationary;
    b. whether the data are seasonal;
    c. the length of seasonality, if it exists; and
    d. whether the data (or residuals) are random.

The concepts covered in Figures 3-2 through 3-9 are discussed following the presentation of these figures. The reader is encouraged to look at the figures very carefully in order to understand the pattern of the data from their autocorrelations and vice versa. As an overview, the following points are included:

- Figure 3-4 (a) Seasonal data series with slight trend
  Figure 3-4 (b) Autocorrelations of seasonal data with slight trend in 3-4 (a)
  Figure 3-5 Autocorrelations of 3-4 (a) after taking first differences
- Figure 3-6 (a) Seasonal data series with dominant trend
  Figure 3-6 (b) Autocorrelations of seasonal data with dominant trend in 3-6 (a)
  Figure 3-7 Autocorrelations of 3-6 (a) after taking first differences
- Figure 3-8 (a) Data series with trend and randomness
  Figure 3-8 (b) Autocorrelations of series with trend and randomness in 3-8 (a)
  Figure 3-9 Autocorrelations of 3-8 (a) after taking first differences

The number of data points of the series is limited to forty. This size is rather small to produce accurate results when considerable randomness exists in the series as is the case with many observed time series. In the examples used, forty data are sufficient because the series chosen have little or no randomness.

---

*Because the concept of autocorrelation is a statistical one, it can be defined and used much more rigorously than is being done here. However, to minimize the mathematical and statistical background needed to use this concept, the remainder of this and subsequent chapters will illustrate the use of autocorrelations through examples rather than through statistical theory. The reader who would like to go deeper into the topic can consult the references given at the end of this chapter.

FIGURE 3-2(a)   Graph of the Data Points Used in Figure 3-2(b) (Random Data)

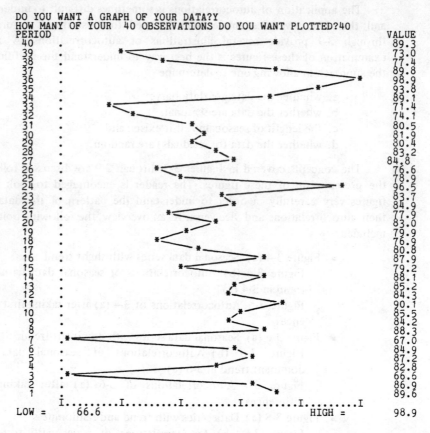

```
DO YOU WANT A GRAPH OF YOUR DATA?Y
HOW MANY OF YOUR  40 OBSERVATIONS DO YOU WANT PLOTTED?40
PERIOD                                                        VALUE
 40                                              *            89.3
 39                       *                                   73.0
 38                           *                               77.4
 37                                         *                 89.8
 36                                             *             98.9
 35                                        *                  93.8
 34                                      *                    89.1
 33           *                                              71.4
 32              *                                            74.1
 31                      *                                    80.5
 30                     *                                     81.9
 29                     *                                     80.4
 28                       *                                   83.2
 27                        *                                  84.8
 26              *                                            76.6
 25               *                                           78.9
 24                                                 *         96.5
 23                  *                                        83.7
 22                    *                                      84.9
 21            *                                              77.9
 20                   *                                       83.0
 19             *                                             79.9
 18         *                                                 76.9
 17                    *                                      80.8
 16                        *                                  87.9
 15             *                                             79.2
 14                          *                                88.1
 13                     *                                     85.2
 12                     *                                     84.3
 11                           *                               90.1
 10                     *                                     85.5
  9                    *                                      84.2
  8                          *                                88.3
  7     *                                                     67.0
  6                     *                                     84.9
  5                       *                                   87.2
  4                    *                                      82.8
  3     *                                                     66.6
  2                      *                                    86.9
  1                        *                                  89.6
     I.........I.........I.........I.........I.........I
LOW =     66.6                                    HIGH =    98.9
```

FIGURE 3-2(b)   Autocorrelations of a Completely Random Series

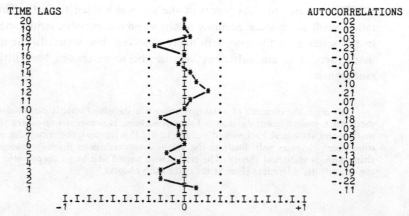

```
DO YOU WANT A GRAPH OF THE AUTOCORRELATIONS?Y

TIME LAGS                                          AUTOCORRELATIONS
  20              .          *        .                 -.02
  19              .        *          .                 -.02
  18              .          *        .                  .03
  17              .   *                .                -.23
  16              .        I           .                -.01
  15              .       I            .                -.07
  14              .         I          .                 .06
  13              .          I   *     .                 .10
  12              .          I      *  .                 .21
  11              .          I*        .                 .07
  10              .        * I         .                -.01
   9              .     *    I         .                -.18
   8              .       *I           .                -.03
   7              .      * I            .               -.05
   6              .         I*          .                .01
   5              .    *     I          .               -.13
   4              .        *I           .               -.04
   3              .   *      I           .              -.19
   2              .  *       I           .              -.22
   1              .          I*          .               .11
     I.I.I.I.I.I.I.I.I.I.I.I.I.I.I.I.I.I.I.I.I.I.I.I
    -1                      0                      +1
```

FIGURE 3-3(a)   Graph of Data Points Used in Figure 3-3(b) (Seasonal Data Only)

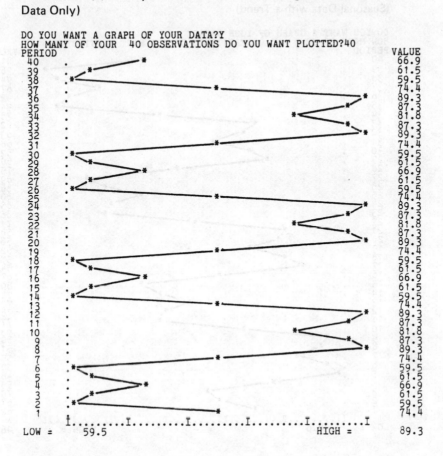

```
DO YOU WANT A GRAPH OF YOUR DATA?Y
HOW MANY OF YOUR  40 OBSERVATIONS DO YOU WANT PLOTTED?40
PERIOD                                                    VALUE
  40                                                       66.9
  39                                                       61.5
  38                                                       59.5
  37                                                       74.4
  36                                                       89.3
  35                                                       87.3
  34                                                       81.8
  33                                                       87.3
  32                                                       89.3
  31                                                       74.4
  30                                                       59.5
  29                                                       61.5
  28                                                       66.9
  27                                                       61.5
  26                                                       59.5
  25                                                       74.4
  24                                                       89.3
  23                                                       87.3
  22                                                       81.8
  21                                                       87.3
  20                                                       89.3
  19                                                       74.4
  18                                                       59.5
  17                                                       61.5
  16                                                       66.9
  15                                                       61.5
  14                                                       59.5
  13                                                       74.4
  12                                                       89.3
  11                                                       87.3
  10                                                       81.8
   9                                                       87.3
   8                                                       89.3
   7                                                       74.4
   6                                                       59.5
   5                                                       61.5
   4                                                       66.9
   3                                                       61.5
   2                                                       59.5
   1                                                       74.4
     I........I........I........I........I........I
LOW =    59.5                          HIGH =        89.3
```

FIGURE 3-3(b)   Autocorrelation Coefficients of a Monthly Seasonal Series

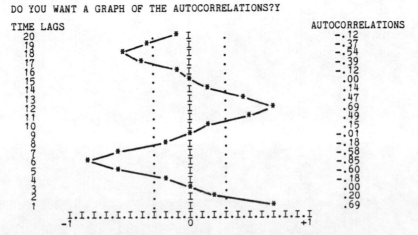

```
DO YOU WANT A GRAPH OF THE AUTOCORRELATIONS?Y
TIME LAGS                                   AUTOCORRELATIONS
  20                                             -.12
  19                                             -.37
  18                                             -.54
  17                                             -.39
  16                                             -.12
  15                                              .00
  14                                              .14
  13                                              .47
  12                                              .69
  11                                              .49
  10                                              .15
   9                                             -.01
   8                                             -.18
   7                                             -.58
   6                                             -.85
   5                                             -.60
   4                                             -.18
   3                                              .00
   2                                              .20
   1                                              .69
    I.I.I.I.I.I.I.I.I.I.I.I.I.I.I.I.I.I.I.I.I
   -1              0              +1
```

FIGURE 3-4(a)   Graph of Data Points Used in Figure 3-4(b) and Figure 3-5 (Seasonal Data with a Trend)

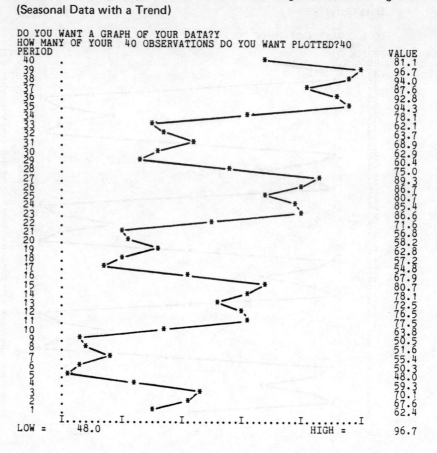

```
DO YOU WANT A GRAPH OF YOUR DATA?Y
HOW MANY OF YOUR  40 OBSERVATIONS DO YOU WANT PLOTTED?40
PERIOD                                                          VALUE
  40                                              *              81.1
  39                                                *            96.7
  38                                              *              94.0
  37                                        *                    87.6
  36                                            *                92.8
  35                                              *              94.3
  34                              *                              78.1
  33                          *                                  62.1
  32                            *                                63.7
  31                                *                            68.9
  30                          *                                  62.9
  29                        *                                    60.4
  28                                  *                          75.0
  27                                        *                    89.3
  26                                      *                      86.7
  25                                  *                          80.7
  24                                    *                        85.4
  23                                        *                    86.6
  22                                *                            71.6
  21                      *                                      56.8
  20                      *                                      58.2
  19                        *                                    62.8
  18                      *                                      57.2
  17                  *                                          54.8
  16                          *                                  67.9
  15                                *                            80.7
  14                              *                              78.1
  13                            *                                72.5
  12                              *                              76.5
  11                                *                            77.5
  10                      *                                      63.8
   9                *                                            50.5
   8                *                                            51.6
   7                  *                                          55.4
   6              *                                              50.3
   5            *                                                48.0
   4                    *                                        59.3
   3                          *                                  70.1
   2                                                             67.6
   1                *                                            62.4
     I.........I.........I.........I.........I.........I
LOW =      48.0                              HIGH =       96.7
```

FIGURE 3-4(b)   Autocorrelations of a Series Consisting of Monthly Seasonal Data and a Trend (The Seasonality is More Dominant than the Trend)

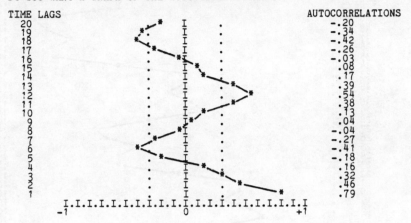

DO YOU WANT A GRAPH OF THE AUTOCORRELATIONS?Y

TIME LAGS ... AUTOCORRELATIONS

TIME LAGS	AUTOCORRELATIONS
20	-.20
19	-.34
18	-.42
17	-.26
16	-.03
15	.08
14	.17
13	.39
12	.54
11	.38
10	.13
9	.04
8	-.04
7	-.27
6	-.41
5	-.18
4	.16
3	.32
2	.46
1	.79

FIGURE 3-5   Autocorrelations of the Series in Figure 3-4(b), After the First Difference Has Been Taken

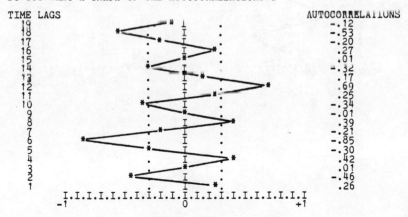

DO YOU WANT A GRAPH OF THE AUTOCORRELATIONS?Y

TIME LAGS	AUTOCORRELATIONS
19	-.12
18	-.53
17	-.20
16	.27
15	.01
14	-.32
13	.17
12	.69
11	.25
10	-.34
9	-.01
8	.39
7	-.21
6	-.85
5	-.30
4	.42
3	.01
2	-.46
1	.26

FIGURE 3-6(a) Graph of Data Points Used in Figure 3-6(b) and 3-7 (Seasonal Data with a Trend)

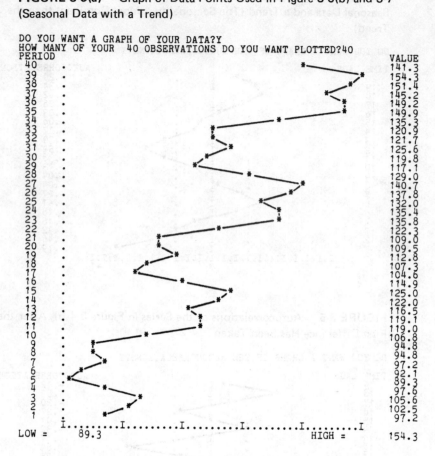

```
DO YOU WANT A GRAPH OF YOUR DATA?Y
HOW MANY OF YOUR  40 OBSERVATIONS DO YOU WANT PLOTTED?40
PERIOD                                                      VALUE
```

PERIOD	VALUE
40	141.3
39	154.3
38	151.4
37	145.2
36	149.2
35	149.9
34	135.3
33	120.9
32	121.7
31	125.6
30	119.8
29	117.1
28	129.0
27	140.7
26	137.8
25	132.0
24	135.4
23	135.8
22	122.3
21	109.0
20	109.5
19	112.8
18	107.3
17	104.6
16	114.9
15	125.0
14	122.0
13	116.5
12	119.1
11	119.0
10	106.8
9	94.8
8	94.8
7	97.2
6	92.1
5	89.3
4	97.6
3	105.6
2	102.5
1	97.2

```
       I.........I.........I.........I.........I.........I
LOW =     89.3                           HIGH =    154.3
```

FIGURE 3-6(b) Autocorrelations of a Series with Seasonality and Trend (Where the Trend is Dominant)

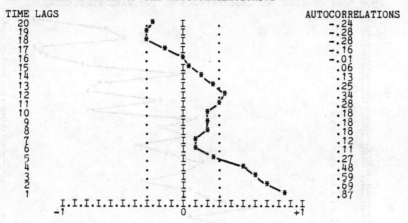

```
DO YOU WANT A GRAPH OF THE AUTOCORRELATIONS?Y
TIME LAGS                                      AUTOCORRELATIONS
  20                                               -.24
  19                                               -.28
  18                                               -.28
  17                                               -.16
  16                                               -.01
  15                                                .06
  14                                                .13
  13                                                .25
  12                                                .34
  11                                                .28
  10                                                .18
   9                                                .18
   8                                                .18
   7                                                .12
   6                                                .11
   5                                                .27
   4                                                .48
   3                                                .59
   2                                                .69
   1                                                .87
     I.I.I.I.I.I.I.I.I.I.I.I.I.I.I.I.I.I.I.I
     -1              0              +1
```

FIGURE 3-7 Autocorrelations of the Series in Figure 3-6(b), After the Trend has been Removed by the Method of First Differences

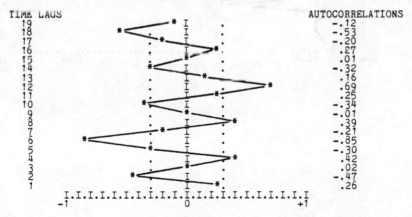

```
DO YOU WANT A GRAPH OF THE AUTOCORRELATIONS?Y
TIME LAGS                                      AUTOCORRELATIONS
  19                                               -.12
  18                                               -.53
  17                                               -.20
  16                                                .27
  15                                                .01
  14                                               -.32
  13                                                .16
  12                                                .69
  11                                                .25
  10                                               -.34
   9                                               -.01
   8                                                .39
   7                                               -.21
   6                                               -.85
   5                                               -.30
   4                                                .42
   3                                                .02
   2                                               -.47
   1                                                .26
     I.I.I.I.I.I.I.I.I.I.I.I.I.I.I.I.I.I.I.I
     -1              0              +1
```

FIGURE 3-8(a)   Graph of Data Points Used in Figure 3-8(b) and 3-9 (Series with Trend and Randomness)

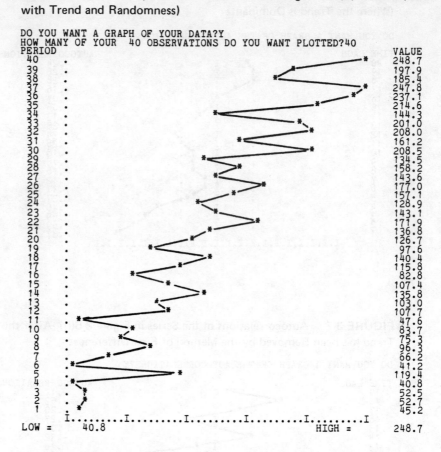

```
DO YOU WANT A GRAPH OF YOUR DATA?Y
HOW MANY OF YOUR  40 OBSERVATIONS DO YOU WANT PLOTTED?40
PERIOD                                                    VALUE
  40                                                      248.7
  39                                                      197.9
  38                                                      185.4
  37                                                      247.8
  36                                                      237.1
  35                                                      214.6
  34                                                      144.3
  33                                                      201.0
  32                                                      208.0
  31                                                      161.2
  30                                                      208.5
  29                                                      134.5
  28                                                      158.2
  27                                                      143.6
  26                                                      177.0
  25                                                      157.1
  24                                                      128.9
  23                                                      143.1
  22                                                      171.9
  21                                                      136.8
  20                                                      126.7
  19                                                       97.6
  18                                                      140.4
  17                                                      115.8
  16                                                       82.8
  15                                                      107.4
  14                                                      135.8
  13                                                      103.0
  12                                                      107.7
  11                                                       47.6
  10                                                       83.7
   9                                                       75.3
   8                                                       96.3
   7                                                       66.2
   6                                                       41.2
   5                                                      119.4
   4                                                       40.8
   3                                                       52.5
   2                                                       52.7
   1                                                       45.2
LOW =     40.8                              HIGH =      248.7
```

FIGURE 3-8(b)   Autocorrelations of a Non-Seasonal and Non-Stationary (Trend) Series

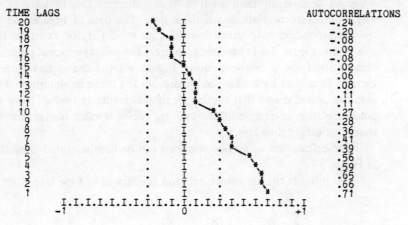

FIGURE 3-9   Autocorrelations of the Series in Figure 3-8(b) after the Trend has been Removed through First Differences

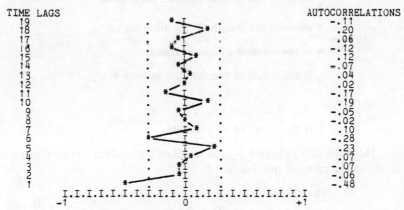

It is very important to understand that a graph of the autocorrelation coefficient is not the same as a graph of the data. The former is a summary of all data points. Thus, instead of having to examine the graph of all data in order to understand their pattern, it is sufficient just to look at the corresponding autocorrelations for these data. The data of Figures 3-3 (a), and the corresponding autocorrelations of Figure 3-3 (b), for example, illustrate this point. Figure 3-3 (a) presents a perfect (noiseless) seasonal pattern which repeats itself every twelve periods. Figure 3-3 (b) shows this pattern very clearly. Thus, just by looking at Figure 3-3 (b) it can be determined that the data are seasonal and that the length of seasonality is twelve. Note that the autocorrelation corresponding to time lag twelve is much higher than those of shorter or longer time lags.)

The calculation of autocorrelations can be demonstrated using the data in Figure 3-1.

The formula for the autocorrelation coefficient of time lage k is:

$$r_k = \frac{\sum\limits_{t=1}^{n-k} (X_t - \overline{X})(X_{t+k} - \overline{X})}{\sum\limits_{t=1}^{n} (X_t - \overline{X})^2}$$

where

r_k denotes the autocorrelation coefficient

k denotes the length of the time lag

n is the number of observations

X_t is the value of the variable at time t

and

$\overline{X}$ is the mean of all the data

Using the data in Figure 3-1 the calculations involved can be illustrated.
The mean of this data is

$$\frac{13 + 8 + 15 + 4 + 4 + \ldots + 7 + 14 + 12}{10} = \frac{100}{10} = 10.$$

$$r_1 = \frac{(13 - 10)(8 - 10) + (8 - 10)(15 - 10) +}{(13 - 19)^2 \quad + \quad (8 - 10)^2 +}$$

$$\frac{+\ (15 - 10)\ (4 - 10) + \ldots \ldots + (14 - 10)\ (12 - 10)}{+\ (15 - 10)^2 + \ldots \ldots + 14 - 10)^2 + (12 - 10)^2}$$

$$r_1 = \frac{-27}{144} = -0.188$$

When k = 3 the calculation is as follows:

$$r_3 = \frac{\displaystyle\sum_{t=1}^{7} (X_t - 10)\ (X_{t+3} - 10)}{\displaystyle\sum_{t=1}^{10} (X_t - 10)}$$

$$= \frac{(13 - 10)\ (15 - 10) + (8 - 10)\ (4 - 10)}{(13 - 10)^2 \quad + \quad (8 - 10)^2}$$

$$+ \frac{\ldots + (7 - 10)\ (12 - 10)}{\ldots + \quad (12 - 10)^2}$$

$$= \frac{26}{144} = 0.18$$

If the autocorrelations drop slowly to zero, and more than two or three differ significantly from zero, it indicates the presence of a trend in the data. This trend can be removed by differencing, as described in the next section. If the autocorrelations oscillate around zero and differ significantly from zero for some time lags, it indicates a seasonal pattern. The length of seasonality can be determined either from the number of periods it takes for the autocorrelations to make a complete cycle or by the time lag giving the largest autocorrelation. (See Figures 3-3 (b) and 3-4 (b).) (Note that the maximum number of lags, k for which the autocorrelations are computed is a matter of judgment. Generally, it's desirable to compute them for lags covering two or more complete cycles, but keeping k ≤ n/3.)

DIFFERENCING A SERIES

When a trend is present in a data series, it can often obscure the presence of a seasonal pattern in the autocorrelations.* (See Figures 3-6 (b) and 3-7.) Once the presence of a trend is determined, it is common to remove it and to then examine the autocorrelations of the de-trended (stationary) series to identify any seasonality. The most frequently used procedure for eliminating trends (removing non-stationarity) is that of differencing.

The method of differencing a series consists of subtracting the first value of a series from the second, the second from the third, the third from the fourth, and so on. The purpose of differencing is the elimination of a possible trend. This can be seen in the simple linear series: 2, 4, 6, 8, 10, 12, 14. Subtracting successive values from each other gives

$$4 - 2 = 2$$
$$6 - 4 = 2$$
$$8 - 6 = 2$$
$$10 - 8 = 2$$
$$12 - 10 = 2$$
$$14 - 12 = 2$$

The new series 2,2,2,2,2,2 is stationary. If the outcome of the first differencing does not produce a horizontal or stationary series, the data should be differenced again. This repeat of the process provides the second differences:

$$2 - 2 = 0$$
$$2 - 2 = 0$$
$$2 - 2 = 0$$
$$2 - 2 = 0$$
$$2 - 2 = 0$$

*In this book the concept of trend is often used synonimously with non-stationarity. Even though, statistically speaking, this is not strictly correct (there are non-stationarities, for example, in levels that are not the results of a trend) it is a convenient and practical way to convey the idea of non-stationarity to those with a non-statistical background.

In most practical situations, the first or second difference will remove any non-stationarity that exists. Since randomness is also present in most real situations, some variation will generally be present in the differenced series, unlike the differences obtained for the randomless series 2,4,6,8,10,12,14. (This series was used to illustrate the idea that differencing produces stationarity.)

In addition to facilitating identification of seasonality, differencing is also useful because there are several methods of forecasting that can handle only stationary data. Thus, if the data is non-stationary it must be transformed to a stationary level using differencing before such a method of time series can be applied.

Non-stationary data can be seen in Figures 3-4 (a), 3-6 (a) and 3-8 (a), while stationary data are shown in the remaining figures. The difference between the two is that the autocorrelations for stationary data are scattered around zero, while for non-stationary data with a trend they have a diagonal direction from right to left.

The level at which the data are stationary can be found:

a. By examining the graphical plots of the autocorrelation coefficients of the original and differenced data.

b. By comparing the means of the autocorrelations. (An indication of non-stationarity can be obtained by comparing the means of the first and second half of the autocorrelations. If the data are non-stationary, usually the two means will be different.)

c. By comparing the coefficients of variation (the smallest coefficient of variation will correspond to the level of differencing at which the data are stationary).

d. By examining the computed χ^2, Box-Pierce Q statistic. (The smallest computed χ^2 usually corresponds to the level of differencing at which the data are stationary.)

A typical sequence in the application of autocorrelation analysis to a new series would be first to examine the autocorrelations of the original data for the presence of non-stationarity. If it is present, the series is differenced, and the autocorrelations of the differenced series are examined for trend. This differencing is repeated until stationarity is achieved. With a stationary (trendless or horizontal) series, the autocorrelations are examined for seasonality as indicated by values that oscillate in sign and are periodically significantly different from zero.

This same procedure of autocorrelation analysis can be applied to the series of errors (residuals) obtained from a forecasting model as well as to original data. If the model is a good one, the errors should be random, containing no pattern. The autocorrelations for such a random error series will be like those shown in Figure 3-2 (b). However, if the errors are not random,— indicating that the model has not removed all of the pattern in forecasting – their autocorrelations will indicate that fact by having some values significantly different from zero. In fact, the series of error values might contain a trend or a seasonal pattern.

INPUT/OUTPUT FOR AUTOCORRELATION PROGRAM (ACORN)

Inputs

1. *Enter number of time lags for autocorrelation*

Input the number of autocorrelation coefficients to be calculated. Since each time lag corresponds to one autocorrelation, if you put 10, the program will calculate 10 autocorrelations corresponding to time lags of 1, 2, ...10.

One aim of autocorrelation analysis is to reveal the existence and length of seasonality. To achieve this one needs to consider two to three times as many autocorrelations as the suspected length of seasonality.

2. *Do you want a graph of the autocorrelations?*

The answer can be YES or NO. YES indicates that you want to have the autocorrelations printed as a graph. (It is a fairly slow process to print a graph on a 10-character-per-second time sharing terminal.) If the autocorrelations of the original data indicate the existence of non-stationarity (they do not drop to zero rapidly – after the second or third autocorrelation), one must rerun the series, after taking the first difference. Similarly, if the autocorrelation coefficients of the first differences do not drop to zero reasonably fast, one must rerun with second differences. This rerun is done automatically. The program can attempt to determine at what level data are stationary, whether the data are seasonal, and if the data are seasonal, their length of seasonality. There are not, however, any statistically-based rules to determine the level of stationarity, the existence of seasonality, and its length. Intuitive judgment is required. The program tries to substitute for this in a way that is not always successful. The user should therefore check the conclusions reached by the program.

Outputs

1. *Mean of all autocorrelations*

2. *Mean of the first half of all autocorrelations*

3. *Mean of the second half of all autocorrelations*

When the data are stationary, the mean of the first half (2) should be about the same as the mean of the second half (3). If this is not so, the program should be rerun with the first differences.

4. *Coefficient of variation*

The coefficient of variation shows how dispersed the data are around the constant mean of the original data. Stationary data have a smaller coefficient of variation than non-stationary data. Thus, as a practical rule, the smaller coefficient of variation corresponds to the degree of differencing at which the data are stationary.

5. *Standard error of the autocorrelations*

It indicates the extent of dispersion of the autocorrelations about their mean value, assuming that the data are completely random. The dotted lines in the autocorrelation graph are from minus two to plus two standard deviations from zero. This represents the 95% confidence interval for the hypothesis: This autocorrelation is not significantly different from zero.

6. *Chi-squared (χ^2)*

The χ^2 computed indicates whether the autocorrelations computed are from a random set of data. If the χ^2 computed is smaller than the corresponding χ^2 from the table (e.g., at a 95% confidence interval) the data are random; otherwise they may have some pattern. Also a smaller value of the computed χ^2 indicates less pattern is present in the data. This may be an indication of the level of differencing at which the data are stationary.

7. *Graph of autocorrelations*

The graph allows one to see the data pattern as summarized by the autocorrelations. The user should be cautious in interpreting this graph because when there is a strong trend, the remaining components of the pattern are dominated by the trend and cannot be seen.

References for further study:

       Box, G. and G. Jenkins (1976, Chapter 2)
       Box, G. and D. Pierce (1970)
       Jenkins, G. and D. Watts (1968)
       Makridakis, S. and S. Wheelwright (1978, Chapter 2)
       Quenouille, M. (1949)
       Wheelwright, S. and S. Makridakis (1977, Chapter 8)

EXERCISE 3.1

The following 36 data points exhibit a strong trend and a seasonal pattern whose length is 12 periods. Use the autocorrelation program to confirm this.

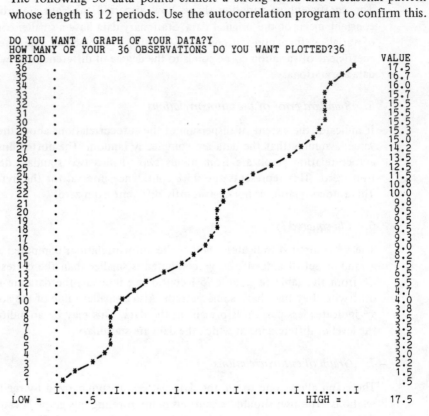

```
DO YOU WANT A GRAPH OF YOUR DATA?Y
HOW MANY OF YOUR  36 OBSERVATIONS DO YOU WANT PLOTTED?36
PERIOD                                                    VALUE
```

Explanation of solution

Figures 3-10 through 3-12 indicate the results obtained using the autocorrelation program, ACORN. It should be noted that in Figure 3-12, the computer program has used a very simple decision rule — identification of the largest absolute autocorrelation — to determine seasonality. In this instance, it has mistakenly determined a seasonal pattern of 6, pointing out the need for judgment on the part of the user because the graphs in Figures 3-11 and 3-12 indicate a complete seasonal pattern that covers 12 periods.

FIGURE 3-10   Autocorrelation Analysis

```
WHICH PROGRAM WOULD YOU LIKE TO RUN?ACORN


*****   AUTOCORRELATION ANALYSIS   *****
NEED HELP (Y OR N)?N
HOW MANY OBSERVATIONS DO YOU WANT TO USE?36
DATA FILENAME?EX2

DO YOU WANT TO ANALYSE YOUR DATA YOURSELF(Y OR N)?N


        TABLE    1   AUTOCORRELATIONS    OTH DIFFERENCE
        -----       -----------------       ----------

MEAN AUTOCORRELATION         .277     STANDARD ERROR=        .167
MEAN OF FIRST   9   VALUES=      .604  CHI SQUARE (COMPUTED)=     138.4
MEAN OF LAST    9   VALUES=    -.051   CHI SQUARE (FROM TABLE)=    8.7
COEFFICIENT OF VARIATION =     .548

DO YOU WANT A GRAPH OF THE AUTOCORRELATIONS?Y
```

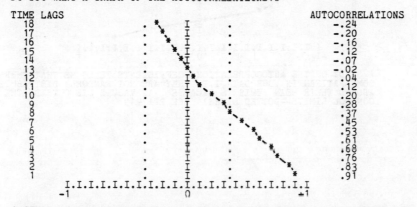

```
A STUDY OF THE AUTOCORRELATION COEFFICIENTS TELLS ME THAT THERE IS
SOME PATTERN IN YOUR DATA,I.E. THEY ARE NOT RANDOMLY DISTRIBUTED
AROUND THEIR MEAN (THIS IS BECAUSE    8 VALUES LIE OUTSIDE THE
CONTROL LIMITS--DOTTED LINES)   SEE PP11-19
```

FIGURE 3-11   Autocorrelations of First Differences

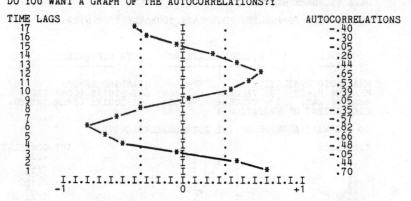

```
     TABLE     2    AUTOCORRELATIONS     1ST DIFFERENCE
     -----          ----------------     --------------

MEAN AUTOCORRELATION       -.014    STANDARD ERROR=       .169
MEAN OF FIRST    8  VALUES=    -.224    CHI SQUARE (COMPUTED)=    134.5
MEAN OF LAST    9  VALUES=      .173    CHI SQUARE (FROM TABLE)=    8.0
COEFFICIENT OF VARIATION =     .036

DO YOU WANT A GRAPH OF THE AUTOCORRELATIONS?Y

TIME LAGS                                          AUTOCORRELATIONS
  17          *            I         .                -.40
  16          *            I         .                -.30
  15          .         *  I         .                -.05
  14          .            I      *  .                 .26
  13          .            I         .*                .44
  12          .            I         . *               .65
  11          .            I         *                 .53
  10          .            I      *  .                 .39
   9          .            I *       .                 .05
   8          .       *    I         .                -.35
   7          .   *        I         .                -.57
   6       *  .            I         .                -.82
   5          . *          I         .                -.66
   4          .      *     I         .                -.48
   3          .         *  I         .                -.05
   2          .            I      *  .                 .44
   1          .            I         . *               .70
     I.I.I.I.I.I.I.I.I.I.I.I.I.I.I.I.I.I.I.I.I
    -1                     0                   +1
```

A STUDY OF THE AUTOCORRELATION COEFFICIENTS TELLS ME THAT THERE IS
SOME PATTERN IN YOUR DATA,I.E. THEY ARE NOT RANDOMLY DISTRIBUTED
AROUND THEIR MEAN (THIS IS BECAUSE   12 VALUES LIE OUTSIDE THE
CONTROL LIMITS--DOTTED LINES)   SEE PP11-19

FIGURE 3-12 Autocorrelations of Second Differences

```
     TABLE    3    AUTOCORRELATIONS      2ND DIFFERENCE
     -----         -----------------      -----------

MEAN AUTOCORRELATION      -.007      STANDARD ERROR=        .171
MEAN OF FIRST   8  VALUES=     -.147   CHI SQUARE (COMPUTED)=       67.3
MEAN OF LAST    9  VALUES=      .117   CHI SQUARE (FROM TABLE)=      8.0
COEFFICIENT OF VARIATION =     .027
```

DO YOU WANT A GRAPH OF THE AUTOCORRELATIONS?Y

```
TIME LAGS                                          AUTOCORRELATIONS
  17                    .        *                      -.00
  16                .  *     I                          -.26
  15                    .     I        *                -.00
  14                          I  *     .                 .31
  13                    .   * I                          -.00
  12                          I        .        *         .66
  11                    .   * I                          -.00
  10                          I        .  *               .34
   9                    .   * I                          -.00
   8                .  *     I                           -.40
   7                    .   * I                          -.00
   6      *     .             I                           .83
   5                    .   * I                          -.00
   4                .  *     I                           -.43
   3                    .   * I                          -.00
   2                          I        .        *         .48
   1                    .   * I                          -.00
    I.I.I.I.I.I.I.I.I.I.I.I.I.I.I.I.I.I.I.I.I.I
   -1                   0                      +1
```

A STUDY OF THE AUTOCORRELATION COEFFICIENTS TELLS ME THAT THERE IS
SOME PATTERN IN YOUR DATA,I.E. THEY ARE NOT RANDOMLY DISTRIBUTED
AROUND THEIR MEAN (THIS IS BECAUSE   6 VALUES LIE OUTSIDE THE
CONTROL LIMITS--DOTTED LINES)   SEE PP11-19

YOUR DATA ARE STATIONARY AT THEIR   1ST DIFFERENCE

THEREFORE I AM LOOKING AT TABLE   2 TO DETERMINE THE LENGTH
OF SEASONALITY,IF ANY EXISTS

AUTOCORRELATIONS SIGNIFICANTLY DIFFERENT THAN ZERO (ORDERED)
TIME LAG   AUTOCORRELATIONS
    6   .82
    1   .70
    5   .66
   12   .65
    7   .57
   11   .53
    4   .48
   13   .44
    3   .44
   17   .40
   10   .39
    8   .35
```

I CONSIDER THE LENGTH OF YOUR SEASONALITY IS =      6
```

CHAPTER 4

Moving Averages

The major objective in using moving averages is to eliminate randomness in a time series. This objective is achieved by averaging several data points together in such a way that positive and negative errors eventually cancel themselves out. The averaging is done over a constant number of observations. The term "moving average" is used because, as each new observation in the series becomes available, the oldest observation is dropped and a new average is computed. The result of calculating the moving average over a set of data points is a new series of numbers with little or no randomness.

The ability of moving averages to eliminate randomness can be used in time series analysis for two main purposes: a) to eliminate trend; and b) to eliminate seasonality.

ELIMINATION OF TREND

The elimination of trend consists of averaging a certain number of data points and then subtracting this average from the middle value of the data that were averaged. The outcome is a stationary series. If the number of terms is not an odd number, then one has the choice of subtracting the average from either the term before or the term after the middle, or applying a two-period moving average on the already averaged data. This two-period moving average is frequently called a *centered* moving average. Figure 4-1 shows a simple series where N, the number of terms in the moving average, is an odd number. On the other hand, Figure 4-2 shows a moving average where N is even. In the latter case, one must center the moving averages already obtained by averaging consecutive pairs of terms, (i.e., making N = 2).

As can be seen from Figure 4-1, averaging three items and subtracting that average from the middle of the three numbers averaged gives a stationary series of values. (In this example it is exactly 0 because there is no randomness.) Similarly, in Figure 4-2 the objective of making the series stationary is achieved, but instead of taking a single moving average, a centered moving average is taken because the number of terms involved is even, and there is not a single middle number in the four terms that are averaged. This requires that a two-term moving average be taken to center the first moving average. The result of subtracting the centered moving average from the middle number is a stationary series.

FIGURE 4-1 A Single Moving Average (Odd Number of Terms)

t Time Period	X_t Time Series	$N-3$	M_t Moving Average	$X_t - M_t$
1	3		—	
2	5	$M_2 = \dfrac{X_1 + X_2 + X_3}{3} =$	5	0
3	7		7	0
4	9		9	0
5	11	$M_5 = \dfrac{X_4 + X_5 + X_6}{3} =$	11	0
6	13		13	0
7	15		15	0
8	17	$M_8 = \dfrac{X_7 + X_8 + X_9}{3} =$	17	0
9	19		19	0
10	21		21	0
11	23	$M_{11} = \dfrac{X_{10} + X_{11} + X_{12}}{3} =$	23	0
12	25		—	

FIGURE 4-2   A Single Centered Moving Average (Even Number of Terms)

t Time Period	X_t Time Series	$N = 4$	M_t Moving Average	$N = 2$	M'_t Moving Average	$X_t - M'_t$
1	3		—		—	
2	5	$M_2 = \dfrac{X_1 + X_2 + X_3 + X_4}{4} =$	6		—	
3	7	.	8	$M'_3 = \dfrac{M_2 + M_3}{2} =$	7	0
4	9	.	10	.	9	0
5	11	$M_5 = \dfrac{X_4 + X_5 + X_6 + X_7}{4} =$	12	$M'_5 = \dfrac{M_4 + M_5}{2} =$	11	0
6	13	.	14	.	13	0
7	15	.	16	.	15	0
8	17	$M_8 = \dfrac{X_7 + X_8 + X_9 + X_{10}}{4} =$	18	$M'_8 = \dfrac{M_7 + M_8}{2} =$	17	0
9	19	.	20	.	19	0
10	21	.	22	.	21	0
11	23	.	—		—	
12	25	.	—		—	

ELIMINATION OF SEASONALITY

One may assume that a series is made of three subcomponents (for more information see Chapter 15): trend, cycle, and randomness. Furthermore, if one assumes that their effect is additive, then the time series, X_t, is equal to

$$X_t = T_t + C_t + R_t \qquad (4\text{-}1)$$

where

T_t denotes trend

C_t denotes cycle

and

R_t denotes randomness.

Averaging X_t eliminates the randomness since negative and positive errors cancel each other. Thus (4-1) becomes:

$$M_t = T_t + C_t \qquad (4\text{-}2)$$

where M_t is the moving average at period t.

Subtracting (4-2) from (4-1) gives:

$$X'_t = T_t + C_t + R_t - (T_t + C_t)$$

$$X'_t = R_t \qquad (4\text{-}3)$$

That is, what remains in (4-3) is randomness illustrating that a moving average is indeed capable of isolating the random component.

If instead of assuming an additive relationship in (4-1), a multiplicative one is assumed, (4-1) becomes:

$$X_t = T_t \times C_t \times R_t \qquad (4\text{-}4)$$

$$M_t = T_t \times C_t \qquad (4\text{-}5)$$

and if (4-4) is divided (instead of subtracted) by (4-5), the same result is obtained:

$$X'_t = R_t \qquad (4\text{-}6)$$

A moving average can also be used to eliminate the seasonality of a series if N, the number of terms averaged, is equal to the length of seasonality. Obviously, with monthly data an average taken on 12 values gives the average for the year. Such an average is by definition free of seasonal effects, since high months are canceled by low months. Chapter 15 and 16 show more precisely how seasonality is removed by the method of moving averages.

A double moving average is a moving average of a moving average. For example, the centered moving average in Figure 4-2 can also be called a double moving average because M_t' is a moving average of the moving average presented in column M_t. In this way, one can calculate any desired level of moving averages whether it be triple, quadruple or even higher. The full range of possible moving averages is unlimited. Those which are more commonly used, however, are the following:

1. Single Moving Average
2. Centered Single Moving Average
3. A 3 × 3 Double Moving Average
4. A 3 × 5 Double Moving Average
5. A 3 × 9 Double Moving Average
6. Spencer's 15-term Weighted Moving Average
7. Henderson's 5-term Weighted Moving Average
8. Henderson's 9-term Weighted Moving Average
9. Henderson's 13-term Weighted Moving Average
10. Henderson's 23-term Weighted Moving Average

The reasons for using the above moving averages and not some others are mainly empirical. Investigators have found that one type of moving average is better than some other for most observed time series. The general rule is that the larger the N or the higher the order of the averaging, the better its ability to remove randomness. In this respect a value of N as large as possible should be sought. However, $(N - 1)/2$ terms are lost in the beginning and $(N - 1)/2$ in the end of the data (see Figure 4-2). This is undesirable, particularly for the observations at the end of the series. Thus a compromise must be made in such a way that N is large enough to eliminate randomness effectively but not so large that too many data are lost at the end of the series. (See Chapters 16 and 20 for more detail.)

The following are the formulas used to calculate these different moving averages where N is the number of terms in the moving average.

1. Single Moving Average

$$M_t = \frac{X_{t-\frac{N-1}{2}} + X_{t-\frac{N-3}{2}} + \ldots + X_{t+\frac{N-1}{2}}}{N}$$

2. Centered Moving Average (used after formula 1)

$$M_t' = \frac{M_{t-1} + M_t}{2}$$

3. A 3 × 3 Moving Average (or a 5-term weighted moving average)

$$M_t = .111X_{t-2} + .222X_{t-1} + .333X_t + .222X_{t+1} + .111X_{t+2}$$

4. A 3 × 5 Moving Average (or a 7-term weighted moving average)

$$M_t = .067X_{t-3} + .133X_{t-2} + .200X_{t-1} + .200X_t$$
$$+ .200X_{t+1} + .133X_{t+2} + .067X_{t+3}$$

5. A 3 × 9 Moving Average (or an 11-term weighted moving average)

$$M_t = .037X_{t-5} + .074X_{t-4} + .111X_{t-3} + .111X_{t-2} + .111X_{t-1}$$
$$+ .111X_t + .111X_{t+1} + .111X_{t+2} + .111X_{t+3} + .074X_{t+4}$$
$$+ .037X_{t+5}$$

6. Spencer 15-term Weighted Moving Average

$$M_t = -.009X_{t-7} - .019X_{t-6} - .016_{t-5} + .009X_{t-4} + .066X_{t-3}$$
$$+ .144X_{t-2} + .209X_{t-1} + .231X_t + .209X_{t+1}$$
$$+ .144X_{t+2} + .066X_{t+3} + .009X_{t+4} - .016X_{t+5}$$
$$- .019X_{t+6} - .009X_{t+7}$$

7. Henderson 5-term Weighted Moving Average

$$M_t = -.073X_{t-2} + .294X_{t-1} + .558X_t + .294X_{t+1} - .073X_{t+2}$$

8. Henderson 9-term Weighted Moving Average

$$M_t = -.041X_{t-4} - .010X_{t-3} + .119X_{t-2} + .267X_{t-1} + .33X_t$$
$$+ .267X_{t+1} + .119X_{t+2} - .010X_{t+3} - .041X_{t+4}$$

9. Henderson 13-term Weighted Moving Average

$$M_t = -.019X_{t-6} - .028X_{t-5} + 0X_{t-4} + .066X_{t-3} + .147X_{t-2}$$
$$+ .214X_{t-1} + .24X_t + .214X_{t+1} + .147X_{t+2} + .066X_{t+3}$$
$$+ 0X_{t+4} - .028X_{t+5} - .019X_{t+6}$$

10. Henderson 23-term Weighted Moving Average

$$M_t = -.004X_{t-11} - .011X_{t-10} - .016X_{t-9} - .015X_{t-8}$$

$$- .005X_{t-7} + .013X_{t-6} + .039X_{t-5} + .068X_{t-4}$$

$$+ .097X_{t-3} + .122X_{t-2} + .138X_{t-1} + .148X_t + .138X_{t+1}$$

$$+ .122X_{t+2} + .097X_{t+3} + .068X_{t+4} + .039X_{t+5} + .013X_{t+6}$$

$$- .005X_{t+7} - .015X_{t+8} - .016X_{t+9} - .011X_{t+10}$$

$$- .004X_{t+11}$$

References for further study:

Kendall, M. (1973, Chapter 2)
Makridakis, S. and S. Wheelwright (1978, Chapter 3)
Shiskin, J. et al. (1967)
Wheelwright, S. and S. Makridakis (1977, Chapter 3)

INPUT/OUTPUT FOR THE MOVING AVERAGE PROGRAM (MUVING)

Input

1. *What kind of moving average would you like to use?*

The user may choose any one of the moving averages mentioned above by typing in the number referencing that average. Except for the single and centered moving averages where the user must specify the number of terms to be included in the moving average, there are no other inputs required to run this program. For single moving averages, the number of terms should be an odd number, and for centered moving averages it *must* be an even number.

Output

The output of the program will be a list of the actual data, the moving averages, and the ratio of the two series. The user has the option of putting either the moving averages or the difference between the actual and the moving average data into a file for later use in any of the other programs in the system. Doing so provides a stationary series to work with.

EXERCISE 4.1

Given the following 28 values, calculate:

    a. a single moving average
    b. a centered moving average
    c. a Henderson's 23-term weighted moving average.

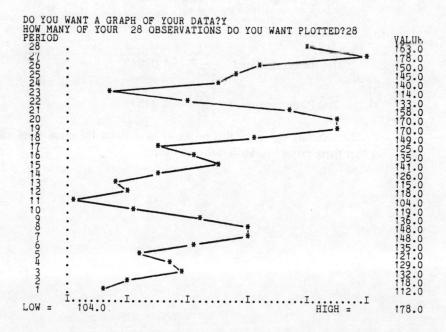

```
DO YOU WANT A GRAPH OF YOUR DATA?Y
HOW MANY OF YOUR  28 OBSERVATIONS DO YOU WANT PLOTTED?28
PERIOD                                                        VALUE
  28    .                                    *                163.0
  27    .                                         *           178.0
  26    .                               *                     150.0
  25    .                              *                      145.0
  24    .                             *                       140.0
  23    .          *                                          114.0
  22    .                  *                                  133.0
  21    .                                 *                   158.0
  20    .                             *                       170.0
  19    .                             *                       170.0
  18    .                      *                              149.0
  17    .             *                                       125.0
  16    .                *                                    135.0
  15    .                   *                                 141.0
  14    .              *                                      126.0
  13    .           *                                         115.0
  12    .            *                                        118.0
  11    .       *                                             104.0
  10    .            *                                        119.0
   9    .                 *                                   136.0
   8    .                     *                               148.0
   7    .                     *                               148.0
   6    .                *                                    135.0
   5    .          *                                          121.0
   4    .            *                                        129.0
   3    .             *                                       132.0
   2    .         *                                           118.0
   1    .       *                                             112.0
        I.........I.........I.........I.........I.........I
LOW =    104.0                                  HIGH =        178.0
```

Explanation of Solutions to the Single Moving Average Exercise (See Figure 4-3)

The numbers printed in the column "Moving Average" were obtained as follows:

$$\text{Moving Average} = \frac{\text{a sum of actual values}}{\text{number of terms in moving average}}$$

Thus, in this series:

$$\text{1st Moving Average calculated} = \frac{112 + 118 + 132 + 129 + 121}{5} = \frac{612}{5} = 122.4$$

$$\text{2nd Moving Average calculated} = \frac{118 + 132 + 129 + 121 + 135}{5} = \frac{635}{5} = 127$$

In the second moving average calculated, the first observation — 112 — is dropped and the sixth observation — 135 — is added. It follows that in the third moving average calculated, the second observation — 118 — is dropped and the 7th observation — 148 — is added.

The "Ratio: Actual/Moving Average" column is obtained as follows:

$$\text{Ratio} = \frac{\text{middle value of data averaged}}{\text{moving average}}$$

Thus, in this series:

$$\text{1st Ratio calculated} = \frac{132}{122.4} = 1.07843$$

$$\text{2nd Ratio calculated} = \frac{129}{127} = 1.01575$$

The resulting ratios for this series all have about the same value, suggesting that these ratios provide a stationary series.

FIGURE 4-3   Single Moving Average

WHICH PROGRAM WOULD YOU LIKE TO RUN?MUVING

********** COMMONLY USED MOVING AVERAGE**********

NEED HELP (Y OR N)?N
HOW MANY OBSERVATIONS DO YOU WANT TO USE?28
DATA FILENAME?ANYNAM
WHAT TYPE OF MOVING AVERAGE WOULD YOU LIKE TO USE
1 - SINGLE MOVING AVERAGE
2 - CENTERED MOVING AVERAGE
3 - A 3X3 DOUBLE MOVING AVERAGE
4 - A 3X5 DOUBLE MOVING AVERAGE
5 - A 3X9 DOUBLE MOVING AVERAGE
6 - SPENCER-S 15-POINTS WEIGHTED MOVING AVERAGE
7 - HENDERSON S 5 TERM WEIGHTED MOVING AVERAGE
8 - HENDERSON S 9 TERM WEIGHTED MOVING AVERAGE
9 - HENDERSON S 13 TERM WEIGHTED MOVING AVERAGE
10 - HENDERSON S 23 TERM WEIGHTED MOVING AVERAGE
ANSWER BY TYPING 1,2,3,4,......,10.
?1

ENTER THE NUMBER OF PERIODS IN THE MOVING AVERAGE?5

PERIOD	ACTUAL	MOVING AVERAGE	RATIO ACTUAL/MOV. AV.
1	112.		
2	118.		
3	132.	122.400	1.078
4	129.	127.000	1.016
5	121.	133.000	.910
6	135.	136.200	.991
7	148.	137.600	1.076
8	148.	137.200	1.079
9	136.	131.000	1.038
10	119.	125.000	.952
11	104.	118.400	.878
12	118.	116.400	1.014
13	115.	120.800	.952
14	126.	127.000	.992
15	141.	128.400	1.098
16	135.	132.200	.???
17	125.	144.000	.868
18	149.	149.800	.995
19	170.	154.400	1.101
20	170.	156.000	1.090
21	158.	149.000	1.060
22	133.	143.000	.930
23	114.	138.000	.826
24	140.	136.400	1.026
25	145.	145.400	.997
26	150.	155.200	.966
27	178.		
28	165.		

DO YOU WANT TO SAVE EITHER THE MOVING AVERAGE OR THE RATIO?N

Explanation of Solution to the Centered Moving Average Exercise (See Figure 4-4)

The numbers printed in the "Moving Average" columns were obtained as follows:

$$\frac{112 + 118 + 132 + 129 + 121 + 135}{6} = \frac{747}{6} = 124.5 \left.\begin{array}{c} \\ \\ \end{array}\right\} \frac{124.5 + 130.5}{2} = \frac{255}{2}$$

$$\frac{118 + 132 + 129 + 121 + 135 + 148}{6} = \frac{783}{6} = 130.5 \left.\begin{array}{c} \\ \\ \\ \end{array}\right\} \qquad = 127.5$$

$$\frac{132 + 129 + 121 + 135 + 148 + 148}{6} = \frac{813}{6} = 135.5 \left.\begin{array}{c} \\ \\ \\ \end{array}\right\} \frac{130.5 + 135.5}{2} = \frac{266}{2}$$

$$\left.\begin{array}{c} \\ \end{array}\right\} \qquad = 133$$

The "Ratio: Actual/Moving Average" column is obtained as before, the computer having chosen the fourth value of data averaged as center:

$$\text{Ratio} = \frac{\text{4th data value included in average}}{\text{centered moving average}}$$

Thus, in this series:

$$\text{1st Ratio calculated} = \frac{129}{127.5} = 1.01176$$

$$\text{2nd Ratio calculated} = \frac{121}{133} = 0.909774$$

Again, the series is stationary.

Explanation of Solution to Henderson's 23-Term Weighted Moving Average (See Figure 4-5)

The values on Figure 4-5 are found by applying formula 10 (Henderson's 23-term average). For example, the value 123.783 is found as follows:
When $t = 12$

TABLE 4-1

Description of Value	Value	Henderson's Coefficient	Product (Result)
X_{t-11}	112	− 0.004 =	−0.0448
X_{t-10}	118	− 0.011 =	−1.298
.	.	.	.
.	.	.	.
.	.	.	.
$X_t - 1$	104	+ 0.138 =	+14.352
X_t	118	+ 0.148 =	+17.464
X_{t+1}	115	+ 0.138 =	+15.87
.	.	.	.
.	.	.	.
.	.	.	.
X_{t+10}	133	− 0.011 =	−1.463
X_{t+11}	114	− 0.004 =	−0.456
		Sum of results	123.783

It can be seen that X_t is the middle term of the 23 data points averaged, i.e., the 12th value. Therefore, the first moving average calculated is placed next to the 12th observation. As in the previous examples, if 118 is divided by 123.783 the resulting ratio is 0.9533, and all of these ratios together form a stationary series.

FIGURE 4-4   Centered Moving Average

```
EXE METHOD
  8K
WHICH PROGRAM WOULD YOU LIKE TO RUN?MUVING

********** COMMONLY USED MOVING AVERAGE**********

NEED HELP (Y OR N)?N
HOW MANY OBSERVATIONS DO YOU WANT TO USE?28
DATA FILENAME?ANYNAM
WHAT TYPE OF MOVING AVERAGE WOULD YOU LIKE TO USE
1 - SINGLE MOVING AVERAGE
2 - CENTERED MOVING AVERAGE
3 - A 3X3 DOUBLE MOVING AVERAGE
4 - A 3X5 DOUBLE MOVING AVERAGE
5 - A 3X9 DOUBLE MOVING AVERAGE
6 - SPENCER-S 15-POINTS WEIGHTED MOVING AVERAGE
7 - HENDERSON S 5 TERM WEIGHTED MOVING AVERAGE
8 - HENDERSON S 9 TERM WEIGHTED MOVING AVERAGE
9 - HENDERSON S 13 TERM WEIGHTED MOVING AVERAGE
10 - HENDERSON S 23 TERM WEIGHTED MOVING AVERAGE
ANSWER BY TYPING 1,2,3,4,.....,10.
?2

ENTER THE NUMBER OF PERIODS IN THE MOVING AVERAGE?6

PERIOD          ACTUAL      MOVING AVERAGE    RATIO ACTUAL/MOV. AV.
   1            112.
   2            118.
   3            132.
   4            129.          127.500              1.012
   5            121.          133.000               .910
   6            135.          135.833               .994
   7            148.          135.333              1.094
   8            148.          133.083              1.112
   9            136.          130.250              1.044
  10            119.          126.083               .944
  11            104.          121.500               .856
  12            118.          120.083               .983
  13            115.          121.833               .944
  14            126.          124.917              1.009
  15            141.          129.250              1.091
  16            135.          136.417               .990
  17            125.          144.667               .864
  18            149.          149.750               .995
  19            170.          151.000              1.126
  20            170.          149.917              1.134
  21            158.          148.250              1.066
  22            133.          145.417               .915
  23            114.          141.667               .805
  24            140.          141.667               .988
  25            145.          145.833               .994
  26            150.
  27            178.
  28            163.

DO YOU WANT TO SAVE EITHER THE MOVING AVERAGE OR THE RATIO?N
```

FIGURE 4-5   Henderson's 23-Term Weighted Moving Average

```
EXE METHOD
 8K
WHICH PROGRAM WOULD YOU LIKE TO RUN?MUVING

********** COMMONLY USED MOVING AVERAGE**********

NEED HELP (Y OR N)?N
HOW MANY OBSERVATIONS DO YOU WANT TO USE?28
DATA FILENAME?ANYNAM
WHAT TYPE OF MOVING AVERAGE WOULD YOU LIKE TO USE
1 - SINGLE MOVING AVERAGE
2 - CENTERED MOVING AVERAGE
3 - A 3X3 DOUBLE MOVING AVERAGE
4 - A 3X5 DOUBLE MOVING AVERAGE
5 - A 3X9 DOUBLE MOVING AVERAGE
6 - SPENCER-S 15-POINTS WEIGHTED MOVING AVERAGE
7 - HENDERSON S 5 TERM WEIGHTED MOVING AVERAGE
8 - HENDERSON S 9 TERM WEIGHTED MOVING AVERAGE
9 - HENDERSON S 13 TERM WEIGHTED MOVING AVERAGE
10 - HENDERSON S 23 TERM WEIGHTED MOVING AVERAGE
ANSWER BY TYPING 1,2,3,4,......,10.
?10
```

PERIOD	ACTUAL	MOVING AVERAGE	RATIO ACTUAL/MOV. AV.
1	112.		
2	118.		
3	132.		
4	129.		
5	121.		
6	135.		
7	148.		
8	148.		
9	136.		
10	119.		
11	104.		
12	118.	123.783	.953
13	115.	125.045	.920
14	126.	127.970	.985
15	141.	132.013	1.068
16	135.	136.372	.990
17	125.	140.477	.890
18	149.		
19	170.		
20	170.		
21	150.		
22	133.		
23	114.		
24	140.		
25	145.		
26	150.		
27	178.		
28	163.		

```
DO YOU WANT TO SAVE EITHER THE MOVING AVERAGE OR THE RATIO?Y

DO YOU WANT TO SAVE THE MOVING AVERAGE ?Y
```

CHAPTER 5

Single Exponential Smoothing

Exponential smoothing methods were developed in the early 1950s. Since then they have become a particularly popular method of forecasting among businessmen because they are easy to use, require very little computer time, and need only a few data points to produce future predictions. These smoothing methods are well suited for short or immediate term predictions of a large number of items. They are suitable for stationary data or when there is a slow growth or decline over time.

The method of exponential smoothing is based on averaging (smoothing) past values of a time series in a decreasing (exponential) manner. This is achieved by formula (5-1) which, if expanded by substituting previous values of F_t [see (5-2)], results in exponentially decreasing weights being given to past observations. Another way of looking at equation (5-1) is to express it in the form of equation (5-8), which is obtained by the simple algebraic manipulation of (5-1). Equation (5-8) indicates that the forecast for the next period is equal to the forecast of the current period plus an adjustment that depends upon the magnitude and sign of the error. This adjustment will be a multiple of the value of alpha. Thus, the error between the actual and the forecasted value for the current period is the means for correcting the forecast for the next period. This is the simplest form of the control principle which is used extensively in more sophisticated methods.

$$F_{t+1} = \alpha X_t + (1 - \alpha) \, F_t \tag{5-1}$$

but

$$F_t = \alpha X_{t-1} + (1 - \alpha) \, F_{t-1} \tag{5-2}$$

$$F_{t-1} = \alpha X_{t-2} + (1 - \alpha) \, F_{t-2} \tag{5-3}$$

Substituting (5-2) into (5-1) gives:

$$F_{t+1} = \alpha X_t + (1 - \alpha) \, (\alpha X_{t-1} + (1 - \alpha) \, F_{t-1})$$

$$F_{t+1} = \alpha X_t + \alpha(1 - \alpha) \, X_{t-1} + (1 - \alpha)^2 \, F_{t-1} \tag{5-4}$$

Similarly, substituting (5-3) into (5-4) will result in:

$$F_{t+1} = \alpha X_t + \alpha(1 - \alpha) \, X_{t-1} + \alpha(1 - \alpha)^2 \, X_{t-2} + (1 - \alpha)^3 \, F_{t-2} \tag{5-5}$$

In general,

$$F_{t+1} = \alpha X_t + \alpha(1 - \alpha) \, X_{t-1} + \alpha(1 - \alpha)^2 \, X_{t-2} + \alpha(1 - \alpha)^3 \, X_{t-3}$$

$$+ \, \alpha(1 - \alpha)^4 \, X_{t-4} + \ldots \tag{5-6}$$

Equation (5-6) illustrates how past values of the time series, X_t, are weighted in an exponentially or geometrically decreasing manner, accounting for the name, exponential smoothing.

Equation (5-1) can be rewritten as:

$$F_{t+1} = \alpha X_t + F_t - \alpha F_t$$

or

$$F_{t+1} = F_t + \alpha(X_t - F_t) \tag{5-7}$$

or

$$F_{t+1} = F_t + \alpha e_t \tag{5-8}$$

where $X_t - F_t$ is the difference between the actual and forecasted value, i.e. the residual error, e_t. Thus, the forecast for period $t + 1$ is the same as that of period t plus a percentage adjustment of the last error. (The forecast is adjusted in the direction indicated by the sign of the error.) This adjustment principle is a powerful one that is used widely in engineering and statistics.

References for further study:

Brown, R. (1959, Chapter 3)
Brown, R. (1961)
Brown, R. (1963, Chapters 4 and 7)
S. Makridakis and S. Wheelwright (1978, Chapter 3)
S. Wheelwright and S. Makridakis (1977, Chapter 3)

INPUT/OUTPUT FOR THE SINGLE EXPONENTIAL SOOTHING PROGRAM (EXPO)

Input

1. *Enter a value for alpha?*

Alpha (or Greek letter α) is a constant whose value is between zero and one. The value of alpha influences the accuracy of the forecast. It must therefore be specified in such a way that the forecast errors will be as small as possible. This can be done by trial and error, starting with a certain value of alpha and then increasing or decreasing it to find the value that minimizes the mean squared error. However, as an empirical rule, it is known that small values of alpha smooth the data to a much greater extent than large values of alpha. Thus, if the data have substantial fluctuations or randomness, one should use a small value for alpha. On the other hand, data with little randomness or with a clear pattern, need a larger value of alpha. (This is not always true for higher order exponential smoothing methods.) The importance of alpha can be seen in the equations (5-7) and (5-8).

Exponential smoothing is a method that is frequently used on an on-going basis by repeated application of (5-1). When a new X_t value becomes available, (5-1) is applied and a forecast for F_{t+1}, i.e., the next period, is found. At the outset of any use of exponential smoothing, however, an initial value for F_1 must be specified. There are many ways of estimating a starting value of F_1. In the program it is simply set at $F_1 = X_1$. The value of F_1 has little importance in an on-going system after the first several periods because of the exponentially decreasing weights of (5-6).

If the program EXPO is run continually for non-teaching purposes, line 1180 should be replaced by:

    1180 F = F(1)

where F(1) is the last forecasted value produced when the program was last run for the same forecasting operation.

Output

The outputs of the single exponential smoothing program are:
1. An optional table of all the actual and forecasted values together with their absolute and percentage errors;
2. Mean squared error, mean absolute percentage error, and the mean percentage error;
3. An optional forecast for as many periods as the user desires.

These outputs are standard for all forecasting programs and they will not be described individually from now on. Only output that differs from the three items above will be described.

EXERCISE 5.1

The following 22 data points are the past history of item E14 used for the production of product E. Could you forecast the demand of item E14 for the next four periods (21-24)?

What is the alpha value that minimizes the mean squared error?

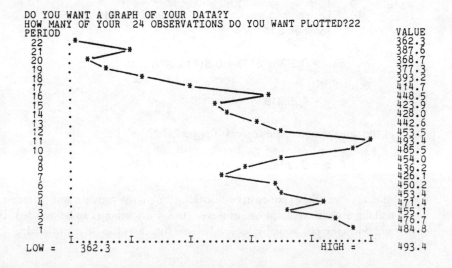

```
DO YOU WANT A GRAPH OF YOUR DATA?Y
HOW MANY OF YOUR  24 OBSERVATIONS DO YOU WANT PLOTTED?22
PERIOD                                                      VALUE
  22    *                                                   362.3
  21    .          *                                        387.6
  20    .       *                                           368.7
  19    .        *                                          377.3
  18    .           *                                       393.2
  17    .               *                                   414.7
  16    .                     *                             448.5
  15    .                  *                                423.9
  14    .                  *                                428.0
  13    .                    *                              442.6
  12    .                      *                            453.5
  11    .                           *                       493.4
  10    .                         *                         485.5
   9    .                     *                             454.0
   8    .                *                                  436.2
   7    .               *                                   426.1
   6    .                  *                                450.2
   5    .                  *                                453.4
   4    .                    *                              471.4
   3    .                 *                                 455.1
   2    .                   *                               476.7
   1    .                      *                            484.8
        I.........I.........I.........I.........I.........I
LOW =   362.3                              HIGH =           493.4
```

Explanation of Solution

Suppose one wants to find the forecast for period 21. Using Formula 5.1 gives:

$$F_{21} = \alpha X_{20} + (1 - \alpha) F_{20}$$

where

$$\alpha = 0.2$$

$$X_{20} = 368.712$$

and

$$F_{20} = 420.784 \text{ which was the forecast obtained during period 20.}$$

Thus

$$F_{21} = 0.2(368.712) + (1 - 0.2) (420.784)$$

$$= 410.369$$

Continuing, the forecast for period 22 will be:

$$F_{22} = 0.2(X_{21}) + (1 - 0.2) (410.369)$$

actual	α	Forecasted value at
value		period 21 (see
i.e. demand		previous result)
for item E14		

$$F_{22} = 0.2(387.617) + 0.8(410.369)$$

$$= 405.819$$

In the same manner, the forecast for period 23 will be:

$$F_{23} = 0.2(362.34) + 0.8(405.819) = 397.123$$

Actually, simple exponential smoothing can only provide one forecast. By making a slight assumption, however, this shortcoming can be avoided. Using 362.34 as the last actual value, additional forecasts can be completed:

$$F_{24} = 0.2(362.34) + 0.8(397.123) = 390.16$$

$$F_{25} = 0.2(362.34) + 0.8(390.16) = 380.15$$

$$F_{26} = 0.2(362.34) + 0.8(384.60) = 380.15$$

In order to find the best alpha (α), one must run the program with alpha at different values and see their effect on the mean squared error (MSE). In the three examples shown in Figures 5-1 through 5-3, for

$$\alpha = 0.2 \text{ the MSE} = 962.845$$

$$\alpha = 0.5 \text{ the MSE} = 533.327$$

$$\alpha = 0.8 \text{ the MSE} = 433.753$$

Therefore, of these three examples, $\alpha = 0.8$ is the best since it gives the minimum MSE of 433.753. It should be noted that in many applications of simple exponential smoothing, a range of $0.01 < \alpha < .30$ is used. When an alpha value close to 1.0 gives the best results, it may indicate that trends or seasonality are present and that simple exponential smoothing is not the best approach.

FIGURE 5-1   Single Exponential Smoothing ($\alpha = 0.2$)

```
WHICH PROGRAM WOULD YOU LIKE TO RUN?EXPO


***      EXPONENTIAL SMOOTHING      ***
YOU CAN CHOOSE BETWEEN 6 PROGRAMS
1. EXPO - SINGLE EXPONENTIAL SMOOTHING
2. EXPOTL - ADAPTIVE RESPONSE RATE EXPONENTIAL SMOOTHING
3. EXPO2 - LINEAR EXPONENTIAL SMOOTHING
4. EXPOH - LINEAR (HOLTS) EXPONENTIAL SMOOTHING
5. EXPOQ - QUADRATIC (BROWN S) EXPONENTIAL SMOOTHING
6. EXPOW - LINEAR AND SEASONAL (WINTERS) EXPONENTIAL SMOOTHING.
HOW MANY OBSERVATIONS DO YOU WANT TO USE?22
DATA FILENAME?NEWNAM

WHAT PROGRAM DO YOU WANT TO RUN - TYPE ITS NAME ?EXPO


***      SINGLE EXPONENTIAL SMOOTHING      ***

NEED HELP (Y OR N)?N

ENTER A VALUE FOR ALPHA (BETWEEN 0 AND 1)?0.2
DO YOU WANT A TABLE OF ACTUAL AND PREDICTED (Y OR N)?Y
```

PERIOD	ACTUAL	FORECAST	ERROR	PCT ERROR
1	484.83			
2	476.71	484.83	-8.12	1.70%
3	455.07	483.21	-28.14	6.18%
4	471.39	477.58	-6.19	1.31%
5	453.40	476.34	-22.94	5.06%
6	450.23	471.75	-21.52	4.78%
7	426.11	467.45	-41.34	9.70%
8	436.18	459.18	-23.00	5.27%
9	454.02	454.58	-.56	.12%
10	485.50	454.47	31.03	6.39%
11	493.40	460.67	32.73	6.63%
12	453.52	467.22	-13.70	3.02%
13	442.62	464.48	-21.86	4.94%
14	428.04	460.11	-32.07	7.49%
15	423.93	453.69	-29.76	7.02%
16	448.50	447.74	.76	.17%
17	414.71	447.89	-33.18	8.00%
18	393.16	441.26	-48.10	12.23%
19	377.34	431.64	-54.30	14.39%
20	368.71	420.78	-52.07	14.12%
21	387.61	410.36	-22.75	5.87%
22	362.34	405.81	-43.47	12.00%

```
MEAN PC ERROR (MPE) OR BIAS     =     -5.24%
MEAN SQUARED ERROR (MSE)        =      962.8
MEAN ABSOLUTE PC ERROR  (MAPE) =        6.5%


FORECAST FOR HOW MANY PERIODS AHEAD (0=NONE,24=MAX)?4
```

PERIOD	FORECAST
23	397.12
24	390.16
25	384.60
26	380.15

```
WOULD YOU LIKE AUTOCORRELATIONS ON THE RESIDUALS ?N

DO YOU WANT TO RUN ANOTHER PROGRAM?Y
```

FIGURE 5-2   Single Exponential Smoothing ($\alpha = 0.5$)

```
HOW MANY OBSERVATIONS DO YOU WANT TO USE?22
DATA FILENAME?NEWNAM

WHAT PROGRAM DO YOU WANT TO RUN - TYPE ITS NAME ?EXPO

***      SINGLE EXPONENTIAL SMOOTHING      ***

NEED HELP (Y OR N)?N

ENTER A VALUE FOR ALPHA (BETWEEN 0 AND 1)?0.5
DO YOU WANT A TABLE OF ACTUAL AND PREDICTED (Y OR N)?Y

PERIOD        ACTUAL        FORECAST        ERROR      PCT ERROR

   1          484.83
   2          476.71        484.83          -8.12         1.70%
   3          455.07        480.77         -25.70         5.65%
   4          471.39        467.92           3.47          .74%
   5          453.40        469.66         -16.26         3.59%
   6          450.23        461.53         -11.30         2.51%
   7          426.11        455.88         -29.77         6.99%
   8          436.18        440.99          -4.81         1.10%
   9          454.02        438.59          15.43         3.40%
  10          485.50        446.30          39.20         8.07%
  11          493.40        465.90          27.50         5.57%
  12          453.52        479.65         -26.13         5.76%
  13          442.62        466.59         -23.97         5.41%
  14          428.04        454.60         -26.56         6.21%
  15          423.93        441.32         -17.39         4.10%
  16          448.50        432.63          15.87         3.54%
  17          414.71        440.56         -25.85         6.23%
  18          393.16        427.64         -34.48         8.77%
  19          377.34        410.40         -33.06         8.76%
  20          368.71        393.87         -25.16         6.82%
  21          387.61        381.29           6.32         1.63%
  22          362.34        384.45         -22.11         6.10%
MEAN PC ERROR (MPE) OR BIAS     =    -2.70%
MEAN SQUARED ERROR (MSE)        =     533.3
MEAN ABSOLUTE PC ERROR  (MAPE)  =      4.9%

FORECAST FOR HOW MANY PERIODS AHEAD (0=NONE,24=MAX)?4

PERIOD        FORECAST
  23           373.39
  24           367.87
  25           365.10
  26           363.72

WOULD YOU LIKE AUTOCORRELATIONS ON THE RESIDUALS ?N

DO YOU WANT TO RUN ANOTHER PROGRAM?Y
```

FIGURE 5-3 Single Exponential Smoothing ($\alpha = 0.8$)

```
HOW MANY OBSERVATIONS DO YOU WANT TO USE?22
DATA FILENAME?NEWNAM

WHAT PROGRAM DO YOU WANT TO RUN - TYPE ITS NAME ?EXPO

***      SINGLE EXPONENTIAL SMOOTHING      ***

NEED HELP (Y OR N)?N

ENTER A VALUE FOR ALPHA (BETWEEN 0 AND 1)?0.8
DO YOU WANT A TABLE OF ACTUAL AND PREDICTED (Y OR N)?Y
```

PERIOD	ACTUAL	FORECAST	ERROR	PCT ERROR
1	484.83			
2	476.71	484.83	-8.12	1.70%
3	455.07	478.33	-23.26	5.11%
4	471.39	459.72	11.67	2.48%
5	453.40	469.06	-15.66	3.45%
6	450.23	456.53	-6.30	1.40%
7	426.11	451.49	-25.38	5.96%
8	436.18	431.19	4.99	1.14%
9	454.02	435.18	18.84	4.15%
10	485.50	450.25	35.25	7.26%
11	493.40	478.45	14.95	3.03%
12	453.52	490.41	-36.89	8.13%
13	442.62	460.90	-18.28	4.13%
14	428.04	446.28	-18.24	4.26%
15	423.93	431.69	-7.76	1.83%
16	448.50	425.48	23.02	5.13%
17	414.71	443.90	-29.19	7.04%
18	393.16	420.55	-27.39	6.97%
19	377.34	398.64	-21.30	5.64%
20	368.71	381.60	-12.89	3.50%
21	387.61	371.29	16.32	4.21%
22	362.34	384.35	-22.01	6.07%

```
MEAN PC ERROR (MPE) OR BIAS      =    -1.80%
MEAN SQUARED ERROR (MSE)         =   433.7
MEAN ABSOLUTE PC ERROR   (MAPE)  =     4.4%


FORECAST FOR HOW MANY PERIODS AHEAD (0=NONE,24=MAX)?4
```

PERIOD	FORECAST
23	366.74
24	363.22
25	362.52
26	362.38

```
WOULD YOU LIKE AUTOCORRELATIONS ON THE RESIDUALS ?N

DO YOU WANT TO RUN ANOTHER PROGRAM?Y
```

CHAPTER 6

Adaptive Response Rate Exponential Smoothing

In the previous chapter the method of single exponential smoothing was discussed. The application of the method is based on expression (5-1) which requires only a value for the smoothing constant α before it can be applied. However, specifying α may give rise to difficulties, particularly if it must be done for a large number of series, such as in inventory forecasting, for example. To avoid this difficulty several methods have been proposed in the literature that allow for the automatic specification of the parameter α. These methods eliminate the need for the user to specify α and in addition, they allow α to vary from one period to another so that the forecast can respond to the pattern of the data as closely as possible. When the forecasting error increases – that is, the forecasts are consistently under or over the actual values – a smaller or larger value for α, respectively, is specified.

In this manner, forecasting methods truly responsive to changes in the pattern of data and completely automatic, requiring no input from the user, can be obtained. The only shortcoming of "adaptive" methods is the fact that they might overreact to changes in the data in such a way that random fluctuations are mistakenly identified as changes in the pattern of data, giving poor forecasting results.

There are several adaptive methods of forecasting available. (See references at the end of this chapter.) The authors feel that the one proposed by Trigg and Leach is representative, probably most widely applied, and performs extremely well. It will be described next.

Adaptive response rate exponential smoothing is based on the same equations as single exponential smoothing [see (5-1) and (6-1)], but α varies and is calculated according to (6-2).

$$F_{t+1} = \alpha_t X_t + (1 - \alpha_t) F_t \qquad (6\text{-}1)$$

$$\alpha_{t+1} = \left| \frac{E_t}{M_t} \right| \qquad (6\text{-}2)$$

where

$$E_t = \beta e_t + (1 - \beta) E_{t-1} \qquad (6\text{-}3)$$

$$M_t = \beta \left| e_t \right| + (1 - \beta) M_{t-1} \qquad (6\text{-}4)$$

and

$$e_t = X_t - F_t \qquad (6\text{-}5)$$

β is usually set at .1 or .2. Finally, α_{t+1} is computed instead of α_t to allow the system to "settle" a little by not being too responsive to changes. α_t could, however, have been used also.

As can be seen from (6-2), the value of α_t will vary with the ratio of actual to absolute error values in a way that deserves some further consideration. Expression (6-3) smooths the actual errors while (6-4) smooths the absolute error values. If E_t and M_t are about equal this means that (6-1) forecasts in such a way that no obvious bias of under or overestimating occurs. Thus, a value of α_t close to one will result. The same high value of α_t will result when the magnitude of errors (e_{t-1}) are small since in this case M_t will not be too different than E_t. But most importantly α_t will vary, according to (6-2), based on variations in the pattern of the data.

References for further study:

Brown, R. G. (1963)
Chow, W. M. (1965)
Farley, J. and M. Hinich (1970)
McClain, J. and L. Thomas (1973)
Makridakis, S. and S. Wheelwright (1978, Chapter 3)
Trigg, D. (1964)
Trigg, D. and D. Leach (1967)

INPUT/OUTPUT FOR THE ADAPTIVE RESPONSE RATE EXPONENTIAL SMOOTHING PROGRAM (EXPOTL)

Input

None.

If the program EXPOTL is run continually for non-teaching purposes the last values for F_t, E_t and M_t should be printed each time the program is run. Line 210 should be inserted as follows:

    210 PRINT S(NØ), E1, E2

These three printed values should be input as follows as initial values the next time the program is run for the same forecasting operation:

    1010 S(1) = value of S(NØ)
    1040 E1   = value of E1
    1050 E2   = value of E2

Output (as with Single Exponential Smoothing. Chapter 5).

1. Optional table of actual, forecasted and error values
2. MSE, MAPE and MPE
3. Optional forecasts for future periods.

EXERCISE 6.1

Using the data in Chapter 5, obtain forecasts for the next four periods. Compare the forecasts obtained using the adaptive response rate exponential smoothing with those of the single exponential smoothing. How would you compare the accuracy of the two methods?

Explanation of Solution (See Figure 6-1)

The mean squared error for the adaptive response rate is much better than for single exponential smoothing with $\alpha = 0.2$, it is about the same with $\alpha = 0.5$, and it is worse with $\alpha = 0.8$.

FIGURE 6-1 Adaptive Response Rate Exponential Smoothing

```
HOW MANY OBSERVATIONS DO YOU WANT TO USE?22
DATA FILENAME?NEWNAM

WHAT PROGRAM DO YOU WANT TO RUN - TYPE ITS NAME ?EXPOTL


***      ADAPTIVE RESPONSE RATE EXPONENTIAL SMOOTHING      ***

NEED HELP (Y OR N)?N
DO YOU WANT A TABLE OF ACTUAL AND PREDICTED (Y OR N)?Y

PERIOD          ACTUAL          FORECAST          ERROR        PCT ERROR

   1            484.83
   2            476.71          484.83            -8.12          1.70%
   3            455.07          483.21           -28.14          6.18%
   4            471.39          477.58            -6.19          1.31%
   5            453.40          476.34           -22.94          5.06%
   6            450.23          453.40            -3.17           .70%
   7            426.11          450.23           -24.12          5.66%
   8            436.18          426.11            10.07          2.31%
   9            454.02          436.18            17.84          3.93%
  10            485.50          447.72            37.78          7.78%
  11            493.40          454.67            38.73          7.85%
  12            453.52          467.11           -13.59          3.00%
  13            442.62          459.48           -16.86          3.81%
  14            428.04          453.55           -25.51          5.96%
  15            423.93          450.52           -26.59          6.27%
  16            448.50          446.37             2.13           .48%
  17            414.71          447.14           -32.43          7.82%
  18            393.16          436.48           -43.32         11.02%
  19            377.34          413.20           -35.86          9.50%
  20            368.71          388.26           -19.55          5.30%
  21            387.61          373.11            14.50          3.74%
  22            362.34          384.84           -22.50          6.21%
MEAN PC ERROR (MPE) OR BIAS       =         -2.54%
MEAN SQUARED ERROR (MSE)          =        595.5
MEAN ABSOLUTE PC ERROR   (MAPE) =          5.0%


FORECAST FOR HOW MANY PERIODS AHEAD (0=NONE,24=MAX)4


PERIOD          FORECAST
  23             371.64
  24             365.44
  25             363.38
  26             362.69

WOULD YOU LIKE AUTOCORRELATIONS ON THE RESIDUALS ?N
```

The advantage of adaptive response rate exponential smoothing, however, is that one does not have to worry about finding the optimal alpha. The mean squared error may not be the minimum possible but it will be very close to it. Furthermore, even if the data change drastically, the value of alpha will be adjusted automatically to account for it.

Obtaining forecasts is a little more involved in the case of adaptive response rate exponential smoothing. Values have to be found for equations (6-2), (6-3), and (6-4). This is done as follows:

Assume one is at period 21. The values of E_t and M_t when $\beta = .2$ will be

$$E_{20} = -13.8458, \quad M_{20} = 23.5993 \text{ and } \alpha_{20} = .809$$

Substituting these values into (6-2), the value for α is

$$\alpha_{21} = \left| \frac{-13.8458}{23.5993} \right| = .5867$$

Using this value of α in formula (6-1) gives a forecast for period 22 of

$$F_{22} = \alpha_{21} X_{21} + (1 - \alpha_{21}) F_{21}$$

$$= 0.5867 \, (387.617) + 0.4219 \, (373.109)$$

$$F_{22} = 384.841$$

One could then update the value of α for use in forecasting the next period. Formula (6-5) becomes:

$$e_{22} = X_{22} - F_{22} = 362.34 - 384.84 = -22.501$$

$$E_{22} = 0.2(e_{22}) + (1 - 0.2)(E_{21}) \text{ where } \beta = 0.2$$

$$= 0.2 \, (-22.501) + (0.8) \, (-13.8458)$$

where E_{21} was calculated above

$$= -15.5767$$

Similarly Formula (6-4) becomes:

$$M_{22} = 0.2 \, e_{21} + (1 - 0.2) \, M_{21} \text{ where } \beta = 0.2$$

$$= 0.2 \left| -22.501 \right| + (0.8) \, (23.5993)$$

where M_{21} was calculated above

$$= 23.3796$$

Therefore the value of alpha becomes:

$$\alpha_{23} = \left| \frac{E_{22}}{M_{22}} \right| = \left| \frac{-15.5767}{23.3796} \right| = 0.6663$$

If the value of α is used in formula (6-1) one obtains a forecast for period 23, 24, 25 and 26 as follows (the last actual value, X_{22}, is used in forecasting beyond period 22):

$$F_{23} = .5867(362.34) + .4133(384.842) = 371.639$$

$$F_{24} = .6663(362.34) + .3337(371.639) = 365.443$$

$$F_{25} = .6663(362.34) + .3337(365.443) = 363.375$$

$$F_{26} = .6663(362.34) + .3337(363.375) = 362.686$$

Cases

EXERCISES ON SIMPLE EXPONENTIAL SMOOTHING

1. The Chamber of Commerce and Industry of Paris has been asked by several of its members to include a forecast of the French index of industrial production in its monthly newsletter. Using the monthly data given below

   a. Compute a forecast for periods 13 through 29 using the method of moving averages with 12 months of observations in each average.

   b. Compute the error in each forecast. How accurate would you say these forecasts are?

   c. Now compute a new series of moving averages using six months of observations in each average and again compute the errors.

   d. How do these two moving average forecasts compare? Which seems most accurate? Why do you think this is so?

Period	French Index of Industrial Production	Period	French Index of Industrial Production
1	108	15	98
2	108	16	97
3	110	17	101
4	106	18	104
5	108	19	101
6	108	20	99
7	105	21	95
8	100	22	95
9	97	23	96
10	95	24	96
11	95	25	97
12	92	26	98
13	95	27	94
14	95	28	92

2. The treasurer of a publicly-traded firm has been asked by the directors to include a forecast of the trading price of the company's stock in his monthly report to the board.

Using the method of moving averages and the data given below, answer questions (a) through (d) asked in Question 1.

e. Does this method of moving averages seem to work better in this case than in the situation given in Question 1? Explain why or why not.

Period	Average Stock Price	Period	Average Stock Price
1	72	13	73
2	71	14	75
3	73	15	75
4	72	16	68
5	70	17	68
6	74	18	71
7	76	19	72
8	76	20	69
9	78	21	72
10	71	22	75
11	73	23	72
12	70	24	74

3. Do the following:
   a. Apply the forecasting method of exponential smoothing to the data given in Problem 1 using $\alpha = .3$ and $\alpha = .6$.
   b. Compute the mean squared error for each of these two forecasts.
   c. Which value of α gives the best results? What does this imply about the basic pattern in the data?
   d. Apply the forecasting method of adaptive response rate exponential smoothing, compute the mean squared error and compare it with the result you found in (b) above.

4. Answer the three parts of Question 3 using the data given in Question 2.

5. Now compare the two moving average forecasts and the two exponential smoothing forecasts prepared for the data given in Question 1. Which of these forecasting approaches would you recommend be used in this situation? Why?

6. Answer Question 5 for the forecasts prepared on the data contained in Question 2.

7. The production manager of a firm that manufacturers small appliances has asked you to help him prepare a monthly forecast of the demand for toasters. The following data are available:

Month:	1/69	2/69	3/69	4/69	5/69	6/69	7/69	8/69	9/69	10/69	11/69
Demand:	2000	1350	1950	1975	3100	1750	1550	1300	2200	2775	2350

Questions

a. Use the method of moving averages to compute a forecast of monthly demand for toasters.

(1) Use a 3-month moving average.

(2) Use a 5-month moving average.

b. Compare the accuracy of these forecasts by

(1) Computing the mean absolute deviation of the errors.

(2) Developing a probability distribution of the errors.

c. What point forecast would you make for the demand for toasters for December 1969?

d. Suppose you had a decision tree that contained an event fan representing "demand for toasters in December 1969." How would you use the answers to Question 1-3 and the bracket median technique to convert that event fan into an event with four branches? What would be the values of each of the four branches?

e. Answer Questions 1-4, but rather than using the method of moving averages, use exponential smoothing with $\alpha = .1$, $\alpha = .5$, and $\alpha = .9$.

f. Answer Questions 1-4 using adaptive response rate exponential smoothing.

g. Of the six forecasting equations you used above, which one do you think is most appropriate for this situation? Why do you think this one works best?

8. Expression (5-6), if expanded further, yields

$$F_{t+1} = \alpha X_t + \alpha(1 - \alpha)X_{t-2} + \alpha(1 - \alpha)^2 X_{t-2} + \alpha(1 - \alpha)^3 X_{t-3}$$

$$+ \alpha(1 - \alpha)^4 X_{t-4} + \alpha(1 - \alpha)^5 X_{t-5} + \ldots + \alpha(1 - \alpha)^{n-1} X_{t-n-1} \quad (1)$$

where $e_{t+1} = X_{t+1} - F_{t+1}$.

Similarly, if (5-6) is expressed for F_t it will be

$$F_t = \alpha X_{t-1} + \alpha(1 - \alpha)X_{t-2} + \alpha(1 - \alpha)^2 X_{t-3} + \alpha(1 - \alpha)^3 X_{t-4}$$

$$+ \alpha(1 - \alpha)^4 X_{t-5} + \alpha(1 - \alpha)^5 X_{t-6} + \ldots + \alpha(1 - \alpha)^{n-2} X_{t-n-1} \quad (2)$$

But

$$e_{t+1} = X_{t+1} - F_{t+1}, \text{ and } e_t = X_t - F_t$$

Thus,

$$X_{t+1} = F_{t+1} + e_{t+1}, \quad \text{and} \quad X_t = F_t + e_t$$

Expressions (1) and (2) could be expressed in terms of X_t therefore:

$$X_{t+1} = \alpha X_t + \alpha(1-\alpha)X_{t-1} + \alpha(1-\alpha)^2 X_{t-2} + \alpha(1-\alpha)^3 X_{t-3}$$

$$+ \alpha(1-\alpha)^4 X_{t-4} + \ldots + \alpha(1-\alpha)^{n-1} X_{t-n-1} + e_{t+1} \qquad (3)$$

and

$$X_t = \alpha X_{t-1} + \alpha(1-\alpha)X_{t-2} + \alpha(1-\alpha)^2 X_{t-3} + \alpha(1-\alpha)^3 X_{t-4}$$

$$+ \alpha(1-\alpha)^4 X_{t-5} + \ldots + \alpha(1-\alpha)^{n-2} X_{t-n-1} + e_t \qquad (4)$$

Multiplying both sides of (4) by $(1-\alpha)$ yields

$$(1-\alpha)X_t = \alpha(1-\alpha)X_{t-1} + \alpha(1-\alpha)^2 X_{t-2} + \alpha(1-\alpha)^3 X_{t-3} + \alpha(1-\alpha)^4 X_{t-4}$$

$$+ \alpha(1-\alpha)^5 X_{t-5} + \ldots + \alpha(1-\alpha)^{n-1} X_{t-n-1} + (1-\alpha)e_t \qquad (5)$$

Subtracting (5) from (3), gives:

$$X_{t+1} - (1-\alpha)X_t = \alpha X_t + [\alpha(1-\alpha)X_{t-1} - \alpha(1-\alpha)X_{t-2}] + [\alpha(1-\alpha)^2 X_{t-2}$$

$$- \alpha(1-\alpha)^2 X_{t-2}] + \ldots + [\alpha(1-\alpha)^{n-1} X_{t-n-1} \qquad (6)$$

$$- \alpha(1-\alpha)^{n-1} X_{t-n-1}] + e_{t+1} - (1-\alpha)e_t$$

Expression (6) is considerably simplified since all terms except the first one of (3) and the last one of (3) and (6) cancel each other. Thus, (6) becomes

$$X_{t+1} - (1-\alpha)X_t = \alpha X_t + e_{t+1} - (1-\alpha)e_t$$

or

$$X_{t+1} - X_t + \alpha X_t = \alpha X_t + e_{t+1} - (1-\alpha)e_t$$

or

$$X_{t+1} - X_t = \alpha X_t - \alpha X_t + e_{t+1} - (1-\alpha)e_t$$

$$X_{t+1} - X_t = e_{t+1} - (1-\alpha)e_t \qquad (7)$$

Expression (7) is an alternative form for stating simple exponential smoothing. It indicates that the difference between two successive *actual* values is equal to the error in the present period minus a percentage of the error of the previous period. Expression (7) is related to several concepts that will be discussed in Chapters 17 and 18 and is an important way of looking at simple exponential smoothing.

a. Relate and explain the three alternative ways of expressing simple exponential smoothing, namely:

$$F_{t+1} = \alpha X_t + (1 - \alpha)F_t \tag{8}$$

$$F_{t+1} = F_t + \alpha(X_t - F_t) = F_t + \alpha e_t \tag{9}$$

and

$$X_{t+1} - X_t = e_{t+1} - (1 - \alpha)e_t \tag{10}$$

b. If there is a trend in the data how well do you think simple exponential smoothing will do? Why?

c. How do (8), (9), and (10) relate to adaptive response rate exponential smoothing? What is the primary difference between simple exponential smoothing and adaptive response rate exponential smoothing?

CHAPTER 7

Brown's One Parameter Linear Exponential Smoothing

The exponential smoothing methods examined in Chapters 5 and 6 are most appropriate for series that have a constant mean or a mean that changes very gradually with time. The linear exponential smoothing methods examined in this and subsequent chapters attempt to deal directly with non-stationary time series that exhibit a significant trend. Their only difference from single exponential smoothing is that they introduce additional formulas that estimate the trend so that it can be subsequently used to forecast. (This is frequently called double exponential smoothing.)

The equations used in Brown's model are:

$$S'_t = \alpha X_t + (1 - \alpha)S'_{t-1} \tag{7-1}$$

$$S''_t = \alpha S'_t + (1 - \alpha)S''_{t-1} \tag{7-2}$$

where S'_t is the single exponentially smoothed value for time t, and S''_t is the double exponentially smoothed value for that time period.

$$b_t = \frac{\alpha}{1 - \alpha} (S'_t - S''_t) \tag{7-3}$$

(See Exercise 1 on Smoothing, Averaging and Trends for explanation as to why (7-3) is multiplied by $\alpha/1 - \alpha$.)

where b_t is an estimate of trend;

$$a_t = S'_t + (S'_t - S''_t) = 2S'_t - S''_t \qquad (7\text{-}4)$$

where a_t is an estimate of the intercept; finally the forecasts are found using:

$$F_{t+m} = a_t + b_t m \qquad (7\text{-}5)$$

where m is the number of periods ahead to be forecast.

Equation (7-1) is exactly the same as the formula used for single exponential smoothing. Formula (7-2) is also similar, except it smooths the values of the single smoothing of (7-1). It is introduced to estimate the trend.

The basic idea behind equation (7-2) is that the double exponential smoothing values (S''_{t+1}) will lag behind the single exponential smoothing values (S'_{t+1}) by as much as the single exponential smoothing values will lag the original data. Thus, by subtracting the double exponential smoothing values from the single exponential smoothing values, one obtains an estimate of the trend. This is shown in equation (7-3). The factor α, divided by $1 - \alpha$, is multiplied by the difference between the single and double exponential smoothing values. This results in a trend for a single period because $\alpha/(1 - \alpha)$ is the time delay between the period when smoothing is applied and the current period.

In addition to estimating the trend, linear exponential smoothing makes an estimate of the present level intercept of the data. This is given by equation (7-4), whose basic concept is similar to equation (7-3), i.e., the single exponential smoothing values (S'_t) are below or behind the data by as much as the difference between the single exponential smoothing values and the double exponential smoothing values. Thus, if their difference is added to the single exponential smoothing value, the result will be an update bringing the data to their current level. The value of a_t is different from the corresponding value of X_t in the sense that a_t is a smoothed value and therefore does not include randomness, as does X_t. This is why a_t is preferred to X_t. If randomness was of no concern, (7-5) could alternatively be written as $F_{t+m} = X_t + (X_t - X_{t-1})m$. With randomness present, (7-5) does much better than this. Finally, in order to forecast, one must use formula (7-5) which starts from the current level, a, and adds as many times the trend, b, as the number of periods ahead one wants to forecast. This is therefore a direct adjustment for the trend factor which may exist in the data.

References for further study:

Brown, R. (1959, Chapter 4)
Brown, R. (1963, Chapter 9)
Makridakis, S. and S. Wheelwright (1978, Chapter 3)
Wheelwright, S. and S. Makridakis (1977, Chapter 3)

INPUT/OUTPUT FOR BROWN'S ONE PARAMETER LINEAR EXPONENTIAL SMOOTHING (EXPO2)

Input

1. *Enter a value for α.*

The parameter, alpha, is analogous in function to that used with exponential smoothing. Its value, however, should be considerably smaller. Brown suggests a value of $\alpha = 0.1$ even though he says that the forecasts may not respond fast enough to changes in the data when $\alpha = 0.1$. It seems that values between 0.1 and 0.3 are the most appropriate. However, any value between 0 and 1 can, under some circumstances, be employed in such a way as to minimize the mean squared error.

Output

Standard Output only. (See Chapter 5.)
   If the program EXPO2 is run continually for non-teaching purposes, then the values for S'_t and S''_t should be printed each time the program is run. Line 1515 should be inserted in the program as follows:

   1515 PRINT S(NØ); R(NØ)

The two printed values should be inserted as follows as initial values the next time the program is run for the same forecasting operation:

   1390 S(1) = value of S (NØ)

   1400 R(1) = value of R (NØ)

As it is currently done in the program $S'_1 = X_1$ and $S''_1 = S'_1 = X_1$.

EXERCISE 7.1

The data below show the weekly demand (in thousands of units) of transistors. Forecast their demand for the next four weeks (26-29).

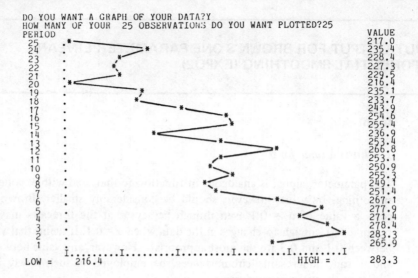

```
DO YOU WANT A GRAPH OF YOUR DATA?Y
HOW MANY OF YOUR   25 OBSERVATIONS DO YOU WANT PLOTTED?25
PERIOD                                                        VALUE
  25    *                                                     217.0
  24    :                         *                           235.4
  23    :                  *                                  228.4
  22    :                *                                    227.3
  21    :              *                                      229.5
  20    *                                                     216.4
  19    :                  *                                  235.1
  18    :               *                                     233.7
  17    :                      *                              243.9
  16    :                         *                           254.6
  15    :                        *                            255.4
  14    :                 *                                   236.9
  13    :                          *                          253.4
  12    :                             *                       266.8
  11    :                       *                             253.1
  10    :                      *                              250.9
   9    :                         *                           255.3
   8    :                       *                             249.1
   7    :                    *                                251.4
   6    :                             *                       267.1
   5    :                                 *                   277.9
   4    :                                *                    271.4
   3    :                               *                     278.4
   2    :                                    *                283.3
   1    :                            *                        265.9
        i........I.........I.........I........I.......I
  LOW =    216.4                             HIGH =    283.3
```

Solution

Assuming a value of $\alpha = 0.2$, the forecasted demands for the next four periods are 220.333, 218.19, 216.047 and 213.904.

In order to forecast, one needs the values from single and double exponential smoothing. [See (7-3), (7-4) and (7-5).] To illustrate how forecasts are found consider a forecast for period 25. The value of the single and double smoothing for period 24 are:

$$S'_{24} = 234.561 \qquad S''_{24} = 241.763$$

Using (7-3) gives:

$$b_{24} = \frac{0.2}{1 - 0.2} \, (234.561 - 241.763)$$

$$= -1.8005$$

Using (7-4) the value of a can be determined as:

$$a_{24} = 2(234.561) - 241.763$$

$$= 227.359$$

Using (7-5) one can now forecast period 25:

$$F_{25} = a_{24} + b_{24}(1)$$

$$= 227.359 - 1.8005$$

$$= 225.558$$

To forecast the next period, one has to update the values of S'_t and S''_t.

$$S'_{25} = \alpha X_{25} + (1 - \alpha) S'_{24}$$

$$= 231.048$$

and

$$S''_{25} = .2(231.048) + .8(241.763)$$

$$= 239.62$$

From the values of S'_{25} and S''_{25}, one can calculate the values of a_{25} and b_{25}

$$a_{25} = 2(231.048) - 239.62$$

$$= 222.476$$

and

$$b_{25} = \frac{0.2}{1 - 0.2} (231.048 - 239.62)$$

$$= -2.143$$

Thus the forecasts for periods 26 to 29 are:

$$F_{26} = a_{25} + b_{25}(1) = 222.476 - 2.143 = 220.333$$

$$F_{27} = a_{25} + b_{25}(2) = 222.476 - 4.286 = 218.19$$

$$F_{28} = a_{25} + b_{25}(3) = 222.476 - 6.429 = 216.407$$

$$F_{29} = a_{25} + b_{25}(4) = 222.476 - 8.572 = 213.904$$

```
WHICH PROGRAM WOULD YOU LIKE TO RUN?EXP

***     EXPONENTIAL SMOOTHING      ***
YOU CAN CHOOSE BETWEEN 6 PROGRAMS
1. EXPO - SINGLE EXPONENTIAL SMOOTHING
2. EXPOTL - ADAPTIVE RESPONSE RATE EXPONENTIAL SMOOTHING
3. EXPO2 - LINEAR EXPONENTIAL SMOOTHING
4. EXPOH - LINEAR (HOLTS) EXPONENTIAL SMOOTHING
5. EXPOQ - QUADRATIC (BROWN S) EXPONENTIAL SMOOTHING
6. EXPOW - LINEAR AND SEASONAL (WINTERS) EXPONENTIAL SMOOTHING.
HOW MANY OBSERVATIONS DO YOU WANT TO USE?25
DATA FILENAME?TSISTR

WHAT PROGRAM DO YOU WANT TO RUN - TYPE ITS NAME ?EXPO2


***  LINEAR EXPONENTIAL SMOOTHING  ***

NEED HELP (Y OR N)?N

ENTER A VALUE FOR ALPHA (BETWEEN 0 AND 1)?0.2
DO YOU WANT A TABLE OF ACTUAL AND PREDICTED (Y OR N)?Y

    PERIOD         ACTUAL          FORECAST        PCT ERROR
      1            265.89
      2            283.35          271.16             4.30%
      3            278.36          272.87             1.97%
      4            271.45          275.77             1.59%
      5            277.95          274.96             1.08%
      6            267.11          276.90             3.67%
      7            251.38          273.85             8.94%
      8            249.05          265.33             6.54%
      9            255.31          258.40             1.21%
     10            250.90          256.08             2.07%
     11            253.11          252.81              .12%
     12            266.83          251.52             5.74%
     13            253.42          256.25             1.12%
     14            236.85          254.33             7.38%
     15            255.36          246.44             3.49%
     16            254.65          248.41             2.45%
     17            243.85          249.67             2.39%
     18            233.75          246.35             5.39%
     19            235.10          240.09             2.12%
     20            216.35          236.37             9.25%
     21            229.50          226.43             1.34%
     22            227.30          224.93             1.04%
     23            228.43          223.28             2.26%
     24            235.40          222.83             5.34%
     25            216.99          225.55             3.95%
MEAN PC ERROR (MPE) OR BIAS       =        -1.10%
MEAN SQUARED ERROR (MSE)          =         111.1
MEAN ABSOLUTE PC ERROR    (MAPE) =          3.5%


FORECAST FOR HOW MANY PERIODS AHEAD (0=NONE,24=MAX)?4

    PERIOD      FORECAST
     26          220.33
     27          218.18
     28          216.04
     29          213.90

WOULD YOU LIKE AUTOCORRELATIONS ON THE RESIDUALS ?N

DO YOU WANT TO RUN ANOTHER PROGRAM?N
```

CHAPTER 8

Holt's Two Parameters Linear Exponential Smoothing

As with Brown's One Parameter Linear Exponential Smoothing, Holt's Two Parameters Smoothing estimates the trend and uses it in forecasting.

The three equations used in Holt's model are:

$$S_t = \alpha X_t + (1 - \alpha)(S_{t-1} + T_{t-1}) \tag{8-1}$$

$$T_t = \beta(S_t - S_{t-1}) + (1 - \beta) T_{t-1} \tag{8-2}$$

$$F_{t+m} = S_t + mT_t \tag{8-3}$$

where m is the number of periods ahead to be forecast.

Equation (8-1) is similar to the original single exponential smoothing equation except that a term for the trend (T_{t-1}) is added. The value of this term is calculated using equation (8-2). The difference between two successive exponential smoothing values is used as an estimate of the trend. Since successive values have been smoothed for randomness, their difference constitutes the trend in the data. This is equivalent to the procedure described in the last chapter — namely, subtracting the double from the single exponential smoothing values. The estimate of the trend is smoothed by multiplying it by β and then multiplying the old estimate of the trend by $(1 - \beta)$. That is, equation (8-2) is similar to equation (8-1), or the general form of the smoothing equation, except that the smoothing is not done for the actual data but rather for the trend. The final result of (8-2) is a smoothed trend that does not, we hope, include much randomness.

In order to forecast, the trend is multiplied by the number of periods ahead that one desires to forecast and then the product is added to S_t (the current level of the data that has been smoothed to eliminate randomness).

The equations for Holt's model are not much different than those used in Chapters 5, 6, and 7. Equation (8-2) is similar to (5-1), except that it involves an additional term, T_{t-1}, for the trend. The term $S_{t-1} + T_{t-1}$ is the equivalent of F_{t-1} in (5-2), which is added to adjust for the trend in the data. The difference between two successive S_t values is, almost by definition the trend. It is "smoothed" by averaging the most recent as well as past values using (8-2). Beta, the smoothing constant for (8-2), is analogous to alpha used in (8-1). The principle of smoothing combines the most recent value, multiplied by some constant (α or β), with the previous smoothed value, multiplied by one minus the smoothing constant.

In comparison to Brown's One Parameter Exponential Smoothing, Holt's model has the disadvantage of requiring the specification of two parameters (α and β) whose values have to be optimized if the mean squared error is to be minimized. On the other hand, one has the opportunity of applying different weights to randomness and trend depending upon the specific type of data involved. In practice, it appears that Brown's model is preferable to Holt's because of the simplicity of having a single parameter. However, little work has been done concerning the accuracy of the forecasts obtained by these two alternative methods of forecasting.

References for further study:

Holt, C. (1957)
Makridakis, S. and S. Wheelwright (1978, Chapter 3)
Winters, P. (1960)

INPUT/OUTPUT FOR HOLT'S TWO PARAMETERS LINEAR EXPONENTIAL SMOOTHING PROGRAM (EXPOH)

Input

1. *Input a value for alpha and beta.*

The value for alpha is the same as that used in the smoothing models in the two previous chapters. In Holt's model it smooths the data to eliminate randomness. Beta is similar to alpha except that it smooths the trend in the data (see equation (8-2)). Both smoothing constants remove randomness in the same manner, i.e., by weighting past values.

Output

1. *Standard output (see Chapter 5).*

If the program EXPOH is run continually for non-teaching purposes then the last values for S_t and T_t should be printed each time the program is run. Line 1140 should be inserted as follows:

> 1140 PRINT S(NØ); T(NØ)

The next time the program is run for the same forecasting operation, the two printed values should be input as initial values as follows:

> 1021 S(1) = value of S(NØ)

> 1030 T(1) = value of T(NØ)

The program initializes

$$S_1 = X_1$$

and

$$T_1 = \frac{X_2 - X_1}{2} + \frac{X_4 - X_3}{2}$$

Alternative specification for initial values are possible.

EXERCISE 8

The following data are the dollar sales of electronic calculators in a large department store. Forecast the sales for the next six weeks. Use different values for the smoothing constants α and β to determine their effect on the mean squared error.

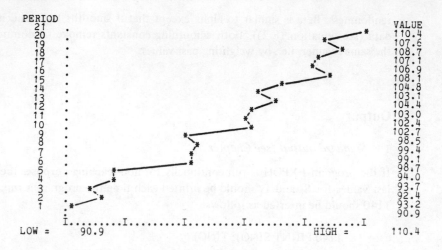

```
PERIOD                                                              VALUE
  21    .                                              *            110.4
  20    .                                        *  *               107.6
  19    .                                      *                    108.7
  18    .                                     *                     107.1
  17    .                                   *                       106.9
  16    .                                  *                        108.1
  15    .                                 *                         104.8
  14    .                          *  *                             103.1
  13    .                         *                                 104.4
  12    .                         *                                 103.0
  11    .                      * *                                  102.4
  10    .                     *                                     102.7
   9    .              *                                             98.5
   8    .             * *                                            99.4
   7    .            *                                               99.1
   6    .            *                                               98.7
   5    .        * *                                                 94.0
   4    .       *                                                    93.7
   3    .     *  *                                                   92.4
   2    .   *                                                        93.2
   1    . *                                                          90.9
        I.........I.........I.........I.........I.........I
LOW =     90.9                                      HIGH =          110.4
```

Solution

How well the model fits the data is measured by the mean squared error (MSE) whose magnitude is sensitive to different values of the smoothing constants. For example, the following MSE values result for different values for α and β. The detailed results for $\alpha = .1, \beta = .1$ are shown in Figure 8.1)

TABLE 8-1   Mean Squared Errors

	$\alpha = .1$	$\alpha = .3$
$\beta = .1$	1600.63	373.89
$\beta = .5$	343.83	225.34

Obviously, if $\alpha = .3$ and $\beta = .5$, than the MSE will be the minimum of the four cases run because of a rather strong trend in the data and a smaller week-to-week randomness.

Equation (8-1) can be used to illustrate the calculations. At period 20, the smoothed value for randomness, S_{20}, is 1134.598 while that of the trend, T_{20}, is 11.052. Using (8-3) one can obtain a forecast for period 21.

$$F_{21} = 1134.598 + 1 (11.052) = 1145.65$$

Values S_t and T_t can then be updated using (8-1) and (8-2).

$$S_{21} = \alpha X_{21} + (1 - \alpha) (S_{20} + T_{20})$$

$$= 0.1(1104.45) + 0.9(1134.598 + 11.052)$$

$$= 1141.53$$

$$T_{21} = \beta(S_{21} - S_{20}) + (1 - \beta) T_{20}$$

$$= 0.1(1141.53 - 1134.598) + 0.9(11.052)$$

$$= 10.64$$

Using (8-3) a forecast can then be prepared

$$F_{22} = 1141.53 + 1(10.64) = 1152.17$$

$$F_{23} = 1141.53 + 2(10.64) = 1162.81$$

$$F_{24} = 1141.53 + 3(10.64) = 1173.45$$

$$\cdot \qquad \cdot \qquad \cdot \qquad \cdot$$

$$\cdot \qquad \cdot \qquad \cdot \qquad \cdot$$

$$\cdot \qquad \cdot \qquad \cdot \qquad \cdot$$

$$F_{27} = 1141.53 + 6(10.64) = 1205.37$$

FIGURE 8-1   Holt's Exponential Smoothing ($\alpha = 0.1$, $\beta = 0.1$)

```
HOW MANY OBSERVATIONS DO YOU WANT TO USE?21
DATA FILENAME?SALES
ALL INPUT DATA DIVIDED BY 10** 1

WHAT PROGRAM DO YOU WANT TO RUN - TYPE ITS NAME ?EXPOH

***  LINEAR (HOLTS) EXPONENTIAL SMOOTHING  ***

NEED HELP (Y OR N)?N

INPUT A VALUE FOR ALPHA AND BETA ?0.1 0.1
DO YOU WANT A TABLE OF ACTUAL AND PREDICTED (Y OR N)?Y
```

PERIOD	ACTUAL	FORECAST	ERROR	PCT ERROR
1	90.91			
2	93.20	92.68	.52	.56%
3	92.41	94.51	-2.10	2.27%
4	93.66	96.05	-2.39	2.56%
5	94.02	97.54	-3.53	3.75%
6	98.75	98.89	-.14	.14%
7	99.05	100.57	-1.52	1.53%
8	99.44	102.10	-2.66	2.67%
9	98.52	103.49	-4.96	5.03%
10	102.70	104.59	-1.89	1.84%
11	102.36	105.99	-3.63	3.54%
12	103.00	107.17	-4.17	4.05%
13	104.42	108.26	-3.84	3.68%
14	103.13	109.35	-6.22	6.03%
15	104.77	110.13	-5.36	5.12%
16	108.05	110.95	-2.90	2.68%
17	106.89	111.98	-5.09	4.76%
18	107.06	112.75	-5.69	5.31%
19	108.75	113.39	-4.64	4.27%
20	107.64	114.10	-6.46	6.00%
21	110.44	114.56	-4.12	3.73%

```
MEAN PC ERROR (MPE) OR BIAS      =      -3.42%
MEAN SQUARED ERROR (MSE)         =       16.0
MEAN ABSOLUTE PC ERROR   (MAPE) =        3.5%


FORECAST FOR HOW MANY PERIODS AHEAD (0=NONE,24=MAX)?6
```

PERIOD	FORECAST
22	115.21
23	116.27
24	117.34
25	118.40
26	119.46
27	120.53

```
WOULD YOU LIKE AUTOCORRELATIONS ON THE RESIDUALS ?N

DO YOU WANT TO RUN ANOTHER PROGRAM?Y
```

CHAPTER 9

Brown's One Parameter Quadratic Exponential Smoothing

Quadratic Exponential Smoothing, frequently called triple exponential smoothing, is an extension of Linear Exponential Smoothing. It aims at dealing with trend of a higher order than linear trend. This is achieved by introducing triple exponential smoothing which, in addition to the double exponential smoothing, is used to remove quadratic trends.

The seven equations used in Brown's model are:

$$S'_t = \alpha X_t + (1 - \alpha)S'_{t-1} \tag{9-1}$$

$$S''_t = \alpha S'_t + (1 - \alpha)S''_{t-1} \tag{9-2}$$

$$S'''_t = \alpha S''_t + (1 - \alpha)S'''_{t-1} \tag{9-3}$$

where

S'_t is a single exponential smoothing

S''_t is a double exponential smoothing

S'''_t is a triple exponential smoothing.

$$a_t = 3S'_t - 3S''_t + S'''_{t-1} \tag{9-4}$$

$$b_t = \frac{\alpha}{2(1-\alpha)^2} \left((6 - 5\alpha)S'_t - (10 - 8\alpha)S''_t + (4 - 3\alpha)S'''_t \right) \tag{9-5}$$

$$c_t = \frac{\alpha^2}{(1-\alpha)^2} (S'_t - 2S''_t + S'''_t) \tag{9-6}$$

$$F_{t+m} = a_t + b_t m + \frac{1}{2} c_t m^2 \tag{9-7}$$

Equation (9-1) is the original single exponential smoothing equation. Equation (9-2) is the double exponential smoothing of Chapter 7, while (9-3) is the triple exponential smoothing equation, new to this. Equation (9-4) determines the current value of the data, (9-5) determines the linear trend, and (9-6) determines the quadratic trend. Finally, equation (9-7) is used for predicting future periods. As can be seen from equation(9-7), both the linear trend and the quadratic trend are used to forecast. The last term of equation (9-7) is mulitiplied by a factor of 1/2 that is derived by differentiation.

Quadratic exponential smoothing is more complicated than linear smoothing, which is more complex than single smoothing. However, the level of complexity is not too great, and the principle of smoothing is always the same. Linear smoothing used the equation of a line ($a_t + b_t m$) while quadratic smoothing uses that of a parabola ($a_t + b_t m + c_t m^2$). It is the parameters that differ in each case and which are estimated in such a way as to reduce the amount of randomness involved. Single and linear exponential smoothings cannot predict turning points well. Quadratic smoothing can do better, but, at the same time, might overreact to random fluctuations in the data. This is a general problem facing the forecaster who on the one hand wants to follow changes in the pattern of the data but on the other wants a method that can distinguish random fluctuations from changes in the basic pattern of the data.

References for further study:

Brown, R. (1963, Chapter 9)
Makridakis, S. and S. Wheelwright (1978, Chapter 3)

INPUT/OUTPUT FOR BROWN'S ONE PARAMETER QUADRATIC EXPONENTIAL SMOOTHING PROGRAM (EXPOQ)

Input

1. *This program requires the user to input a value for alpha which is used in the same manner as in the other exponential smoothing programs.*

Output

Standard outputs include

1. an optional table of actual and forecasted values and their errors;
2. the mean squared error and the mean percentage errors (both absolute and arithmetic);
3. any number of forecasts for future periods.

If the program EXPOQ is run continually for non-teaching purposes, then the last values for S(1), R(1), and Q(1) should be printed each time the program is run. Line 1250 should be inserted as follows:

```
1250 PRINT S(NØ), R(NØ), Q(NØ)
```

The next time the program is run for the same forecasting operation, the three printed values should be input as initial values as follows:

```
line 1100 S(1) = value of S(NØ)

line 1110 R(1) = value of R(NØ)

line 1120 Q(1) = value of Q(NØ)
```

As with all the smoothing methods, this program provides "simple" initial estimates to start with. The sophisticated user should use lines 1100, 1110, and 1120 to input his or her own initial estimates if not satisfied with those of the program.

EXERCISE 9.1

The following data show the closing stock market value, at the end of each month, of Enterprise, Inc. Forecast the price of the stock for the next four months using alpha values of 0.2, 0.5, and 0.8. Determine which value gives the best forecast, i.e., which has the minimum mean squared error.

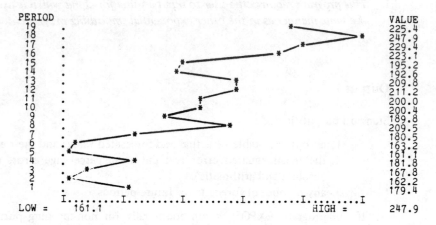

```
PERIOD                                                          VALUE
 19    .                                    *                   225.4
 18    .                                      *                 247.9
 17    .                                *                       229.4
 16    .                              *                         223.1
 15    .                  *                                     195.2
 14    .                 *                                      192.6
 13    .                    *                                   209.8
 12    .                    *                                   211.2
 11    .               *                                        200.0
 10    .               *                                        200.4
  9    .             *                                          189.8
  8    .                    *                                   209.5
  7    .         *                                              180.5
  6    .     *                                                  163.2
  5    .   *                                                    161.1
  4    .        *                                               181.8
  3    .     *                                                  167.8
  2    . *                                                      162.2
  1    .        *                                               179.4
       I.........I.........I.........I.........I.........I
LOW =     161.1                                  HIGH =         247.9
```

Solution

Of the three smoothing values tested, the minimum mean squared error occurs for $\alpha = 0.2$. This should have been expected since the quadratic exponential smoothing requires small values for alpha; otherwise it overreacts to random changes in the data, assuming that they are caused by trend. (In this instance smaller values of alpha give even better results.)

TABLE 9-1   Mean Squared Errors

x = 0.8	855.5
x = 0.5	437.2
x = 0.2	248.7

The calculations can be illustrated using periods 18 and 19. At period 18, the values for the single, double and triple smoothing values are:

$$S'_{18} = 214.996 \qquad S''_{18} = 198.813 \qquad S'''_{18} = 189.394$$

From those values one can calculate a_{18}, b_{18}, and c_{18} using (9-4), (9-5) and (9-6) respectively.

Substituting the appropriate values in (9-4) gives:

$$a_{18} = 3(214.996) - 3(198.813) + 189.394 = 237.792$$

Substituting in (9-5) yields:

$$b_{18} = \frac{0.2}{(1 - 0.2)^2} \Big((6 - (5)0.2)\, 214.996 - (10 - (8)0.2)\, 198.813$$

$$+ (4 - (3)0.2)\, 189.394 \Big)$$

$$= 0.15625 \Big(5(214.996) - 8.4(198.813) + 3.4(189.394) \Big)$$

$$= 7.639$$

Similarly, substituting in (9-6) gives:

$$c_{18} = \frac{0.2^2}{(1 - 0.2)^2} \Big(214.996 - 2(198.813) + 189.394 \Big)$$

$$= 0.0625(6.764)$$

$$= 0.4227$$

The values of a_{18}, b_{18}, and c_{18}, can be used in (9-7) to forecast for the next period (s). For period 19, for example, the forecast will be:

$$F_{19} = 237.942 + 7.639(1) + \frac{1}{2}(0.4227)\,(1)$$

$$F_{19} = 245.793$$

If a new actual value becomes available for period 19, one updates the smoothed values (9-1), (9-2) and (9-3), calculates a_{19}, b_{19}, and c_{19} and then prepares a forecast.

Using (9-1), one obtains:

$$S'_{19} = 0.2X_{19} + (1 - 0.2)S'_{18}$$

$$= 0.2(225.366) + 0.8(214.996)$$

$$= 217.07$$

From (9-2) the double smoothing value is

$$S''_{19} = 0.2(217.07) + 0.8(198.813)$$

$$= 202.465$$

and finally, using (9-3) gives:

$$S'''_{19} = 0.2(202.465) + 0.8(189.394)$$

$$= 192.009$$

If the values of S'_{19}, S''_{19}, and S'''_{19} are substituted in (9-4), (9-5) and (9-6) one obtains:

$$a_{19} = 235.824$$

$$b_{19} = 5.856$$

$$c_{19} = 0.2593$$

Thus, one can now forecast for future periods. For example the predictions for the next four periods can be found using (9-7):

$$F_{20} = 235.824 + 5.856(1) + \frac{1}{2}(0.2593)(1)^2 = 241.801$$

$$F_{21} = 235.824 + 5.856(2) + \frac{1}{2}(0.2593)(2)^2 = 248.054$$

$$F_{22} = 235.824 + 5.856(3) + \frac{1}{2}(0.2593)(3)^2 = 254.558$$

$$F_{23} = 235.824 + 5.856(4) + \frac{1}{2}(0.2593)(4)^2 = 261.322$$

Figure 9-1 shows the results of using EXPOQ for these data with $\alpha = .2$.

FIGURE 9-1   Brown's Quadratic Exponential Smoothing ($\alpha = 0.2$)

```
HOW MANY OBSERVATIONS DO YOU WANT TO USE?19
DATA FILENAME?PRICE

WHAT PROGRAM DO YOU WANT TO RUN - TYPE ITS NAME ?EXPOQ

*** QUADRATIC (BROWNS) EXPONENTIAL SMOOTHING ***

NEED HELP (Y OR N)?N

ENTER A VALUE FOR ALPHA (BETWEEN 0 AND 1)?0.2
DO YOU WANT A TABLE OF ACTUAL AND PREDICTED (Y OR N)?Y
```

PERIOD	ACTUAL	FORECAST	ERROR	PCT ERROR
1	179.39			
2	162.24	162.24	0	0%
3	167.83	169.10	-1.27	.76%
4	181.81	166.28	15.53	8.54%
5	161.07	173.25	-12.18	7.56%
6	163.17	165.31	-2.14	1.31%
7	180.46	161.91	18.55	10.28%
8	209.51	170.55	38.96	18.60%
9	189.83	193.52	-3.69	1.94%
10	200.41	195.59	4.82	2.41%
11	199.99	202.64	-2.65	1.33%
12	211.21	206.08	5.13	2.43%
13	209.82	214.20	-4.38	2.09%
14	192.59	217.55	-24.96	12.96%
15	195.19	208.37	-13.18	6.75%
16	223.14	203.58	19.56	8.77%
17	229.44	216.97	12.47	5.44%
18	247.88	228.46	19.42	7.83%
19	225.36	245.79	-20.43	9.07%

```
MEAN PC ERROR (MPE) OR BIAS      =      1.14%
MEAN SQUARED ERROR (MSE)         =    248.7
MEAN ABSOLUTE PC ERROR  (MAPE)   =      6.0%

FORECAST FOR HOW MANY PERIODS AHEAD (0=NONE,24=MAX)?4
```

PERIOD	FORECAST
20	241.80
21	248.05
22	254.55
23	261.32

```
WOULD YOU LIKE AUTOCORRELATIONS ON THE RESIDUALS ?N

DO YOU WANT TO RUN ANOTHER PROGRAM?Y
```

CHAPTER 10

Winters' Three Parameters Linear and Seasonal Exponential Smoothing

Winters' exponential smoothing is an extension of Holt's linear exponential smoothing. It includes an extra equation that is used to estimate seasonality.

The three equations used by Winters' method are given by equations (10-1), (10-2) and (10-3), while equation (10-4) is used to forecast.

$$I_t = \beta \frac{X_t}{S_t} + (1 - \beta)I_{t-L} \tag{10-1}$$

$$S_t = \alpha \frac{X_t}{I_{t-L}} + (1 - \alpha)(S_{t-1} + T_{t-1}) \tag{10-2}$$

$$T_t = \gamma(S_t - S_{t-1}) + (1 - \gamma)T_{t-1} \tag{10-3}$$

$$F_{t+m} = (S_t + mT_t)I_{t-L+m} \tag{10-4}$$

The estimate of seasonality is given as an index, fluctuating around 1, and is calculated with equation (10-1). The form of equation (10-1) is similar to that of all other exponential smoothing equations. That is, a value — in this case X_t divided by S_t — is multiplied by a constant, β, and is then added to its previous smoothed estimate which has been multiplied by $1 - \beta$. The reason that X_t/S_t is multiplied by the smoothing constant, β, rather than simply X_t or S_t, is to express the dividend as an index rather than in absolute terms.

In addition to equation (10-1), Winters' smoothing uses the last three equations of Holt's model but introduces the seasonal index, I_t, into the formulas. Thus equations (10-2), and (10-3) and (10-4) are used to obtain estimates of the present level of the data, the trend, and the forecast for some future period, t + m. There is a slight difference in equation (10-2) and the corresponding one in Holt's model. In formula (10-2), X_t is divided by I_{t-L}, which adjusts X_t for seasonality. This removes the seasonal effects which may exist in the original data X_t.

Similarly, in equation (10-4) a forecast is obtained which is equal to $[S_t + m(T_t)]$ I_{t-L+m}. It is exactly the same as the corresponding formula used to obtain a forecast in Holt's model. However, this estimate for the future period, t + m, is multiplied by I_{t-m+L}. This is the last seasonal index available and therefore is used to re-adjust the forecast for seasonality. In other words, this formula introduces seasonal effects into the predictions. Multiplying the forecast by I_{t-L+m} has the opposite effect of dividing X_t by I_{t-L} in equation (10-2).

Adjusting data for seasonality is used frequently and in a variety of situations. Government statistics (unemployment, inflation, and even GNP) are very often given on a seasonal adjusted bases. This means that the raw data is divided by the seasonal index (i.e., X_t/I_{t-L}). This is exactly what is done in (10-2). Similarly to take seasonality into account in forecasting, the non-seasonal estimate must be seasonalized. This is done with (10-4).

References for further study:

Holt, C. (1957)
Makridakis, S. and S. Wheelwright (1978, Chapter 3)
Wheelwright, S. and S. Makridakis (1977, Chapter 3)
Winters, P. (1960)

INPUT/OUTPUT FOR WINTERS' THREE PARAMETERS LINEAR AND SEASONAL EXPONENTIAL SMOOTHING PROGRAM (EXPOW)

Input

1.   *Values for alpha, beta and gamma are required.*

There is no difference in their behavior because their use and effects are similar. They are used in the same manner as alpha in the other exponential smoothing programs. Alpha is used to smooth randomness, beta to smooth seasonality, and gamma to smooth trend. Usually, the values of beta and gamma are smaller than alpha.

Output

1. *Standard Output (See Chapter 5).*

If the program EXPOW is run continually for non-teaching purposes, then the last smoothed values of S_t and T_t should be printed together with the last L values of I_t (where L is the length of seasonality) each time the program is run. Line 1612 should be inserted as follows:

1612 PRINT S(NØ), T(NØ), M(NØ), M(NØ – 1), M(NØ) . . . . . . M(NØ – L)

The next time the program is run for the same forecasting operation the values of S_t, T_t, and I_t should be input as initial values as follows:

line   1360 S(NØ) = value of S(NØ)

1380 T(NØ) = value of T(NØ)

1400 I(NØ) = value of I(NØ)

EXERCISE 10.1

The following series is the monthly number of car accidents (in thousands) in Germany. Predict the continuation of the series during the next five periods. The length of seasonality is twelve, the number of months in the year.

Solution

Winters' model is more difficult to optimize because it has three parameters. Through trial and error, however, one can find values for α, β and γ which minimize the mean squared error. For lack of space only a few runs are included. The best model is for values of $\alpha = 0.5$, $\beta = 0.5$ and $\gamma = 0.5$. As can be seen from these results, the values of the smoothing constants do not have to be very small as with Brown's linear and quadratic smoothing models.

In order to illustrate the calculations involved, one can use equation

```
DO YOU WANT A GRAPH OF YOUR DATA?Y
HOW MANY OF YOUR  36 OBSERVATIONS DO YOU WANT PLOTTED?36
```

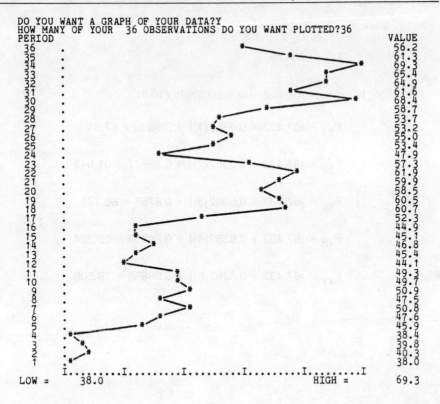

PERIOD		VALUE
		56.2
		61.3
		69.3
		65.4
		64.9
		61.6
		68.4
		58.7
		53.7
		53.2
		55.0
		53.4
		47.9
		57.3
		61.9
		59.9
		58.5
		60.5
		60.7
		52.3
		44.9
		45.1
		46.8
		45.4
		44.1
		49.3
		49.7
		50.9
		47.5
		50.8
		47.6
		45.9
		38.4
		39.8
		40.3
		38.0

```
LOW =     38.0                              HIGH =     69.3
```

(10-1) and assume $\beta = 0.1$ and time is at period 35. Equation (10-2) can be used to find the smoothed value, S_{36}, assuming that $\alpha = 0.1$.

$$S_{36} = 0.1 \frac{56.2158}{0.95475} + 0.9(67.653 + 0.7303)$$

$$= 5.888 + 61.545 = 67.432$$

In order to use (10-2) one needs the value of I_{24} which is 0.95475. This is the last available smoothed index of seasonality.

The trend can be found using (10-3) and assuming $\gamma = 0.1$.

$$T_{36} = 0.1(67.432 - 67.653) + 0.9(0.7303) = 0.6352$$

The smoothed trend at period 35 was 0.7303.

Finally one can update the seasonal index, I_t, using (10-1).

$$I_{36} = 0.1 \frac{56.2158}{67.432} + 0.9(0.95475) = 0.94264$$

Since

$$I_{35} = .95475.$$

In order to forecast, one must apply (10-4):

$$F_{37} = (67.432 + 0.6352(1)) + 0.9893 = 67.34$$

$$F_{38} = (67.432 + 0.6352(2)) + 0.89577 = 61.542$$

$$F_{39} = (67.432 + 0.6352(3)) + 0.8765 = 60.774$$

$$F_{40} = (67.432 + 0.6352(4)) + 0.84638 = 59.224$$

$$F_{41} = (67.432 + 0.6352(5)) + 0.99858 = 70.508$$

FIGURE 10-1   Solution to Exercise 10 Using Winter's Linear Exponential Smoothing (A, B, C = 0.1)

```
EXE METHOD
  8K
WHICH PROGRAM WOULD YOU LIKE TO RUN?EXPOW


***      EXPONENTIAL SMOOTHING      ***
YOU CAN CHOOSE BETWEEN 6 PROGRAMS
1. EXPO - SINGLE EXPONENTIAL SMOOTHING
2. EXPOTL - ADAPTIVE RESPONSE RATE EXPONENTIAL SMOOTHING
3. EXPO2 - LINEAR EXPONENTIAL SMOOTHING
4. EXPOH - LINEAR (HOLTS) EXPONENTIAL SMOOTHING
5. EXPOQ - QUADRATIC (BROWN S) EXPONENTIAL SMOOTHING
6. EXPOW - LINEAR AND SEASONAL (WINTERS) EXPONENTIAL SMOOTHING.
HOW MANY OBSERVATIONS DO YOU WANT TO USE?36
DATA FILENAME?CARACC

WHAT PROGRAM DO YOU WANT TO RUN - TYPE ITS NAME ?EXPOW


***  LINEAR AND SEASONAL (WINTERS) EXPONENTIAL SMOOTHING ***

NEED HELP (Y OR N)?N

WHAT IS THE LENGTH OF SEASONALITY?12

ENTER VALUES FOR A,B AND C (0,0,0=HELP)?0.1 0.1 0.1
DO YOU WANT A TABLE OF ACTUAL AND PREDICTED (Y OR N)?Y

PERIOD          ACTUAL          FORECAST          ERROR       PCT ERROR
  14            46.76            39.76             7.00         14.96%
  15            45.09            44.44              .65          1.45%
  16            44.89            45.79             -.90          2.00%
  17            52.32            45.90             6.42         12.27%
  18            60.73            56.67             4.06          6.68%
  19            60.46            61.03             -.57           .94%
  20            58.52            66.94            -8.42         14.38%
  21            59.94            64.21            -4.27          7.13%
  22            61.86            69.98            -8.12         13.13%
  23            57.34            69.59           -12.25         21.37%
  24            47.90            69.72           -21.82         45.54%
  25            53.36            62.41            -9.05         16.95%
  26            55.02            64.62            -9.60         17.45%
  27            53.24            58.82            -5.58         10.48%
  28            53.65            58.44            -4.79          8.93%
  29            58.73            57.11             1.62          2.76%
  30            68.40            68.24              .16           .23%
  31            61.58            72.20           -10.62         17.25%
  32            64.95            76.10           -11.15         17.17%
  33            65.44            71.64            -6.20          9.47%
  34            69.27            76.21            -6.94         10.02%
  35            61.26            74.99           -13.73         22.41%
  36            56.21            73.41           -17.20         30.60%
MEAN PC ERROR (MPE) OR BIAS       =        -10.31%
MEAN SQUARED ERROR (MSE)          =          87.6
MEAN ABSOLUTE PC ERROR    (MAPE) =          13.8%


FORECAST FOR HOW MANY PERIODS AHEAD (0=NONE,24=MAX)?5


PERIOD        FORECAST
  37            67.33
  38            61.54
  39            60.77
  40            59.21
  41            70.50

WOULD YOU LIKE AUTOCORRELATIONS ON THE RESIDUALS ?N

DO YOU WANT TO RUN ANOTHER PROGRAM?Y
```

FIGURE 10-2 Solution to Exercise 10 Using Winter's Linear Exponential Smoothing (A = 0.2, B = 0.2, C = 0.1)

```
HOW MANY OBSERVATIONS DO YOU WANT TO USE?36
DATA FILENAME?CARACC

WHAT PROGRAM DO YOU WANT TO RUN - TYPE ITS NAME ?EXPOW

***  LINEAR AND SEASONAL (WINTERS) EXPONENTIAL SMOOTHING ***

NEED HELP (Y OR N)?N

ENTER VALUES FOR A,B AND C (0,0,0=HELP)?0.2 0.2 0.1
DO YOU WANT A TABLE OF ACTUAL AND PREDICTED (Y OR N)?Y

PERIOD          ACTUAL          FORECAST          ERROR      PCT ERROR
  14            46.76            39.76             7.00        14.96%
  15            45.09            44.94              .15          .34%
  16            44.89            46.35            -1.46         3.24%
  17            52.32            46.45             5.87        11.22%
  18            60.73            56.95             3.78         6.22%
  19            60.46            61.48            -1.02         1.69%
  20            58.52            66.82            -8.30        14.19%
  21            59.94            63.65            -3.71         6.19%
  22            61.86            68.47            -6.61        10.68%
  23            57.34            67.59           -10.25        17.88%
  24            47.90            66.70           -18.80        39.25%
  25            53.36            58.48            -5.12         9.60%
  26            55.02            59.94            -4.92         8.94%
  27            53.24            55.23            -1.99         3.73%
  28            53.65            54.63             -.98         1.83%
  29            58.73            53.71             5.02         8.55%
  30            68.40            63.63             4.77         6.97%
  31            61.58            68.27            -6.69        10.86%
  32            64.95            70.59            -5.64         8.69%
  33            65.44            66.81            -1.37         2.09%
  34            69.27            70.59            -1.32         1.91%
  35            61.26            70.18            -8.92        14.56%
  36            56.21            67.91           -11.70        20.82%
MEAN PC ERROR (MPE) OR BIAS       =      -5.81%
MEAN SQUARED ERROR (MSE)          =      49.4
MEAN ABSOLUTE PC ERROR   (MAPE) =      10.2%

FORECAST FOR HOW MANY PERIODS AHEAD (0=NONE,24=MAX)?5

PERIOD          FORECAST
  37            62.57
  38            58.12
  39            56.79
  40            55.33
  41            64.99

WOULD YOU LIKE AUTOCORRELATIONS ON THE RESIDUALS ?N

DO YOU WANT TO RUN ANOTHER PROGRAM?Y
```

FIGURE 10-3   Solution to Exercise 10 Using Winter's Linear Exponential Smoothing (A, B, C = 0.5)

```
HOW MANY OBSERVATIONS DO YOU WANT TO USE?36
DATA FILENAME?A_CARACC

WHAT PROGRAM DO YOU WANT TO RUN - TYPE ITS NAME ?EXPOW
```

`***`   LINEAR AND SEASONAL (WINTERS) EXPONENTIAL SMOOTHING `***`

```
NEED HELP (Y OR N)?N

ENTER VALUES FOR A,B AND C (0,0,0=HELP)?0.5 0.5 0.5
DO YOU WANT A TABLE OF ACTUAL AND PREDICTED (Y OR N)?Y
```

PERIOD	ACTUAL	FORECAST	ERROR	PCT ERROR
14	46.76	39.76	7.00	14.96%
15	45.09	47.35	-2.26	5.01%
16	44.89	48.34	-3.45	7.68%
17	52.32	47.63	4.69	8.97%
18	60.73	55.75	4.98	8.20%
19	60.46	61.13	-.67	1.10%
20	58.52	63.68	-5.16	8.82%
21	59.94	59.52	.42	.70%
22	61.86	61.41	.45	.72%
23	57.34	60.94	-3.60	6.29%
24	47.90	58.18	-10.28	21.46%
25	53.36	48.28	5.08	9.51%
26	55.02	50.72	4.30	7.82%
27	53.24	52.04	1.20	2.26%
28	53.65	53.75	-.10	.18%
29	58.73	55.39	3.34	5.69%
30	68.40	63.57	4.83	7.06%
31	61.58	69.92	-8.34	13.54%
32	64.95	65.97	-1.02	1.58%
33	65.44	63.43	2.01	3.08%
34	69.27	65.49	3.78	5.46%
35	61.26	67.48	-6.22	10.15%
36	56.21	62.46	-6.25	11.12%

```
MEAN PC ERROR (MPE) OR BIAS        =        -.57%
MEAN SQUARED ERROR (MSE)           =        23.0
MEAN ABSOLUTE PC ERROR   (MAPE) =          7.3%

FORECAST FOR HOW MANY PERIODS AHEAD (0=NONE,24=MAX)?5
```

PERIOD	FORECAST
37	59.07
38	54.48
39	50.08
40	47.27
41	53.23

```
WOULD YOU LIKE AUTOCORRELATIONS ON THE RESIDUALS ?N

DO YOU WANT TO RUN ANOTHER PROGRAM?Y
```

CHAPTER 11

Single and Linear Moving Averages

The single and linear moving averages of this chapter are in principle exactly the same as those examined in Chapter 4. The main difference is that in Chapter 4 the outcome of the averaging process is centered in the middle of the data averaged, while in the methods of this chapter the averages obtained are projected and used for forecasting. They are extrapolated several periods ahead and used for predicting what is basically an historical average.

Moving average methods used for forecasting purposes are usually less attractive than corresponding exponential smoothing methods because they require more data points, need more calculations, and are not necessarily any more accurate than exponential smoothing methods. They should be used only when the randomness in the data is very high and the autocorrelation among successive values extremely low. Otherwise, their overall advantages will not match that of exponential smoothing models.

The formulas for the two major forms of moving average forecasting models are:

1. Single Moving Average

$$F_{t+1} = \frac{X_t + X_{t-1} + X_{t-2} + \ldots + X_{t-n+1}}{n} \qquad (11\text{-}1)$$

That is, F_{t+1} is the mean (average of n values). It differs from the population mean in the sense that n remains constant as time t varies. Expression (11-1) simplifies to:

$$F_{t+1} = F_t + \frac{X_t}{n} - \frac{X_{t-n}}{n} \qquad (11\text{-}2)$$

where n is the number of periods to be included in the moving average.

2. Linear Moving Average

This approach employs the idea of a double moving average in the same way that double exponential smoothing was used to forecast linear trends in Chapter 7.

$$S'_t = \frac{X_t + X_{t-1} + X_{t-2} + \ldots + X_{t-n+1}}{n}$$

$$= S'_{t-1} + \frac{X_t}{n} - \frac{X_{t-n}}{n} \qquad (11\text{-}3)$$

$$S''_t = \frac{S'_t + S'_{t-1} + S'_{t-2} + \ldots + S'_{t-n+1}}{n}$$

$$= S''_{t-1} + \frac{S'_t}{n} - \frac{S'_{t-n}}{n} \qquad (11\text{-}4)$$

In order to forecast one should first calculate a_t and b_t (as in Chapter 7). This is done as follows:

$$a_t = 2S'_t - S''_t \qquad (11\text{-}5)$$

$$b_t = \frac{2}{n-1}(S'_t - S''_t) \qquad (11\text{-}6)$$

(See Exercise 1 on Smoothing, Averaging and Trends for an explanation of why the denominator on (11-6) is n − 1.)

Forecasts can finally be made by applying

$$F_{t+m} = a_t + b_t m \qquad (11\text{-}7)$$

where m is the number of periods ahead to be forecast.

References for futher study:

Brown, R. (1959, Chapter 3)
Brown, R. (1963, Chapter 7)
Makridakis, S. and S. Wheelwright (1978, Chapter 3)

INPUT/OUTPUT FOR THE SINGLE AND LINEAR MOVING AVERAGE FORECASTING PROGRAMS (MAVE AND MAVE2).

Input

1.   *Enter a value for n.*

The parameter n is the number of periods in the moving average. A high value of n averages many data points and therefore smooths the data more than a small value of n. In this respect n behaves in exactly the opposite manner of α. In fact, the two can be approximately related by

$$n = \frac{1}{\alpha}. \tag{11-8}$$

A more precise relationship can be found (see exercise 12-2) to be

$$\frac{1 - \alpha}{\alpha} = \frac{n - 1}{2} \tag{11-9}$$

Thus, the more random or fluctuating the data are, the larger n should be, while for data with little randomness a small value of n should be used. The user can try different values for n until one is found that minimizes the mean squared error.

Output

1.   *Standard Outputs* (see Chapter 5).

EXERCISE 11.1

The following data represent the weekly inventory demand for item YB24. What demand would you forecast in period 21? Use a 3, 5 and 7 month single moving average.

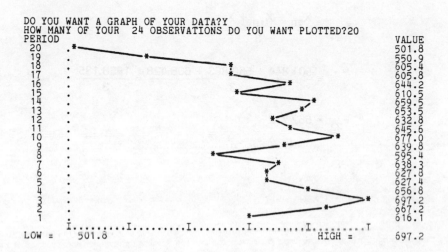

```
DO YOU WANT A GRAPH OF YOUR DATA?Y
HOW MANY OF YOUR  24 OBSERVATIONS DO YOU WANT PLOTTED?20
PERIOD                                                      VALUE
20                                                          501.8
19                                                          550.9
18                                                          605.4
17                                                          605.8
16                                                          644.2
15                                                          610.5
14                                                          659.5
13                                                          653.5
12                                                          632.8
11                                                          645.6
10                                                          677.0
 9                                                          639.8
 8                                                          595.4
 7                                                          638.3
 6                                                          627.3
 5                                                          627.4
 4                                                          656.8
 3                                                          697.2
 2                                                          667.2
 1                                                          616.1
LOW =    501.8                              HIGH =          697.2
```

Solution

The mean squared errors (MSE) for n = 3, n = 5 and n = 7 are 1352.83, 1784.76 and 2178.7, respectively, This means that n = 3 gives the smallest error. The forecast for this n is:

F_{21} = 552.712, which should be used as a prediction for next period (see Figure 11-1).

In order to obtain a forecast one must average the latest terms and use their mean as the prediction for next period. At period 19, for example, if one assumes n = 3 (see Figure 11-1) and desires a forecast for period 20, the following results are obtained:

$$F_{20} = \frac{X_{19} + X_{18} + X_{17}}{3}$$

$$= \frac{550.863 + 605.428 + 605.811}{3} = \frac{1762.102}{3}$$

$$F_{20} = 587.37$$

The actual value, X_{20}, for period 20 is 501.844 which gives an error of $e_{20} = 501.844 - 587.37 = -85.53$, or -17%.

Similarly, the forecast for period 21 is:

$$F_{21} = \frac{X_{20} + X_{19} + X_{18}}{3}$$

$$= \frac{501.844 + 550.863 + 605.428}{3} = \frac{1658.135}{3}$$

$$F_{21} = 552.712$$

FIGURE 11-1   Solution to Exercise 11.1 Using MAVE (Using n = 3)

```
EXE METHOD
  8K
WHICH PROGRAM WOULD YOU LIKE TO RUN?MAVE


      *** FORECASTING USING THE IDEA ***
      ***          OF AVERAGE          ***

YOU CAN CHOOSE ONE OF THESE METHODS
1. MAVE - SINGLE MOVING AVERAGE
2. MAVE2 - LINEAR MOVING AVERAGE
3. MEAN

HOW MANY OBSERVATIONS DO YOU WANT TO USE?20
DATA FILENAME?YB24

WHICH METHOD DO YOU WANT TO USE - ANSWER BY TYPING 1,2,OR 3?1

      *** MAVE ***

NEED HELP (Y OR N)?N

ENTER THE NUMBER OF PERIODS IN THE MOVING AVERAGE?3
DO YOU WANT A TABLE OF ACTUAL AND PREDICTED (Y OR N)?Y


PERIOD        ACTUAL          FORECAST         ERROR      PCT ERROR
   4          656.78          660.17           -3.39         .52%
   5          627.36          673.73          -46.38        7.39%
   6          627.81          660.45          -32.64        5.20%
   7          638.25          637.31             .94         .15%
   8          595.42          631.14          -35.72        6.00%
   9          639.75          620.50           19.26        3.01%
  10          676.96          624.48           52.48        7.75%
  11          645.57          637.38            8.20        1.27%
  12          632.77          654.10          -21.33        3.37%
  13          653.46          651.77            1.69         .26%
  14          659.48          643.93           15.55        2.36%
  15          610.46          648.57          -38.11        6.24%
  16          644.17          641.13            3.04         .47%
  17          605.81          638.04          -32.22        5.32%
  18          605.43          620.15          -14.72        2.43%
  19          550.86          618.47          -67.61       12.27%
  20          501.84          587.37          -85.52       17.04%
MEAN PC ERROR (MPE) OR BIAS     =     -2.97%
MEAN SQUARED ERROR (MSE)        =    1352.8
MEAN ABSOLUTE PC ERROR  (MAPE) =       4.8%


FORECAST FOR HOW MANY PERIODS AHEAD (0=NONE,24=MAX)?6


PERIOD        FORECAST
  21          552.71
  22          518.18
  23          518.80
  24          524.25
  25          512.94
  26          514.96

WOULD YOU LIKE AUTOCORRELATION ON THE RESIDUALS?N

DO YOU WANT TO USE ANOTHER PROGRAM?Y
```

FIGURE 11-2   Solution to Exercise 11.1 Using MAVE (Using n = 5)

```
HOW MANY OBSERVATIONS DO YOU WANT TO USE?20
DATA FILENAME?YB24

WHICH METHOD DO YOU WANT TO USE - ANSWER BY TYPING 1,2,OR 3?1

          *** MAVE ***

NEED HELP (Y OR N)?N

ENTER THE NUMBER OF PERIODS IN THE MOVING AVERAGE?5
DO YOU WANT A TABLE OF ACTUAL AND PREDICTED (Y OR N)?Y

PERIOD         ACTUAL          FORECAST          ERROR      PCT ERROR
   6           627.81           652.93          -25.12         4.00%
   7           638.25           655.27          -17.02         2.67%
   8           595.42           649.48          -54.06         9.08%
   9           639.75           629.12           10.63         1.66%
  10           676.96           625.72           51.24         7.57%
  11           645.57           635.64            9.93         1.54%
  12           632.77           639.19           -6.43         1.02%
  13           653.46           638.09           15.36         2.35%
  14           659.48           649.70            9.78         1.48%
  15           610.46           653.65          -43.19         7.07%
  16           644.17           640.35            3.82          .59%
  17           605.81           640.07          -34.25         5.65%
  18           605.43           634.67          -29.25         4.83%
  19           550.86           625.07          -74.21        13.47%
  20           501.84           603.35         -101.50        20.23%
MEAN PC ERROR (MPE) OR BIAS     =        -3.52%
MEAN SQUARED ERROR (MSE)        =       1784.8
MEAN ABSOLUTE PC ERROR   (MAPE) =          5.5%

FORECAST FOR HOW MANY PERIODS AHEAD (0=NONE,24=MAX)?6

PERIOD         FORECAST
  21           581.62
  22           553.16
  23           548.32
  24           537.87
  25           537.36
  26           544.56

WOULD YOU LIKE AUTOCORRELATION ON THE RESIDUALS?N

DO YOU WANT TO USE ANOTHER PROGRAM?Y
```

FIGURE 11-3   Solution to Exercise 11.1 Using MAVE (Using n = 7)

```
HOW MANY OBSERVATIONS DO YOU WANT TO USE?20
DATA FILENAME?YB24

WHICH METHOD DO YOU WANT TO USE - ANSWER BY TYPING 1,2,OR 3?1

        *** MAVE ***

NEED HELP (Y OR N)?N

ENTER THE NUMBER OF PERIODS IN THE MOVING AVERAGE?7
DO YOU WANT A TABLE OF ACTUAL AND PREDICTED (Y OR N)?Y

PERIOD          ACTUAL          FORECAST          ERROR      PCT ERROR
   8            595.42          647.24           -51.82         8.70%
   9            639.75          644.29            -4.54          .71%
  10            676.96          640.37            36.59         5.41%
  11            645.57          637.48             8.10         1.25%
  12            632.77          635.88            -3.11          .49%
  13            653.46          636.65            16.81         2.57%
  14            659.48          640.31            19.17         2.91%
  15            610.46          643.34           -32.89         5.39%
  16            644.17          645.49            -1.32          .21%
  17            605.81          646.12           -40.31         6.65%
  18            605.43          635.96           -30.53         5.04%
  19            550.86          630.22           -79.36        14.41%
  20            501.84          618.52          -116.68        23.25%
MEAN PC ERROR (MPE) OR BIAS     =      -4.05%
MEAN SQUARED ERROR (MSE)        =      2178.7
MEAN ABSOLUTE PC ERROR  (MAPE)  =       5.9%


FORECAST FOR HOW MANY PERIODS AHEAD (0=NONE,24=MAX)?6

PERIOD          FORECAST
  21            596.86
  22            574.35
  23            572.40
  24            562.43
  25            557.66
  26            551.51

WOULD YOU LIKE AUTOCORRELATION ON THE RESIDUALS?N

DO YOU WANT TO USE ANOTHER PROGRAM?N
```

EXERCISE 11.2

The data below are the demand of YB24 for a longer time period. Use the
method of linear moving average to forecast the next 4 periods using an n of
3, 5 and 7 periods.

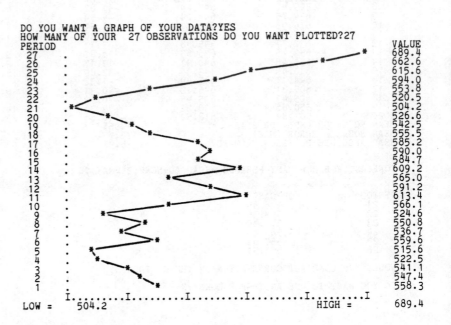

```
DO YOU WANT A GRAPH OF YOUR DATA?YES
HOW MANY OF YOUR  27 OBSERVATIONS DO YOU WANT PLOTTED?27
PERIOD                                                      VALUE
```

Solution

The mean squared errors for n = 3, n = 5 and n = 7 is 1122.23, 2085.26 and
3954.04, respectively. Thus, n = 3 provides the best results as was true with
single moving average.

In order to forecast with the method of linear moving averages, one must
first calculate a_t and b_t using (11-5) and (11-6). These values are a function of
the double moving averages of the last n periods.

In order to forecast period 26 (assuming an n = 3; see Figure 11-4) one
must determine S'_t and S''_t.

$$S'_{24} = \frac{X_{24} + X_{23} + X_{22}}{3} = \frac{594.002 + 553.841 + 520.466}{3} = 556.103$$

$$S'_{25} = \frac{X_{25} + X_{24} + X_{23}}{3} = \frac{615.552 + 594.002 + 553.841}{3} = 587.80$$

$$S'_{26} = \frac{X_{26} + X_{25} + X_{24}}{3} = \frac{662.593 + 615.552 + 594.002}{3} = 624.049$$

$$S'_{27} = \frac{X_{27} + X_{26} + X_{25}}{3} = \frac{689.366 + 662.593 + 615.552}{3} = 655.837$$

$$S''_{26} = \frac{S'_{27} + S'_{25} + S'_{24}}{3} = \frac{624.049 + 587.8 + 556.103}{3} = 589.32$$

$$S''_{27} = \frac{S'_{27} + S'_{26} + S'_{25}}{3} = \frac{655.837 + 624.049 + 587.8}{3} = 622.562$$

Once the values of the single and double moving averages are known, one can find a_t and b_t through (11-5) and (11-6). Thus

$$a_{26} = 2S'_{26} - S''_{26}$$

$$= 2(624.049) - 589.32$$

$$= 658.778$$

$$b_{26} = \frac{2}{n-1}(S'_{26} - S''_{26})$$

$$= \frac{2}{3-1}(624.049 - 589.32)$$

$$= 34.729$$

Applying (11-7) in order to forecast period 27 gives:

$$F_{27} = a_{26} + b_{26}(1)$$

$$= 658.778 + 34.729$$

$$= 693.513$$

One can continue and find a_{27} and b_{27}

$$a_{27} = 2S'_{27} - S''_{27} = 2(655.837) - 622.562 - 622.562 = 689.112$$

and

$$b_{27} = \frac{2}{3-1}(655.837 - 622.562) = 33.275.$$

Thus,

$$F_{28} = a_{27} + b_{27}(1) = 689.112 + 33.275 = 722.388$$

$$F_{29} = a_{27} + b_{27}(2) = 689.112 + 33.275(2) = 755.664$$

$$F_{30} = a_{27} + b_{27}(3) = 689.112 + 33.275(3) = 788.939$$

$$F_{31} = a_{27} + b_{27}(4) = 689.112 + 33.275(4) = 822.215$$

FIGURE 11-4 Solution to Exercise 11.2 Using MAVE 2 (Using n = 3)

```
WHICH PROGRAM WOULD YOU LIKE TO RUN?MAVE2


    *** FORECASTING USING THE IDEA ***
    ***          OF AVERAGE          ***

YOU CAN CHOOSE ONE OF THESE METHODS
1. MAVE - SINGLE MOVING AVERAGE
2. MAVE2 - LINEAR MOVING AVERAGE
3. MEAN

HOW MANY OBSERVATIONS DO YOU WANT TO USE?27
DATA FILENAME?YB24NEW

WHICH METHOD DO YOU WANT TO USE - ANSWER BY TYPING 1,2,OR 3?2

        *** MAVE2 ***

NEED HELP (Y OR N)?N

ENTER THE NUMBER OF PERIODS IN THE MOVING AVERAGES (  13=MAX)?3
DO YOU WANT A TABLE OF ACTUAL AND PREDICTED (Y OR N)?Y

PERIOD          ACTUAL          FORECAST          ERROR      PCT ERROR
  6             559.62          504.27            55.35        10.74%
  7             536.73          533.72             3.01          .54%
  8             550.77          547.78             2.99          .56%
  9             524.64          567.85           -43.21        -7.85%
 10             566.08          529.65            36.43         6.94%
 11             613.41          552.43            60.98        10.77%
 12             591.25          602.41           -11.16        -1.82%
 13             564.96          633.77           -68.81       -11.64%
 14             609.19          604.18             5.01          .89%
 15             584.72          586.34            -1.62         -.27%
 16             589.99          582.45             7.54         1.29%
 17             585.21          604.31           -19.10        -3.24%
 18             555.48          581.54           -26.06        -4.45%
 19             542.49          558.57           -16.08        -2.89%
 20             526.61          533.45            -6.84        -1.26%
 21             504.22          504.93             -.71         -.13%
 22             520.46          488.64            31.82         6.31%
 23             553.84          495.91            57.93        11.13%
 24             594.00          533.38            60.62        10.95%
 25             615.55          602.05            13.50         2.27%
 26             662.59          650.01            12.50         2.04%
 27             689.36          693.51            -4.15         -.63%
MEAN PC ERROR (MPE) OR BIAS     =       1.13%
MEAN SQUARED ERROR (MSE)        =      1122.2
MEAN ABSOLUTE PC ERROR  (MAPE) =       4.4%


FORECAST FOR HOW MANY PERIODS AHEAD (0=NONE,24=MAX)?6


PERIOD          FORECAST
 28             722.38
 29             755.66
 30             788.93
 31             822.21
 32             855.48
 33             888.75

WOULD YOU LIKE AUTOCORRELATION ON THE RESIDUALS?N

DO YOU WANT TO USE ANOTHER PROGRAM?Y
```

FIGURE 11-5 Solution to Exercise 11.2 Using MAVE 2 (Using n = 5)

```
HOW MANY OBSERVATIONS DO YOU WANT TO USE?227_
*D*
27
DATA FILENAME?YB24NEW

WHICH METHOD DO YOU WANT TO USE - ANSWER BY TYPING 1,2,OR 3?2

        *** MAVE2 ***

NEED HELP (Y OR N)?N

ENTER THE NUMBER OF PERIODS IN THE MOVING AVERAGES (  13=MAX)?5
DO YOU WANT A TABLE OF ACTUAL AND PREDICTED (Y OR N)?Y

PERIOD          ACTUAL          FORECAST          ERROR       PCT ERROR
  10            566.08          538.54            27.54          5.25%
  11            613.41          560.61            52.80          9.33%
  12            591.25          581.17            10.08          1.64%
  13            564.96          598.19           -33.23         -5.62%
  14            609.19          594.77            14.42          2.55%
  15            584.72          621.59           -36.87         -6.05%
  16            589.99          617.37           -27.38         -4.68%
  17            585.21          596.75           -11.54         -1.96%
  18            555.48          588.46           -32.98         -5.64%
  19            542.49          579.86           -37.37         -6.73%
  20            526.61          551.73           -25.12         -4.63%
  21            504.22          532.50           -28.28         -5.37%
  22            520.46          503.18            17.28          3.43%
  23            553.84          487.90            65.94         12.67%
  24            594.00          503.70            90.30         16.30%
  25            615.55          538.98            76.57         12.89%
  26            662.59          584.15            78.44         12.74%
  27            689.36          649.39            39.97          6.03%
MEAN PC ERROR (MPE) OR BIAS       =          2.00%
MEAN SQUARED ERROR (MSE)          =        2085.2
MEAN ABSOLUTE PC ERROR  (MAPE) =             6.7%


FORECAST FOR HOW MANY PERIODS AHEAD (0=NONE,24=MAX)?5

PERIOD          FORECAST
  28            705.87
  29            733.48
  30            761.08
  31            788.68
  32            816.28

WOULD YOU LIKE AUTOCORRELATION ON THE RESIDUALS?N

DO YOU WANT TO USE ANOTHER PROGRAM?Y
```

FIGURE 11-6 Solution to Exercise 11.2 Using MAVE 2 (Using n = 7)

```
HOW MANY OBSERVATIONS DO YOU WANT TO USE?27
DATA FILENAME?YB24NEW

WHICH METHOD DO YOU WANT TO USE - ANSWER BY TYPING 1,2,OR 3?2

        *** MAVE2 ***

NEED HELP (Y OR N)?N

ENTER THE NUMBER OF PERIODS IN THE MOVING AVERAGES (  13=MAX)?7
DO YOU WANT A TABLE OF ACTUAL AND PREDICTED (Y OR N)?Y

PERIOD      ACTUAL          FORECAST        ERROR       PCT ERROR
  14        609.19          585.64          23.55          4.17%
  15        584.72          603.29         -18.57         -3.05%
  16        589.99          606.97         -16.98         -2.90%
  17        585.21          618.72         -33.51         -5.68%
  18        555.48          615.22         -59.74        -10.21%
  19        542.49          590.09         -47.60         -8.57%
  20        526.61          571.40         -44.79         -8.26%
  21        504.22          557.37         -53.15        -10.09%
  22        520.46          525.96          -5.50         -1.09%
  23        553.84          510.79          43.05          8.27%
  24        594.00          507.76          86.24         15.57%
  25        615.55          519.98          95.57         16.09%
  26        662.59          546.09         116.50         18.93%
  27        689.36          587.61         101.75         15.36%
MEAN PC ERROR (MPE) OR BIAS     =        1.62%
MEAN SQUARED ERROR (MSE)        =       3953.9
MEAN ABSOLUTE PC ERROR   (MAPE) =        9.0%

FORECAST FOR HOW MANY PERIODS AHEAD (0=NONE,24=MAX)?6

PERIOD          FORECAST
  28            637.88
  29            649.50
  30            661.11
  31            672.72
  32            684.33
  33            695.95

WOULD YOU LIKE AUTOCORRELATION ON THE RESIDUALS?N

DO YOU WANT TO USE ANOTHER PROGRAM?N
```

CHAPTER 12

Trend Analysis

Trend analysis is a special case of simple regression analysis. The only difference between the two methods is that with trend analysis the value of the independent variable is restricted to time, whereas with simple regression it can be any variable of interest. The concept behind trend analysis is to draw a straight line in such a way that it passes throught the middle of the data. The mathematical equation describing this straight line is given by (12-1). It is fitted in such a way that the sum of the squared deviations (the differences between actual values of the time series and those estimated by equation (12-1)) are as small as possible. This is done by calculating the values of a and b (needed to specify the line) using the method of least squares.

The equation used to calculate the trend line is

$$\hat{X}_t = a + bt \tag{12-1}$$

where

    a   is an estimate of the intercept

    b   is an estimate of the slope

    t   denotes time (i.e., 1, 2, 3, . . . , t)

    $\hat{X}_t$ is the estimate of the data point corresponding to period t.

The quantity $X_t - \hat{X}_t$ is the residual or error, e_t. The squared sum of all the errors will be a minimum as long as (12-2) and (12-3) are used to estimate the values of a and b.

The values of a and b can be found using the following two formulas:

$$b = \frac{n\Sigma tX - \Sigma t\Sigma X}{n\Sigma t^2 - (\Sigma t)^2} \tag{12-2}$$

$$a = \frac{\Sigma X}{n} - b\frac{\Sigma t}{n} \tag{12-3}$$

where all the summations go from 1 to n, and n is the number of data points.

Forecasting requires use of equation (12-4)

$$F_{t+m} = a + b(t + m) \tag{12-4}$$

where t + m is the number of the period for which a forecast is to be prepared.

It can be seen that there is considerable similarity between (12-2), (12-3), (12-4) and the linear smoothing methods of Chapters 7 and 8. Both trend analysis and exponential smoothing require estimates for a and b. However, in regression these estimates are constant throughout the entire range of data. In smoothing methods they vary from period to period in a way that is more responsive to changes in the pattern of the data. The final object, however, is the same — both types of methods look for an intercept point (a) and a trend (b) in order to forecast.

Unlike smoothing models, trend analysis is used mostly for medium- and longer-term forecasts. It generally requires more data points and is more difficult to update when new data become available. If one wants to obtain new estimates for a and b, one must rerun the entire set of data whereas with smoothing models new forecasts can be found simply by updating the smoothing values. The relative merits of regression and smoothing models for a specific application cannot be precisely specified, even though smoothing models are preferable almost always for short term forecasts. Finally, it should be pointed out that trend analysis, like exponential smoothing methods with the exception of Winters' model, is incapable of handling seasonal data.

References for further study:

Johnston, J. (1972, Chapters 2 and 3)
Makridakis, S. and S. Wheelwright (1978, Chapter 5)
Wheelwright, S. and S. Makridakis (1977, Chapter 5)

INPUT/OUTPUT TO THE TREND ANALYSIS PROGRAM (SREG)

Input

None.

Output

In addition to the standard outputs, the program provides the following information:

1. the value of a and b
2. the t test of a and b
3. F-test
4. R^2 and R
5. standard deviation of regression
6. ΣX_i, ΣX_i^2, and $\overline{X}$

Except for the values of a and b that are used directly for forecasting through equation (12-1), the remainder of the output provides information about the significance or the trend equation — how good the fit is and how reliable the forecasts will be.

For example, R^2 varies between 0 and 1. An R^2 of 0 means that the regression equation does not describe the data at all, and a value of 1 means that the fit is perfect, although the regression equation may still be a poor predictor. This is checked using the t-tests and the F-test, which indicate whether the parameters, a and b and the overall trend relationship between the forecast variable and time are significant. In general, if the t-test corresponding to b is greater than 2, or the F-test is greater than 5, the trend relationship is significant. When the opposite is true, that is, the t-test < 2 or the F-test < 5, one should use the mean as an estimate rather than the trend line. Finally, the standard deviation of regression and the summation of X_i and X_i^2 are used to construct confidence intervals.

In order to forecast one must add to a (the intercept), (t + m) times the value of b (the slope), where m is the number of periods ahead to be forecast and t is the current period.

EXERCISE 12.1

The following data represent the GNP of France in constant 1969 prices for the years 1950-1970. Forecast the GNP values for the next four years.

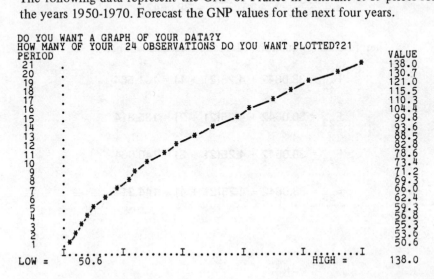

```
DO YOU WANT A GRAPH OF YOUR DATA?Y
HOW MANY OF YOUR  24 OBSERVATIONS DO YOU WANT PLOTTED?21
PERIOD                                                        VALUE
 21   .                                              *         138.0
 20   .                                        *--*--          130.7
 19   .                                     *--                121.0
 18   .                                  *--                   115.5
 17   .                               *--                      110.3
 16   .                             *--                        104.4
 15   .                          *--                            99.8
 14   .                       *--                               93.6
 13   .                    *--                                  88.5
 12   .                 *--                                     82.8
 11   .              *--                                        78.6
 10   .            *                                            73.4
  9   .          *                                             71.2
  8   .         *                                              69.3
  7   .       *                                                66.0
  6   .      *                                                 62.4
  5   .    *--                                                 59.3
  4   .   *                                                    56.8
  3   .  *                                                     55.3
  2   . *                                                      53.6
  1   .*                                                       50.6
LOW =    50.6  I........I........I........I........I........I
                                                      HIGH =   138.0
```

Solution (See Figure 12-1)

In order to forecast one must first calculate the values of a and b. This is done using (12-2) and (12-3), where the different summations needed are shown below:

Actual (X_i)	Time (t_i)	Actual $\times$ time $(X_i t_i)$	Time squared (t_i^2)
50.6	1	50.6 $\times$ 1	1^2
53.6	2	53.6 $\times$ 2	2^2
55.3	3	55.3 $\times$ 3	3^2
.			
.			
138	21	138 $\times$ 21	21^2
$\Sigma X_i = 1781.1$	$\Sigma t_i = 231$	$\Sigma X_i t_i = 22864.6$	$\Sigma t_i^2 = 3311$

Thus

$$b = \frac{21(22864.6) - 231(1781.1)}{21(3311) - (231)^2} = 4.25$$

$$a = \frac{1781.1}{21} - 4.25 \; \frac{231}{21} = 38.0642$$

Therefore using equation (12-4) the forecasts will be:

$$F_{22} = 38.0642 + 4.25(21 + 1) = 131.564$$

$$F_{23} = 38.0642 + 4.25(21 + 2) = 135.814$$

$$F_{24} = 38.0642 + 4.25(21 + 3) = 140.064$$

$$F_{25} = 38.0642 + 4.25(21 + 4) = 144.314$$

FIGURE 12-1   Forecasts for French GNP Using Simple Regression

```
EXE METHOD
 8K
WHICH PROGRAM WOULD YOU LIKE TO RUN?SREG
HOW MANY OBSERVATIONS DO YOU WANT TO USE?21
DATA FILENAME?FRANCE
WHAT PROGRAM DO YOU WANT TO RUN(SCURVE,SREG,OR EXGROW)?SREG


    *** SREG ***

NEED HELP (Y OR N)?N
DO YOU WANT A TABLE OF ACTUAL AND PREDICTED (Y OR N)?Y

PERIOD        ACTUAL          FORECAST          ERROR      PCT ERROR
  1           50.60            42.31             8.29         16.37%
  2           53.60            46.56             7.04         13.13%
  3           55.30            50.81             4.49          8.11%
  4           56.80            55.06             1.74          3.06%
  5           59.30            59.31             -.01           .02%
  6           62.40            63.56            -1.16          1.87%
  7           66.00            67.81            -1.81          2.75%
  8           69.30            72.06            -2.76          3.99%
  9           71.20            76.31            -5.11          7.18%
 10           73.40            80.56            -7.16          9.76%
 11           78.60            84.81            -6.21          7.91%
 12           82.80            89.06            -6.26          7.57%
 13           88.50            93.31            -4.81          5.44%
 14           93.60            97.56            -3.96          4.24%
 15           99.80           101.81            -2.01          2.02%
 16          104.40           106.06            -1.66          1.59%
 17          110.30           110.31             -.01           .01%
 18          115.50           114.56              .94           .81%
 19          121.00           118.81             2.19          1.81%
 20          130.70           123.06             7.64          5.84%
 21          138.00           127.31            10.69          7.74%
                COEFFICIENT      STD. ERROR       T-TEST
INTERCEPT (A)     38.06            .25            150.95
SLOPE (B)          4.25            .04            109.71

F-TEST =      495.99      R-SQUARE =    .96      R =     .98

STANDARD DEVIATION OF REGRESSION =    1.16
SUM OF X(I) SQUARED =          165503.6
SUM OF X(I) =          1781.1
X-DAR =      84.81


MEAN PC ERROR (MPE) OR BIAS       =        .12%
MEAN SQUARED ERROR (MSE)          =       25.4
MEAN ABSOLUTE PC ERROR    (MAPE)  =        5.3%


FORECAST FOR HOW MANY PERIODS AHEAD (0=NONE,24=MAX)?4

PERIOD    FORECAST
 22       131.56
 23       135.81
 24       140.06
 25       144.31

WOULD YOU LIKE AUTOCORRELATION ON THE RESIDUALS?N
WOULD YOU LIKE TO USE ANOTHER PROGRAM?N
```

**EXERCISES ON SMOOTHING,
AVERAGING, AND TREND FITTING
EXERCISE 12-2**

Expressions (11-2) and (5-1) can be related by assuming that $F_t = X_{t-n}$. Equation (11-2) then becomes:

$$F_{t+1} = F_t + \frac{X_t}{n} - \frac{F_t}{n} \tag{1}$$

By substituting $\alpha = \frac{1}{n}$, equation (1) becomes

$$F_{t+1} = F_t + \alpha X_t + \alpha F_t$$

and

$$F_{t+1} = \alpha X_t + (1 - \alpha)F_t \tag{2}$$

Thus, equation (11-2) becomes equation (2) — which is exactly the same as (5-1) — if it is assumed that $F_t = X_{t-n}$.

Expression (11-1) can also be viewed from a different perspective. If it is assumed that $t = 0, t - 1 = 1, t - 2 = 2, t - 3 = 3, \ldots, t - n + 1 = n - 1$, the weights applied to $X_t, X_{t-2}, X_{t-3}, \ldots, X_{t-n+2}$ can be expressed in terms of how far back in time they are from the present, i.e., 0. Thus, the equivalent of (11-1) is:

$$F_{t+1} = \frac{X_t}{n} + \frac{X_{t-1}}{n} + \frac{X_{t-2}}{n} + \frac{X_{t-3}}{n} + \ldots + \frac{X_{t-n+1}}{n} \tag{3}$$

Changing the origin to t and considering the equal weights, 1/n, gives:

$$0\frac{1}{n} + 1\left(\frac{1}{n}\right) + 2\left(\frac{1}{n}\right) + 3\left(\frac{1}{n}\right) + \ldots + (n-1)\left(\frac{1}{n}\right) \tag{4}$$

Because the sum of $0, 1, 2, 3, \ldots (n - 1)$ is $n(n - 1)/2$, the sum of the weights shown in (4) is

$$\frac{(n - 1)\,\cancel{(n)}}{2}\left(\frac{1}{\cancel{n}}\right) = \frac{n - 1}{2} \tag{5}$$

If the "aging" of weights in (5-6) is expressed in a similar way as (3) above, the result is:

$$\alpha(0) + \alpha(1 - \alpha)1 + \alpha(1 - \alpha)^2 2 + \alpha(1 - \alpha)^3 3 + \ldots + \alpha(1 - \alpha)^n n$$

$$(6)$$

Equation (6) is a geometric progression whose sum is

$$\frac{1 - \alpha}{\alpha} \tag{7}$$

(a) It should be noted that (7) is the factor by which the trend, $S'_t - S''_t$, of (7-3) is multiplied. Similarly (5) is the term by which (11-6) is multiplied.

What can be said about the similarities of (5) and (7) and of (7-3) and (11-6)?

(b) Assuming that

$$\frac{n - 1}{2} = \frac{1 - \alpha}{\alpha}$$

provides a rough way of relating n and α.

How does this differ from stating that $\alpha = 1/n$?

(c) What can you say about the ability of single exponential smoothing, (5-1) or as it is expanded in (5-6), to keep up with trend changes in the data?

Cases

UTAH INDUSTRIAL PROMOTION BOARD

It was with some surprise that Jeff Miller read the governor's response to the Industrial Promotion Board's budget request that Jeff had submitted early in 1973. As head of that board for the past several years, Jeff had often heard the governor talk about the need to expand the state's activities in attracting new industry to the four major counties of Utah. (Those counties were Salt Lake, Weber, Davis and Utah.) In his budget request for fiscal 1974, Jeff had included a 20% increase in out-of-state travel that he thought was necessary in order to attract more major industries and jobs to the four-county region. While the governor's response clearly had positive overtones to it, the governor had made very clear that he thought Jeff's request was incomplete. The governor indicated that Jeff's office should forecast several items assuming no change in the current budget levels and then forecast the difference that would arise if a 20% increase were given for next year and for several years following.

One of the variables that the governor had indicated was particularly important was that of non-agricultural employment. Thus as a first step Jeff obtained the historical data for this variable for each of the four counties over the past several years (See Figure 1). Since he had as much or more data for this variable as he did for any of the others mentioned by the governor, he thought it might make sense to start with the task of forecasting this one before tackling any of the others.

While Jeff knew that he would want to be completely fair in his presentation to the governor, he also realized that he wanted to consider any changes in the long-term pattern that might cause him to give somewhat lower estimates of future non-agricultural employment than would a simple trend line fitted to the historical data. He certainly did not want to over-estimate the growth in non-agricultural employment (assuming no increase in the budget spent by the Utah Industrial Promotion Board) since that would perhaps jeopardize his own position as head of the board if he were unable to then show an increase when he was given the higher budget level. Therefore Jeff thought it might be useful to try several different approaches in forecasting this data in order to determine which tended to give higher forecasts and which lower, so that he could then select the appropriate combination or single forecast that would most honestly present the future situation to the governor.

In preparing these forecasts, it was unclear in Jeff's mind whether or not he should try to forecast total non-agricultural employment in the four-county region and then break it up according to some percentage in each of the counties or whether he should simply forecast them individually and then add them together in order to get a total. One of the things that argued for the former approach was the fact that the four-county region had no clear-cut geographical boundaries as far as companies and people taking up residency in the state were concerned and thus if one area were low, others might tend to be high just because of short-run changes in preferences.

Jeff had considered briefly the possibility of using some form of regression analysis and obtaining several other data series that could be associated with this one. However, a discussion with a local professor had suggested to him that all of these series would be closely related to time and therefore multicollinearity might create special problems for that technique. Thus, at least for the time being, he decided to limit himself to time series methods of forecasting.

FIGURE 1 Utah Industrial Promotion Board Non-Agricultural Employment in Salt Lake, Weber, Davis and Utah Counties 1950 to 1972

Year	Salt Lake	Weber	Davis	Utah
1950	95,021	24,528	11,720	18,190
1951	99,566	25,874	18,110	19,354
1952	102,549	24,129	21,848	18,540
1953	105,002	25,822	20,533	20,027
1954	106,901	24,116	18,053	19,778
1955	112,913	25,073	18,340	22,186
1956	118,987	26,023	18,857	23,895
1957	121,728	26,272	18,823	24,673
1958	122,804	26,950	19,156	23,438
1959	128,783	29,385	20,418	23,358
1960	134,545	30,578	20,758	25,193
1961	139,704	31,193	20,913	24,943
1962	148,153	31,154	22,197	25,436
1963	154,100	31,363	22,888	26,344
1964	154,595	30,943	22,581	28,097
1965	157,620	31,110	23,668	29,957
1966	162,783	34,447	28,784	30,876
1967	161,923	38,344	32,374	32,474
1968	168,090	38,494	32,309	33,861
1969	177,060	38,636	31,768	36,349
1970	183,659	38,759	31,999	37,433
1971	191,340	33,473	32,899	38,858
1972ᵖ	204,210	40,000	34,740	41,700

ᵖPreliminary

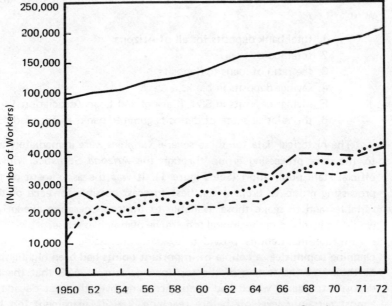

Source: Utah Rom Department of Employment Security.

ARIZONA SECURITY BANK

With the increased level of interest in long-range planning, the Arizona Security Bank (ASB) in early 1974 had almost overnight decided to initiate an annual long-range planning exercise. The purpose of this exercise was to review important trends in the banking business and in various characteristics of ASB's performance and to project those into the future assuming no changes in management policies. It was then felt that a group of the top operating executives in the bank could spend two to three days reveiwing these data and determining what changes in policy would be required in coming months and years if the bank were going to achieve its long-term goals.

Since the bank had a substantial data processing group to handle all of its accounting and financial data, top management thought it would be a relatively simple task for this group to prepare forecasts of several important variables that could serve as input to this planning exercise. Six of the specific items relating to ASB's area of competition that were felt to be important were

1. total bank deposits for all of Arizona
2. total loans
3. the ratio of loans to deposits
4. savings deposits in banks
5. savings deposits in S&L (Savings and Loan Associations)
6. the relative share of deposits going to banks as opposed to S&L's.

The historical data for these several variables were immediately available to the data processing group through the *Arizona Statistical Review,* an official state publication (See Figure 1). It was the assignment of the data processing group to prepare the best forecast possible for each of these six variables and to make those available to the top operating executives who would be involved in the annual long-range planning exercise.

In the instructions provided to data processing from the long-range planning committee, a couple of important points had been highlighted. One of these was the feeling of the top management at ASB that the ratio of deposits to loans, which had been increasing over the past decade, would almost certainly level off before reaching 80%. (Management felt that the

Federal Government would not allow that figure to rise above 80% in the future.) It was also the feeling of the bank's management that the banking segment of the industry would continue to grow more rapidly than the savings and loan segment and thus banks would gradually increase their share of the total market.

One of the questions raised by the long-range planning committee's memo was whether these six variables should be forecast independently of one another or if some approach that forecasted one variable and then prepared a forecast for another by simply taking the complement might be more accurate. For example, total bank and S&L deposits might be forecast, the bank's share of those might be forecast and then the actual dollar figures for both the banks and the S&L's could be projected using those two forecast items. The long-range planning committee had made it quite clear in their directives to the data processing group that it was up to data processing to determine which forecasting techniques would be most appropriate for this situation.

FIGURE 1   Arizona Security Bank

DEPOSITS AND LOANS OF ALL ARIZONA BANKS

Dec. 31	Deposits	Loans	Ratio
1963 (Dec. 20)	$1,732,374,000	$1,201,556,000	69.4%
1964	1,970,219,000	1,358,105,000	68.9
1965	2,096,020,000	1,452,733,000	69.3
1966	2,238,692,000	1,566,465,000	70.0
1967	2,484,924,000	1,708,776,000	68.5
1968	2,856,225,000	1,971,650,000	69.0
1969	3,110,704,000	2,324,095,000	74.7
1970	3,552,079,000	2,585,854,000	72.8
1971	4,322,060,000	3,163,440,000	73.3
1972	5,248,274,000	3,997,988,000	76.2
1973 (June 30)	5,523,727,000	4,193,845,000	75.9

Source: Arizona Statistical Review

SAVINGS DEPOSITS IN ARIZONA

Dec. 31	Banks	Savings & Loan Association	Total
1963 (Dec. 20) . . .	$ 728,191,000	$ 488,202,000	$1,216,393,000
1964	872,848,000	582,876,000	1,455,724,000
1965	992,231,000	619,418,000	1,611,649,000
1966	1,131,334,000	597,318,000	1,728,652,000
1967	1,293,367,000	684,833,000	1,978,200,000
1968	1,461,418,000	741,016,000	2,202,434,000
1969	1,521,213,000	791,084,000	2,312,297,000
1970	1,790,884,000	919,172,000	2,710,056,000
1971	2,260,126,000	1,231,714,000	3,491,840,000
1972	2,807,359,000	1,667,807,000	4,475,166,000
1973 (June 30) . . .	2,996,978,000	1,862,995,000	4,859,973,000

Source: Arizona Statistical Review

PHOENIX CITY COUNCIL

In spite of the fact that it was a sunny, warm day in early April 1974, Paul Marshall was visibly concerned about the task he had recently been given by the Phoenix City Council. As a staff assistant to that group for the past three years he had been involved in generating a number of ideas and programs that would help the Council make better decisions. He also supplied them with any analytical and data support that they requested. In one of their recent meetings, an outside consultant had suggested that it might be very helpful to them in planning their tax levies and estimating their future tax base to have projections concerning the population of the city of Phoenix. The City Council had felt that was a great idea and had immediately assigned it to Paul and asked him to prepare annual forecasts for three, five and ten years. Their thought on the various number of time periods had been that some people in the city government might only be concerned with one time horizon while others might deal with very different time horizons depending on the responsibilities involved and the decisions in question.

Paul was anxious to meet the needs of the City Council but was not sure exactly where he should begin. He knew that there were several state agencies that seemed to pull population projections out of thin air whenever they needed them to support their position. However, he was not aware of any group that had systematically been projecting Phoenix city population with any consistency or any level of accuracy. Thus he felt he really needed to start from scratch in pursuing his assignment.

As a first step he gathered together the historical data published by the city's planning department (See Figure 1). Upon reviewing this data, Paul decided that there were at least three questions that he needed to answer before preparing his forecast. The first revolved around the specific technique most suitable for this situation. Not having had much experience with forecasting, he really was not sure which approach should be tested and how to go about doing that.

The second question concerned whether or not he should supply identical forecasts for the first three years of the three- five- and ten-year projections. Since these were going to different people who would use them for different things, he thought there might be some argument for having them somewhat different. Of particular concern was the fact that if one of them was to be used in making projections concerning the tax base in the city, that it would be much better to underestimate future population figures rather than to overestimate them and thus create financial problems for the city in the future.

The third question concerned the use of the population data available for the county of Maricopa. While he had historical data for the county, he was not sure whether or not that would be useful to him in projecting future population in the city of Phoenix. The City Council had not expressed any particular interest in having population projections for the county as a whole since there was an entirely different government organization concerned with the county's problems.

FIGURE 1   Phoenix City Council Population Growth of Phoenix and Maricopa County

As of July 1	Phoenix City Limits	Maricopa* County
1950 (April Census)	106,818	331,770
1951	109,000	364,000
1952	119,000	390,000
1953	130,000	417,000
1954	140,000	446,000
1955	155,000	477,000
1956	170,000	510,000
1957	179,000	545,000
1958	242,000	583,000
1959	364,000	625,000
1960 (April Census)	439,170	663,510
1961	452,000	740,000
1962	468,000	775,000
1963	483,000	808,000
1964	494,000	833,000
1965	504,000	852,000
1966	511,000	870,000
1967	519,000	890,000
1968	528,000	914,000
1969	546,000	946,000
1970 (April Census)	582,500	969,425
1971	621,000	1,017,000
1972	674,000	1,060,000
1973	724,000	1,105,000

Source: Inter-census year figures for Maricopa County are by the Arizona Department of Economic Security in cooperation with the U.S. Bureau of the Census; all Phoenix inter-census year figures are by the Phoenix City Planning Department.

*Totals for the county including Phoenix

FIGURE 1 (continued)

POPULATION

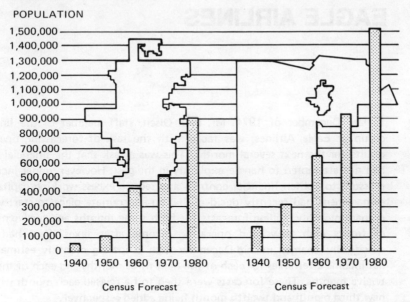

Census Forecast   Census Forecast

Phoenix Square Mile Area — 269   Maricopa County Square Mile Area — 9,155

EAGLE AIRLINES

In late December of 1974, Mr. Paul Olsen, staff member of the financial group at Eagle Airlines, was faced with the task of forecasting operating results for the next several months. This was a task that the financial group had not attempted to handle explicitly in the past. However, under increased pressure to tighten financial controls and reduce excess working capital, the management had recently decided that an appropriate place to start was to prepare monthly, rolling forecasts for both three months and twelve months of future cash flows and operating income after taxes. Paul had been specifically charged in mid-December with preparing monthly estimates of net income after taxes for each of the next three months and each of the next twelve months. These forecasts were then to be revised each month with the new third month and twelfth month being added respectively.

Upon receiving the assignment, Paul had initially considered the possibility of relating the variable, net income after taxes, to other financial and operating statistics available in the company. However, a couple of days of investigation of this possibility led him to conclude that it would be very difficult at the present time to obtain data on related series soon enough to be useful in this assignment. He also concluded that any relationships that might be identified as existing among net income after taxes and these other variables might be very unstable over time. Therefore Paul had decided that it would be most appropriate to consider some time series forecasting method for use in this assignment.

As background for preparing his forecast for December, January and February and for the coming twelve months as well, Paul obtained the historical data shown in Figure 1. He then set about to determine what forecasting technique would be most suitable for his task and how he could best apply it. In the back of his mind was the thought that once he had prepared these forecasts, he would then want to check to see whether or not the results were sufficiently reliable (for the historical data) that they could form a basis from which to eventually predict cash flows.

Paul thought it might be the case that he should just stick with trying to predict the seasonal factor related to each month of the year and then use some more subjective approach to estimating net income for the next several months. However, he realized that the preference of those to whom he would supply these forecasts would be for a complete dollar forecast for each of the required months rather than simply a seasonal factor. He also knew that one of the first questions they would ask him would deal with the reliability of his forecasts and how much faith they should put in the numbers he had actually developed.

FIGURE 1   Eagle Airlines — Net Income (Loss) After Taxes ($000)

	1974	1973	1972	1971	1970
January	$ 4,296	$ (253)	$ 2,781	$ (1,937)	$ 1,201
February	1,517	(1,969)	2,514	(2,720)	1,035
March	6,917	(522)	6,656	5,681	2,688
April	7,162	3,835	3,205	1,791	2,246
May	(178)	(844)	991	491	(843)
June	1,463	212	791	(916)	1,051
July	3,437	2,655	1,080	(470)	274
August	4,105	3,684	1,771	2,387	2,154
September	(5,547)	(7,445)	(7,586)	(6,579)	(7,234)
October	(1,853)	(4,478)	(8,559)	(2,754)	(6,981)
November	687	2,697	(2,903)	(1,106)	(7,424)
December		8,118	4,720	3,809	(106)
Total		$ 5,690	$ 5,461	$ (2,323)	$ (11,030)

Source: Company records.

FERRIC PROCESSING

Richard Hansen, the Assistant to the General Manager of Ferric Processing, was preparing his presentation for the upcoming Executive Committee meeting. He wanted to present the results of his recent cost study, and propose a new way of predicting processing costs for prospective plant sites. He also wanted to suggest several possible extensions of his study which he felt would help the company better understand their long-run costs, and the way these costs were influenced by several key factors.

COMPANY BACKGROUND

Ferric Processing was a relatively new subsidiary of the Metzger Machinery Company. It was organized in 1956 to capitalize upon a newly-patented process for recovering iron from the waste slag of steel mills. In the last eight years, Ferric had built ten new slag processing plants, and compiled an enviable record of both sales and earnings growth.

As a largely unwanted by-product of steel-making in an open hearth furnace, great quantities of slag are produced. Particularly in the vicinity of the open hearth pouring pits, this slag contains small amounts of iron and iron oxides which would be potentially valuable if they could be economically recovered from the slag. Several steel companies had experimented with recovery processes which re-cycled the recovered iron to the blast furnaces, but these processes seemed to be of doubtful utility.

In the middle 1950's, an executive of the Metzger Machinery Company who was familiar with the steel industry developed and patented a new process for this recovery which promised great potential. Basically, the slag was carried along conveyor belts through a continuous cycle of alternating crushers, magnets, and screens. The crushers broke the slag into smaller and smaller pieces, the magnets recovered the iron, and the screens sorted the remaining slag with a very low iron content into various particle sizes which could be used for other industrial purposes.

Once the patents for the process seemed secure, the Metzger Machinery Company formed a new subsidiary named Ferric Processing, Inc., and negotiated contracts for two of these slag processing plants. Although there were many unexpected problems in the initial break-in period of about two years, these plants were eventually very successful. They were followed in succeeding years by eight other plants, all of a similar design, but modified slightly because of Ferric's experience with earlier plants.

In every instance, Ferric's plant processed the slag from a particular steel mill, and each of the plants was located as close as possible (usually within five or ten miles) to the steel mill which it served. Ferric's personnel would dig the slag from the deposits in the open hearth area, and load it into large heavy duty trucks. These trucks would deliver it to the intake conveyor at Ferric's plant, where it would be fed into the recovery process. After processing, the recovered iron would be returned to the steel mill to be charged into the blast furnace, and the remaining slag would be returned to the slag operations at the steel mill. Ferric Processing was paid a flat processing charge per ton of slag it handled, and this fee was always negotiated at the time the original contract with the steel company was signed. Generally, a rather lengthy series of negotiations preceded the decision to build any new plant. As the final outcome of these negotiations, the steel company and Ferric would agree upon a contract covering at least three to five years, which stipulated both the processing charge per ton and several minimum guarantees to Ferric in terms of total tonnage processed. After the contract, Ferric would begin constructing the new plant, whose capacity was matched to the size of the particular steel mill.

Over the years, Ferric had attempted to make sure that the negotiated rate per ton was high enough to cover all their expenses, including labor, transportation, depreciation of the capital equipment, etc. They found, however, that on several of their plants they had seriously underestimated the total processing costs per ton, and in several instances they had renegotiated a second contract at higher rates after the first contract expired. On the other hand, they also had overestimated these costs in several plant locations which were now producing efficiently at cost/ton rates significantly below the negotiated processing charge.

THE COST STUDY

Currently, Ferric was considering three additional plant sites, and in one case negotiations were proceeding rapidly to a final agreement. Because the difference between the negotiated charge per ton and the average processing

costs per ton was the critical variable for Ferric's profitability, it was important to estimate future processing costs early in the stages of negotiation for
each prospective plant site.

Ferric's management had, on several earlier occasions, asked its industrial
engineer to provide detailed cost estimates for various parts of their overall
process. Based upon the auditing of an individual plant's cost records and
time-and-motion studies of various functions within the plant, these studies
had provided what was considered to be a very detailed and accurate breakdown of all costs within a plant, as well as some suggestions for cost
reduction. On the whole, however, they did not provide an adequate comparison of costs across different plants; and it was very difficult to extrapolate
cost estimates for a new plant from them.

In a recent discussion with the General Manager, his Assistant, Richard
Hansen, had volunteered to make a new kind of cost study. Mr. Hansen
believed that some of the key factors influencing Ferric's costs were the
economies of scale inherent in the different plant sizes, and he believed that
the effect of these economies of scale were acknowledged but not well
understood by top management. Indeed, it seemed that there could be a
substantial disagreement about the way plant size affected total costs per ton,
particularly among several of Ferric's management group responsible for
making cost projections. This disagreement could be inferred from the different cost estimates these executives had submitted for the smallest of the three
prospective plant sites, although none of them had ever directly addressed the
problem of economies of scale as such. With the concurrence of the General
Manager, Mr. Hansen set aside his other duties for several days and concentrated upon this new study of average costs per ton across different plants.

As the basic data to study this problem of costs, Mr. Hansen compiled a
list of all of Ferric's plants, their processing capacity, and an estimate of their
individual average costs/ton for the last year. These data are shown below in
Table 1. The cost estimates were furnished by the Controller's office. The
total yearly costs attributable to each plant, including all direct and indirect
labor, materials, transportation, depreciation of capital equipment, super-
visors' salaries, etc., were already compiled as an important part of Ferric's
management control system. These costs for last year, divided by the number
of tons actually processed last year by the plant, provided the average
costs/ton figure which the Controller's office furnished.

As a first look at his problem, Mr. Hansen constructed the graph shown
in Exhibit 1. On the basis of this plot of the data, he was convinced that
economies of scale were indeed important, and he decided to use regression
analysis to study them. Employing the services of a statistical analyst and a
computer, he first obtained a simple linear regression analysis of the average
costs/ton versus plant size (capacity). The input data and results of this
regression are shown in Exhibit 2. Mr. Hansen and the analyst then jointly
decided to run another regression analysis, using (1/capacity) as the revised
independent variable. The input data and results of this regression are shown
in Exhibit 3.

TABLE 1

Plant #	First Began Operations	Capacity	Average Costs/Ton in 1964
1	1956	900 tons/mo.	$21.95
2	1956	500 tons/mo.	27.18
3	1958	1750 tons/mo.	16.90
4	1959	2000 tons/mo.	15.37
5	1960	1400 tons/mo.	16.03
6	1960	1500 tons/mo.	18.15
7	1961	3000 tons/mo.	14.22
8	1962	1100 tons/mo.	18.72
9	1962	2600 tons/mo.	15.40
10	1963	1900 tons/mo.	14.69

Intrigued with the progress of his cost study, Mr. Hansen discussed the results in general terms with several managers in the company. They were all quite interested, but many of them objected to drawing any conclusions from studies across different plants. There were so many other important differences from plant to plant, they argued, that it was very risky to draw conclusions using only plant capacity as a discriminating factor. In particular, several of these managers pointed out that the costs of hauling the slag between the steel mill and processing plant was a major factor in Ferric's overall costs. Because some of the plants were built adjacent to the steel mills and others were up to nine miles away, the differences in transportation mileage could cause major differences in total costs which would be reflected in the average costs/ton figures. Several plant managers argued that these differences could be seriously distorting the results of Hansen's cost study.

In addition, the manager of one of the older plants argued that the age of the plants could be one of the principal determinants of costs. More recent plants incorporated several modifications made in light of previous plants' experiences, and thus they tended to be somewhat more efficient. Because the early plants tended to be smaller and later plants tended to be larger, the increased efficiency of the later plants could be distorting the results of the cost study. Furthermore, the older plants might well have higher maintenance costs, although the newer plants might have initially higher costs before management could get the "bugs" shaken out of the processing unit. The manager of one of the older plants was concerned that Hansen's study was overlooking these possible effects.

Richard Hansen sat down to prepare a brief presentation of his results for the upcoming Executive Committee meeting. He wanted to briefly summarize and interpret the results, and discuss possible problems with the interpretation. He also wanted to suggest several extensions of the study which he felt would help management better understand their average costs, and the way these costs were related to various key factors.

EXHIBIT 1 Plot of "Costs/Ton" Versus "Capacity" for 10 Plants

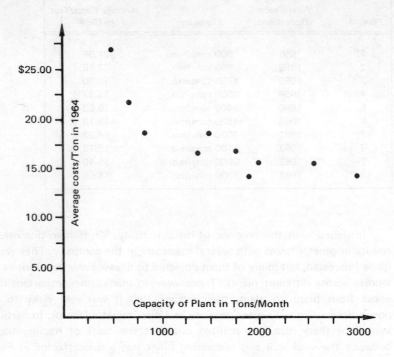

Capacity of Plant in Tons/Month

EXHIBIT 2 Ferric Processing Data

PART 1 The Input Data

y (Average Costs/Ton)	x (Capacity) in 000 tons/month
$27.18	.50
21.95	.90
18.72	1.10
16.03	1.40
18.15	1.50
16.90	1.75
14.69	1.90
15.37	2.00
15.40	2.60
14.22	3.00

EXHIBIT 2 (continued)

PART 2   The Results of the Linear Regression Analysis

VARIABLE	B	STD. ERROR	T-TEST	PCT VARIATION EXPLAINED
CONSTANT	85.193	1.85876	13.5536	
2	-4.40357	1.02457	-4.29796	.697798

```
R-SQUARED=0.698   R=0.835
F-TEST=  18.47   STD. OF REGR.=   2.33
DEGREES OF FREEDOM FOR NUMBER.= 1    FOR DENUMBER.=8

DO YOU WANT THE RESIDUALS TO BE PRINTED?Y
ACTUAL          PREDICTED        RESIDUALS        % ERROR
37.18           22.9912          4.18884          15.4115
21.95           21.2287          .720266          8.28139
19.72           20.843           -1.62902         -8.70302
16.03           19.0279          -2.99795         -18.7021
18.15           18.5876          -.437588         -2.41086
16.9            17.4867          -.586695         -3.47157
14.69           16.8862          -2.13616         -14.5416
15.77           16.3859          -1.0158          -6.60999
15.4            18.7437          1.65634          10.7555
14.22           11.9822          2.23777          15.7858

DURBIN-WATSON STAT. = .851989
```

EXHIBIT 3   Ferric Processing Data

y (Average Costs/Ton)	x (1/Capacity in 000's tons/month)
$27.18	2.00
21.95	1.11
18.72	.909
16.03	.714
18.15	.666
16.90	.571
14.69	.526
15.37	.500
15.40	.385
14.22	.333

EXHIBIT 3 (continued)

PART 2 The Results of the Regression Analysis

VARIABLE	B	STD. ERROR	T-TEST	PCT. VARIATION EXPLAINED
CONSTANT	11.7478	.602338	19.5036	
2	7.92466	.562293	11.8534	.94517

```
R-SQUARED=0.845    8=0.873
F-TEST= 140.62   STD. OF REGR.=   0.99
DEGREE OF FREEDOM FOR NUMBER.= 1     FOR DENUMBER.= 8
```

```
DO YOU WANT THE RESIDUALS TO BE PRINTED?Y
ACTUAL         PREDICTED      RESIDUALS      % ERROR
27.18          27.5875        -.417486       -1.53601
21.95          20.=444         1.40564        6.40383
13.72          18.9515        -.231462       -1.23644
16.03          17.4061        -1.37611       -3.58462
18.15          17.0257         1.12438        6.19438
16.9           16.2729         .627144        3.71091
14.69          15.9162        -1.22624       -8.34744
15.37          15.7102        -.340192       -2.21335
15.4           14.7988         .601166        3.90368
14.22          14.3867        -.166741       -1.17253
```

```
DURBIN-WATSON STAT.=2.5128
```

LENEX CORPORATION

Richard Daron, having been the controller at the Lenex Corporation for less than a month, had started to realize the magnitude and seriousness of the problems Lenex faced. During the past six months Lenex had undergone the worst liquidity crisis of its history. A number of events had combined to bring Lenex to the brink of disaster. It was a rather peculiar situation since Lenex was technically far from bankruptcy. Profits had been satisfactory (except for the past year), the company had a ratio of total debt to total assets of .4 and a rather good growth rate which had slowed only during the last recession. Daron recalled that the company's "liquidity squeeze" (not enough cash to meet short term obligations) had created a substantial shock wave through management when it was discovered. The stock market had then administered a final blow when the price of a share of common stock fell 8 points (18%) in a single week. Daron had been appointed to the controller's job after his predecessor had resigned for "age" reasons and after a group of three banks had guaranteed a loan of up to $10,000,000 to meet short and medium term obligations of the company.

During the few weeks that had passed since the bank loan had been arranged, the problems had eased a little, and the stock price of the company had recovered somewhat, but things were still far from good. The same basic conditions that had caused the liquidity crisis continued to exist. The recession had caused the company to lose certain revenues and had led to increased competition in the form of price reductions and more aggressive sales promotions.

COMPANY BACKGROUND

As a manufacturer of electrical appliances, Lenex was considered medium-sized, employing 1500 people. This had led the company to be a price follower, often aiming at specialized markets, such as customized products that could be sold through specialty gift stores and customized orders designed to meet the demands of large industrial customers. With the recession, the industry price leader had cut prices to maintain the level of revenue it required to cover its high fixed costs. While Lenex had a cost structure with somewhat lower fixed costs and higher variable costs, the need to match competitors by reducing prices 8% had been painful.

On top of the price cuts, the level of demand had fallen as a result of the recession. This had a double effect: first it resulted in unused man-power and equipment, and second it led to increased inventory levels and increased costs for carrying those inventories. At the same time competition from abroad, mainly Japan, had increased, driving prices and revenues even lower. Unfortunately costs were still running at their pre-recession levels since nobody at Lenex had predicted the downturn. It had taken Lenex four months to adjust production and inventory schedules to take account of the reduced level of business.

Lenex had managed to meet all of its obligations for about one year following the downturn, but eventually the pressure had become too great and creditors and financial analysts had become aware that something was wrong. The Board of Directors had been obliged to take drastic measures to remedy the situation. One of the corrective actions had been Daron's appointment as controller. His assignment had been not only to find solutions for the short term crisis, but also to develop a system which would help to avoid a similar crisis in the future.

As a first step Daron had started a series of studies aimed at identifying the company's major problems. Two of his assistants were in charge of these studies, and an outside consultant had worked with them for several months. The results of the studies were not yet ready, but Daron had been talking regularly with those conducting them and with others in the organization. As a result of these conversations and the data that had been gathered, he thought he could now identify the roots of the problems.

LENEX OPERATIONS

Lenex produced about 200 basic products. These products came in about 2,000 variations in terms of quality, major technical characteristics and, of course, price. When minor variations in shape and differences in color were considered, Lenex actually provided a minimum of 5,000 items which required raw material, semi-finished, and finished goods inventories of about 100,000 different items.

Many other factors, such as production scheduling, distribution to either wholesaler or retailer, special orders, assignment of personnel, etc., also had to be considered in the production and sales of the 5,000 items. This was done through different forms of planning.

Lenex manufactured products for both the consumer market and large industrial customers. The first group consisted of a large number of individual customers whose buying behavior was difficult to understand but followed certain probabilistic laws due to the large numbers. The second group was less predictable since its demands varied widely, but planning for this group was easier since orders were large and the lead times were typically much longer than those required by the consumer market.

Lenex produced a wide variety of small and medium sized electrical appliances for household use. These included radios, air conditioners, plugs, switches, electic heaters, can-openers, tooth-brushes and mixers. For the industrial market, it produced dynamos and other electrical apparatus ordered in sufficient quantity to compensate for the set-up costs.

Daron had several questions in his mind as he contemplated analyzing his available data. Would it be possible to design a forecasting system capable of averting future liquidity crises in addition to providing forecasting inputs to meet planning requirements? What factors should be considered in this design? What techniques of forecasting should be considered?

TABLE 1   Shipments of Air Conditioners

	1967	*1968*	*1969*	*1970*	*1971*
January	210	215	211	187	201
February	223	225	210	196	205
March	204	230	214	195	235
April	244	214	208	246	243
May	274	276	276	266	250
June	246	261	269	228	234
July	237	250	265	257	256
August	267	248	253	233	231
September	212	229	244	227	229
October	211	221	202	188	185
November	188	209	221	195	187
December	188	214	210	191	189
	2704	2792	2783	2609	2645

Year 1967

— Average price per air conditioner 100 dollars
— 29.08% of total dollar sales

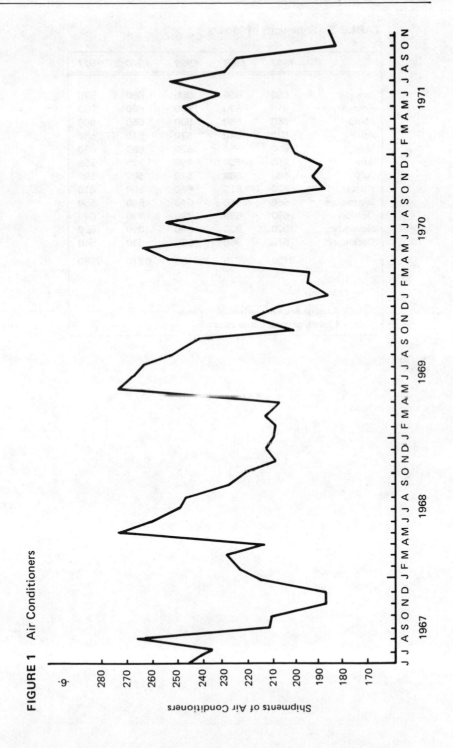

FIGURE 1   Air Conditioners

TABLE 2   Shipments of Mixers

	1967	*1968*	*1969*	*1970*	*1971*
January	650	600	600	600	570
February	610	620	590	650	580
March	650	650	590	650	600
April	620	540	570	640	580
May	600	580	620	660	630
June	950	830	990	920	850
July	580	560	570	580	590
August	620	610	660	560	610
September	560	550	640	640	550
October	690	690	790	730	660
November	1020	980	980	1090	890
December	570	960	580	550	540
	8120	7770	8180	8270	7650

Year 1967

— Average price per mixer 20 dollars
— 17.46% of total dollar sales

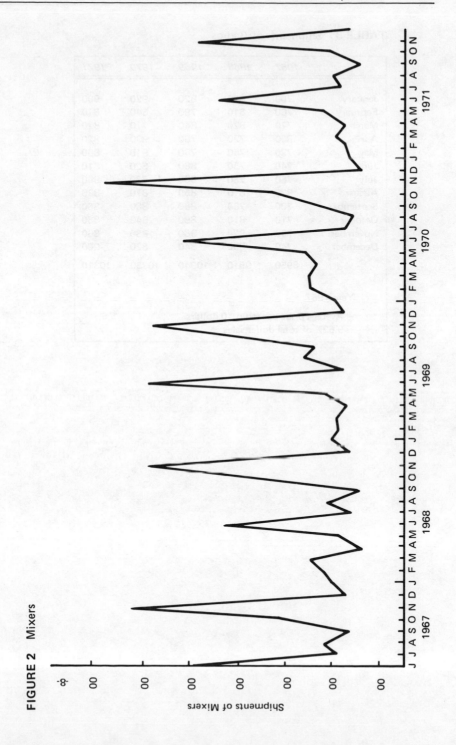

FIGURE 2 Mixers

TABLE 3   Shipments of Radios

	1967	1968	1969	1970	1971
January	700	810	820	920	900
February	760	810	790	840	810
March	770	820	840	770	820
April	800	720	760	930	820
May	720	750	760	910	860
June	740	750	880	820	810
July	760	780	880	770	960
August	820	810	850	870	830
September	700	780	860	880	860
October	710	810	890	940	880
November	730	820	880	880	860
December	740	850	840	820	900
	8950	9510	10940	10350	10310

Year 1967

— Average price per radio 10 dollars
— 9.62% of total dollar sales

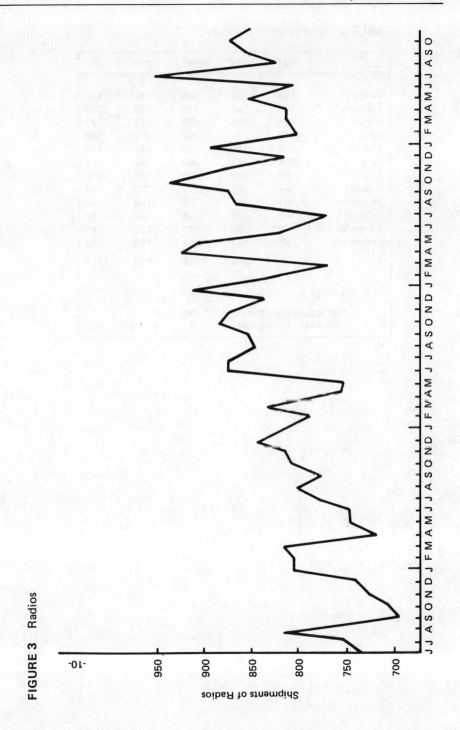

FIGURE 3 Radios

TABLE 4 Shipments of Blenders

	1967	1968	1969	1970	1971
January	153	164	155	140	143
February	159	153	135	150	136
March	138	165	189	161	171
April	142	147	163	161	160
May	145	150	140	141	141
June	154	161	160	140	161
July	160	158	155	147	159
August	144	161	137	152	140
September	152	147	136	183	155
October	155	137	158	144	159
November	146	139	145	159	145
December	158	137	163	153	140
	1806	1819	1836	1670	1810

Year 1967

— Average price per blender 15 dollars
— 2.9% of total dollar sales

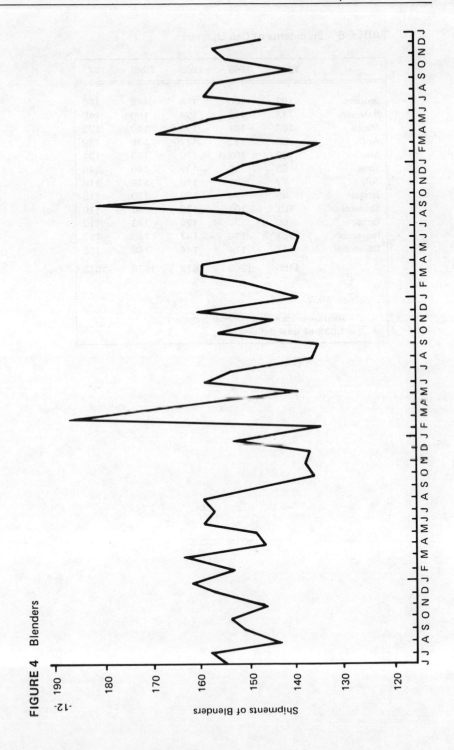

FIGURE 4   Blenders

-12-

TABLE 5   Shipments of Can Openers

	1967	*1968*	*1969*	*1970*	*1971*
January	103	100	114	149	146
February	113	116	124	150	148
March	105	101	115	130	122
April	99	112	111	135	139
May	102	102	120	138	131
June	93	96	132	149	140
July	77	91	116	119	110
August	89	94	126	120	118
September	101	104	139	148	110
October	95	120	125	143	112
November	107	110	147	146	122
December	111	114	144	149	115
	1195	1260	1513	1676	1513

Year 1967

— Average price per can opener 8 dollars

— 1.03% of total dollar sales

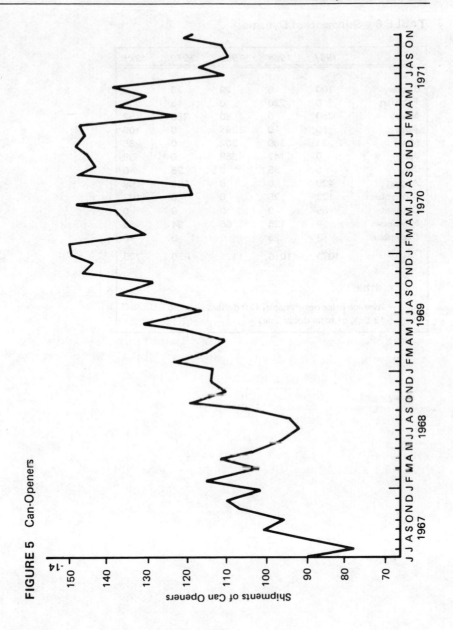

FIGURE 5   Can-Openers

TABLE 6   Shipments of Dynamos

	1967	1968	1969	1970	1971
January	100	0	29	35	0
February	0	220	0	45	0
March	250	0	80	285	152
April	50	0	387	0	102
May	33	389	302	0	35
June	0	147	258	0	98
July	0	85	0	28	0
August	420	0	0	163	169
September	123	25	0	0	107
October	85	0	0	0	0
November	9	125	65	94	35
December	3	25	0	0	10
	1073	1016	1121	650	708

Year 1967

— Average price per dynamo 120 dollars
— 13.85% of total dollar sales

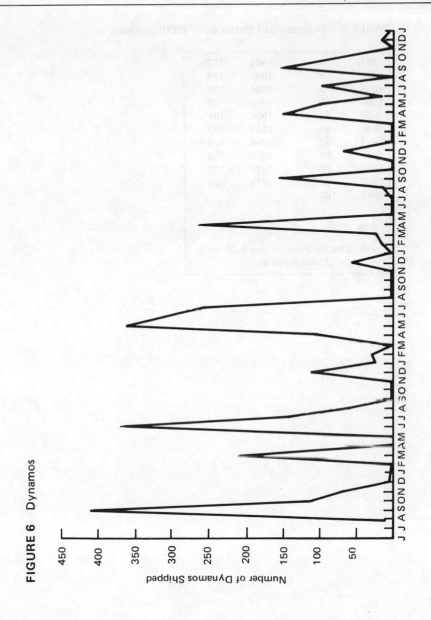

FIGURE 6   Dynamos

TABLE 7   Shipments of Batteries in (000) of Units

1951	3	1962	112
1952	8	1963	115
1953	13	1964	120
1954	15	1965	128
1955	30	1966	135
1956	35	1967	142
1957	51	1968	160
1958	62	1969	166
1959	70	1970	172
1960	82	1971	175
1961	100		

Year 1970

— Average price per battery 0.10 dollar
— 1.45% of total dollar sales

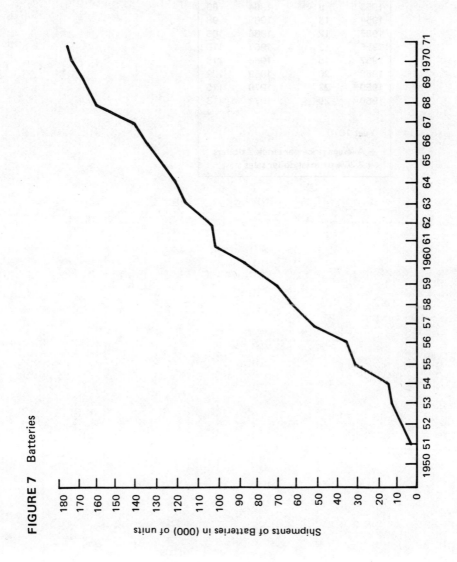

FIGURE 7 Batteries

TABLE 8   Shipments of Electric Clocks in (000) of Units

1950	11	1961	42
1951	10	1962	50
1952	9	1963	63
1953	8	1964	69
1954	13	1965	98
1955	12	1966	105
1956	13	1967	110
1957	15	1968	118
1958	20	1969	129
1959	23	1970	115
1960	29	1971	118

Year 1970

— Average price per clock 2 dollars
— 2.26% of total dollar sales

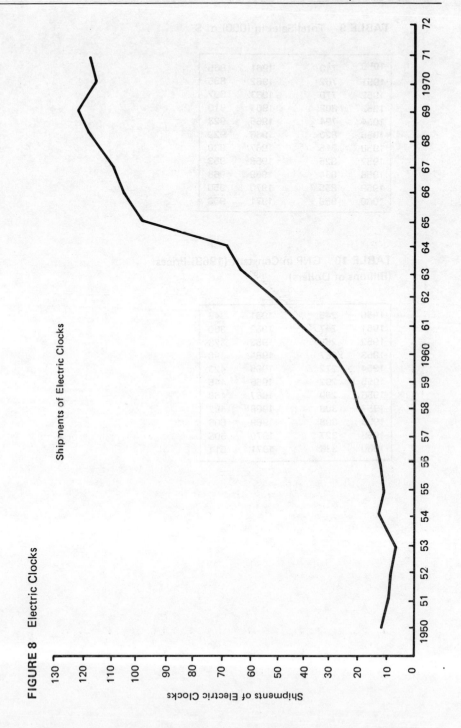

FIGURE 8   Electric Clocks

TABLE 9   Total Sales in (000) of $

1950	748	1961	865
1951	762	1962	885
1952	779	1963	897
1953	802	1964	910
1954	794	1965	923
1955	820	1966	929
1956	815	1967	930
1957	825	1968	953
1958	834	1969	968
1959	858	1970	950
1960	859	1971	953

TABLE 10   GNP in Constant (1963) Prices
(Billions of Dollars)

1950	245	1961	343
1951	247	1962	360
1952	255	1963	376
1953	267	1964	398
1954	272	1965	423
1955	292	1966	445
1956	299	1967	458
1957	306	1968	482
1958	308	1969	506
1959	327	1970	508
1960	336	1971	511

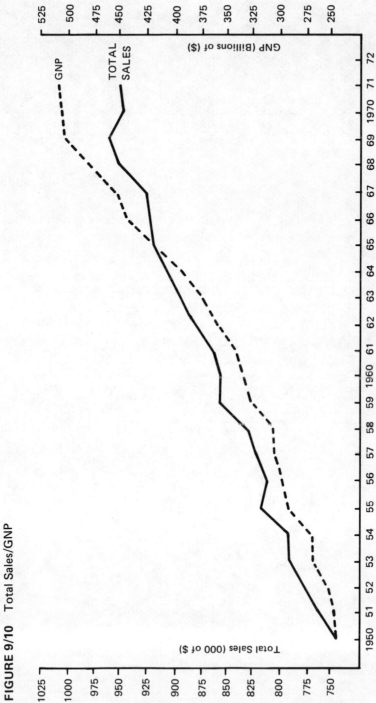

FIGURE 9/10   Total Sales/GNP

CHAPTER 13

S-Curve and Exponential Growth Models

1. S-CURVE FITTING

S-curve fitting is appropriate for long-term forecasting. It is used to estimate the life cycle of technologies and products. An S-curve has a slow start, a rather steep growth, and a saturation that comes after some period in time. A typical S-curve is shown in Figure 13-1

There are several alternative mathematical expressions for an S-shaped curve. Perhaps the simplest and most widely applied is:

$$X_t = e^{a + \frac{b}{t}}$$

where X_t is the S-curve estimate

      e   is a constant equal to 2.71828
      a   is the intercept
      b   is the equivalent to the slope in the linear case (the sign of b will always be negative)
      t   is time

FIGURE 13-1   Typical S-Curve

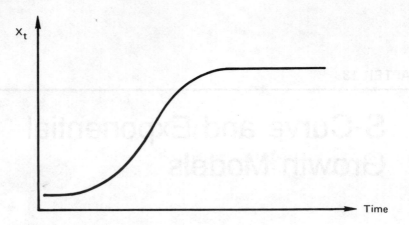

Because an S-curve is a nonlinear function, the standard estimation procedure of linear least squares that is used in trend analysis and the other regression methods cannot be applied directly on (13-1). However, Equation (13-1) can be transformed into a linear equation by taking logarithms of the time series involved (see (13-2)) and replacing 1/t by t as can be seen in (13-4). The resulting transformed equation, (13-5), is linear, and the classical least squares method can then be applied to estimate the values of a and b. When the values of a and b are known, they can be substituted into equation (13-6) to obtain forecasts for as many periods as desired.

$$\log_e X_t = \left(a + \frac{b}{t} \right)(\log_e e) \qquad\qquad (13\text{-}2)$$

or

$$\log_e X_t = a + \frac{b}{t} \qquad\qquad (13\text{-}3)$$

since

$$\log_e e = 1$$

Defining

$$X_t' = \log_e X_t$$

and replacing

$$t' = \frac{1}{t} \tag{13-4}$$

equation (13-2) becomes:

$$X'_t = a + bt' \tag{13-5}$$

In preparing a forecast, equation (13-6) is used:

$$F_{(t+m)} = e^{a + \frac{b}{(t+m)}} \tag{13-6}$$

where m is the number of periods ahead to be forecast.

---

2. EXPONENTIAL GROWTH MODELS

Exponential growth models are appropriate for long-term forecasting when growth rates are constant over time. This is characteristic of the sales of several products and companies and has been the growth pattern of most industrialized economies during recent decades. An exponential growth curve has a constant rate of growth at any point in time, resulting in staggering increases over longer periods of time as the increase is compounded exponentially. With an annual growth rate of 5%, the series will double in fourteen years and will become more than five times its initial value in thirty-four years time. This concept of exponential growth has been the focus of much discussion among scientists concerned with future requirements for food, raw materials, and other resources. There are those who argue that there are limits to growth because exponential increases are bound by finite space and resources. On the other hand, there are others who argue that stabilizing social and market mechanisms will slow down or stop exponential growth. Whatever one's position, exponential growth may take place for extended periods of time, and some model is needed to account for it. This model can be approximated by the exponential curve of the following form:

$$y_t = e^{a+bt} \tag{13-7}$$

where

y_t is the time series at time t

$e = 2.71828$

a is the intercept

b is the equivalent of the slope and denotes the growth rate.

As with the S-curve, equation (13-7) can be transformed to make it linear, by taking natural logarithms of both sides of the equation. This gives

$$\log_e y_t = (a + bt) \log_e (e)$$

or

$$\log_e y_t = a + bt \tag{13-8}$$

since $\log_e (e) = 1$

Defining

$$y'_t = \log_e y_t \tag{13-9}$$

equation (13-8) becomes

$$y'_t = a + bt \tag{13-10}$$

Applying classical least squares to (13-10) allows one to estimate the values of a and b and then use them in (13-11) to forecast future values.

$$F_{(t+m)} = e^{a+bt} \tag{13-11}$$

where m is the number of periods ahead to be forecast.

References for further study:

Makridakis, S. and S. Wheelwright (1978, Chapter 14)

INPUT/OUTPUT TO THE S—CURVE COMPUTER PROGRAM (SCURVE)

Input

None.

Output

The same as with trend analysis (see Chapter 12).

INPUT/OUTPUT TO THE EXPONENTIAL GROWTH PROGRAM (EXGROW)

Input

None.

Output

The same as with trend analysis.

EXERCISE 13.1

The following fourteen data points show the yearly overseas shipments (in thousands) of a deodorant product. Estimate the number of shipments for the next five years using an S-curve extrapolation and trend extrapolation. Compare the predictions obtained using the two methods.

```
***** LISTS THE DATA OF A FILE *****
PERIOD    OBSERVATION  PERIOD   OBSERVATION  PERIOD    OBSERVATION
   1          750.00      5        2210.00      9         2510.00
   2         1000.00      6        2380.00     10         2570.00
   3         1640.00      7        2400.00     11         2615.00
   4         2060.00      8        2500.00     12         2640.00
                                               13         2910.00
                                               14         3050.00
```

Solution (See Figures 13-3 and 13-4)

To be able to forecast using an S-curve, one must transform the data into a linear form and then calculate the values of a and b. These values can then be used to make forecasts. The steps are shown below.

TABLE 13.1

Original Time Series	Transformed Time Series	Time	Transformed Time
X_t	$X'_t = \log_e X_t$	t	$t' = \frac{1}{t}$
750	6.62007	1	1
1000	6.90776	2	0.5
1640	7.40245	3	0.3333
.	.	.	.
.	.	.	.
.	.	.	.
2910	7.97591	13	0.07692
3050	8.02290	14	0.07143
X'_t	t'	$X't'$	t'^2
6.62007	1	6.62007 × 1	1
6.90776	0.5	6.90776 × 0.5	0.25
7.40245	0.3333	.	.
.	.	.	.
.	.	.	.
.	.	.	.
7.97591	0.07692	7.97591 × 0.07692	0.00510
8.02290			
$\Sigma X't = 107.07$	$\Sigma t' = 3.252$	$\Sigma X't' = 23.5884$	$\Sigma t'^2 = 1.5$

The values for a and b are calculated as in trend analysis (see Chapter 12) using formulas (12-2) and (12-3):

$$b = \frac{14(23.5884) - (3.252)(107.07)}{14(1.576) - (3.252)^2} = -1.558$$

$$a = \frac{107.07}{14} - (-1.558)\frac{3.252}{14} = 8.0097$$

Using equation (13-4) where t is the current period (14) the forecasts for the following periods are:

$$F_{15} = 2.71828^{8.0097} - \frac{1.558}{15} = 2.71828^{7.905} = 2713.16$$

$$F_{16} = 2.71828^{8.0097} - \frac{1.558}{16} = 2730.83$$

$$F_{19} = 2.71828^{8.0097} - \frac{1.558}{19} = 2773.14$$

COMPARISON OF THE USE OF TREND ANALYSIS AND S—CURVE

If the same data are used in the trend analysis and S-curve programs the forecasts for the next five periods are:

TABLE 13.2

Trend Analysis	S-curve	Difference
3320.38	2713.16	607.22
3465.63	2730.83	734.80
3610.87	2746.52	864.35
3756.11	2760.54	995.57
3901.35	2773.14	1128.21

From the forecasts above and the data graph (Figure 13-2) it can be seen that forecasting with different methods gives very different results. The question of which method of forecasting is better can only be answered by judging how well each method fits. In this example, the R^2 of the S-curve is 0.924 while that of trend analysis is 0.82. However, with only fourteen data points, the comparison is not completely reliable. One can say that in the face of historical evidence, the S-curve analysis fits the data better and therefore should be used to predict the future. Nevertheless, one must realize that if the data do not really follow an S-curve pattern, the errors may be quite substantial.

EXERCISE 13.2

The following table contains IBM (world-wide) sales data, in millions of dollars, for the years 1947 to 1968. Use these to forecast for the years 1968 through 1977. Compare the forecasting results using trend analysis and exponential growth.

```
***** LISTS THE DATA OF A FILE *****
PERIOD    OBSERVATION   PERIOD   OBSERVATION   PERIOD   OBSERVATION
1           144.54        8        461.35        15       1694.20
2           161.98        9        563.54        16       1925.20
3           183.46       10        734.34        17       2059.60
4           214.91       11       1000.40        18       3239.30
5           266.79       12       1171.70        19       3572.80
6           333.72       13       1309.70        20       4247.70
7           409.98       14       1436.00        21       5345.20
```

Solution

The table below shows the actual values, the forecasts, and the percentage errors for both the exponential growth and trend fitting models. It can be seen that the exponential growth model provides the better fit. The percentage errors for trend fitting indicate a definite pattern cycling from high positive, to zero, to negative, etc. This suggests that trend fitting does not adequately describe the data. This is not true with the exponential growth model, whose errors are considerably smaller and show no pattern. The same conclusion that the exponential growth model is better than the trend one can be reached by looking at the R^2 and the F-test values. These are .801 and 80.03, and .995 and 3549.34, respectively, for the trend and exponential

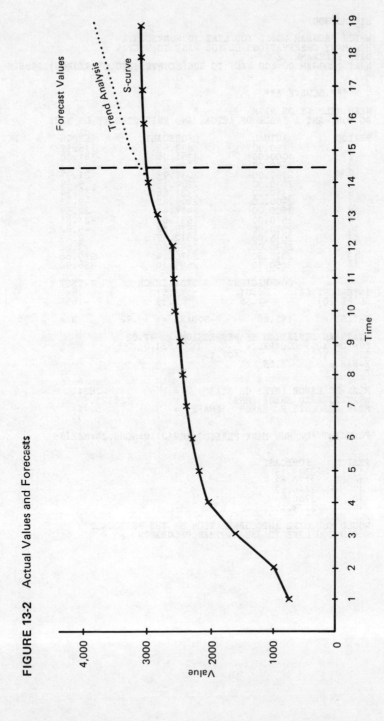

FIGURE 13-2 Actual Values and Forecasts

FIGURE 13-3   Solution to Exercise 13.1 Using SCURVE

```
EXE METHOD
 8K
WHICH PROGRAM WOULD YOU LIKE TO RUN?SCURVE
HOW MANY OBSERVATIONS DO YOU WANT TO USE?14
DATA FILENAME?DEOD
WHAT PROGRAM DO YOU WANT TO RUN(SCURVE,SREG,OR EXGROW)?SCURVE


        *** SCURVE ***

NEED HELP (Y OR N)?N
DO YOU WANT A TABLE OF ACTUAL AND PREDICTED (Y OR N)?Y

PERIOD          ACTUAL          FORECAST          ERROR      PCT ERROR
   1            750.00           633.63          116.37        15.52%
   2           1000.00          1381.06         -381.06        38.11%
   3           1640.00          1790.63         -150.63         9.18%
   4           2060.00          2038.93           21.07         1.02%
   5           2210.00          2204.15            5.85          .26%
   6           2380.00          2321.66           58.34         2.45%
   7           2400.00          2409.42           -9.42          .39%
   8           2500.00          2477.41           22.59          .90%
   9           2510.00          2531.61          -21.61          .86%
  10           2570.00          2575.82           -5.82          .23%
  11           2615.00          2612.57            2.43          .09%
  12           2640.00          2643.60           -3.60          .14%
  13           2910.00          2670.14          239.86         8.24%
  14           3050.00          2693.10          356.90        11.70%

                COEFFICIENT     STD. ERROR       T-TEST
INTERCEPT (A)       8.01          12.57            .64
SLOPE (B)          -1.56           2.39           -.65

F-TEST =      145.68      R-SQUARE =    .92        R =     .96

STANDARD DEVIATION OF REGRESSION =  47.02
SUM OF X(I) SQUARED =              821.0
SUM OF X(I) =           107.1
X-BAR =        7.65


MEAN PC ERROR (MPE) OR BIAS     =         -.62%
MEAN SQUARED ERROR (MSE)        =      26525.5
MEAN ABSOLUTE PC ERROR  (MAPE) =          6.4%


FORECAST FOR HOW MANY PERIODS AHEAD (0=NONE,24=MAX)?5

PERIOD    FORECAST
  15      2713.15
  16      2730.83
  17      2746.52
  18      2760.54
  19      2773.15

WOULD YOU LIKE AUTOCORRELATION ON THE RESIDUALS?N
WOULD YOU LIKE TO USE ANOTHER PROGRAM?N
```

FIGURE 13-4   Solution to Exercise 13.1 Using SREG

```
EXE METHOD
 8K
WHICH PROGRAM WOULD YOU LIKE TO RUN?SREG
HOW MANY OBSERVATIONS DO YOU WANT TO USE?14
DATA FILENAME?DEOD
WHAT PROGRAM DO YOU WANT TO RUN(SCURVE,SREG,OR EXGROW)?SREG


        *** SREG ***

NEED HELP (Y OR N)?N
DO YOU WANT A TABLE OF ACTUAL AND PREDICTED (Y OR N)?Y

PERIOD        ACTUAL        FORECAST        ERROR      PCT ERROR
  1            750.00        1287.00        -537.00        71.60%
  2           1000.00        1432.24        -432.24        43.22%
  3           1640.00        1577.48          62.52         3.81%
  4           2060.00        1722.73         337.27        16.37%
  5           2210.00        1867.97         342.03        15.48%
  6           2380.00        2013.21         366.79        15.41%
  7           2400.00        2158.45         241.55        10.06%
  8           2500.00        2303.69         196.31         7.85%
  9           2510.00        2448.93          61.07         2.43%
 10           2570.00        2594.18         -24.18          .94%
 11           2615.00        2739.42        -124.42         4.76%
 12           2640.00        2884.66        -244.66         9.27%
 13           2910.00        3029.90        -119.90         4.12%
 14           3050.00        3175.14        -125.14         4.10%
               COEFFICIENT    STD. ERROR      T-TEST
INTERCEPT (A)  1141.76           21.14         54.01
SLOPE (B)       145.24             .59        246.33

F-TEST =       54.80      R-SQUARE =   .82      R =    .91

STANDARD DEVIATION OF REGRESSION = 79.09
SUM OF X(I) SQUARED =     75537625.0
SUM OF X(I) =        31235.0
X-BAR =    2231.07


MEAN PC ERROR (MPE) OR BIAS    =       -4.76%
MEAN SQUARED ERROR (MSE)       =      75068.4
MEAN ABSOLUTE PC ERROR   (MAPE) =       15.0%


FORECAST FOR HOW MANY PERIODS AHEAD (0=NONE,24=MAX)?5

PERIOD     FORECAST
 15         3320.38
 16         3465.63
 17         3610.87
 18         3756.11
 19         3901.35

WOULD YOU LIKE AUTOCORRELATION ON THE RESIDUALS?N
WOULD YOU LIKE TO USE ANOTHER PROGRAM?N
```

growth models. Clearly in this case, the exponential growth model provides the better forecasts.

Substituting the summation values in formulas (12-2) and (12-3) will yield:

$$b = \frac{21(1695.6) - (231)(141.376)}{21(3311) - (231)^2}$$

$$= .1825$$

and

$$a = \frac{141.376}{21} - .1825 \frac{231}{21}$$

$$= 4.7249$$

TABLE 13-3 IBM Sales - Linear Trend and Exponential Growth Models

	IBM Sales	Exponential Growth Model		Trend-Fitting	
PERIOD	ACTUAL	FORECAST	PCT ERROR	FORECAST	PCT ERROR
1	144.54	135.28	6.00%	-701.58	585.39%
2	161.98	162.37	-0%	-486.29	400.22%
3	183.46	194.87	-6.00%	-271.01	247.72%
4	214.91	233.88	-8.00%	-55.73	125.93%
5	266.79	280.70	-5.00%	159.56	40.19%
6	333.72	336.89	-0%	374.84	12.32%
7	409.98	404.34	1.00%	590.12	43.94%
8	461.35	485.28	-5.00%	805.41	74.58%
9	563.54	582.43	-3.00%	1020.69	81.12%
10	734.34	699.02	4.00%	1235.97	68.31%
11	1000.40	838.96	16.00%	1451.26	45.07%
12	1171.70	1006.91	14.00%	1666.54	42.23%
13	1309.70	1208.48	7.00%	1881.82	43.68%
14	1436.00	1450.40	-1.00%	2097.11	46.04%
15	1694.20	1740.75	-2.00%	2312.39	36.49%
16	1925.20	2089.23	-8.00%	2527.68	31.29%
17	2059.60	2507.46	-21.00%	2742.96	33.18%
18	3239.30	3009.43	7.00%	2958.24	8.68%
19	3572.80	3611.87	-1.00%	3173.53	11.18%
20	4247.70	4334.93	-2.00%	3388.81	20.22%
21	5345.20	5202.73	2.00%	3604.09	32.57%

The forecasts obtained by each method are as follows:

TABLE 13-4

PERIOD	Year	Exponential Growth FORECAST*	Trend-Fitting FORECAST	Actual
22	1968	6244.25	3819.38	6888.549
23	1969	7494.27	4034.66	7197.295
24	1970	8994.53	4249.94	7503.96
25	1971	10795.12	4465.23	8273.603
26	1972	12956.17	4680.51	9532.593
27	1973	15549.83	4895.79	10993.242
28	1974	18662.72	5111.08	12675.292
29	1975	22398.76	5326.36	14436.541
30	1976	26882.72	5541.64	11785.623 (ending 9/30/76)
31	1977	32264.30	5756.93	

In order to forecast, one must use equation (13-11). The prediction for 1968 (period 22) is

$$F_{22} = e^{4.7249+.1825(22)}$$

$$= 6244.41$$

For period 23,

$$F_{23} = e^{4.7249+.1825(23)}$$

$$= 7494.49$$

while for 1977 (period 31),

$$F_{31} = e^{4.7249+1.825(23)}$$

$$= 32,265 \text{ million dollars}$$

If the forecast is correct, the implication is that the sales of IBM in 1977 will be more than twenty-three times their 1947 level.

---

*The actual sales of IBM for the years 1968 through 1976 (which were treated as unknown when the exponential model was estimated) were as indicated in the last column of Figure 13-4. The exponential growth predictions are generally higher suggesting that IBM's growth rate slowed in the years 1968-1976 in comparison to the years 1947-1967.

The values for a and b in (13-7) and the related statistics of the exponential growth model can be seen below:

```
                 COEFFICIENT     STD. ERROR         T-TEST
INTERCEPT (A)      4.72            .10              47.10
SLOPE (B)           .18            .02               7.47

F-TEST =     3544.12       R-SQUARE =    .99        R =    1.00

STANDARD DEVIATION OF REGRESSION =        .46
SUM OF X(I) SQUARED =              977.5
SUM OF X(I) =           141.4
X-BAR =        6.73


MEAN PC ERROR (MPE) OR BIAS       =           -.33%
MEAN SQUARED ERROR (MSE)          =         18028.2
MEAN ABSOLUTE PC ERROR   (MAPE) =             6.1%
```

The values of a and b can be found by first transforming y_t as in (13-9) and then applying the least squares method to (13-10). For the IBM data this is done as follows:

TABLE 13.5

Period	Original IBM data y_t	Transformed IBM data $y'_t = \log_e y_t$	Time t	$y'_t t$	t^2
1	144.543	4.9736	1	4.9736	1
2	161.983	5.0875	2	10.175	4
3	183.465	5.2120	3	15.636	9
.	.	.	.	.	.
.	.	.	.	.	.
20	4247.706	8.3541	20	167.082	400
21	5345.291	8.5840	21	180.2634	441
		$\Sigma y'_t = 141.37\dot{5}$	$\Sigma t = 231$	$\Sigma y'_t t = 1695.6$	$\Sigma t^2 = 3311$

CHAPTER 14

Harrison's Harmonic Smoothing

Harrison's Harmonic Smoothing (HHS) falls between the exponential smoothing models and the decomposition and autoregressive/moving average methods. It requires limited data, as do exponential smoothing models, but it may predict seasonal turning points better than exponential smoothing models. However, when adequate past information is available, HHS does not seem to perform as well as the more accurate methods of decomposition or autoregressive/moving average schemes.

HHS is a seasonal method that uses the principle of Fourier analysis to transform the data into sine and cosine terms. These are then used to estimate the seasonality in the series. The computational procedure used by the HHS method is much more complicated than that of the exponential smoothing models. This has hindered the wider application of the approach among practitioners. However, the HHS method provides some advantages when limited data exist or the data include considerable randomness.

The procedure involved in applying the HHS method is based on the premise that actual values represent a multiplicative combination of seasonal, trend, cyclical, and random components. The HHS method consists of first removing the trend-cycle and calculating crude seasonal estimates. Second, these estimates are smoothed using Fourier analysis and keeping only those Fourier coefficients that are significantly different from zero. Third, any outlying data values — values in the data which are unusually high or low —

are replaced. Fourth, the adequacy of the seasonal fit is tested and, if it is good, the seasonal indices are used to forecast future periods. A more detailed description of the computational procedure involved is given below.

1. REMOVAL OF TREND-CYCLE

A standard procedure for removing the trend-cycle is to calculate an L-period moving average (where L is the length of seasonality) and to divide it into the time series, X_t (see Chapter 4). The effect of the moving average is to eliminate the seasonality and randomness in the series, thus highlighting the trend-cycle. That is,

$$M_t = T_t C_t \tag{14-1}$$

where

T_t denotes trend

and

C_t denotes cycle.

Depending upon the type of moving average (single, double, or weighted), one can remove different types of trend — linear, quadratic, or cubic. Generally, HHS assumes a linear trend and uses an L-period single moving average.

2. CALCULATION OF CRUDE SEASONAL ESTIMATES

Once M_t in (14-1) is known, it can be divided into X_t yielding:

$$R_t^* = \frac{X_t}{M_t} = \frac{S_t T_t C_t R_t}{T_c C_t} = S_t R_t \tag{14-2}$$

where

S_t denotes seasonality

and

R_t denotes randomness.

The R_t^*'s are estimates of seasonality and randomness.

In order to eliminate randomness, R_t^* is arranged in such a way that similar months can be averaged together. Letting r_{ij} denote the ith year and jth month, one could average all the same months together as follows:

$$r_j = \frac{\sum_{i=1}^{m_j} r_{ij}}{m_j} \qquad (14\text{-}3)$$

where m_j is the number of observations available for the jth month.

The value r_j will therefore be the crude seasonal estimates for the data since the randomness of (14-2) has been averaged out through (14-3).

Finally, one can calculate the standard deviation of the crude seasonal estimates, as

$$\sigma = \sqrt{\frac{\sum_{i=1}^{m_j} \sum_{j=1}^{L} (r_{ij} - r_j)^2}{\sum_{j=1}^{L} m_j - L}} \qquad (14\text{-}4)$$

Equation (14-4) is a measure of the dispersion of each of the seasonal factors. The greater the difference between the two values, r_{ij} and r_j, the larger the standard deviation will be.

3. OBTAINING SMOOTHED SEASONAL ESTIMATES THROUGH FOURIER ANALYSIS

The smoothed seasonal estimates, f_j, can be obtained by applying (14-5):

$$\hat{f}_j = 1 + \sum_{k=1}^{L} [a_k \cos(kd_j) + b_k \sin(kd_j)] \qquad j = 1, 2, \ldots, L \qquad (14\text{-}5)$$

where

$$d_j = \frac{2(j-1)\pi}{L/2} - \pi \qquad\qquad j = 1, 2, \ldots, L \qquad (14\text{-}6)$$

where

$$a_k = \frac{1}{L/2} \sum_{j=1}^{L} \hat{f}_j \cos(kX_j) \qquad\qquad\qquad (14\text{-}7)$$

$$b_k = \frac{1}{L/2} \sum_{j=1}^{L} \hat{f}_j \sin(kX_j) \qquad\qquad\qquad (14\text{-}8)$$

$$\text{for } k = 1, 2 \ldots L/2$$

In (14-7) and (14-8) the crude seasonal estimates $\hat{f}_j$, and the actual values X_j are used to estimate the Fourier coefficients, which in turn are used in (14-5) to calculate the smoothed seasonal estimates. Only those $\hat{f}_j$, in (14-5), that are significantly different from zero are included in the final calculations, the remainder being dropped.

4. REPLACEMENT OF THE OUTLIERS

If the difference between r_{ij}, the original estimates of seasonality, and f_j, the smoothed estimates, is greater than 2.5σ (see (14-4) for a definition of σ), the corresponding X_t value is replaced by some more likely value (for example $\hat{f}_j R_t^*$). The result of replacing outliers is that the effect of unusual events, such as strikes, wars, or total breakdowns is eliminated from the series, which allows a more realistic estimate.

Once there are some replaced outliers, the steps 1, 2, and 3 above are recomputed with the replaced values. Thus, new smoothed seasonal estimates are calculated using (14-5).

5. MEASUREMENT OF THE ADEQUACY OF THE SEASONAL FIT

The adequacy of the seasonal fit is tested by an F-test. The F-value is computed as the ratio of the mean squared errors between the variation due to seasonal effects, M_s, and that due to unexplained residual-factors, M_r.

$$F\text{-test} = \frac{M_s}{M_r}$$

If the value of the F-test is not statistically significant (usually greater than 5), it indicates that the seasonality in the data is not statistically important and the seasonal indeces are set equal to 1.

6. FORECASTING

In order to forecast, there must be some estimate of the trend-cycle that will be multiplied by the corresponding value of r_j for the period(s) corresponding to those of the forecasts.

References for further study:

Harrison, P. (1965)
Makridakis, S. and S. Wheelwright (1978, Chapter 3)

INPUT/OUTPUT FOR THE HHS PROGRAM (HARM)

Input

What is the length of seasonality? The length of seasonality, if not known, can be found by using the ACORN program.

Output

1. *Significant Harmonics*

The significant harmonics are those values of (14-7) and (14-8) that are significantly different from zero. The statistical significance is established by calculating the amplitude, c_k, and by finding the variance of the cosine and sine terms. The amplitude is found as:

$$c_k = \sqrt{a_k^2 + b_k^2}$$

and the quantity:

$$t = \frac{c_k \sqrt{n}}{2\sigma}$$

is distributed as a t-distribution (the value of σ is given by (14-4); thus when its value is greater than about 2 the corresponding sine and cosine values are significant; otherwise they are assumed to be zero and are not printed.

2. *Replaced Outliers*

If there are some extreme values, i.e., greater than 2.5σ (see equation (14-4)), they are replaced by the seasonally adjusted 12-months moving average corresponding to the period.

3. *Seasonal Indices*

The seasonal indices correspond to each season (e.g., month), and indicate the extent of seasonality. A seasonal index of less than 100 means that the seasonality of that period is below the average for the year, while one above 100 means the opposite.

4. *Analysis of Variance*

The analysis of variance provides an F-value that indicates whether or not overall seasonality of the data is significant. If the F-value computed is less than about 5, this implies that one should not use a seasonal model of forecasting like the HHS, but rather try a nonseasonal method.

EXERCISE 14-1

The following data show the monthly road casualties in the United Kingdom between the years 1964 and 1970. Estimate the seasonal indexes for each of the 12-months.

```
***** LISTS THE DATA OF A FILE *****
PERIOD      OBSERVATION   PERIOD    OBSERVATION   PERIOD    OBSERVATION
  1            26.00        29         35.10         57         31.20
  2            24.50        30         34.40         58         31.40
  3            27.90        31         35.70         59         30.80
  4            29.10        32         33.60         60         30.60
  5            34.70        33         31.90         61         27.60
  6            33.10        34         35.10         62         23.40
  7            36.00        35         33.40         63         25.00
  8            37.50        36         37.60         64         26.30
  9            34.80        37         27.50         65         31.00
 10            35.50        38         27.20         66         29.30
 11            33.40        39         30.20         67         31.70
 12            32.90        40         28.60         68         32.00
 13            29.00        41         34.10         69         30.00
 14            24.70        42         30.90         70         31.80
 15            31.30        43         34.70         71         33.60
 16            32.40        44         33.70         72         31.60
 17            33.90        45         33.60         73         26.90
 18            35.00        46         31.00         74         26.60
 19            36.40        47         28.90         75         27.60
 20            36.50        48         29.70         76         27.10
 21            34.40        49         23.90         77         29.80
 22            33.90        50         24.70         78         29.10
 23            33.90        51         27.50         79         32.60
 24            38.40        52         26.70         80         31.60
 25            27.00        53         28.70         81         31.10
 26            26.30        54         30.30         82         33.20
 27            29.80        55         31.30         83         33.60
 28            32.60        56         32.10         84         34.00
```

Solution (See Figures 14-1 through 14-3)

1. *Computing a 12-months Moving Average (12-MMA)*

The purpose of the 12-MMA is to remove the seasonality and randomness of the data (see (14-1)). The 12-MMA should be centered but since 12 is an even number, it is put in the middle plus one period. Thus:

$$M_7 = \frac{26 + 24.5 + 27.9 + 29.10 + \ldots + 35.5 + 33.4 + 32.9}{12}$$

$$= 32.117$$

$$M_8 = \frac{24.5 + 27.90 + 29.10 + 34.7 + \ldots + 33.4 + 32.9 + 29.0}{12}$$

$$= 32.367$$

$$M_9 = \frac{27.9 + 29.10 + 34.7 + 33.1 + \ldots 32.9 + 29.0 + 24.7}{12}$$

$$= 32.383$$

etc. (for a table of all 12-MMA values see Chapter 15).

In order to avoid losing the last five values because of the moving averaging, the method of Harmonic Smoothing estimates the last five values by extrapolating the last twelve values of the moving average that are available (i.e., periods 67 to 79). If there is a statistically significant trend, then the extrapolation includes the trend; otherwise the trend is zero and thus the extrapolation is horizonatal, or the value for period 79 is used as an estimate for periods 80 through 84.

Thus:

$$M_{78} = \frac{31.6 + 26.9 + 26.6 + 27.6 + \ldots + 31.1 + 33.2 + 33.6}{12}$$

$$= 30.067$$

$$M_{79} = \frac{26.9 + 26.6 + 27.6 + 27.1 + \ldots + 33.2 + 33.6 + 34}{12}$$

$$= 30.267$$

Then:

$$M_{80} = 30.267$$

$$M_{81} = 30.267$$

$$M_{82} = 30.267$$

$$M_{83} = 30.267$$

$$M_{84} = 30.267$$

2. *Calculating Crude Seasonal Estimates*

Dividing the data by the 12-MMA computed above gives a series which includes seasonality and randomness only (see (14-2)). The values obtained

are as follows: (All values for January are calculated to show how the crude
seasonal index for the month can be found.)

$$R_7^* = \frac{X_7}{M_7} = \frac{36}{32.117} = 1.1209$$

$$R_8^* = \frac{X_8}{M_8} = \frac{37.5}{32.367} = 1.1586$$

$$R_9^* = \frac{X_9}{M_9} = \frac{34.8}{32.367} = 1.07463$$

.

.

.

$$R_{13}^* = \frac{X_{13}}{M_{13}} = \frac{29}{33.033} = .8779$$

.

.

.

$$R_{25}^* = \frac{X_{25}}{M_{25}} = \frac{27}{33.058} = .8167$$

.

.

.

$$R_{37}^* = \frac{X_{37}}{M_{37}} = \frac{27.5}{32.15} = .8554$$

$$R_{49}^* = \frac{X_{49}}{M_{49}} = \frac{23.9}{29.45} = .8115$$

.

.

.

$$R_{61}^* = \frac{X_{61}}{M_{61}} = \frac{27.6}{29.167} = .9463$$

.

.

.

$$R_7^* = \frac{X_{73}}{M_{73}} = \frac{26.9}{29.817} = .90218$$

.

.

.

$$R_{78}^* = \frac{X_{78}}{M_{78}} = \frac{29.1}{30.067} = .9678$$

$$R_{79}^* = \frac{X_{79}}{M_{79}} = \frac{32.6}{30.267} = 1.0771$$

$$R_{80}^* = \frac{X_{80}}{M_{80}} = \frac{31.6}{30.267} = 1.0441$$

$$R_{81}^* = \frac{X_{81}}{M_{81}} = \frac{31.1}{30.267} = 1.0275$$

$$R_{82}^* = \frac{X_{82}}{M_{82}} = \frac{33.2}{30.267} = 1.0969$$

$$R^*_{83} = \frac{X_{83}}{M_{83}} = \frac{33.6}{30.267} = 1.1101$$

$$R^*_{84} = \frac{X_{84}}{M_{84}} = \frac{34}{30.257} = 1.1233$$

Summing the values of R^*_t for each month (i.e., all January, all February, all March, etc.) and dividing by the number observations for that month gives the average seasonal estimates, since the randomness in (14-2) is eliminated through the averaging process. For January, for example,

$$r_1 = \frac{.8779 + .8167 + .8554 + .8115 + .9463 + .90218}{6}$$

$$= .8683$$

The value .8683 is the crude seasonal estimate for the month of January. In a similar manner, one can calculate crude seasonal estimates for the other months. The values of these estimates, after they are adjusted to sum to 1200, are:

$\hat{f}_1 = .8676$	$\hat{f}_5 = 1.0371$	$\hat{f}_9 = 1.0416$
$\hat{f}_2 = .8210$	$\hat{f}_6 = 1.0176$	$\hat{f}_{10} = 1.0638$
$\hat{f}_3 = .9206$	$\hat{f}_7 = 1.094$	$\hat{f}_{11} = 1.0456$
$\hat{f}_4 = .9338$	$\hat{f}_8 = 1.0875$	$\hat{f}_{12} = 1.0698$

3. *Obtaining Smoothed Seasonal Estimates*

Once the values for $\hat{f}_j$ are found, they can be used in (14-6) and (14-7) and transformed into Fourier coefficients that, when substituted in (14-5), provide smoothed estimates of seasonality. This gives (see (14-6)):

$$d_1 = \frac{2(1 - 1)\ 3.14159}{6} - 3.14159 = -3.14159$$

$$d_2 = \frac{2(2-1)\ 3.14159}{6} - 3.14159 = -2.09444$$

.

.

.

$$d_3 = \frac{2(3-1)\ 3.14159}{6} - 3.14159 = -1.0472$$

The Fourier coefficient can then be computed by (14-7) and (14-8). Thus:

$$a_2 = \frac{1}{6}\ (.8676\ \cos(-3.14159) + .8210\ \cos(-2.0944)$$

$$+ .9206(-1.0472) + \ldots + 1.0698(8.3776)$$

$$= .0781$$

$$a_2 = \frac{1}{6}\ (.8676\ \cos(2x - 3.14159) + .8210\ \cos(2x - 2.0944)$$

$$+ .0206(2x - 1.0472) + \ldots + 1.0098(2x - 8.3776)\ )$$

$$= -.01006$$

Similarly the remaining a_k and b_k can be calculated. The values of all of them are:

	a_k	b_k
1	.0781	.0669
2	-.0101	-.0432
3	.0190	.0314
4	-.0102	-.0084
5	.0162	.0295
6	.0021	-.0000

Except for the sixth harmonic, the remaining are significant and their values are shown in Figure 14-1 where a_k is the cosine coefficient, b_k is the sine, and the amplitude, c_k, is found as:

$$c_k = \sqrt{a_k^2 + b_k^2}$$

For the first harmonic, for example, c_1 is:

$$c_1 \sqrt{.078^2 + .067^2} = .1028$$

The value of c_k is needed to exclude those harmonics that are not significantly different from zero. With the sine, cosine, and amplitude, one can analyze the series in terms of its frequency characteristic if so desired; however, in this example only the sine and cosine values will be used in (14-5) to calculate smoothed estimates of seasonality.

$$\hat{r}_1 = 1 + .078 \cos(1x - .314159) + .067 \sin(1x - 3.14159)$$

$$- .010 \cos(2x - 3.14159) - .043 \sin(2x - 3.14159)$$

$$+ \ldots + .016 \cos(5x - 3.14159) + .03 \sin(5x - 3.14159)$$

$$= .86652$$

$$\hat{r}_2 = 1 + .078 \cos(1x - 2.0944) + .067 \sin(1x - 2.0944)$$

$$- .010 \cos(2x - 2.0944) - .043 \sin(2x - 2.0944)$$

$$+ \ldots + .016 \cos(5x - 2.0944) + .03 \sin(5x - 2.0944)$$

$$= .8221$$

.

.

.

$$\hat{r}_{12} = 1 + .078 \cos(1x\ 8.3776) + .067 \sin(1x\ 8.3776)$$

$$- .010 \cos(2x\ 8.3776) - .043 \sin(2x\ 8.3776)$$

$$+ \ldots + .016 \cos(5x\ 8.3776) + .03 \sin(5x\ 8.3776)$$

$$= 1.07083$$

FIGURE 14-1 Significant Harmonics

```
EXE METHOD
 8K
WHICH PROGRAM WOULD YOU LIKE TO RUN?HARM

   *** HARRISONS HARMONIC SMOOTHING ***

NEED HELP (Y OR N)?N
HOW MANY OBSERVATIONS DO YOU WANT TO USE?84
DATA FILENAME?UKROAD

WHAT IS THE LENGTH OF YOUR SEASONALITY (E.G. 12 IF MONTHLY)?12
          SIGNIFICANT HARMONICS
HARMONIC      COSINE           SINE          AMPLITUDE
   1           .078            .067            .1028
   2          -.010           -.043            .0444
   3           .019            .031            .0367
   4          -.010           -.008            .0132
   5           .016            .030            .0337


     R E P L A C E D  O U T L I E R S
PERIOD            OUTLIER           REPLACEMENT
```

FIGURE 14-2 Seasonal Indices

```
OUTLIERS REPLACED NONE
   BEST SEASONAL ESTIMATES ARE THE CRUDE ESTIMATES

PERIOD     SEASONAL INDICES
   1            86.87
   2            81.99
   3            92.17
   4            93.27
   5           103.81
   6           101.65
   7           109.51
   8           108.65
   9           104.26
  10           106.28
  11           104.67
  12           106.87

      A N A L Y S I S   O F   V A R I A N C E

    MEAN SQUARED ERRORS
SEAS. =      .0575    RES. =        .0018
F-TEST =   32.5696
```

FIGURE 14-3 Analysis of Variance

```
MEAN PC ERROR (MPE) OR BIAS     =      -.09%
MEAN SQUARED ERROR (MSE)        =       1.6
MEAN ABSOLUTE PC ERROR  (MAPE)  =       3.4%

FORECAST FOR HOW MANY PERIODS AHEAD (0=NONE,24=MAX)?3

PERIOD      FORECAST
  85          28.09
  86          26.51
  87          29.80

WOULD YOU LIKE AUTOCORRELATION ON THE RESIDUALS?N
```

The Classical Decomposition Method

Decomposition methods, as the name implies, "break down" a time series into the four components — seasonality, trend, cycle, and randomness — that frequently are present in economic and business time series. The principle of decomposition is intuitive and easy to understand. Practitioners often prefer to use decomposition methods even though they require some human inputs (predicting the trend-cycle) rather than more statistically complete approaches, such as AutoRegressive/Moving Average (ARMA) methods, that may be more difficult to understand. Statisticians, on the other hand, prefer the ARMA methods, popularized by Box and Jenkins, because they provide a wide choice of forecasting models that can theoretically fit any type of data. With seasonal data, decomposition approaches may be as accurate in practice as the more statistically-oriented methods. In addition they provide information (such as the state of the trend-cycle) which is not available with other time series methods.

In the decomposition category there are two main methods — Classical Decomposition, and the Census II method. Both are similar in principle, even though the Census II is much more refined and elaborate. Both the methods assume that economic- or business-oriented time series are made up of four components — seasonality, trend, cycle and randomness. Furthermore, they usually assume that the relationship between these four components is multiplicative (Census II allows for an additive relationship if the user so desires) as shown in equation (15-1).

$$X_t = S_t T_t C_t R_t \qquad\qquad (15\text{-}1)$$

where

X_t is the time series;

S_t denotes seasonality;

T_t denotes trend;

C_t denotes cycle;

and

R_t denotes randomness.

(Alternatively, one could assume an additive relationship of the form

$$X_t = S_t + T_t + C_t + R_t$$

But additive models are not commonly encountered in practice.)

If a moving average of L periods (where L is the length of seasonality) is calculated, it will represent the mean value for the year. Such a value will obviously be free of seasonal effects, since high months will be offset by low ones. If M_t denotes the moving average of (15-1), it will be free of seasonality and will contain little randomness. Thus, it can be denoted:

$$M_t = T_t C_t \qquad\qquad (15\text{-}2)$$

Equation (15-2) excludes randomness because it has been eliminated through the averaging process. Thus, the equation only includes the trend-cycle component. One can decompose (15-2) further by assuming some form of trend. For example, deciding that the trend is linear (see Chapter 12), gives:

$$T_t = a + bt \qquad\qquad (15\text{-}3)$$

If (15-3) is divided into (15-2) the cycle will be isolated from the trend.

$$\frac{M_t}{a + bt} = \frac{T_t C_t}{T_t} = C_t \qquad\qquad (15\text{-}4)$$

If a linear trend is not adequate, one may wish to specify a nonlinear one. If one is willing to assume a pattern for the trend, this can be used to separate it from the cycle. In practice, however, it is often difficult to separate the two, and often one may prefer to work with the trend-cycle figures of (15-2). The isolation of the trend will add little to the overall ability to forecast.

To isolate seasonality one simply divides (15-2) into (15-1). This yields:

$$\frac{S_t}{M_t} = \frac{S_t T_t C_t R_t}{T_t C_t} = S_t R_t \tag{15-5}$$

Thus, dividing the original series by the moving averages gives the seasonal pattern and whatever randomness exists. Finally, randomness can be eliminated by averaging the different values of (15-5). The averaging is done on the same months or seasons of different years. (For example, the average of all Januaries, all Februaries, ..., all Decembers.) The result is a set of seasonal values free of randomness, called seasonal indices, and widely used in practice.

In order to forecast, one must reconstruct each of the components of (15-1). The seasonality is known through averaging the values obtained through (15-5), the trend through (15-3), the randomness cannot be predicted, and the cycle of (15-4) must be estimated by the user. (See Chpater 21 for more information on how to predict the cycle.)

References for further study:

> Macauley, F. (1930)
> Makridakis, S. and S. Wheelwright (1978, Chapter 4)
> Wheelwright, S. and S. Makridakis (1977, Chapter 6)

INPUT/OUTPUT FOR THE CLASSICAL DECOMPOSITION PROGRAM (DCOMP)

Input

1. Length of seasonality (use the autocorrelation program to find it, if it is not known).
2. Input estimates of the cyclical factors (optional).

The user should input estimates of the cyclical factors. To make a good forecast, it is important that these be predicted correctly. It is not easy to prepare an estimate since no quantitative method can be used because cycles vary in length. Rather, it is a judgmental input which depends upon intuitive feel and a knowledge of economic conditions. It is clearly influenced by a boom or a recession, and that is why it does not repeat itself at constant intervals.

The graph of the cyclical factors output by the program before the above question is asked can only be used as an aid to estimating the cycle. There are $L/2 - 1$ values of cyclical factors missing at the end of the graph because the moving averages calculated by equation (15-2) are required before the cyclical factors can be calculated in equation (15-4). This means the user must estimate the cyclical factors for the last $L/2 - 1$ data points before predicting future values.

The program assumes a linear trend of the form (15-3). If the trend is not linear, then the program must be changed accordingly.

Outputs

In addition to the standard output there are:

1. A table of the moving averages and the trend-cycle ratios (see (15-2).)
2. The seasonal indices for L seasons. (The difference between the medial average, i.e., the average after the largest and smallest values have been excluded) and the seasonal indices is that the sum of the former does not add up to 1200 while the sum of the latter does. A seasonal index of 100 is the average while anything below indicates that the corresponding month is below the average, and an index above 100 indicates the opposite. (Government figures given on a seasonal adjusted basis have been found by dividing the actual value by the seasonal index.) Seasonal indices are important for planning and scheduling purposes.
3. The graph of the cyclical factors. If the trend is linear, this is merely a graph of equation (15-4). This graph can be used as an aid to predicting the cycle (see Chapter 21 for more information).

SOLVED EXERCISE 15.1

Use the data from Chapter 14 to estimate the seasonal indices for each month, and predict road casualties for the year 1971 assuming a linear trend.

Solution

1. *Calculating the Seasonal Factors* (see Figure 15-1)

    a. 12-Months Moving Average (12-MMA) of the data

    The purpose of this average is to eliminate the seasonality and randomness in the data. The average should be centered, but since twelve is an even number it is put next to the middle, plus one period. Thus:

    $$M_7 = \frac{26 + 24.5 + 27.9 + 29.10 + \ldots + 35.50 + 33.40 + 32.90}{12}$$

    $$= \frac{385.4}{12}$$

    $$M_7 = 32.117$$

    $$M_8 = \frac{24.5 + 27.9 + 29.10 + 34.7 + \ldots + 33.4 + 32.90 + 29.0}{12}$$

    $$= \frac{388.4}{12}$$

    $$M_8 = 32.367$$

    .

    .

    .

    $$M_{79} = \frac{26.9 + 26.6 + 27.60 + 27.10 + \ldots + 33.2 + 33.6 + 34.0}{12}$$

    $$= \frac{363.2}{12}$$

    $$M_{79} = 30.267$$

    b. dividing the 12-MMA into the data

    The moving averages are the equivalent of (15-2). If they are divided into the original data, their ratio represents seasonality and randomness (see equation (15-5)). Thus:

$$\frac{X_7}{M_7} = \frac{36.0}{32.117} = 1.1209$$

$$\frac{X_8}{M_8} = \frac{37.5}{32.367} = 1.1586$$

.

.

.

$$\frac{X_{79}}{M_{79}} = \frac{32.6}{30.267} = 1.0771$$

The ratios (preferably multiplied by 100) are the equivalent of (15-5) and include seasonality and randomness only. The next step in the classical decomposition method is to exclude randomness from the ratios of seasonal factors just computed.

2. *Calculating the Seasonal Indices* (see Figure 15-2)

One can construct a table where the ratios of seasonal factors for all the months of January, all the months of February, ... , all the months of December are together. For January, for example, one would have:

	January		January
1964	—		—
1965	87.79		87.79
1966	81.674		81.674
1967	85.537		85.537
1968	81.154	smallest value – excluded	—
1969	94.629	highest value – excluded	—
1970	90.218		90.218
			345.319 ÷ 4 =
		Medial Average for January	86.305

where the Medial Average is the average of all values of the same month (e.g., January) after the smallest and largest values have been excluded.

Once all seasonal factors (ratios) for January are together, the smallest and largest values are excluded. This is done to eliminate the possible effects

FIGURE 15-1   Ratios Representing Seasonality and Randomness

```
EXE METHOD
8K
WHICH PROGRAM WOULD YOU LIKE TO RUN?DCOMP

    *** CLASSICAL  DECOMPOSITION ***

NEED HELP (Y OR N)?N
HOW MANY OBSERVATIONS DO YOU WANT TO USE?84
DATA FILENAME?UKROAD

WHAT IS THE LENGTH OF YOUR SEASONALITY ?12

DO YOU WANT A TABLE OF MOVING AVERAGES AND RATIOS?Y.
```

SEASON	DATA	MOVING AVERAGE	RATIO (× 100)
1	26.00		
2	24.50		
3	27.90		
4	29.10		
5	34.70		
6	33.10		
7	36.00	32.12	112.09
8	37.50	32.37	115.86
9	34.80	32.38	107.46
10	35.50	32.67	108.67
11	33.40	32.94	101.39
12	32.90	32.87	100.08
13	29.00	33.03	87.79
14	24.70	33.07	74.70
15	31.30	32.98	94.90
16	32.40	32.95	98.33
17	33.90	32.82	103.30
18	35.00	32.86	106.52
19	36.40	33.15	109.80
20	36.50	32.98	110.66
21	34.40	33.12	103.88
22	33.90	32.99	102.75
23	33.90	33.01	102.70
24	36.40	33.11	109.94
25	27.00	33.06	81.67
26	26.30	33.00	79.70
27	29.80	32.76	90.97
28	32.60	32.55	100.15
29	35.10	32.65	107.50
30	34.40	32.61	105.49
31	35.70	32.71	109.15
32	33.60	32.75	102.60
33	31.90	32.82	97.18
34	35.10	32.86	106.82
35	33.40	32.53	102.69
36	37.60	32.44	115.90
37	27.50	32.15	85.54
38	27.20	32.07	84.82
39	30.20	32.07	94.15
40	28.60	32.22	88.77
41	34.10	31.87	106.98
42	30.90	31.50	98.10
43	34.70	30.84	112.51
44	33.70	30.54	110.34
45	33.60	30.33	110.77
46	31.00	30.11	102.96
47	28.90	29.95	96.49
48	29.70	29.50	100.68
49	23.90	29.45	81.15
50	24.70	29.17	84.69
51	27.50	29.03	94.72
52	26.70	28.83	92.60
53	28.70	28.87	99.42
54	30.30	29.02	104.39
55	31.30	29.10	107.56
56	32.10	29.41	109.15
57	31.20	29.30	106.48
58	31.40	29.09	107.93
59	30.80	29.06	105.99
60	30.60	29.25	104.62
61	27.60	29.17	94.63

FIGURE 15-1 (Continued)

62	23.40	29.20	80.14
63	25.00	29.19	85.64
64	26.30	29.09	90.40
65	31.00	29.12	106.44
66	29.30	29.36	99.80
67	31.70	29.44	107.67
68	32.00	29.38	108.91
69	30.00	29.65	101.18
70	31.80	29.87	106.47
71	33.60	29.93	112.25
72	31.60	29.83	105.92
73	26.90	29.82	90.22
74	26.60	29.89	88.99
75	27.60	29.86	92.44
76	27.10	29.95	90.48
77	29.80	30.07	99.11
78	29.10	30.07	96.78
79	32.60	30.27	107.71
80	31.60		
81	31.10		
82	33.20		
83	33.60		
84	34.00		

of unusual events. Averaging the remaining seasonal factors eliminates the randomness in them (see (15-5)). These averages are called the medial averages for each month.

The only difference between the medial averages and the seasonal indices is that the former do not add up to 1200 (100 times 12 months). Thus some adjustment is still needed. This is accomplished by multiplying each month's medial average by the correction ratio of 1200/1198.979 = 1.000852. The standardized medial average for January then becomes:

$$86.305(1.000852) = 86.378,$$

for February

$$82.336(1.000852) = 82.406,$$

etc.

FIGURE 15-2 Seasonal Indices and Trend Values

```
**** SEASONAL INDEX ****
   FOR PERIOD 1 THROUGH    7

SEASON    MEDIAL AVERAGE     NORMALIZED INDEX
   1          86.30               86.38
   2          82.34               82.41
   3          93.07               93.15
   4          92.95               93.03
   5         104.04              104.12
   6         101.95              102.03
   7         109.28              109.38
   8         109.77              109.86
   9         104.75              104.84
  10         106.05              106.14
  11         103.19              103.28
  12         105.29              105.38

TOTAL       1198.98             1200.00

        **** TREND ****
INTERCEPT (A) =     32.57
SLOPE (B) =       -.03
```

3. Calculating a Linear Trend

One can assume that the data have a linear trend and therefore calculate the parameters of a straight line, a and b (see Chapter 13 for details). This will be the equivalent of finding (15-3). If the trend is non-linear, one must estimate the appropriate parameters. In the road casualty data, the values for a and b are a = 32.5666 and b = −.03476. These imply there is a gradual decline in the number of casualties with time.

4. Calculating the Cyclical Factors (see Figure 15-3)

Once a and b are estimated, the trend for any period can be found using (15-3). For example, for period 7, the trend is:

$$T_7 = 32.5666 - .034761(7) = 32.3233$$

If T_7 is divided into the corresponding 12-MMA, one obtains a measure of the cycle, or the cyclical factor for period 7.

$$C_7 = \frac{M_7}{T_7} = \frac{32.117}{32.323} = .9936$$

Similarly,

$$T_8 = 32.5666 - .034761(8) = 32.2885$$

and

$$C_8 = \frac{M_8}{T_8} = \frac{32.367}{32.2885} = 1.0024$$

Similarly,

$$T_{79} = 32.5666 - .034761(79) = 29.8205$$

and

$$C_{79} = \frac{M_{79}}{T_{79}} = \frac{30.267}{29.8205} = 1.015$$

From the above cyclical factors (preferably multiplied by 100), one can attempt to forecast future values of the cycle. Unfortunately, as can be seen from Figure 15-3 the last 5 values for the cyclical factors are missing, which makes the job of estimating the cyclical factors beyond period 84 more difficult.

5. *Forecasting* (see Figure 15-4)

In order to forecast one must reconstruct the seasonal trend and cyclical components. For period 85 (January), for example, the seasonal index is .86378, the trend is $T_{85} = 32.5666 - .034761(85) = 29.612$ and assuming that the cycle is 1.00 (or 100), gives

$$\hat{X}_{85} = .86378(29.612)(1.0) = 25.578$$

For period 90 (June), seasonal index $= 1.0203$,

$$\text{trend } T_{90} = 32.5666 - .034761(90) = 29.438$$

and assuming a cycle of 1. (100), the forecast is

$$\hat{X}_{90} = 1.0203(29.438)(1) = 30.036$$

FIGURE 15 3 Graph of Cyclical Factors

```
DO YOU WANT A LIST OF CYCLICAL FACTORS AND FORECASTS ?Y
      **** GRAPH OF CYCLICAL FACTOR ****

        50        75       100       125       150
SEASON I.........I.........I.........I.........I          C.F.     FORECAST
   7   .                   *                   .          99.4       35.1
   8   .                   *                   .         100.2       35.6
   9   .                   *                   .         100.4       34.0
  10   .                   *                   .         101.4       34.7
  11   .                   *                   .         102.4       34.0
  12   .                   *                   .         102.3       34.6
  13   .                   *                   .         102.9       28.5
  14   .                   *                   .         103.1       27.2
  15   .                   *                   .         102.9       30.7
  16   .                   *                   .         102.9       30.7
  17   .                   *                   .         102.6       34.2
  18   .                   *                   .         102.9       33.5
  19   .                    *.                 .         103.9       36.3
  20   .                    *                  .         103.5       36.2
  21   .                     *                 .         104.0       34.7
  22   .                    *                  .         103.7       35.0
  23   .                     *                 .         103.9       34.1
  24   .                      *                .         104.3       34.9
  25   .                      *                .         104.3       28.6
  26   .                      *                .         104.2       27.2
  27   .                    *                  .         103.6       30.5
  28   .                   *                   .         103.0       30.3
  29   .                    *                  .         103.5       34.0
  30   .                    *                  .         103.4       33.3
  31   .                     *                 .         103.9       35.8
  32   .                      *                .         104.1       36.0
  33   .                      *                .         104.5       34.4
  34   .                      *                .         104.7       34.9
  35   .                    *                  .         103.7       33.6
  36   .                    *                  .         103.6       34.2
```

FIGURE 15-3 (Continued)

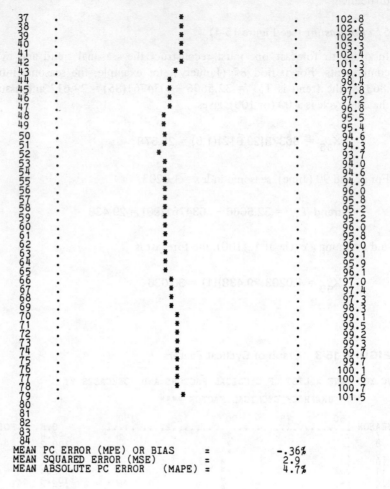

37	*	.	102.8	27.8
38	*	.	102.6	26.4
39	*	.	102.8	29.9
40	*	.	103.3	30.0
41	*	.	102.4	33.2
42	*	.	101.3	32.1
43	*	.	99.3	33.7
44	*	.	98.4	33.6
45	*	.	97.8	31.8
46	*	.	97.2	32.0
47	*	.	96.8	30.9
48	*	.	95.5	31.1
49	*	.	95.4	25.4
50	*	.	94.6	24.0
51	*	.	94.3	27.0
52	*	.	93.7	26.8
53	*	.	94.0	30.1
54	*	.	94.6	29.6
55	*	.	94.9	31.8
56	*	.	96.0	32.3
57	*	.	95.8	30.7
58	*	.	95.2	30.9
59	*	.	95.2	30.0
60	*	.	96.0	30.8
61	*	.	95.8	25.2
62	*	.	96.0	24.1
63	*	.	96.1	27.2
64	*	.	95.9	27.1
65	*	.	96.1	30.3
66	*	.	97.0	30.0
67	*	.	97.4	32.2
68	*	.	97.3	32.3
69	*	.	98.3	31.1
70	*	.	99.1	31.7
71	*	.	99.5	30.9
72	*	.	99.2	31.4
73	*	.	99.3	25.8
74	*	.	99.7	24.6
75	*	.	99.7	27.8
76	*	.	100.1	27.9
77	*	.	100.6	31.3
78	*	.	100.7	30.7
79	*	.	101.5	33.1
80				32.6
81				32.7
82				31.2
83				31.5
84				30.6

```
MEAN PC ERROR (MPE) OR BIAS    =      -.36%
MEAN SQUARED ERROR (MSE)       =       2.9
MEAN ABSOLUTE PC ERROR  (MAPE) =       4.7%
```

FIGURE 15-4   Forecasts

```
FORECAST FOR HOW MANY PERIODS AHEAD (0=NONE,24=MAX)?12

DO YOU WANT TO ESTIMATE THE    12 CYCLICAL FACTORS FOR YOUR
FORECASTS (IF NOT THE PROGRAM WILL ASSIGN A VALUE OF 100 TO EACH)?N


      **** FORECAST ****

PERIOD     FORECAST
85         25.58
86         24.37
87         27.52
88         27.45
89         30.69
90         30.04
91         32.16
92         32.26
93         30.75
94         31.10
95         30.22
96         30.80

WOULD YOU LIKE AUTOCORRELAIONS ON THE RESIDUALS ?N
```

CHAPTER 16

Census II Decomposition Method

In principle Census II is similar to other decomposition methods in general and the classical decomposition of Chapter 15 in particular. Its aim is to decompose and at the same time identify each component of a time series. The process for doing this, however, is more elaborate than that of other decomposition approaches. In fact, there are three main differences between the Census and the Classical Decomposition methods:

1. The Census method calculates preliminary estimates of seasonality and trend-cycle and then final estimates. The result is that the influence of each component can be removed separately. Classical Decomposition, on the other hand, attempts to decompose the series for more than one component at a time. The importance of isolating each component separately has both statistical and practical consequences.
2. The Census method removes outliers (i.e., values which are abnormally high or low) and irons out irregular fluctuations to a much greater extent than does Classical Decomposition.
3. The Census method provides several measures, or tests, which allow the user to determine how well the process of decomposition has been achieved. These measures, or tests, are discussed in the exercise.

Census II, as is true with many forecasting procedures, can frequently predict seasonality variations but can do little to forecast the cycle and turning points caused by changes in the general level of economic activity. The main reason is that the length of the cycle is not constant, which makes its prediction by mechanical approaches practically impossible.

The major advantage of the Census II method is that it provides information about the trend-cycle and allows the user to make intelligent estimates of the continuation of the trend cycle. (It does not attempt to separate the trend from the cycle.) It is often said that Census can predict the trend-cycle components. That is not correct. Census can provide the user with only enough information to estimate the trend-cycle more accurately. The user must identify and input estimates of the trend-cycle before the method can provide forecasts. (More information on estimating the trend-cycle values is included in Chapters 20, 21, and 22.)

It is beyond this chapter to provide a complete description of the Census II method. A summary of the procedure followed by the Census II method can be seen in Figure 16-1. The interested reader is encouraged to look at the references below.

References for further study:

> McLaughlin, R. (1962)
> Makridakis, S. and S. Wheelwright (1978, Chapter 4)
> Shiskin, J. (1967)

INPUT/OUTPUT FOR THE CENSUS PROGRAM (CENSUS)

Input

1. *The Length of Seasonality*

If the length is not known it can be found by running the autocorrelation analysis program (see Chapter 3).

Output

Only the most important output will be discussed. A description and explanation of the full output of the Census II program can be found in the

FIGURE 16-1   Major tasks in Census II

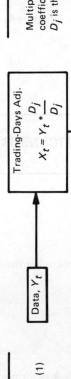
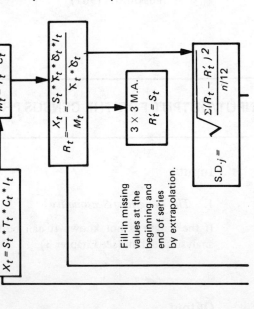

Steps	Description
(1)	Multiply the data of each month by the Trading-Days (T-D) adjustment coefficient. D_j D_j where D_j is the T-D for each of the months, and D_j is the average, over all years, for this particular month.
(2)	Calculate a 12-month, centered, Moving Average (12-MMA), its purpose is to eliminate the seasonality and some of randomness. Thus the Trend-cycle will remain.
(3)	Divide the M.A. into the original data. (1)/(2). The result is: Ratios of seasonal — irregular components.
(a)	Calculate a 3 × 3 M.A. (a 3-term Moving Average — double M-A) to each month separately. Its effect will be to eliminate the randomness in (3).
(4) (b)	Calculate the standard deviation (S.D.) for each month separately (12 in all) by summing up the square difference between R_t (M.A. with randomness ± (3)) and R'_t (M.A. without randomness (4.a)).

Steps boxes:

Data, Y_t

Trading-Days Adj. $X_t = Y_t * \dfrac{D_j}{D_j}$

$M_t = T_t * C_t$

Original Data $X_t = S_t * T_t * C_t * I_t$

$R_t = \dfrac{X_t}{M_t} = \dfrac{S_t * T_t * \delta_t * I_t}{Y_t * \delta_t}$

3×3 M.A. $R'_t = S_t$

Fill-in missing values at the beginning and end of series by extrapolation.

$S.D._j = \sqrt{\dfrac{\Sigma(R_t - R'_t)^2}{n/12}}$

FIGURE 16-1 (Continued)

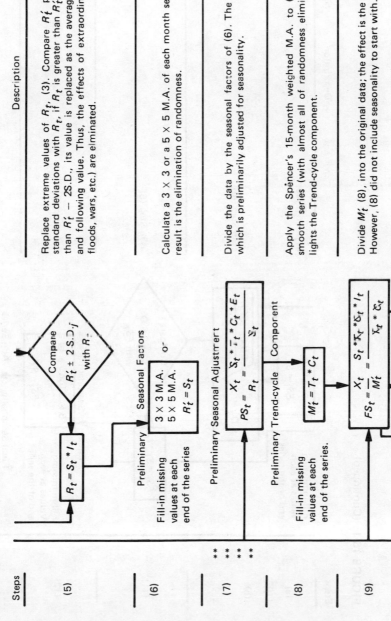

Steps		Description
(5)	Compare $R'_t \pm 2\,S.D._j$ with R_- $R_t = S_t * I_t$	Replace extreme values of R_t, (3). Compare R'_t plus, or minus, two standard deviations with R_t, if R_t is greater than $R'_t + 2S.D.$ or smaller than $R'_t - 2S.D.$, its value is replaced as the average of the preceding and following value. Thus, the effects of extraordinary events (strikes, floods, wars, etc.) are eliminated.
(6)	Preliminary Seasonal Factors 3×3 M.A. or 5×5 M.A. $R'_t = S_t$ Fill-in missing values at each end of the series	Calculate a 3×3 or a 5×5 M.A. of each month separately, in (5); the result is the elimination of randomness.
(7)	Preliminary Seasonal Adjustment $PS_t = \dfrac{X_t}{R_t} = \dfrac{S_t * T_t * C_t * E_t}{S_t}$	Divide the data by the seasonal factors of (6). The outcome is a series which is preliminarily adjusted for seasonality.
(8)	Preliminary Trend-cycle Component $M'_t = T_t * C_t$ Fill-in missing values at each end of the series.	Apply the Spencer's 15-month weighted M.A. to (7). The result is a smooth series (with almost all of randomness eliminated) which highlights the Trend-cycle component.
(9)	$FS_t = \dfrac{X_t}{M'_t} = \dfrac{S_t * T_t * C_t * I_t}{X_t * C_t}$	Divide M'_t, (8), into the original data; the effect is the same as in Step (3). However, (8) did not include seasonality to start with.

FIGURE 16-1 (Continued)

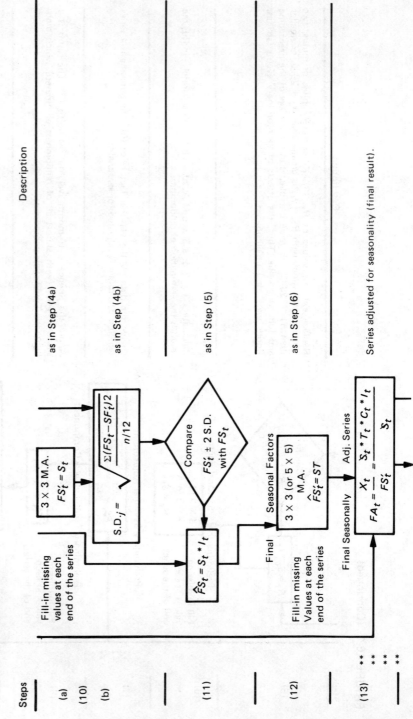

FIGURE 16-1 (Continued)

(14)
- ** Fill-in missing
- ** values at each
- ** end of the series
- **

Apply Spencer's 15-point weighted M.A. on (13); the outcome is to eliminate the randomness, thus achieving a refined estimate of Trend-cycle (final result).

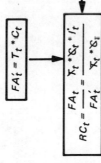

$$FA_t' = T_t * C_t$$

$$RC_t = \frac{FA_t}{FA_t'} = \frac{\bar{x}_t * \&_t * I_t}{\bar{x}_t * \&_t}$$

(15)
- **
- **
- **

RC_t is an estimate of the random component. It is found by dividing (13) by (14).

(16)
- Perform Different Test to make sure that the estimation of the Seasonal, Trend-cycle, and random components are correct.
- Compute Months for Cyclical Dominance (MCD).
- Calculate Summary Statistics.

Use original data and those of steps (13), (14), and (15) to obtain results.

(17)
- Compute a Moving Average whose length is equal to the months for cyclical dominance. Update this average to obtain estimates of trend-cycle.
- Use the seasonal factors projected in (12) as seasonal indices.

Use the seasonally adjusted data of step (13). Update this moving average to easily obtain current estimates of Trend-cycles (i.e., it is used instead of FA_t' in (14)).

references. (In the CENSUS program every effort has been made to put complete headings at the top of each output table so that they will be as intelligible as possible.)

1.   *Stable Factors – Seasonal Indices*

The stable factors are indices fluctuating around 100 and indicating the seasonality of each month involved. For example, a stable factor of 110 corresponding to the month of July will mean that July's figures are 10% above the average for the year. The stable factors are obtained after excluding extreme values and averaging the seasonal indices of all the same months (i.e., all Januaries, Februaries, etc.). Thus, they give a picture of the average seasonality of each month.

2.   *Final Seasonal Adjustment Factors*

The final seasonal adjustment factors are obtained if randomness is eliminated from seasonality. A times series, X_t, is composed of the following four components:

$$X_t = S_t T_t C_t R_t \qquad\qquad (16\text{-}1)$$

where

X_t is the time series:

S_t denotes seasonality:

T_t denotes trend;

C_t denotes cycle;

and

R_t denotes randomness.

An L month moving average of (16-1) will yield:

$$M_t = T_t C_t \qquad\qquad (16\text{-}2)$$

If (16-2) is divided into (16-1), one obtains:

$$\frac{X_t}{M_t} = \frac{S_t T_t C_t R_t}{T_t C_t} = S_t R_t \qquad\qquad (16\text{-}3)$$

Equation (16-3) isolates both seasonality and randomness. The randomness of (16-3) can be eliminated by applying a moving average (usually a 3 × 3 or a 5 × 5, see Chapter 4 for definition) to similar months — all Januaries, Februaries, etc. — so that only seasonality is left in (16-3). From the final seasonal adjustment factors, Census II calculates forecasts for seasonality one year ahead. As can be seen below, this is done by multiplying the last year's factors by 3, subtracting the preceding year's factor and dividing the result by 2. For example, the figure for April is 98.1; this is calculated by multiplying 98.3 × 3 – 98.5 and dividing their differences by 2 (see below).

Final Seasonal Adjustment Factors

91.8 87.4 101.4 98.5 98.2 111.1 123.8 120.3 105.5 91.9 79.9 90.2
91.8 87.1 100.9 98.3 98.1 111.6 124.6 120.3 105.6 91.7 79.7 90.2

One Year Ahead Forecasted Seasonal Factors

91.9 87.0 100.7 98.1 98.0 111.8 125.0 120.4 105.6 91.6 79.6 90.2
             ↑
          April

3. *Percentage Change from Previous Month*

There are several percentage change tests which are calculated by finding the percentage change of each value from that of the previous month. Specifically there are 4 percentage change tests: one for the original data, and one for each one of the three major components of a time series-seasonality, trend-cycle, and random or irregular.

(a) *Original data*   The percentage change of the original data is used as a guide for comparison with the rest of the percentage change tables. Its magnitude should be the highest of all the remaining percentage change tables. It indicates something about the fluctuation in the original data. The difference between the percentage change in the original data and the percentage change in the seasonally adjusted data indicates the success of the seasonal adjustment. The higher the difference, the better the seasonal adjustment. Finally, the difference between the percentage change in the original data and that of the random or irregular component represents the maximum possible improvement that forecasting can achieve. In other words, it gives an idea of the minimum error level one can ever expect to achieve.

(b) *Final Seasonally-Adjusted Series*   The seasonally-adjusted series does not include seasonal effects. It can, therefore, reveal the extent of the adjustment if compared to the percentage change of the original data.

(c) *Random components*   Like the previous percentage change tests, this one indicates the extent of the randomness in the data. The overall average percentage change is the minimum absolute percentage error when forecasting. By definition, one cannot reduce the error below the values of the percentage change of the random component.

(d) *Trend-cycle component*   This percentage change refers to the trend-cycle component and determines one of the most important measures of the Census II method – the Months for Cyclical Dominance (MCD). The MCD is calculated by finding the percentage changes of trend-cycle over the random component, when the time span goes from one, two, three, four, to n months. The MCD is chosen for the time span corresponding to when the ratio is below 1.

4. *Ratios of Preceding and Following Month, (Original Data and Seasonally Adjusted Series)*

Calculating the ratio of a month to that of the average of the preceding and following months gives an idea of how a particular month varies in relation to the preceding and following months. Where the data are not seasonal, such variations will be small. However, when strong seasonality exists, the variations would be considerable because there will exist a pattern between successive months. Averaging the ratios of all the years, for each one of the months separately, indicates the variability of the series. The ratios of the preceding and following months of original data usually vary considerably, a fact suggesting that the original series is seasonal. The same calculation, however, performed on the seasonally-adjusted data (because the seasonality has been removed) usually results in ratios that are much lower than those of the original data. As a matter of fact, the ratios of the seasonally adjusted data indicate how successful the seasonal adjustment has been in removing seasonality. Where the average ratios are below 95 or above 105, there will be a strong suggestion that the seasonal adjustment process is not adequate and that some part of the seasonal variation has not been removed through the various steps followed by Census II.

5. *Ratios of Preceding January* (Final Adjusted Series)

Dividing each value of the final seasonally-adjusted series by its corresponding value of each preceding January provides a set of standardized values with January as the base. By examining (plotting) these standardized ratios, one can determine whether or not there is any constant pattern of longer than one month duration. Where this is so, it will appear that the seasonality has not been removed properly from the data. The January test, therefore, will reveal if there is some intra-year seasonality left, while the adjacent month test (preceding and following month) reveals if there is seasonality between years. Both the adjacent month and January tests must be used in conjunction with each other to make sure that the removal of seasonality has been successful.

6. *Ratios* (12-Month Moving Averages of Seasonally-Adjusted Series/ 12-Month Moving Average of Original Data)

It should be understood that one not only worries about adjusting properly but also about the possibility of over-adjusting. A test to help determine if this has happened can be obtained by dividing a twelve-month moving average of the original data into a twelve-month moving average of the seasonally-adjusted data. The twelve-month moving average of the original data eliminates the seasonality without altering the volume of the data. However, the twelve-month moving average of the final seasonally-adjusted data eliminates seasonality but at the same time includes other adjustments, such as elimination of randomness, replacement of extremes, etc. The end result is that it may include some of the random component as part of the seasonality. Thus, by finding the ratio between the two averages one can determine whether or not any over-adjustment has taken place. If the ratios are close to 100, then there is no over-adjustment. However, if there are ratios below 90 or above 110, one must seriously question the success of the seasonal adjustment because it has been overzealous in eliminating fluctuations in the data.

7. *Spencer's 15-Point Weighted Average* (Seasonally-Adjusted Series)

The Spencer's 15-point weighted moving average aims at eliminating the randomness from the data. If it is done on the seasonally-adjusted series that already excludes seasonality, the result will be the elimination of randomness in such a way that the trend-cycle will become evident. Because it included fifteen terms in the averaging process, Spencer's average is powerful and can eliminate up to cubic trends. Its disadvantage is that there are seven terms lost at the end of the series that are needed to predict the trend-cycle. Even though these values have been calculated in the Census II output, what is shown can be very misleading and should be used with utmost caution. In summary, the Spencer's average is used to highlight the trend-cycle in an ex-post fashion. Item 9 below (MCD Moving Average of Adjusted Series) is used for the purpose of predicting the trend cycle.

8. *Months for Cyclical Dominance* (MCD)

The Months for Cyclical Dominance, or MCD, indicate the time span it takes for the trend-cycle component to become more important than (dominate) randomness. The longer the MCD, the less important is the cyclical component over the random component, and vice-versa. The MCD are found by calculating the average percentage change for one, two, three, four, five, etc., months duration for each of the random and trend-cycle components. Once this is done the ratio between the two is calculated and the time span for which the ratio is less than one indicates the MCD.

9. *MCD Moving Average of Adjusted Series* (Centered)

Once the MCD is known, one would like to calculate a measure that would enable prediction of the trend-cycle component. If one finds, for example, that the number of months for cyclical dominance is three, one could calculate a 3-months MA of the final seasonally-adjusted data. These data are free of seasonality, they include only trend-cycle and randomness, and their randomness dominates the data for a period of less than three months. Calculation of a 3-months MA would eliminate most of the randomness involved. In addition there would be the advantage that only one value would be missing at the end of the series through the averaging. This makes this type of moving average more desirable than the corresponding type found using Spencer's method, in which 7 values are missing in the end of the series. Where the MCD is five there will be two values missing at the end. For all practical purposes the MCD is rarely more than five.

10. *Summary Statistics*

The final output of the Census II method is a number of summary statistics that relate to three aspects of the data:

  a. the average percentage change of all components and their ratios;
  b. the average duration of positive and negative signs (that is, the average time that it takes until a change in sign takes place), for all components of the series;
  c. the average duration of the percentage change of the random over the trend-cycle ratios for time spans of one, two, three, four and five months. This reveals the month for critical dominance.

EXERCISE 16.1

Use the data of Chapter 14 to estimate the seasonal indices for each month of 1971. In addition, calculate the one-year-ahead forecast of the seasonal indices. Assume that the data will decrease in the linear trend shown between the years 1964 to 1970 (i.e., a = 32.5666 and b = −.03476 — see Figure 15-2) then make forecasts for 1971 using:

  a. Harrison Harmonic Smoothing — see seasonal indices in Chapter 14;
  b. Classical Decomposition — see exercise of Chapter 15;
  c. Census II using the seasonal indices; and
  d. Census II using the one-year-ahead forecast of seasonal factors.

Solution

Using a = 32.5666 and b = −.03476 the trend for the periods 85 to 96 (7th year or 1971) is:

$$T_{85} = 29.612 \qquad T_{89} = 29.473 \qquad T_{93} = 29.334$$

$$T_{86} = 29.577 \qquad T_{90} = 29.438 \qquad T_{94} = 29.299$$

$$T_{87} = 29.542 \qquad T_{91} = 29.403 \qquad T_{95} = 29.264$$

$$T_{88} = 29.508 \qquad T_{92} = 29.369 \qquad T_{96} = 29.230$$

(for example, $T_{88} = 32.5666 - .03476(88) = 29.508$)

Since the seasonal indices are already known from Harrison's method and Classical Decomposition, one could run the Census program to find the seasonal indices and the one-year-ahead forecast of the seasonal factors. These are shown in Figure 16-2. Their calculations are similar to those of Classical Decomposition, except that the process of decomposition is more elaborate and is done by isolating one component of the time series at a time. Figure 16-1 shows the steps followed to obtain the seasonal indices and their forecast for one year ahead.

Figure 16-3 below shows the seasonal indices calculated by each of the methods. As can be seen, they are similar even though the computational procedure used by each of the methods is different. The last column of Figure 16-3 shows the one year forecast of the seasonal factors. These are different than the seasonal indices mainly because they contain the most recent information available on seasonality. (They use the seasonal factors of the last two years.)

Figure 16-3 shows the seasonal indices needed to adjust the trend values calculated above. In addition, the last column is the predictions using the one-year-ahead forecast of seasonal factors. The actual numbers of road casualties for 1971 are: 28.4, 25.5, 26.6. 26.2, 29.3. 28.8, 31.2, 31.9, 28.5, 32.2, 31.7 and 31.8 for each of the 12 months of 1971. One can estimate the errors by each method by subtracting the forecasts shown in Figure 16-4 from the actual values just mentioned.

Figure 16-5 shows the percentage error by period and for each of the methods. The best forecast, in this specific set of data, is that based on the one-year-ahead forecast of the seasonal factors obtained by Census II.

The average ratios of the preceding and following months of the seasonally-adjusted series are very close to 100 (see Figure 16-5). This means

FIGURE 16-2 Centered Ratios and Final Seasonal Adjustment Factors

```
   CENTERED RATIOS (ORIG./15-MONTHS MOV. AVER.)
STABLE FACTORS -SEASONAL INDICES-
  86.5   81.8   92.0   93.6 104.6 102.3 109.4 108.6 104.0 106.2 104.3 106.9

   FINAL SEASONAL ADJUSTMENT FACTORS

ONE YEAR AHEAD FORECASTED SEASONAL FACTORS
  89.1   84.8   91.0   91.1 102.5 100.4 108.6 107.8 103.5 107.1 108.0 106.1
```

FIGURE 16-3   Seasonal Indices

Year 1971	Classical Decomposition	Harrison Harmonic Smoothing	CENSUS II	
			Seasonal Indices	One-Year-Ahead Forecast of Seasonal Factors
J	86.305	96.652	86.5	89.1
F	82.336	82.21	81.8	84.8
M	93.07	91.555	92	91
A	92.955	93.486	93.6	91.1
M	104.035	103.598	104.6	102.5
J	101.946	101.869	102.3	100.4
J	109.284	109.299	109.4	108.6
A	109.765	108.861	108.6	107.8
S	104.751	104.049	104	103.5
O	106.048	106.49	106.2	107.1
N	103.194	104.452	104.3	108
D	105.289	107.083	106.9	106.1

FIGURE 16-4   Forecasts

Year 1971	Classical Decomposition	Harrison Harmonic Smoothing	CENSUS II	
			Seasonal Indices	One-Year-Ahead Forecasts Seasonal Factors
J	25.5566	25.6593	25.6143	26.3842
F	24.3526	24.3154	24.1941	25.0814
M	27.4951	27.1657	27.179	26.8836
A	27.4288	27.5855	27.6191	26.8815
M	30.6621	30.5333	30.8286	30.2097
J	30.011	29.9883	30.1152	29.5559
J	32.1332	32.1376	32.1673	31.932
A	32.2364	31.9709	31.8943	31.6593
S	30.7275	30.5216	30.5072	30.3605
O	31.0711	31.2006	31.1156	31.3793
N	30.199	30.5672	30.5227	31.6055
D	30.7755	31.2999	31.2464	31.0126

FIGURE 16-5   Percentage Errors

Year 1971	Classical Decomposition	CENSUS II		
		Harrison Harmonic Smoothing	Seasonal Indices	One-Year-Ahead Forecast of Seasonal Factors
J	10.01	9.65	9.8	7.09
F	4.49	4.64	5.12	1.64
M	−3.37	−2.12	−2.18	−1.07
A	−4.7	−5.29	−5.42	−2.61
M	−4.65	−4.21	−5.22	−3.11
J	−4.21	−4.13	−4.57	−2.63
J	−3	−3.01	−3.11	−2.35
A	−1.06	−0.23	.01	.75
S	−7.82	−7.1	−7.05	−6.53
O	3.5	3.1	3.36	2.54
N	4.73	3.57	3.71	.29
D	3.22	1.57	1.74	2.47
Mean absolute Percentage Error	4.56	4.05	4.27	2.76

that the seasonal adjustment has been successful in removing seasonality. On the other hand, the same ratios for the original data (see Figure 16-6) indicate wide variation around 100, meaning that the original road casualty data are indeed seasonal.

The ratios to the preceding January (see Figure 16-7) do not show any consistent pattern among the months of the year. (There are no months whose ratios are constantly above or below 100). This means that there is no intra-year seasonality remaining in the data.

The ratios of the 12-month moving averages of the seasonally-adjusted series over the original series (see Figure 16-8) aims at revealing if any over-adjustment has taken place. This is not the case because most values

FIGURE 16-6   Census II Ratios for Preceding and Following Months

```
RATIOS OF PRECEDING AND FOLLOWING MONTH (FINAL SEAS. ADJ. SERIES)
AVERAGES
100.9 100.1  99.9  99.8 100.3  99.8 100.1 100.1  99.9 100.1 100.2  99.2

RATIOS OF PRECEDING AND FOLLOWING MONTH (ORIGINAL DATA)
AVERAGES
 92.4  91.8 104.8  95.0 107.1  95.4 103.9 101.8  96.9 102.0  98.1 111.7
```

FIGURE 16-7   Census II Ratios for Preceding January

```
RATIOS TO PRECEDING JANUARY (FINAL SEAS. ADJ. SERIES)
100.0 101.8  99.1  99.4 107.2 104.4 107.3 112.3 110.3 110.2 107.6 100.3
100.0  91.3  99.4  99.5  93.9  99.0  97.1  98.2  97.3  94.3  97.7  99.3
100.0 103.0 101.5 108.4 104.9 105.1 102.4  97.4  96.5 104.7 102.8 110.3
100.0 103.4 101.9  95.3 101.5  94.0  98.7  96.9 100.1  91.2  87.0  86.5
100.0 107.8 108.7 105.1 100.4 108.4 104.3 107.9 108.5 107.3 106.4 104.5
100.0  88.8  87.3  91.7  95.9  92.7  93.0  94.6  92.0  95.1 100.5  95.1
100.0 103.8  99.9  98.0  95.7  95.4  99.0  96.7  98.9 102.4 103.0 105.8
```

FIGURE 16-8   Census II 12-Month Moving Average Ratios

```
12-MONTH MOV. AVER. SEAS. ADJ./12-MONTH MOV. AVER. ORIG.
109.9 107.6 105.5 104.0 102.5 101.3  99.5  99.2 100.2 100.5 101.4 101.3
101.8 101.8 101.3 101.5 101.1 101.6  99.6  99.2 100.1 100.4 101.3 101.2
101.8 101.7 101.4 101.5 101.2 101.6  99.7  99.3 100.2 100.4 101.3 101.2
101.9 101.8 101.4 101.6 101.3 101.8 100.0  99.4 100.0 100.2 101.2 101.1
101.9 101.9 101.6 101.7 101.5 101.6 100.0  99.7 100.2 100.3 101.3 101.1
101.8 101.8 101.5 101.7 101.6 101.5  99.9  99.6 100.3 100.4 101.4 101.2
101.9 101.8 101.5 101.8 101.8 101.6 100.1  98.3  97.1  95.9  95.7  95.3
```

(excluding the first and last seven which are distorted since they are replaced artifically) are very close to 100 (which means no over-adjustment at all).

The summary statistics shown in Figure 16-9 indicate that on the average the data are well-behaved, with no wide fluctuations among months. The average percentage change in the original time series of road casualties is 7.04%. From this 7.04% about half is due to randomness, and the other half to seasonality with average percentage changes of 3.53 and 3.73, respectively. This is consistent with the average absolute percentage error of the forecasts found in Figure 16-5 of about 4%. Actual forecast errors are always above the average percentage change of the random component and should be lower than the average percentage change in the original data. As a matter of fact, the difference between the average percentage change in the original data and that achieved by each method shows the extent of improvement achieved by formal forecasting. In comparison to Figure 16-5 the improvements are 2.49, 3.00, 2.78 and 4.29, respectively.

Finally, Figure 16-9 indicates that the influence of the trend-cycle is small. This can be seen from its average percentage change, the ratios of random over trend-cycle, seasonal over trend-cycle, and trend-cycle over original. All show that the trend-cycle is dominated by the other components. Also, the trend-cycle will take more than five months to dominate the random component; this implies again that its influence is weak.

FIGURE 16-9   Census II Ratios Involving Trend Cycle

```
                **** SUMMARY STATISTICS ****
                AVERAGE PC. CHANGE
  ORIGINAL        TREND-CYCLE      5-M.M.A.      RANDOM   SEAS ADJ SERIES
  7.04538           .74327          .91782       3.52784      3.73170


   RANDOM        RANDOM      SEASON        RANDOM   TREND-CYCLE    SEASON
  ----------    --------   -----------    --------  -----------   --------
  TREND-CYCLE   SEASON     TREND-CYCLE    ORIGINAL   ORIGINAL     ORIGINAL
    4.75         .95          5.02          .50         .11         .53


  AVERAGE DURATION OF POSITIVE AND NEGATIVE SIGNS
  ORIGINAL         RANDOM      TREND-CYCLE     3-MONTHS M.A.
   1.80435        1.43103        6.91667         1.59615

                  RANDOM/TREND-CYCLE
                    SPAN IN MONTHS
      1               2             3              4              5
  4.746356        2.355974      1.627046       1.420536       1.067391
```

CHAPTER 17

Generalized Adaptive Filtering

A basic principle in time series forecasting techniques is that future values of the series are a function of (determined by or dependent upon) past values. Though not always obvious, this principle has been implicit in all forms of forecasting discussed so far. In Chapter 5, equation (5-6) showed how exponential smoothing determined a forecast by weighting past values in a decreasing fashion as they became further removed from the present. The logic, of course, was that most recent observations should be weighted more heavily in determining future predictions. The single moving average in Chapter 11 (see equation (11-1)) also weighted past values, but weighted them equally. In both cases a combination of past values was used in forecasting, but the user had little control over the exact weights. In (5-6) for example, the weights were always decreasing exponentially, while in (11-1) they were equal for the last n observations, where n was the number of terms in the moving average.

The methods examined in this and the next chapter will explicitly attempt to express forecasts as functions of past values and at the same time assume no fixed weighting scheme. Rather, they are aimed at discovering an optimal set of weights.

The method of Generalized Adaptive Filtering and the Box-Jenkins methodology are similar, except that they use different procedures in selecting the most appropriate model and optimizing its parameters. In the opinion of the authors, Generalized Adaptive Filtering is easier to use, requires less knowledge, and needs less computer time than similar models used in the Box-Jenkins methodology. (See Chapter 18.) The major difference between the two methods is that the weights (parameters) of Generalized Adaptive Filtering change with every observation while those of the Box-Jenkins models are fixed. This may be an advantage with some data series and a disadvantage with others.

AUTOREGRESSIVE (AR) AND MOVING AVERAGE (MA) PROCESSES

Wold (1938) showed that any series could be expressed as (17-1), (17-2) or (17-3), a combination of the first two, in such a way that whatever remains as e_t, on the right-hand side of the equation, will be randomness.

$$X_t = \phi_1 X_{t-1} + \phi_2 X_{t-2} + \phi_3 X_{t-3} + \ldots + \phi_p X_{t-p} + e_t \qquad (17\text{-}1)$$

$$X_t = e_t - \theta_1 e_{t-1} - \theta_2 e_{t-2} - \theta_3 e_{t-3} - \ldots - \theta_q e_{t-q} \qquad (17\text{-}2)$$

$$X_t = \phi_1 X_{t-1} + \phi_2 X_{t-2} + \ldots + \phi_p X_{t-p} + e_t - \theta_1 e_{t-1}$$

$$- \theta_2 e_{t-2} - \ldots - \theta_q e_{t-q} \qquad (17\text{-}3)$$

More than twenty years elapsed before Wold's theorem was put into practice, largely because equations (17-1), (17-2) and (17-3) provided an infinite number of models, making it difficult to make an appropriate selection. In addition, once a model was chosen, the computational efforts involved in finding the best values for the parameters were not insignificant, especially for those of the form (17-2) or (17-3). Advances in mathematics and the advent of computers made the choice of model and the necessary computations manageable.

Before proceeding further, it is useful to see what equations (17-1), (17-2) and (17-3) mean and how the concepts they express can be used in forecasting.

Equation (17-1) indicates that future values are linear combinations of past values. Equation (17-1) can be thought of as similar to equation (5-6) with the following substitutions:

$$\phi_1 = \alpha$$

$$\phi_2 = \alpha(1 - \alpha)$$

$$\phi_3 = \alpha(1 - \alpha)^2$$

$$\phi_4 = \alpha(1 - \alpha)^3$$

etc.

Equation (17-1) is called an autoregressive scheme, since it is similar to a regression equation except that the variables on the right-hand side are previous values of the variable on the left-hand side (hence the name auto). Furthermore p can take different values.

If p = 1, equation (17-1) becomes:

$$X_t = \phi_1 X_{t-1} + e_t \tag{17-4}$$

If p = 2, equation (17-1) becomes:

$$X_t = \phi_1 X_{t-1} + \phi_2 X_{t-2} + e_t \tag{17-5}$$

and if p = 3, equation (17-1) becomes:

$$X_t = \phi_1 X_{t-1} + \phi_2 X_{t-2} + \phi_3 X_{t-3} + e_t \tag{17-6}$$

etc.

Equation (17-4) is called an *autoregressive* (AR) process of degree 1, or AR (1).

Equation (17-5) is called an *autoregressive* (AR) process of degree 2, or AR(2).

Equation (17-6) is an AR(3), etc.

As can be seen, all AR models express future values as a combination of past ones. Depending upon the degree of the AR process, the number of past values included will vary. This is flexible, allowing the most appropriate combination of past values to be selected.

Equation (17-2) is called a moving average (MA) model. The parameter q (as with p in AR models) can take on different values. If q = 1, equation (17-2) becomes:

$$X_t = e_t - \theta_1 e_{t-1} \tag{17-7}$$

The term e_t is the forecasting error, i.e., the difference between actual and forecasted values. When q = 2, equation (17-2) becomes:

$$X_t = e_t - \theta_1 e_{t-1} - \theta_2 e_{t-2} \tag{17-8}$$

When q = 3, equation (17-2) becomes:

$$X_t = e_t - \theta_1 e_{t-1} - \theta_2 e_{t-2} - \theta_3 e_{t-3} \qquad (17\text{-}9)$$

Equation (17-7) is called a *moving average* process of first degree, or MA(1). Equation (17-8) is called a *moving average* process of second degree, or MA(2). Equation (17-9) is an MA(3), etc.

Moving average models are different than autoregressive ones, as they assume that future values of X_t are linear combinations of past values of the forecasting errors.

Finally, equation (17-3) is called a *mixed autoregressive - moving average* scheme, or ARMA. It is a combination of AR and MA equations. Depending upon the values of p and q, an ARMA model can be of p and q degree. Thus:

$$X_t = \phi_1 X_{t-1} + e_t - \theta_1 e_{t-1} \qquad (17\text{-}10)$$

is an ARMA (1, 1), and

$$X_t = \phi_1 X_{t-1} + \phi_2 X_{t-2} + e_t - \theta_1 e_{t-1} - \theta_2 e_{t-2} \qquad (17\text{-}11)$$

is an ARMA (2, 2).

The advantage of an ARMA scheme is that it includes different AR models and uses whatever error remains in an MA equation in attempting to further improve forecasting. This can be done until the errors have been reduced to randomness (white noise), at which point no further improvements in the model fitted are possible since whatever is left is randomness (which by definition cannot be predicted).

It should be recognized that AR(p), MA(q) and their combination ARMA(p,q) are extremely general and can be highly useful because of their flexibility. With the exception of decomposition and harmonic smoothing methods, they are the general case of all other forecasting methods discussed so far. The reason that they are general is that the user has the choice of the type of process — AR, MA or ARMA — and can select the degree of that process — p = 1,2,3, . . . , q = 1,2,3, ... With the correct choice, an appropriate model can be fitted to almost any set of data. That is exactly what Wold showed almost forty years ago: any time series can be expressed by some ARMA scheme. Furthermore, the approach is powerful because the parameters ϕ_i and θ_i of an ARMA model can vary depending upon the data. They are not fixed as in the smoothing models, nor is the estimation restricted to linear relationships as in the regression method.

The problem of the ARMA model is in another direction — the user has difficulty selecting correctly from the wide range of models available. However, the Box-Jenkins methodology has contributed significantly in this regard by providing guidelines that help make ARMA models relevant and applicable in real life situations. This is achieved through a few steps that guide the user in correctly choosing the right ARMA model (see Figure 18.1).

As an example, suppose one has the simple randomless series, 5, 10, 5, 10, 5, 10, ... One could express it as:

$$X_t = \phi_1 X_{t-1} + \phi_2 X_{t-2} + e_t$$

If the values of ϕ_1 and ϕ_2 are estimated, they are found to be $\phi_1 = 0, \phi_2 = 1$. Estimating period 3, gives:

$$\hat{X}_3 = 0X_2 + 1X_1$$

$$= 0(10) + 1(5) = 5$$

and period 4 becomes:

$$\hat{X}_4 = 0X_3 + 1X_2$$

$$= 0(5) + 1(10) = 10$$

Distinguishing the estimated values of the 3rd and 4th period as $\hat{X}_3$ and $\hat{X}_4$, their differences from the actual values, X_3 and X_4, are:

$$e_3 = X_3 - \hat{X}_t = 5 - 5 = 0$$

$$e_4 = X_4 - \hat{X}_t = 10 - 10 = 0$$

Thus, one can estimate with zero error ($e_t = 0$) as many periods as desired. This same principle applies to series with randomness in which case the weights will be determined in such a way that the sum of the squared errors (e_t) are as small as possible.

Given the above background information, the method of Generalized Adaptive Filtering can be examined, and in the next chapter a similar and more complete analysis of the Box-Jenkins methodology for ARMA models can be made.

GENERALIZED ADAPTIVE FILTERING (GAF)

Both this method and the one in the next chapter use the concept of autocorrelation to a great extent. The reader is, therefore, encouraged to review Chapter 3 before continuing.

1. Achieving Stationarity

It might be desirable to transform the data into a stationary form before proceeding with fitting an ARMA model, even though this is not necessary. Stationarity exists when the data are horizontal, or fluctuate around a constant mean. As explained in Chapter 3, this can be achieved through differencing the time series. The user can use the graph of the autocorrelation coefficients to determine stationarity. If the autocorrelations drop quickly to zero (i.e., after the third or fourth value they are not significantly different than zero), or if they fluctuate around zero, the data are stationary. If this is not the case, the data must be differenced.

In the GAF program, the first difference can be taken by using one weight, while a second difference requires two weights. If the user does not know whether one or two weights is required, it is better to start with one, examine the new graph of the autocorrelations, and then decide if the series must be differenced again. (Using one weight twice is equivalent to using two weights initially.) An example of how a series can be made stationary is shown in the exercise at the end of this chapter.

2. Choosing the Appropriate Degree of the AR Process

Choosing the appropriate degree of the AR process requires the examination of the autocorrelation coefficients and another similar statistic (to be discussed in the next chapter), the partial autocorrelations. The exact process used in selection will be examined in detail in the next chapter. At this point an empirical rule of thumb that can serve as a guideline will be given to make the use of Adaptive Filtering easier.

As a practical rule, one can choose the degree of the AR process – the number of weights – to equal the time lag corresponding to the largest positive autocorrelations coefficient after a time lag of 3. This is equivalent to setting p equal to the length of seasonality. If the data are not seasonal, p can be set equal to 1, 2, or 3. Setting p equal to, say, 12 if monthly seasonality is present may result in specifying too many parameters (a non-parsimonious model) but this is a disadvantage to be compared with the advantage of considerably less complexity in using the method. Similarly, for non-seasonal data, p can be set equal to 3, thus providing a simple yet general approach for dealing with seasonal and non-seasonal data. Again, doing so will mean that too many parameters may be used. When this is the case, their values will be close to zero. For forecasting purposes, therefore, there will be little harm.

3.   Choosing the Appropriate Degree of MA Process

Once the AR model has been fitted to the data, one can look at the autocorrelation of the residuals. The residuals e_t, are the difference between the actual values and those predicted by the AR model. As a simple rule of thumb, if they are random noise, their autocorrelation will not be significantly different than zero, nor will it have any pattern. If this is the case, no further steps are needed. One can use an MA(O) process. If this is not the case, some moving average model must be fitted.

The choice of the degree of the MA model is similar to the AR choice. The degree should be equal to the time lag of the largest autocorrelation coefficient. Usually, following this rule will result in an adequate model and in random residuals.

References for further study:

Kalman, R. (1960)
Makridakis, S. and S. Wheelwright (1978, Chapter 9)
Wheelwright, S. and S. Makridakis (1977, Chapter 4)
Widrow, B. (1966)
Wilde, D. and C. Beightler (1967)

INPUT/OUTPUT FOR THE GENERALIZED ADAPTIVE FILTERING PROGRAM (GAF)

Input

1.   *How Many Autocorrelations Do You Want to See?*

The program produces a graph of the autocorrelations. Any number less than sixteen is appropriate as long as there are enough data points. If there are fewer than thirty observations, a number should be chosen which is slightly less than half the total number of data points available.

2. *How Many Weights (Parameters) Do You Want?*

If the data are stationary (see (1) above to find out), then use as many weights as the time lag corresponding to the largest autocorrelation coefficient. If the data are not stationary, use one weight.

3. *Are Your Data Stationary?*

By looking at (1) above, answer YES or NO.

4. *Do You Want to Specify Initial Values for Your Weights?*

The answer should be NO, if the data are being run for the first time. Otherwise, one can answer YES and input the values of the optimal weights from the last run. This will reduce the program's running time considerably.

5. *Enter A Value for the Learning Constant*

The value for the learning constant should be somewhat smaller than $1/m$, where m is the number of weights (this is done when the user enters -1). The larger the value of the learning constant, the more quickly the optimal weight values will be found, the resulting MSE may be larger, too. On the other hand, the smaller the learning constant the greater the computer time required to run and the lower the MSE will be.

6. *Do You Want to Continue the Training?*

The objective is to continue until the autocorrelation of the residuals has been reduced to white noise. If the user does not know, YES is the appropriate response and the program will cycle to (1),(2),(3),(4),(5), and (6). (With (1) it can be discovered whether the residuals have any pattern.)

Output

1. *Graph of Autocorrelations*

Its usage was explained above.

2. *Learning Performance*

The program provides information on how the Mean Squared Error (MSE), the Mean Absolute Percentage Error (MAPE), and Mean Percentage Error (MPE) and the error reduction in the MSE behave at each iteration.

3.   *Optimal Weights*

These are the final weights which correspond to the minimum (optimal) MSE. There is one weight for every degree of AR or MA process decided upon.

EXERCISE 17.1

In Chapter 2 a series of twenty numbers was given and the question was to forecast the next eight values. The numbers in the series were: 19, 24, 32, 63, 99, 120, 144, 191, 243, 280, 320, 383, 451, 504, 560, 639, 723, 792, 864, 959. Forecast the next eight numbers using the method of Generalized Adaptive Filtering. (Note: this data series does not include any randomness.)

Solution

Assuming that one knows nothing about the series, the GAF program can be run as follows:

1.   *Achieving Stationarity*

Figure 17-1 shows the autocorrelations for the series. They exhibit a definite trend going from right to left. This trend must be eliminated before continuing. This is done using one weight. The new series is 5 (24-19), 8 (32-24), 31 (63-32), . . ., 72, 95. This series is shown in Figure 17-2 under the column of ERROR. The program then treats whatever is left (i.e., the ERROR column) as the new series to forecast.

Figure 17-3 shows the autocorrelations of the differenced series (5,85,8, 31, . . . , 72, 95). It can be seen that there is still a trend in the autocorrelations but it is not as strong as that of Figure 17-1. In addition, one can see some seasonal pattern emerging (the autocorrelation of three time lags is larger than one of two or four, but it is not very clear. Since there is still trend it must be removed using one weight again. This is done by taking the first difference of the already differenced series. The resulting series is: 3 (i.e., 8-5), 23 (i.e., 31-8), 5 (i.e., 36-31), . . . 3,23. The outcome is equivalent to having taken a second difference to begin with. These numbers can be seen in Figure 17-4 under ERROR which is the new series to be forecast. (It can also be seen that this series contains no randomness.)

FIGURE 17-1   Autocorrelations for the Data

```
EXE METHOD
 8K
WHICH PROGRAM WOULD YOU LIKE TO RUN?GAF


***** GENERALIZED ADAPTIVE FILTERING *****

NEED HELP (Y OR N)?N
TO RUN THIS PROGRAM YOU MUST HAVE YOUR DATA IN A FILE. IF YOU
DO NOT HAVE THEM ALREADY IN A FILE, STOP THE PPROGRAM AND
USE FINPUT.
HOW MANY OBSERVATIONS DO YOU WANT TO USE?20
DATA FILENAME?CH17
HOW MANY AUTOCORRELATIONS DO YOU WANT TO SEE (0 IF NONE)?9
```

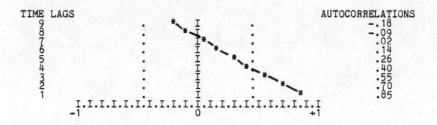

```
TIME LAGS                                              AUTOCORRELATIONS
     9              .    *  I       .        .              -.18
     8              .       I*      .        .              -.09
     7              .       I*-     .        .               .02
     6              .       I *     .        .               .14
     5              .       I  *    .        .               .26
     4              .       I    *- .        .               .40
     3              .       I      *-*       .               .55
     2              .       I         *-*    .               .70
     1              .       I            *-* .               .85
        I.I.I.I.I.I.I.I.I.I.I.I.I.I.I.I.I.I.I.I.I
       -1                    0                  +1
```

FIGURE 17-2   Data with Trend Eliminated Using 1 Weight

```
HOW MANY WEIGHTS (PARAMETERS) DO YOU WANT (16 = MAX)?1
IS YOUR DATA STATIONARY ?N

********    LEARNING PERFORMANCE   ********

LEARNING CONSTANT =      0
ITERATION          MSE          MPE          MAPE    % ERROR REDUCTION
   1            3055.26        17.91%       17.91%          0%


OPTIMAL WEIGHTS

   1      1.000
DO YOU WANT A TABLE OF ACTUAL AND PREDICTED (Y OR N)?Y

PERIOD        ACTUAL        FORECAST        ERROR      PCT ERROR
   2          24.00          19.00          5.00        20.83%
   3          32.00          24.00          8.00        25.00%
   4          63.00          32.00         31.00        49.21%
   5          99.00          63.00         36.00        36.36%
   6         120.00          99.00         21.00        17.50%
   7         144.00         120.00         24.00        16.67%
   8         191.00         144.00         47.00        24.61%
   9         243.00         191.00         52.00        21.40%
  10         280.00         243.00         37.00        13.21%
  11         320.00         280.00         40.00        12.50%
  12         383.00         320.00         63.00        16.45%
  13         451.00         383.00         68.00        15.08%
  14         504.00         451.00         53.00        10.52%
  15         560.00         504.00         56.00        10.00%
  16         639.00         560.00         79.00        12.36%
  17         723.00         639.00         84.00        11.62%
  18         792.00         723.00         69.00         8.71%
  19         864.00         792.00         72.00         8.33%
  20         959.00         864.00         95.00         9.91%
MEAN PC ERROR (MPE) OR BIAS        =        17.91%
MEAN SQUARED ERROR (MSE)           =      3055.3
MEAN ABSOLUTE PC ERROR  (MAPE)     =        17.9%
```

FIGURE 17-3   Autocorrelations of the Differenced Series

```
FORECAST FOR HOW MANY PERIODS AHEAD (0=NONE,24=MAX)?0


DO YOU WANT TO CONTINUE THE TRAINING (Y OR N)?Y
HOW MANY AUTOCORRELATIONS DO YOU WANT TO SEE (0 IF NONE)?9
```

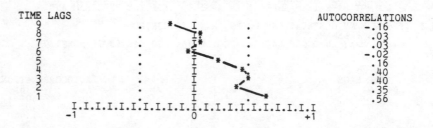

```
TIME LAGS                                           AUTOCORRELATIONS
     9                 .     *   I      .                  -.16
     8                 .         I*.    .                   .03
     7                 .         *.     .                   .03
     6                 .       *I.      .                  -.02
     5                 .         I  *                       .16
     4                 .         I      *.                  .40
     3                 .         I        *                 .40
     2                 .         I      *                   .35
     1                 .         I            *             .56
       I.I.I.I.I.I.I.I.I.I.I.I.I.I.I.I.I.I.I.I.I
      -1               0                        +1
```

FIGURE 17-4   Second Differences of the Data

```
HOW MANY WEIGHTS (PARAMETERS) DO YOU WANT (16 = MAX)?1
IS YOUR DATA STATIONARY ?N

********   LEARNING PERFORMANCE   ********

LEARNING CONSTANT =        0
ITERATION        MSE          MPE         MAPE     % ERROR REDUCTION
1              205.00        3.33%        5.86%            0%


OPTIMAL WEIGHTS

  1      1.000-
DO YOU WANT A TABLE OF ACTUAL AND PREDICTED (Y OR N)?Y

PERIOD        ACTUAL         FORECAST        ERROR       PCT ERROR
   3          32.00           29.00           3.00          9.38%
   4          63.00           40.00          23.00         36.51%
   5          99.00           94.00           5.00          5.05%
   6         120.00          135.00         -15.00         12.50%
   7         144.00          141.00           3.00          2.08%
   8         191.00          168.00          23.00         12.04%
   9         243.00          238.00           5.00          2.06%
  10         280.00          295.00         -15.00          5.36%
  11         320.00          317.00           3.00           .94%
  12         383.00          360.00          23.00          6.01%
  13         451.00          446.00           5.00          1.11%
  14         504.00          519.00         -15.00          2.98%
  15         560.00          557.00           3.00           .54%
  16         639.00          616.00          23.00          3.60%
  17         723.00          718.00           5.00           .69%
  18         792.00          807.00         -15.00          1.89%
  19         864.00          861.00           3.00           .35%
  20         959.00          936.00          23.00          2.40%
MEAN PC ERROR (MPE) OR BIAS      =         3.33%
MEAN SQUARED ERROR (MSE)         =         205.0
MEAN ABSOLUTE PC ERROR  (MAPE) =         5.9%
```

2. *Determining the Number of Weights (Degree of AR Process)*

The autocorrelations of Figure 17-5 are distributed around zero which means that the data are stationary at the second level of difference. The highest autocorrelation is at a lag of two, but at a time lag of four it is almost as large at .767. Thus, one might choose four weights. (One might alternatively have chosen six or eight weights although the fact that the errors repeat themselves every four periods suggests the use of four weights.)

The same results obtained in Figures 17-1 through 17-5 can be obtained using two weights initially (see Figure 17-8) if one knows that the nonstationarity is of second degree. A second application of GAF with four weights (see Figure 17-9) then gives the same results obtained in Figures 17-6 and 17-7.

3. *Calculating Optimal Weights*

The autoregressive equation to be fitted into the data is:

$$\hat{X}_t = \phi_1 X_{t-1} + \phi_2 X_{t-2} + \phi_3 X_{t-3} + \phi_4 X_{t-4} \qquad (17\text{-}12)$$

To find values for ϕ_1^*, ϕ_2^*, ϕ_3^*, and ϕ_4^* in such a way that the MSE will be as small as possible, one starts with some values of ϕ_1, ϕ_2, ϕ_3, and ϕ_4 and then modifies them by applying:

$$\phi'_{it} = \phi_{it} + 2Ke_t X_{t-i} \qquad i - m + 1, m + 2, \dots, n \qquad (17\text{-}13)$$

where

ϕ'_i is the new, "better," weight

FIGURE 17-5   Autocorrelations of Second Differences

FORECAST FOR HOW MANY PERIODS AHEAD (0=NONE,24=MAX)?0

DO YOU WANT TO CONTINUE THE TRAINING (Y OR N)?Y
HOW MANY AUTOCORRELATIONS DO YOU WANT TO SEE (0 IF NONE)?9

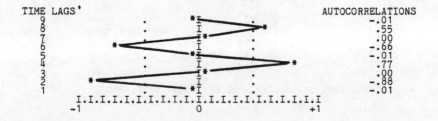

TIME LAGS	AUTOCORRELATIONS
9	-.01
8	.55
7	.00
6	-.66
5	-.01
4	.77
3	.00
2	-.88
1	-.01

FIGURE 17-6   Learning Performance

```
HOW MANY WEIGHTS (PARAMETERS) DO YOU WANT (16 = MAX)?4

DO YOU WANT TO SPECIFY INITIAL VALUES FOR YOUR WEIGHTS (Y OR N)?N

ENTER A VALUE FOR THE LEARNING CONSTANT (IF -1 IT WILL BE SET BY THE PROGRAM)
?-1

********   LEARNING PERFORMANCE   ********

LEARNING CONSTANT =   .142
ITERATION       MSE         MPE        MAPE     % ERROR REDUCTION
2             31.19        1.52%       1.52%         1.1720%
3             15.00        1.04%       1.04%         1.0794%
4              7.22         .72%        .72%         1.0768%
5              3.48         .50%        .50%         1.0767%
6              1.67         .35%        .35%         1.0767%
7               .81         .24%        .24%         1.0767%
8               .39         .17%        .17%         1.0767%
9               .19         .12%        .12%         1.0767%
10              .09         .08%        .08%         1.0767%
11              .04         .06%        .06%         1.0767%
12              .02         .04%        .04%         1.0767%
13              .01         .03%        .03%         1.0767%
14              .00         .02%        .02%         1.0767%
15              .00         .01%        .01%         1.0767%
16              .00         .01%        .01%         1.0767%

OPTIMAL WEIGHTS

1       .544
2       .454
3      -.457
4       .453
```

FIGURE 17-7   Forecasting Performance Using GAF

```
DO YOU WANT A TABLE OF ACTUAL AND PREDICTED (Y OR N)?Y

PERIOD       ACTUAL        FORECAST        ERROR     PCT ERROR
7           144.00        143.97           .03        .02%
8           191.00        190.97           .03        .01%
9           243.00        242.95           .05        .02%
10          280.00        279.95           .05        .02%
11          320.00        319.97           .03        .01%
12          383.00        382.97           .03        .01%
13          451.00        450.96           .04        .01%
14          504.00        503.96           .04        .01%
15          560.00        559.97           .03        .01%
16          639.00        638.97           .03        .00%
17          723.00        722.97           .03        .00%
18          792.00        791.97           .03        .00%
19          864.00        863.97           .03        .00%
20          959.00        958.97           .03        .00%
MEAN PC ERROR (MPE) OR BIAS       =       .01%
MEAN SQUARED ERROR (MSE)          =       .0
MEAN ABSOLUTE PC ERROR   (MAPE) =        .0%


FORECAST FOR HOW MANY PERIODS AHEAD (0=NONE,24=MAX)?8
PERIOD          FORECAST
21             1058.97
22             1143.90
23             1231.80
24             1342.68
25             1458.50
26             1559.24
27             1662.92
28             1789.56

DO YOU WANT TO CONTINUE THE TRAINING (Y OR N)?N

CRUS:       76.52
```

FIGURE 17-8   Data with Trend Eliminated Using 2 Weights

```
HOW MANY WEIGHTS (PARAMETERS) DO YOU WANT (16 = MAX)?2
IS YOUR DATA STATIONARY ?N

********   LEARNING PERFORMANCE   ********

LEARNING CONSTANT =       0
ITERATION          MSE          MPE          MAPE      % ERROR REDUCTION
1                205.00        3.33%         5.86%           0%

OPTIMAL WEIGHTS

  1       -1.000
  2        2.000
DO YOU WANT A TABLE OF ACTUAL AND PREDICTED (Y OR N)?Y

PERIOD          ACTUAL        FORECAST           ERROR         PCT ERROR
  3             32.00          29.00             3.00            9.38%
  4             63.00          40.00            23.00           36.51%
  5             99.00          94.00             5.00            5.05%
  6            120.00         135.00           -15.00           12.50%
  7            144.00         141.00             3.00            2.08%
  8            191.00         168.00            23.00           12.04%
  9            243.00         238.00             5.00            2.06%
 10            280.00         295.00           -15.00            5.36%
 11            320.00         317.00             3.00             .94%
 12            383.00         360.00            23.00            6.01%
 13            451.00         446.00             5.00            1.11%
 14            504.00         519.00           -15.00            2.98%
 15            560.00         557.00             3.00             .54%
 16            639.00         616.00            23.00            3.60%
 17            723.00         718.00             5.00             .69%
 18            792.00         807.00           -15.00            1.89%
 19            864.00         861.00             3.00             .35%
 20            959.00         936.00            23.00            2.40%
MEAN PC ERROR (MPE) OR BIAS         =        3.33%
MEAN SQUARED ERROR (MSE)            =      205.0
MEAN ABSOLUTE PC ERROR   (MAPE) =          5.9%

FORECAST FOR HOW MANY PERIODS AHEAD (0=NONE,24=MAX)?0
```

ϕ_i is the old weight

K   is the learning constant

e_t is the error at period t (standardized)

X_{t-i} is the value of the time series at period t – i; this value corresponds to the ϕ_i weight. (Standardized)

m   is the number of weights

and

n   is the number of data points adjusted for the level of differencing.

FIGURE 17-9   Forecasting Performance Using GAF

```
DO YOU WANT TO CONTINUE THE TRAINING (Y OR N)?Y
HOW MANY AUTOCORRELATIONS DO YOU WANT TO SEE (0 IF NONE)?9
```

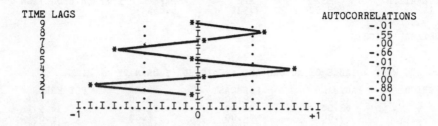

```
HOW MANY WEIGHTS (PARAMETERS) DO YOU WANT (16 = MAX)?4

DO YOU WANT TO SPECIFY INITIAL VALUES FOR YOUR WEIGHTS (Y OR N)?N

ENTER A VALUE FOR THE LEARNING CONSTANT (IF -1 IT WILL BE SET BY THE PROGRAM)
?-1
```

```
********   LEARNING PERFORMANCE   ********
LEARNING CONSTANT =   .142
```

ITERATION	MSE	MPE	MAPE	% ERROR REDUCTION
2	31.19	1.52%	1.52%	1.1720%
3	15.00	1.04%	1.04%	1.0794%
4	7.22	.72%	.72%	1.0768%
5	3.48	.50%	.50%	1.0767%
6	1.67	.35%	.35%	1.0767%
7	.81	.24%	.24%	1.0767%
8	.39	.17%	.17%	1.0767%
9	.19	.12%	.12%	1.0767%
10	.09	.08%	.08%	1.0767%
11	.04	.06%	.06%	1.0767%
12	.02	.04%	.04%	1.0767%
13	.01	.03%	.03%	1.0767%
14	.00	.02%	.02%	1.0767%
15	.00	.01%	.01%	1.0767%
16	.00	.01%	.01%	1.0767%

```
OPTIMAL WEIGHTS

  1       .544
  2       .454
  3      -.457
  4       .453
DO YOU WANT A TABLE OF ACTUAL AND PREDICTED (Y OR N)?N
MEAN PC ERROR (MPE) OR BIAS      =        .01%
MEAN SQUARED ERROR (MSE)         =        .0
MEAN ABSOLUTE PC ERROR   (MAPE) =        .0%


FORECAST FOR HOW MANY PERIODS AHEAD (0=NONE,24=MAX)?0


DO YOU WANT TO CONTINUE THE TRAINING (Y OR N)?N
```

The application of (17-13) is done in the program and the result can be seen in Figure 17-7. The final forecasts are given there and are almost the same as the continuation of the series.

The balance of the description of the solution is more technical for the rest of this chapter. It requires that the reader have some knowledge of matrix algebra. However, what follows is not essential except for those who would like to know the computational details of the GAF method. The other readers are encouraged to go to the next chapter.

One can start by assigning equal weights to equation (17-12).

$$\phi_1 = \phi_2 = \phi_3 = \phi_4 = .25 \qquad (17\text{-}14)$$

and then use equation (17-13) to modify those until the MSE becomes a minimum.

Substituting the weights of equation (17-14) into equation (17-12) and starting at t = 5, equation (17-12) becomes:

$$\hat{X}_5 = .25X_4 + .25X_3 + .25X_2 + .25X_1$$

$$= .25(-15) + .25(5) + .25(23) + .25(3)$$

$$\hat{X}_5 = 4$$

Thus the estimate for period 5 is calculated as 4. The actual value is $X_5 = 3$. Therefore, the error is:

$$e_5 = X_5 - \hat{X}_5$$

$$e_5 = 3 - 4 = -1$$

With similar calculations one can find the error for periods 6, 7, 8, . . . , 20. The mean squared error of all errors using the weights of equation (17-14) is 181. There is an alternative to (17-14). Instead of initializing the weights with equal values, one can obtain their values from a well-known set of equations: the Yule-Walker equations. The result will be that the weights can be calculated as follows:

$$\begin{bmatrix} \phi_1 \\ \phi_2 \\ \phi_3 \\ \phi_4 \end{bmatrix} = \begin{bmatrix} 1 & \rho_1 & \rho_2 & \rho_3 \\ \rho_1 & 1 & \rho_1 & \rho_2 \\ \rho_2 & \rho_1 & 1 & \rho_1 \\ \rho_3 & \rho_2 & \rho_1 & 1 \end{bmatrix}^{-1} \begin{bmatrix} \rho_1 \\ \rho_2 \\ \rho_3 \\ \rho_4 \end{bmatrix} \qquad (17\text{-}15)$$

where ρ_1, ρ_2, ρ_3 and ρ_4 are the autocorrelation coefficients for time lags 1, 2, 3, and 4 respectively. Their values are shown in Figure 17-5. Thus, equation (17-1) becomes:

$$
\begin{bmatrix} \phi_1 \\ \phi_2 \\ \phi_3 \\ \phi_4 \end{bmatrix} =
\begin{bmatrix}
1 & -.0072 & -.8804 & .0042 \\
-.0072 & 1 & -.0072 & -.8804 \\
-.8804 & -.0072 & 1 & -.0072 \\
-.0042 & -.8804 & -.0072 & 1
\end{bmatrix}^{-1}
\begin{bmatrix} -.0072 \\ -.8804 \\ .0042 \\ .767 \end{bmatrix}
$$

(17-16)

To solve equation (17-16) one must find the inverse of the square matrix as shown below. (The reader who has no mathematical knowledge of matrices or no interest in this particular topic can go directly to the solution of equation (17-15) shown in equation (17-18).)

$$
\begin{bmatrix} \phi_1 \\ \phi_2 \\ \phi_3 \\ \phi_4 \end{bmatrix} =
\begin{bmatrix}
4.468 & .3066 & 3.938 & .2795 \\
.3066 & 4.471 & .3305 & 3.938 \\
3.938 & .3305 & 4.471 & .3066 \\
.2795 & 3.938 & .3066 & 4.468
\end{bmatrix}
\begin{bmatrix} -.0072 \\ -.8804 \\ .0042 \\ 767 \end{bmatrix}
$$

(17-17)

$$
\begin{bmatrix} \phi_1 \\ \phi_2 \\ \phi_3 \\ \phi_4 \end{bmatrix} =
\begin{bmatrix} -.0712 \\ -.9172 \\ -.0654 \\ -.0406 \end{bmatrix}
\quad \text{or} \quad
\begin{aligned}
\phi_1 &= -.0712 \\
\phi_2 &= -.9172 \\
\phi_3 &= -.0654 \\
\phi_4 &= -.0406
\end{aligned}
$$

(17-18)

The weights of equation (17-18) provide an MSE of 76.2 in comparison to the error of 181 given by the initial weights, $\phi_1 = \phi_2 = \phi_3 = \phi_4 = .25$. This is a substantial decrease which makes the usage of equations (17-15) worthwhile.

Once some initial weights have been determined, the method of adaptive filtering uses equation (17-13) to modify them in such a way that, on the average, the MSE of the modified weights will be smaller than it was for the old weights. This process continues until the MSE cannot be reduced any further. The next paragraph illustrates how the weights are modified.

Suppose one starts with $\phi_1 = -.0712$, $\phi_2 = -.9172$, $\phi_3 = .0654$ and $\phi_4 = -.0406$. Since there are four weights, one needs four values of the time series to apply them to forecasting the next value. For example, if one is at period 1, the values of ϕ_i and X_i could be substituted into equation (17-12) directly. However, it is more desirable first to standardize the values of the X_i involved. Standardizing adjusts the values of X_i so they are all between zero and one. This can be done by dividing all data by the largest number, or by dividing each set of m values by the square root of the sum of the squared values. Making this transformation assures that, for all practical purposes, the MSE resulting from the new weights will on the average be smaller than the previous MSE as long as K, the learning constant, is less than $1/m$, where m is the number of weights. The four values involved are:

Period	X_i	X_i^2	$X_i \sqrt{\Sigma X_i^2}$	Standardized X_i
1	3	9	.10687	$X_1 = .10687$
2	23	529	.81934	$X_2 = .81934$
3	5	25	.17812	$X_3 = .17812$
4	-15	225	-.53435	$X_4 = -.53435$

$$\sum_{i=1}^{4} X_i^2 = 788$$

$$\sqrt{\Sigma_i^2} = 28.0713$$

Substituting the standardized X_i into equation (17-12) gives:

$$\hat{X}_5 = -.0716(-.53435) - .9172(.17812) - .0654(.81934)$$

$$- .0406(.10687)$$

$$= -.18303$$

The standardized error is:

$$e_5 = X_5/28.0713 - (-.18303)$$

$$= .106870 + .18303 = .29$$

This standardized error of .29 can be used in equation (17-13) to yield new weights.

$$W_1 = -.0716 + 2(.1417)(.29)(-.53435) = -.1155$$

$$W_2 = -.9172 + 2(.1417)(.29)(.17812) = -.9024$$

$$W_3 = -.0654 + 2(.1417)(.29)(.81934) = .00162$$

$$W_4 = -.0406 + 2(.1417)(.29)(.10687) = -.0318$$

The mean squared error of the new weights is 66.277. This is a drop of about 15% from the MSE corresponding to the previous weights.

The training of weights can be repeated using periods 2 to 5 and forecasting for period 6. This gives:

Period	X_i	X_i^2	$\dfrac{X_i \sqrt{\Sigma X_i^2}}{}$ Standardized X_i
2	23	529	.81934
3	5	25	.17812
4	−15	225	−.53435
5	3	9	.10687

$$\Sigma X_i^2 = 788$$

$$\sqrt{\Sigma X_i^2} = 28.0713$$

The standardized error of period 6 is:

$$e_6 = X_6/28.0713 - \hat{X}_6$$

$$= .81934 - .49620 = -.323$$

This new error can be used to train another set of weights and so on, until all twenty observations have been used. This will complete a training iteration, after which one can start again from the beginning of the series and repeat the process until the MSE can be reduced no further. The performance of adaptive filtering is shown in Figure 17-6. For each revision of the weights, the MSE is shown next to the iteration number, as is its reduction from one iteration to the next. At iteration sixteen the program stops because the MSE is almost zero. One could have continued until the MSE had become exactly zero, in which case the forecast error would have also become zero.

Figure 17-7 shows the optimal weights, the table of the actual and forecast values, and the continuation of the series. The actual results are: 1059, 1144, 1232, 1343, 1459, 1560, 1664, 1791, indicating a very accurate set of predictions, as would be expected for a series with no randomness.

CHAPTER 18

The Box-Jenkins Methodology

The Box-Jenkins methodology is an efficient and practical procedure for using autoregressive or moving average schemes for forecasting applications.* The main contribution of Box and Jenkins is the developement of procedures for identifying the ARMA model that best fits a set of data and for testing the adequacy of that model. Figure 18-1 shows the Box-Jenkins methodology and the steps that it includes. If applied correctly, Stage 3 will give an adequate model for forecasting. The different steps of Figure 18-1 entail the following:

1. *Postulate General Class of Models*

The most general class of an autoregressive-moving average (ARMA) model is:

$$X_t = \phi_1 X_{t-1} + \phi_2 X_{t-2} + \ldots + \phi_p X_{t-p} + e_t - \theta_1 e_{t-1} - \theta_2 e_{t-2}$$

$$- \ldots - \theta_q e_{t-q} \qquad (18\text{-}1)$$

Equation (18-1) can fit almost any stationary time series. The task for the user is to determine the degree (p and q) of the ARMA model.

*The reader should review the first part of the last chapter before continuing.

FIGURE 18-1   The Box-Jenkins Methodology

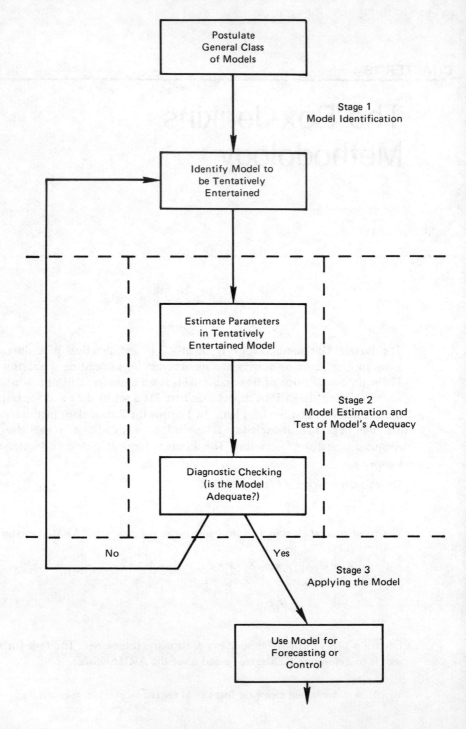

2. *Identify Tentative Model*

It is possible to decide on some specific ARMA (p,q) model by examining the autocorrelation coefficients and a similiar set of parameters, the partial autocorrelation coefficients.* Combinations of autocorrelations and partial autocorrelations, in the absence of randomness, can reveal the exact ARMA (p, q,) model with no possibility of error. Randomness complicates the task because it influences the values of the autocorrelation and partial autocorrelation coefficients and may cause them to deviate from their "true" pattern. This makes the model identification process more difficult, but generally there is enough information available to select an appropriate model. Even if there is a mistake at this point, it can be found later when a test of the model's adequacy is made. Figure 18-2 shows the behavior of the theoretical (assuming the absence of randomness) autocorrelation and partial autocorrelation coefficients. As can be seen, there is a unique set of autocorrelations for each model.

FIGURE 18-2(a)   Examples of Autocorrelation and Partial Autocorrelation Coefficients of AR(1) Model

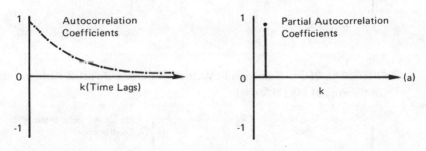

or

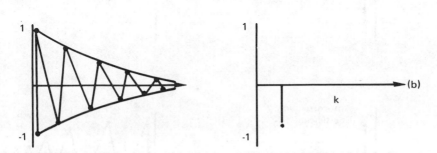

*Partial autocorrelations are analogous to autocorrelations in that they indicate the relationship of the values of a time series to various time lagged values of the same series. However, they differ from autocorrelations in that they are computed for each time lag after removing the effect of all other time lags on the given time lag and on the original series. In essence, they show the relative strength of the relationship that exists for varying time lags.

FIGURE 18-2(b)   Examples of Autocorrelation and Partial Autocorrelation Coefficient of AR(2) Model

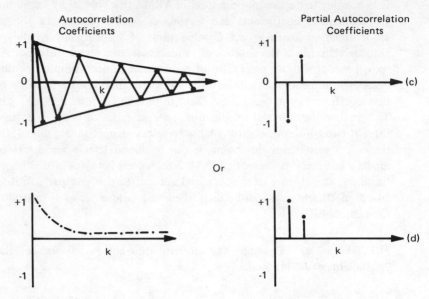

Or

FIGURE 18-2(c)   Examples of Autocorrelation and Partial Autocorrelation Coefficients of MA(1) Model

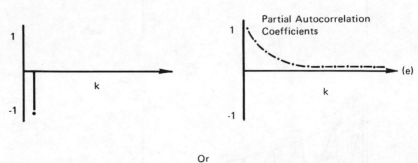

Or

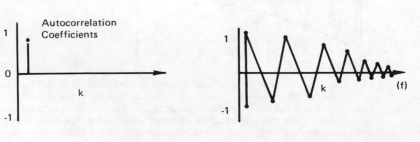

FIGURE 18-2(d)   Examples of Autocorrelation and Partial Autocorrelation
Coefficients of MA(2) Model

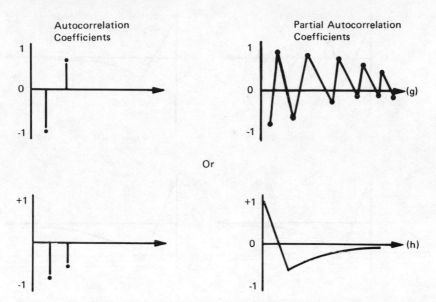

Thus, in order to forecast a time series, the autocorrelations and partial
autocorrelations must be computed first. The graphs of these two items
should be carefully examined and compared with the theoretical patterns of
Figure 18-2. Sometimes the pattern of the computed autocorrelations and
partial autocorrelations can be easily classified as one of the theoretical ones
shown in Figure 18-2. The identification of the model is then easy. Other
times, some free association is required to be able to infer a pattern from the
autocorrelations, or more than one pattern may be implied. In such instances
a judgment must be made. However, most of the time a selection is possible
and, given some experience, not too difficult.

Once the tentative selection of a model has been made (for example,
selection of an ARMA (1,1) model) the next step is to estimate the parameters of that model.

3.   *Estimate the Parameters of Tentative Model*

Assuming that the tentative model is an ARMA (1,1), its mathematical form
is:

$$X_t = \phi_1 X_{t-1} + e_t - \theta_1 e_{t-1} \qquad (18\text{-}2)$$

FIGURE 18-2(e)   Examples of Autocorrelation and Partial Autocorrelation Coefficients of a Mixed ARMA (1,1) Model

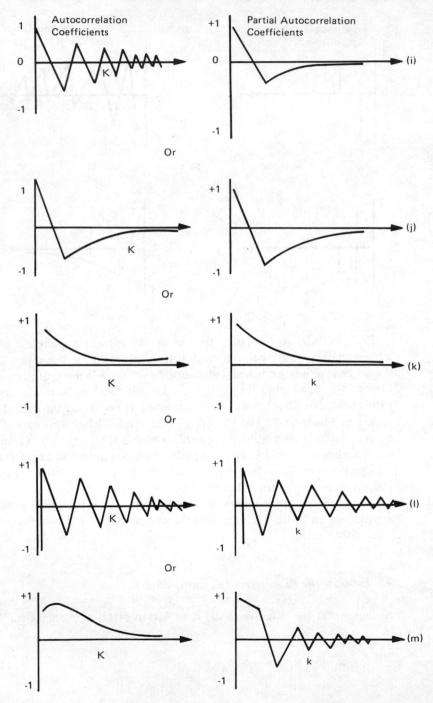

In order to use (18-2), the values of ϕ_1 and θ_1 must be estimated. This is done with standard statistical estimation methods (maximum likelihood estimates or non-linear estimates). These estimation procedures are often computationally cumbersome. However, they can be computed easily with a high speed electronic computer.

The general approach is to start with some initial values for ϕ_1 and θ_1 and then modify them by small steps while observing the mean squared error. This allows the direction of change in ϕ_1 and θ_1 that will result in the smallest MSE to be determined. Eventually, the ϕ_1 and θ_1 corresponding to the minimum mean squared error are found and used as the final estimates of the model. An alternative estimation procedure is to find the MSE for all combinations of ϕ_1 and θ_1 and to specify those parameter values that minimize the MSE. (See Chapters 26 and 27.)

Continuing with the ARMA (1, 1) model, assume that $\phi_1 = .35$ and $\theta_1 = .20$. The tentative model then is:

$$\hat{X}_t = .35X_{t-1} + e_t - .20e_{t-1} \tag{18-3}$$

The next step is to make sure that (18-3) is indeed an adequate model.

4. Testing the Model's Adequacy

If the model represented by (18-3) is an adequate one, the residual differences between the time series values and those estimated by the model must be white noise. These errors can be computed using (18-4).

$$e_t = X_t - \hat{X}_t \tag{18-4}$$

Next, one can find the autocorrelation coefficient of the residuals, e_t. The existence of a pattern in the residuals can be determined from the autocorrelation coefficients. If none of them is significantly different than zero, the errors are assumed to be random or white noise and the model is adequate. It is rather simple, therefore, to test the adequacy of a model by calculating the autocorrelation of the residuals. If the errors are random, then the model is adequate. Otherwise, one must return to step 2, select another model and repeat steps 3 and 4.

5. Applying the Model

If (18-4) indicates random residuals, the model of (18-3) can be used for forecasting purposes. To apply (18-3) one need only know X_{t-1} and e_{t-1}, both easily obtained quantities. (See exercises.)

Stationarity and Seasonality

(a) *Stationarity*   Equation (18-1) is applicable to stationary series only. If the time series (X_t) is non-stationary it must be converted to a stationary series before proceeding with steps 2, 3, 4 and 5. The procedure most widley used by the Box-Jenkins approach to transform a series to a stationary one is the method of differencing.

If the data imply a non-stationary pattern, achieving stationarity is no different from the procedure described in Chapter 17. The autocorrelations of the original data are examined. If they drop to zero quickly (after the third or fourth coefficient), it implies stationarity in the original series. If this is not the case, the first difference is taken, and the autocorrelations of the differenced series are examined. If they drop to zero rapidly, stationarity exists at the first level of differencing. If they do not, the second difference must be taken and the autocorrelations of the resulting series examined. For all practical purposes, stationarity will exist in the original, first, or second differenced series.

One therefore chooses the level at which the data are stationary and uses (18-1) as the general class of models from which to tentatively identify an ARMA (p,q) model which appear to be most appropriate based on an examination of the autocorrelation coefficients. When the data involved are seasonal, there is an extra factor involved in achieving stationarity. This requires stationarity among the same months of different years, that is, it requires stationarity among months (or periods in general) which are L periods apart, where L is the length of seasonality. How this type of non-stationarity can be achieved and the introduction of the most general class of seasonal ARMA models is examined next.

(b) *Seasonal ARMA models*   In addition to the general form of (18-1), ARMA models can be extended to include seasonality by incorporating seasonal coefficients in either the AR or the MA part of the equation. For example, one could make (18-2) seasonal by adding a seasonal coefficient, ϕ_s, in the autoregressive portion.

$$\hat{X}_t = a + \phi_1 X_{t-1} + (\phi_s X_{t-L} + \phi_1 \phi_s X_{t-L-1}) + e_t - \theta_1 e_{t-1} \quad (18\text{-}5)$$

Equation (18-5) is similar to (18-2) except for the term inside the parenthesis which accounts for seasonality. ϕ_s is the seasonal coefficient, while $\phi_1 \phi_s$ exists because the model is multiplicative and therefore shows the combined effects of the AR non-seasonal and the AR seasonal parameters. If one decides to put the seasonal coefficient in the MA, the model becomes:

$$X_t = a + \phi_1 X_{t-1} + e_t - \theta_1 e_{t-1} - (\theta_s e_{t-L} - \theta_1 \theta_s e_{t-L-1}) \quad (18\text{-}6)$$

Again, (18-6) is similar to (18-2) except for the term inside the parenthesis which accounts for the seasonality.

Identifying a model through the program BOXJEN involves choosing one of the fourteen models available. Alternatively any model of up to three parameters can be specified. If none of these models fit the data, the program UNIVBJ should be used so that a higher order model can be specified. (See Chapter 26.) The process of selecting the most appropriate model, once the data have been transformed to a stationary series, is described below.

As always, the starting point is to examine the autocorrelation and partial autocorrelation coefficients. Seasonality exists when some autocorrelations of time lag larger than three or four are significantly different from zero. For a moment, however, the autocorrelations due to seasonality should be ignored. This reduces the examination of the remaining autocorrelations to that of a non-seasonal model. Then by looking at the autocorrelations in conjunction with Figure 18-2, one can choose the most appropriate ARMA model. What remains is deciding whether to put the seasonal addition in the autoregressive portion or the moving average portion. Assuming an ARMA (1,1) non-seasonal model, adding the seasonal to the AR will produce (18-5) while putting it in the MA yields (18-6).

The decision of adding the seasonal to the AR or MA has to be made ignoring the non-seasonal autocorrelation coefficients. Thus, if the length of seasonality is 12 one must examine the 12, 24, 36, 48, etc., autocorrelations and partial autocorrelation coefficients as if they were next to each other and decide what will fit them better — an AR or an MA model. If the autocorrelation coefficients of 12, 24, 36, 48 dampen exponentially and there are only one or two significant partial autocorrelations, this implies an AR seasonal. If the opposite happens — one, or two significant autocorrelations while the partial ones dampen exponentially — it indicates the seasonal on the MA part. The problem is that one never has enough autocorrelations to make this decision accurately and often must decide by trial if the seasonal coefficients fit better in the AR or the MA part of the model. The inconvenience of such a trial-and-error approach is minimal.

A final point to resolve is the level of non-stationarity when the data are seasonal. To find the appropriate level, one has to start by taking short differences (one period apart) and long or seasonal differences (L periods apart). Some combination of these short and long differences can be used to form a stationary data series. Thus, one must start with zero short and zero long differences and examine their autocorrelation coefficients. If they drop to zero quickly and then oscillate around zero, it implies stationarity at the zero-zero level. If this is not the case, one must try different sets of differences. For example, one might try one short and zero long and examine the autocorrelations, and similarly for zero short and one long, one short and one long, two short and one long, one short and two long, etc. Examining the autocorrelation coefficients corresponding to each difference will lead to identification of one that is a more stationary series than the others. (The autocorrelations should be as close to zero as possible in addition to oscillating around zero).

There is a test which can facilitate the choice of the best level of differencing. This is the χ^2-test, or the Box-Pierce test as it is often called. The smaller the value of this test the less significantly different than zero the autocorrelation coefficients are, and the tighter their distribution around zero. Thus, the smallest value of this test will imply that stationarity exists at that level of differencing. Although the χ^2-test works most of the time, there are cases where it does not work. Thus it should be used to supplement, rather than to replace a visual inspection of the autocorrelation coefficients.

Confidence Limits

Statistical forecasting methods such as the Box-Jenkins approach not only provide forecasts for the future but at the same time allow the calculation of upper and lower limits within which future actual values will fall according to some probability. These confidence intervals can be useful because they allow the planner to have an idea of the amount of variation surrounding the forecasts. Thus, even though the forecast is the mostly likely value (assuming that past patterns will continue uninterrupted into the future), the confidence interval provides limits for the forecast. If the method used is not a statistical one, as with most methods examined so far, the mean squared error can be used to estimate confidence limits. These limits are equal to the forecast, plus or minus twice the square root of the mean squared error. Even though not very precise, this will give an indication of the extent of the uncertainty for one period ahead forecasts. (The reader can check this in exercise 18-2.) For more than one period ahead forecasts the limits of the confidence intervals will be a little wider, but still their magnitude can be roughly assessed.

Summary of the Box-Jenkins Approach

1. Decide whether or not the series is stationary.
2. If the time series is not stationary, take the appropriate level of differences to transform it into a stationary series.
3. Examine the autocorrelations and partial autocorrelation coefficients of the stationary series and select the type of ARMA model for tentative use. (See Figure 18-2 for help in identifying an appropriate model.)
4. Estimate the parameters of the ARMA model selected.
5. Find the autocorrelations of the residuals — the difference between the time-series and the model selected in 3 and estimated at 4. If the residuals are white noise, use the model for forecasting

purposes. If they are not, look for a better model in 3. The pattern of the residuals should be of help in determining what was wrong with the model selected initially and therefore be of assistance in finding a more appropriate one.

6. If the residuals are not white noise, repeat steps 3, 4 and 5.
7. If the time-series is seasonal take the appropriate number of short and long differences so that the series will be transformed into a stationary one.
8. Ignore the seasonal autocorrelaton and partial autocorrelation coefficients, and from the remaining coefficients determine the degree of the ARMA model most appropriate for the series. The seasonal autocorrelaton and partial autocorrelation coefficients can be examined to determine whether the seasonal coefficient should be in the AR or the MA portion of the model.
9. Estimate the parameters of the seasonal ARMA model selected above.
10. Find the autocorrelations of the residuals and check to see whether or not they are white noise. If they are random, use the model for forecasting purposes. If they are not repeat steps 8, 9, and 10.
11. Apply the model by obtaining forecasts as well as lower and upper confidence limits for those.

References for further study:

> Box, G. and G. Jenkins (1976)
> Mabert, V. (1975)
> Makridakis, S. and S. Wheelwright (1978, Chapter 10)
> Montgomery, D. and L. Johnson (1976, Chapter 9)
> Nelson, C. (1974)
> Wheelwright, S. and S. Makridakis (1977, Chapter 8)

INPUT/OUTPUT FOR THE BOX-JENKINS PROGRAMS

There are three programs available for the Box-Jenkins method. The program B-J is for time-series data that are non-seasonal. It provides initial estimates, but the estimation of final parameter values must be done manually by the user. The program BOXJEN can deal with seasonal and/or non-seasonal data. It allows a choice of a variety of ARMA models, as long as the number of parameters is three or less. (This is so because of the optimization procedure it uses — a modified simplex method.) It requires no initial estimates and minimizes the MSE by examining many possible combinations of possible parameter values. The third program, UNIVBJ, is examined in Chapter 26. It is the most general of the three and will handle any ARMA model. However, it requires initial estimates of parameter values (these values can be calculated internally, however, in the ID program) which it subsequently optimizes using Marquardt's non-linear estimation algorithm.

NON—SEASONAL PROGRAM B—J

Input

1. *Enter Total Order of Differencing?*

Before any model can be used, the data must be stationary. The question, therefore, requires specification of the level of differencing (0, 1, 2, 3) at which the data are stationary. If the user does not know, start by answering 0, and then see the graph of autocorrelations. If they drop to zero fast, then respond NO in next input; otherwise, the answer should be YES, and the total order of differencing should be chosen as 1. If the first difference makes the series stationary, O.K. If not, the second difference must be taken.

2. *Do You Want to Difference the Series?*

If the graph of autocorrelations shows that the series is stationary (i.e the autocorrelation coefficients are not significantly different than zero after 3 or 4 values) the answer should be NO: otherwise, the user should try another level of differencing. The program fits some ARMA model to the last degree of differencing specified.

3. *What Order of AR Process to be Fitted?*

The order of the autoregressive (AR) process must be decided by looking at Figure 18-2 and the graph of autocorrelation and partial autocorrelation coefficients printed in the program. The idea is to try to find the best resemblance between the autocorrelation coefficients calculated, and those of Figure 18-2. The degree of the AR model should then be specified as that having the highest resemblance with the theoretical autocorrelations shown in Figure 18-3. If the model does not require an AR parameter, the answer should be 0.

4.   *What Order of MA Process to be Fitted?*

This is a similar question to 3 above. The choice should be made by looking at the calculated autocorrelation and partial autocorrelation coefficients and theoretical ones of Figure 18-2.

5.   *Would You Like Another Set of Parameters?*

The B-J program calculates initial estimates of the ARMA parameters which can be improved. This can be done by trying smaller and larger values surrounding the original estimates provided by the program (see 2 and 3 under Output below, see also Exercise 19). The best parameters will correspond to the minimum mean squared error. The user can refine the parameters to any level desired.

6.   *Do You Want Autocorrelations of the Residuals?*

In general, the answer should be YES, unless the user is absolutely sure that the model chosen is the most appropriate one.

Output

1.   *Autocorrelation and Partial Autocorrelations*

Looking at the graph of autocorrelation and partial autocorrelation coefficients, one can decide upon the most appropriate ARMA model for the data.

2.   *Estimate Autoregressive Parameters*

These are the values of $\phi_1, \phi_2, \ldots, \phi_p$ in (17-1). They are initial estimates only. They can be modified (improved) if one so desires.

3.   *Estimate Moving Average Parameters*

These are the values of $\theta_1, \theta_2, \ldots, \theta_q$ in (17-1). They are initial estimates which can be improved.

4.   *Autocorrelation of Residuals*

The graph of the autocorrelation of the residuals indicates whether or not the ARMA model fitted in the data is adequate. If none of the autocorrelation coefficients is significantly different from zero, then the ARMA model fitted is good; otherwise, a new model must be tried.

2.   PROGRAM BOXJEN

Input

1.   *Short Term Differences?*

Level of period-to-period differencing in order to make the series stationary (i.e., 0, 1, 2).

2.   *Long Term Differences?*

Level of L periods apart differencing in order to make the series stationary (i.e., 0, 1, 2).

3.   *Length of Seasonality?*

The length of seasonality, L is needed for the long term differences and the seasonal coefficient. It can be found by looking at a graph of the autocorrelations. The length of seasonality corresponds to the time lag of the highest, after lag of about 3, autocorrelation coefficient.

4.   *Model Number*

The way the program is set, the user has the choice of any one of fourteen different models which, for all practical purposes, are flexible enough to fit most types of data. (Note that none of these uses more than three parameters.)

These models are listed here using a short hand notation frequently used by authors describing the Box-Jenkins approach. The notation makes use of a back shift operator, B^m, where the exponent m indicates the amount of back shifting that is desired. The operator can probably be best understood by examining several examples of its use.

$$(B^m)Y_t \quad \text{is equivalent to } Y_{t-m}$$

$$(B^m)e_t \quad \text{is equivalent to } e_{t-m}$$

$$B^3Y_t \quad \text{is equivalent to } Y_{t-3}$$

$$(1\text{-}B\text{-}B^2)e_t \quad \text{is equivalent to } e_t - e_{t-1} - e_{t-2}$$

It should be noted that standard mathematical operations apply to B^m. Thus

$$(B)(1 - B^2)Y_t = (B - B^3)Y_t = Y_{t-1} - Y_{t-3}$$

The fourteen models available in the BOXJEN program are as follows:

(0) Specify your own model up to three parameters

(1) $Y_t = (1-\theta_1 B-\theta_2 B^2)(1-\theta_s^* B^s)e_t$ MA(2) seasonal in MA

(2) $(1-\phi_1 B)Y_t = (1-\theta_1 B)(1-\theta_s^* B^s)e_t$ ARMA(1,1) seasonal in MA

(3) $(1-\phi_1 B-\phi_2 B^2)Y_t = (1-\theta_s^* B^s)e_t$ AR(2) seasonal in MA

(4) $(1-\phi_s^* B^s)Y_t = (1-\theta_1 B-\theta_2^2 B^2)e_t$ MA(2) seasonal in AR

(5) $(1-\phi_1 B)(1-\phi_s^* B^s)Y_t = (1-\theta_1 B)e_t$ ARMA(1,1) seasonal in AR

(6) $Y_t = (1-\theta_1 B)(1-\theta_s^* B^s)e_t$ MA(1) seasonal in MA

(7) $(1-\phi B)Y_t = (1-\theta_s^* B^s)e_t$ AR(1) seasonal in MA

(8) $(1-\phi_s^* B^s)Y_t = (1-\theta_1 B)e_t$ MA(1) seasonal in AR

(9) $(1-\phi_1 B)(1-\phi_s^* B^s)Y_t = e_t$ AR(1) seasonal in AR

(10) $Y_t = (1-\theta_1 B-\theta_2 B^2)e_t$ Non-seasonal MA(2)

(11) $(1-\phi_1 B)Y_t = (1-\theta_1)e_t$ Non-seasonal ARMA(1,1)

(12) $(1-\phi_1 B-\phi_2 B^2)Y_t = e_t$ Non-seasonal AR(2)

(13) $Y_t = (1-\theta_1 B)e_t$ Non-seasonal MA(1)

(14) $(1-\phi_1 B)Y_t = e_t$ Non-seasonal AR(1)

The model can be chosen by simply entering the number corresponding to the model thought as the most appropriate one. If a model other than the fourteen shown is desired, the user must first type 0 and then input p and q (for ARMA (p,q)) and in the case of seasonal models, specify whether the seasonal parameter is in the AR, MA or both portion. If more than three parameters are desired the program UNIVBJ must be used.

EXERCISE 18.1 (Using B-J)

The following forty-five data values were generated by the DGEN Program. The model specified was a non-stationary (first level), AR(1) with ϕ_1 set equal to .5, and the constant term equal to 15, and a variance of 3. Use these data to test the identification and estimation procedure of the Box-Jenkins approach.

PERIOD	OBSERVATION	PERIOD	OBSERVATION	PERIOD	OBSERVATION
1	2426.411	16	2888.414	31	3317.237
2	2457.842	17	2916.086	32	3342.874
3	2492.729	18	2944.661	33	3370.009
4	2526.982	19	2974.686	34	3399.791
5	2561.318	20	3004.561	35	3434.208
6	2593.570	21	3032.981	36	3465.827
7	2624.803	22	3060.397	37	3491.975
8	2653.578	23	3088.676	38	3522.099
9	2682.360	24	3115.502	39	3550.525
10	2712.987	25	3141.658	40	3578.137
11	2743.418	26	3168.636	41	3606.044
12	2774.085	27	3197.796	42	3633.347
13	2802.829	28	3230.139	43	3662.882
14	2832.128	29	3258.426	44	3693.069
15	2860.593	30	3288.609	45	3725.575

Solution

Figure 18-3 shows the autocorrelation and partial autocorrelations of the original data. As can be seen, the autocorrelation coefficients do *not* drop to zero, even after the fifth value. This means that the data are not stationary. The absence of stationarity can be seen clearly in the upper part of Figure 18-3 by looking at the graph of the autocorrelations. These show a definite trend. The partial autocorrelations are of no use when the data are non-stationary.

Since the autocorrelation coefficients of Figure 18-3 indicate that the data is non-stationary, one must take their first difference. This is done in Figure 18-4.

Figure 18-4 shows the autocorrelations of the first difference. Except for the first and possibly the second values, the autocorrelations drop to a zero, a fact which suggests stationarity. Autocorrelations 7 and 8 are a little high but well below 2(.1508) = .3016 (2 times the standard error). Thus one can assume that they are not significantly different from zero.

Once stationarity is found at some level of differencing, one must decide

FIGURE 18-3   Autocorrelations and Partial Autocorrelations of the
Original Data

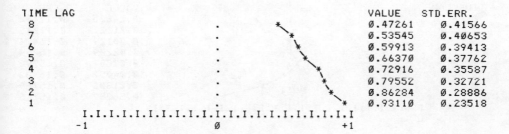

```
ENTER TOTAL ORDER OF DIFFERENCING0
HOW MANY AUTOCORRELATIONS DO YOU WANT?8
```

```
TIME LAG                                          VALUE     STD.ERR.
8                             .      *            0.47261   0.41566
7                             .       *           0.53545   0.40653
6                             .        *          0.59913   0.39413
5                             .         *         0.66370   0.37762
4                             .          *        0.72916   0.35587
3                             .            *      0.79552   0.32721
2                             .              *    0.86284   0.28886
1                             .               *   0.93110   0.23518
        I.I.I.I.I.I.I.I.I.I.I.I.I.I.I.I.I.I.I.I.I
        -1                    0                 +1
```

```
STANDARD ERROR (Q=0) = .149071
```

```
PARTIAL AUTOCORRELATIONS
```

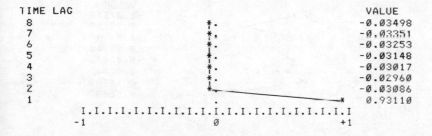

```
TIME LAG                                          VALUE
8                           *.                    -0.03498
7                           *.*                   -0.03351
6                           *.                     -0.03253
5                           *.                    -0.03148
4                           *.                     -0.03017
3                           *.*                    -0.02960
2                           *.                     -0.03086
1                           .               *      0.93110
        I.I.I.I.I.I.I.I.I.I.I.I.I.I.I.I.I.I.I.I.I
        -1                    0                 +1
```

```
DO YOU WANT TO DIFFERENCE SERIES
Y
```

upon the most appropriate model. Looking at Figure 18-4, one sees that the autocorrelations and partial autocorrelations behave as those of Figure 18-2(a) (see Figure 18-2 at the beginning of this chapter). This suggests that the best model should be an AR(1). Obviously the resemblance between Figures 18-4 and 18-2(a) cannot be perfect for two reasons: (1) there is some randomness involved, and (2) there are not enough data. However, Figure 18-4 and Figure 18-2(a) look alike. The autocorrelations do indeed drop exponentially to zero and, except for the first partial, the others do not seem to be significantly different from zero, with the possible exception of 7, which appears to be due to random effects. The examination of Figure 18-4 suggests trying an AR(1) or, equivalently, an ARMA(1,0) model. One should, therefore, answer 1 for AR and 0 for the MA process ot be fitted. The

FIGURE 18-4   First Differences of the Data

```
ENTER TOTAL ORDER OF DIFFERENCING1

HOW MANY AUTOCORRELATIONS DO YOU WANT?8
```

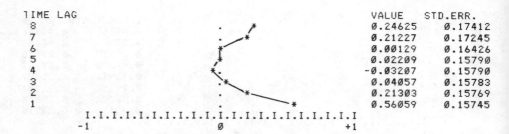

```
TIME LAG                                          VALUE    STD.ERR.
8                                    .       *    0.24625   0.17412
7                                  . .  *         0.21227   0.17245
6                                 *.              0.00129   0.16426
5                                .*.              0.02209   0.15790
4                               *.               -0.03207   0.15790
3                                .*               0.04057   0.15783
2                                .   *            0.21303   0.15769
1                                .        *       0.56059   0.15745
    I.I.I.I.I.I.I.I.I.I.I.I.I.I.I.I.I.I.I.I.I.I.I.I
    -1                           0                +1
```

```
STANDARD ERROR (Q=0) = .150756
```

```
PARTIAL AUTOCORRELATIONS
```

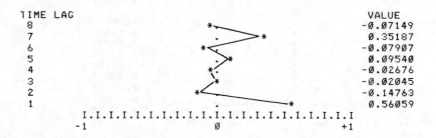

```
TIME LAG                                          VALUE
8                              *.   .            -0.07149
7                                .      *         0.35187
6                              *  .              -0.07907
5                                . *              0.09540
4                               *. .             -0.02676
3                               .*               -0.02045
2                              *.                -0.14763
1                                .        *       0.56059
    I.I.I.I.I.I.I.I.I.I.I.I.I.I.I.I.I.I.I.I.I.I.I.I
    -1                           0                +1
```

```
DO YOU WANT TO DIFFERENCE SERIES
N
WHAT ORDER OF AR PROCESS TO BE FITTED
1

WHAT ORDER OF MA PROCESS TO BE FITTED


0
ESTIMATED AUTOREGRESSIVE PARAMETERS
NO. 1    = .560592



ESTIMATE OF OVERALL CONSTANT TERM= 12.9742
ESTIMATE OF WHITE NOISE VARIANCE= 3.55348
```

program then estimates ϕ_1 = .5593, the constant term = 13.01, and the variance as 3.56, all close to the values specified in the generating program.

One could stop with ϕ_1 = .5593 as the estimate of the AR(1) model; however, there is a possibility of improving the estimate by trying a few values around .5593. Figure 18-5 shows how the MSE behaves. The minimum MSE is for ϕ_1 = .58. The difference, however, between the MSE corresponding to ϕ_1 = .58 and that corresponding to ϕ_1 = .5593 may not be worth the effort. With other data, however, the difference may be larger. In most Box-Jenkins computer programs, including UNIVBJ, the program itself starts from the initial estimates of the parameters and proceeds using a variation of the Gauss-Newton method to arrive at more refined, or final estimates.

The last task in the Box-Jenkins approach is to make sure that the AR(1) model is an adequate one. This is done by making sure that the

FIGURE 18-5   Plot of Mean Squared Error

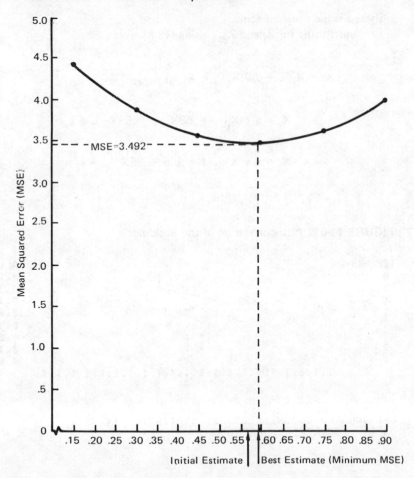

autocorrelations of residuals is white noise. This can be done by checking the autocorrelations of the residuals. Figure 18-6 presents the autocorrelations of the residuals and shows no significant autocorrelations nor any pattern among them.

In order to forecast, one must use equation (18-1), taking into account the fact that the data have been differenced. If the differencing is denoted as:

$$Z_t = X_t - X_{t-1} \tag{18-7}$$

Equation (18-1) then becomes:

$$Z_t = a + .58Z_{t-1} + e_t \tag{18-8}$$

where a is the constant term.

Substituting for Z_t and Z_{t-1} using (18-8) gives

$$X_t - X_{t-1} = .58(X_{t-1} - X_{t-2}) + e_t$$

$$X_t = a + X_{t-1} + .58X_{t-1} - .58X_{t-2} + e_t$$

$$X_t = a + X_{t-1}(1 + .58) - .58X_{t-2} + e_t \tag{18-9}$$

FIGURE 18-6   Autocorrelation of the Residuals

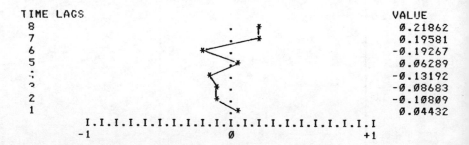

```
TIME LAGS                                                    VALUE
8                              .      *                     0.21862
7                              .     *                      0.19581
6                         *     .                          -0.19267
5                              .   *                        0.06289
:                           *  .                           -0.13192
3                             *.                           -0.08683
2                            * .                           -0.10809
1                             . *                           0.04432
       I.I.I.I.I.I.I.I.I.I.I.I.I.I.I.I.I.I.I.I.I.I.I
       -1                     0                    +1
```

CHI-SQUARED GOODNESS OF FIT STATISTIC= 14.7552
ON 24   DEGREES OF FREEDOM

If t – 1 is period 45 (the last period available), and one wants to forecast for period 46,

$$F_{46} = a + X_{45}(1.58) - .58X_{44}$$

$$F_{46} = 12.401 + 1.58(3725.575) - .58(3693.069) = 3756.83$$

The forecast for period 47 is:

$$F_{47} = 12.401 + 1.58(3756.83) - .58(3725.575) = 3787.36$$

This assumes that F_{46} is the actual value for X_{46} since no actual data point for period 46 is available.

Similarly, the forecast for periods 48, 49, and 50 are:

$$F_{48} = 3817.47$$

$$F_{49} = 3847.33$$

$$F_{50} = 3877.05$$

In order to obtain initial estimates for the AR parameters, one must solve the Yule-Walker equations discussed in the previous chapter [see (17-15)]. However, the AR process is of order 1. Thus, (17-15) is:

$$\phi_1 = (1)^{-1} (\rho_1) = r_1 = .5593$$

ϕ_1 is equal to the autocorrelation coefficient of time lag 1 of the differenced series (see Figure 18-4).

EXERCISE 18.2 (USING BOX JENKINS)

The following data are the industry sales (thousands of Francs) for printing and writing paper in France, between the years 1963 and 1972. Use this data to identify the most appropriate model (s), and check the model's adequacy to fit the data. Then make twelve forecasts for 1973.

```
WHAT PROGRAM DO YOU WANT TO RUN?LIST
DATA FILENAME?BJX20
HOW MANY OBSERVATIONS DO YOU WANT TO USE?120
ALL INPUT DATA DIVIDED BY 10** 1
NAME OF VARIABLE FROM DATA FILE?BJVAR2
```

```
***** LISTS THE DATA OF A FILE *****
```

PERIOD	OBSERVATION	PERIOD	OBSERVATION	PERIOD	OBSERVATION
1	56.27	41	70.11	81	74.20
2	59.90	42	79.01	82	84.72
3	66.85	43	59.46	83	73.17
4	59.78	44	23.07	84	89.85
5	57.99	45	61.72	85	77.81
6	66.82	46	69.14	86	85.61
7	49.92	47	70.11	87	93.88
8	21.52	48	70.58	88	81.30
9	55.58	49	74.76	89	78.34
10	58.69	50	77.34	90	82.81
11	54.61	51	81.38	91	65.73
12	57.11	52	76.67	92	31.00
13	63.47	53	72.89	93	78.00
14	63.93	54	74.92	94	86.00
15	71.22	55	68.10	95	78.00
16	62.16	56	24.14	96	80.80
17	62.10	57	68.02	97	89.52
18	67.60	58	70.83	98	85.61
19	50.13	59	69.42	99	89.33
20	22.03	60	77.21	100	87.50
21	56.07	61	79.53	101	83.51
22	60.25	62	78.84	102	93.46
23	62.64	63	89.00	103	83.25
24	60.55	64	79.74	104	30.00
25	64.68	65	75.10	105	79.14
26	65.84	66	82.13	106	90.00
27	71.29	67	69.16	107	78.17
28	68.77	68	29.07	108	88.00
29	72.39	69	72.71	109	87.50
30	70.72	70	86.84	110	99.30
31	62.90	71	81.24	111	97.68
32	23.75	72	79.96	112	96.87
33	61.33	73	84.30	113	87.17
34	73.04	74	84.70	114	100.69
35	73.49	75	94.20	115	83.20
36	65.18	76	80.43	116	34.56
37	67.62	77	84.03	117	84.95
38	74.82	78	87.15	118	91.39
39	81.07	79	65.63	119	86.87
40	72.94	80	37.05	120	99.37

Solution

Figure 18-7 shows the autocorrelation and partial autocorrelation coefficients of one short- and one long-term difference. At this level of differencing, the data are stationary. This set of differences was arrived at by looking at the graph of autocorrelation coefficients, and examining the χ^2-test. The 1 short and 1 long differences gave one of the lowest χ^2-values in addition to having the autocorrelation coefficients distributed around zero. Once the level of differencing has been set as one short- and one long period, one must look at Figure 18-7 to decide upon the most appropriate model.

Ignoring, for a moment, the twelfth autocorrelation and partial autocorrelation coefficients, Figure 18-7 behaves as Figure 18-2(e). The partial autocorrelations decay exponentially to zero, while there is only one significant autocorrelation coefficient. This suggests an MA(1) model which leaves the option of using model (6) or model (8) depending on whether or not one wants the seasonal coefficient in the AR or MA part. The decision on this is not very clear because of the limited number of autocorrelations and partial autocorrelations. Similarly, the choice of a model is highly uncertain because of randomness and the interaction between seasonal and non-seasonal influences. Thus, quite often the only alternative is to try different models and choose the one which minimizes the MSE.

For the specific data of the French paper industry, one might choose model 6. Its output is shown in Figure 18-8. The first parameter of model 6 is $\theta_1 = .8255$, the second parameter, q, is zero (it does not exist in model 6) and the third is $\theta_s = .61228$. The model then is:

$$Z_t = a + e_t - \theta_1 e_{t-1} - (\theta_s e_{t-12} - \theta_1 \theta_s e_{t-L-1})$$

where

$$Z_t = (X_t - X_{t-1}) + (X_{t-13} - X_{t-12})$$

or

$$Z_t = 0 + e_t - .82553\, e_{t-1} - (.61228\, e_{t-12}$$

$$- .82553(.61228)\, e_{t-13})$$

or

$$(X_t - X_{t-1}) + (X_{t-13} - X_{t-12}) = 0 + e_t - .82553\, e_{t-1}$$

$$- .61228\, e_{t-12} - .50456\, e_{t-13}$$

and solving for X_t yields:

$$X_t = X_{t-1} - X_{t-13} + X_{t-12} + e_t - .82553\, e_{t-1} - .61228\, e_{t-12}$$

$$- .50456\, e_{t-13}$$

Forecasting for period 121 will therefore be:

$$F_{121} = X_{120} - X_{108} + X_{109} + e_{121} - .82553\, e_{120}$$

$$- .61228\, e_{109} - .50456\, e_{108}$$

$$= 993.7 - 880 + 875 + e_{121} - 57.7045 + 8.8168 + 6.206$$

since

$$e_{120} = 69.9, \; e_{109} = -14.4 \text{ and } e_{108} = 12.3$$

$$F_{121} = 946.1 + e_{121}$$

Similarly, forecasts for periods 122 to 132 can be found and their values with 95% confidence limits are shown in Figure 18-9. Very similar results are obtained by using the program UNIVBJ (see Chapter 26). The parameters found for example are: $\theta_1 = .849$ and $\theta_s = .656$. The forecast and the 95% bounds from UNIVBJ are shown below.

PERIOD	FORECAST	LOWER 95% BOUND	UPPER 95% BOUND
121	942.989	853.755	1032.22
122	989.834	899.589	1080.08
123	1027.53	936.291	1118.78
124	974.933	882.701	1067.17
125	927.03	833.819	1020.24
126	1014.93	920.755	1109.11
127	861.715	766.578	956.852
128	424.569	328.484	520.655
129	889.577	792.553	986.602
130	974.008	876.053	1071.96
131	905.776	806.899	1004.65
132	991.586	891.797	1091.38

FIGURE 18-7   Autocorrelation Coefficients for 1 Short and 1 Long
Difference

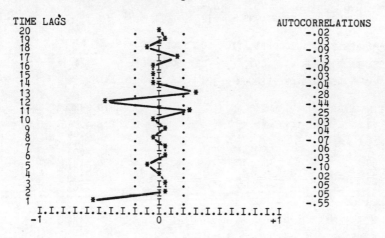

```
          *** AUTOCORRELATION COEFFICIENTS ***
STANDARD ERROR =   .155      CHI SQUARED =        84.09
          DEGREES OF FREEDOM = 32

TIME LAGS                                        AUTOCORRELATIONS
  20                                                  -.02
  19                                                   .03
  18                                                  -.09
  17                                                   .13
  16                                                   .06
  15                                                  -.03
  14                                                  -.04
  13                                                   .28
  12                                                  -.44
  11                                                   .25
  10                                                  -.03
   9                                                   .04
   8                                                  -.07
   7                                                   .06
   6                                                   .03
   5                                                  -.10
   4                                                   .02
   3                                                   .05
   2                                                   .05
   1                                                  -.55
      I.I.I.I.I.I.I.I.I.I.I.I.I.I.I.I.I.I.I.I.I
     -1                  0                  +1
```

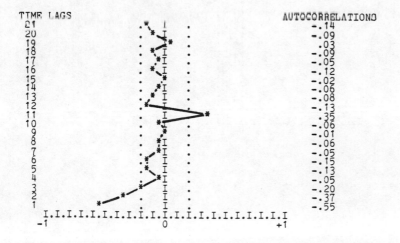

```
     *** PARTIAL AUTOCORRELATION COEFFICIENTS ***
STANDARD ERROR =   .097      CHI SQUARED=        93.65
          DEGREES OF FREEDOM = 32

TIME LAGS                                        AUTOCORRELATIONS
  21                                                  -.14
  20                                                  -.09
  19                                                   .03
  18                                                  -.09
  17                                                  -.05
  16                                                  -.12
  15                                                  -.02
  14                                                  -.06
  13                                                  -.08
  12                                                  -.13
  11                                                   .35
  10                                                  -.06
   9                                                  -.01
   8                                                  -.06
   7                                                  -.05
   6                                                  -.15
   5                                                  -.13
   4                                                  -.05
   3                                                  -.20
   2                                                  -.37
   1                                                  -.55
      I.I.I.I.I.I.I.I.I.I.I.I.I.I.I.I.I.I.I.I.I
     -1                  0                  +1
```

FIGURE 18-8   Output for MA(1) Model, Seasonal in MA

```
SHORT TERM DIFFERENCES =   1  LONG TERM   =   1
LENGTH OF SEASONALITY  =  12  MODEL NUMBER=   6

    1 ST PARAMETER =      .82553
    2 ND PARAMETER =            0
    3 RD PARAMETER =      .61228

    MSE =                 19.7
    NUMBER OF ITERATIONS =   39
```

*** RESIDUAL AUTOCORRELATION COEFFICIENTS *******

```
STANDARD ERROR =   .097      CHI SQUARED =      11.91

        DEGREES OF FREEDOM = 32
```

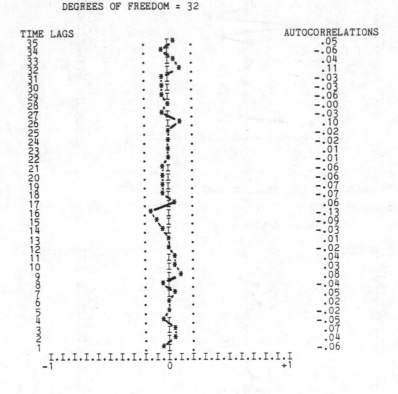

```
TIME LAGS                                       AUTOCORRELATIONS
  35                                                   .05
  34                                                  -.06
  33                                                   .04
  32                                                   .11
  31                                                  -.03
  30                                                  -.03
  29                                                  -.06
  28                                                  -.00
  27                                                  -.03
  26                                                   .10
  25                                                  -.02
  24                                                  -.02
  23                                                   .01
  22                                                  -.01
  21                                                  -.06
  20                                                  -.06
  19                                                  -.07
  18                                                  -.07
  17                                                   .06
  16                                                  -.13
  15                                                  -.09
  14                                                  -.03
  13                                                   .01
  12                                                  -.02
  11                                                   .04
  10                                                   .03
   9                                                   .08
   8                                                  -.04
   7                                                   .05
   6                                                   .02
   5                                                  -.02
   4                                                  -.05
   3                                                   .07
   2                                                   .04
   1                                                  -.06
      I.I.I.I.I.I.I.I.I.I.I.I.I.I.I.I.I.I.I.I.I
     -1                     0                     +1
```

FIGURE 18-9   Forecasts for Periods 122 to 132

```
MEAN PC ERROR (MPE) OR BIAS      =      -1.67%
MEAN SQUARED ERROR (MSE)         =       19.4
MEAN ABSOLUTE PC ERROR   (MAPE) =        5.6%

FORECAST FOR HOW MANY PERIODS AHEAD (0=NONE,24=MAX)?12

PER    FORECAST      95 PC.BOUNDS
121      94.6        85.7        103.5
122      99.7        88.2        111.2
123     103.0        91.5        114.5
124      98.2        86.7        109.7
125      92.9        81.4        104.5
126     102.2        90.7        113.8
127      86.9        75.4         98.4
128      42.3        30.8         53.9
129      89.5        77.9        101.0
130      97.9        86.3        109.4
131      90.7        79.2        102.3
132      99.9        88.4        111.4

DO YOU WANT TO TRY ANOTHER SET OF DIFFERENCES?N

MODEL NO. (0=NO MORE)?0
```

Cases

KOOL KAMP

One cold, snowy day in mid-January of 1975, Mr. Fred Clark, President and chief stockholder of Kool Kamp Inc., sat in his office in Melrose Park, Illinois considering the topics that had been covered in a recent seminar he had attended. While the seminar was aimed at short and long range planning for smaller manufacturing companies, a good part of it had dealt with the problem of forecasting and the available methods that might be used. Fred realized that there was substantial opportunity to use forecasting as part of the planning operation in his own firm and he was anxious to pursue such an application. As a first step, he wanted to focus on the selection and application of one or more forecasting methods to the problem of predicting monthly sales for Kool Kamp Inc. Fred had collected the historical sales data for the company's past four years as shown in Table 1 and Figure 1. However, he was still at ground zero in terms of applying any formal forecasting technique to his situation.

COMPANY BACKGROUND

The Kool Kamp Company had been founded in 1970, shortly after Fred had completed an MBA degree at the Harvard Business School. He had chosen the general Chicago area as a plant location site because of its ready access to a nationwide distribution network and the fact that many of his customers were either headquartered in Chicago or were regularly passing through the area.

Fred had started the business after extensive discussions with a close family friend who owned a small chain of sporting goods stores. This friend had informed Fred that there appeared to be a substantial market for inexpensive styrofoam ice chests and coolers and at that time there were no major manufacturing concerns involved in that segment of the business. Seeing this as an opportunity to get in on the ground floor, Fred had formed the company to manufacture and sell as its first product a styrofoam camp cooler which retailed for less than seven dollars. Since its initial production run, the company had added a couple of sizes and alternative models but had not yet extended its product line to any other segment of the cooler market or the leisure product field. (Fred did hope to be able to expand into other related product lines in the near future.)

Distribution and sales for the company were handled through three manufacturing reps and by Fred himself. Most of the sales went to major discount chains and to retail sporting goods chains. Thus there were only a couple of dozen customers that accounted for ninety percent of the firm's business. Since price seemed to be the most important criteria in sales of this product, the company's emphasis had been placed on maintaining low costs through effective purchasing of materials and efficient manufacturing.

As indicated by the data in Table 1, and Figure 1, sales had grown dramatically since 1970 and Fred now felt that he had a substantial share of the low cost styrofoam cooler market. Unfortunately one of the problems with the demand for the company's product was that it was extremely seasonal both at the final consumer level and in terms of retail store purchases from Kool Kamp's reps. The latter was the case because of the substantial storage space required for holding more than a few dozen units of the product in any single location. While Fred was aware that some steps might be taken to try to spread peak demand over a longer period of time, so far he had not attempted to do so.

The company's general production policy was to work two full shifts in the late winter and spring months and then to work a single shift throughout the remainder of the year. Thus substantial inventories were built up in the late summer and fall periods until demand began to pick up in midwinter. Because of these production requirements, Fred was most anxious to prepare aggregate plans for the company that would be based on the best available information concerning demand. It was in this regard that Fred saw the first use of the sales forecasts whose preparation he was presently contemplating.

Even before Fred had attended the recent seminar, he had asked his three manufacturing reps to help him prepare sales estimates for the first three months of fiscal 1975. These estimates, along with the actual results for those first three months, are shown in Table 1. Fred had found it extremely difficult to obtain these estimates and even when he did, he had been told by all three reps that they were simply ballpark figures and that they (the reps) certainly wouldn't put too much faith in them. The reps had pointed out that one of the problems was that the company's sales were becoming even more seasonal than they had been in the past. This made it extremely difficult to forecast sales for individual months although the reps felt they could do a reasonaly good job in forecasting annual demand.

FORECASTING ISSUES TO BE CONSIDERED

One of the things that Fred had learned at the recent seminar was that there were a number of issues that needed to be considered in applying formal forecasting techniques in any particular situation. One of these which he felt was particularly important in the case of Kool Kamp was selecting one or more techniques that appeared to be appropriate for the situation and then developing a procedure that could be used in testing those to determine which one should be used in the future. From his discussions with the reps, he also felt it would be important to decide whether the initial forecast should be of total annual demand which could then be broken down into forecasts for individual months or whether he should first forecast each month and then simply aggregate them to obtain an estimate of annual demand.

While he wanted accurate forecasts immediately, he realized that there would be a period of time over which he would need to use one or more techniques in order to build up experience with them and to determine their strengths and weaknesses in his situation. One of the things he wanted to do was be sure that he designed his initial procedures in such a way that he could obtain that experience as quickly as possible and build up confidence in his own mind and for others in the organization concerning the reliability of those forecasts.

Fortunately Fred had a close friend who owned a small timesharing bureau and he was certain that he could buy time from that friend at a very reasonable price (perhaps five to eight dollars per terminal hour). In addition he felt that he could impose upon his friend to write him some simple forecasting programs and/or to supply more elaborate programs (such as regression analysis) from the library package supplied by the computer manufacturer. Thus for a rather modest cost Fred was certain that he could gain access to a computer to help him in this forecasting situation.

TABLE 1   Actual Sales (in 000's of dollars) 1971-1974

Month	Fiscal Year	Actual Sales	Month	Fiscal Year	Actual Sales
1. October		$ 1.3	25. October		$ 18.8
2. November		0.1	26. November		13.2
3. December		0.5	27. December		11.6
4. January	(1971)	9.5	28. January	(1973)	53.4
5. February		37.2	29. February		126.2
6. March		81.1	30. March		153.9
7. April		87.3	31. April		188.2
8. May		59.5	32. May		206.1
9. June		72.9	33. June		168.9
10. July		58.6	34. July		97.6
11. August		28.1	35. August		35.2
12. September		14.1	36. September		33.7
13. October		$ 6.4	37. October		$ 20.4
14. November		12.7	38. November		13.5
15. December		13.3	39. December		22.8
16. January	(1972)	20.9	40. January	(1974)	92.7
17. February		65.5	41. February		161.7
18. March		124.5	42. March		316.8
19. April		144.0	43. April		372.8
20. May		137.8	44. May		272.4
21. June		156.0	45. June		181.0
22. July		89.2	46. July		92.1
23. August		36.9	47. August		39.9
24. September		25.2	48. September		34.2

	(Fiscal Year 1975)		*Management Estimates*
49. October		$ 15.8	21.0
50. November		13.1	13.0
51. December		23.0	22.0

FIGURE 1   Graph of Sales Results (from Table 1)

Monthly Sales
(000's of
Dollars)

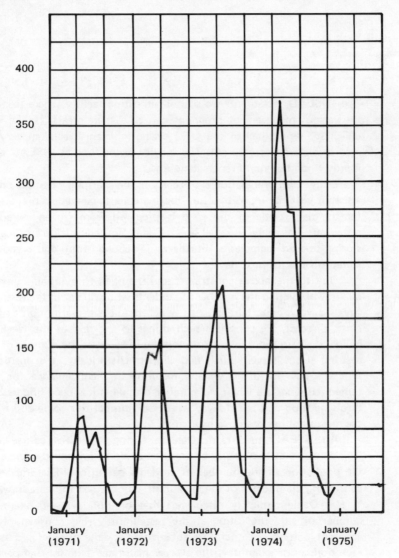

Graph of Sales Results

INSEAD'S RESTAURANT

The INSEAD restaurant is a school-run eating facility where participants of the various management programs can take their meals. Unlike the typical university dining hall in the U.S., eating in France is taken very seriously, even in a student restaurant. The quality of the food at INSEAD's restaurant is pretty good by non-French standards.

Daily food preparation starts early in the morning when the cook and his assistant visit the market to buy the ingredients needed to prepare the meals for the day. Little of the food is prepared from frozen ingredients. The restaurant has only a small freezer and limited refrigerator capacity for storing unused ingredients. Furthermore, cooked food and bread cannot be stored for use the next day.

Due to the above constraints and the need to maintain consistent high quality, predicting the number of students who will eat in the restaurant each day is a very important activity which requires the attention of the restaurant's director. He considers such things as the day of the week, weather, holidays, exam times, etc. in attempting to forecast the number of customers desiring to be served lunch and dinner. Historically, the accuracy of his forecasts has fluctuated widely, resulting in inconvenience for those customers coming late when the number of clients has been underestimated or throwing food and bread away when the number of people desiring to eat in the restaurant has been overestimated.

An INSEAD student, Mr. Dahlem, taking an elective course on forecasting decided that the quality of the food could be improved if the accuracy of the prediction of the number of students eating could be improved. Upon contacting the director of the restaurant, he obtained the data shown in Table 1. Mr. Dahlem quickly discovered that on holidays and exam days, the number of students eating at the restaurant varied considerably with no apparent pattern. He therefore substituted for the number of customers on each holiday or exam day, the average historical value for that particular day of the week. These averages are identified with an * in Table 1 and are as follows:

[1] Institut European D'Administration des Affaires, (Fontainebleu, France)

Mon: 294
Tue: 300
Wed: 291
Thu: 285
Fri: 242
Sat: 182
Sun: 182

By running an autocorrelation program Mr. Dahlem discovered, as expected, that the data showed a strong seasonal pattern of 7 periods (days) duration and were stationary at zero differences. Stationarity was expected because on different days there are different numbers of students eating and because the total number of students at INSEAD did not change over the five months covered by the historical data.

By running a decomposition program, Mr. Dahlem found that Tuesday was the most popular day for eating at the restaurant while Saturday was the least popular. The seasonal coefficients identified had the following values:

Mon: 120.49
Tue: 121.62
Wed: 117.88
Thu: 115.29
Fri: 93.95
Sat: 63.19
Sun: 67.59

The results of this preliminary analysis simply increased Mr. Dahlem's interest in this forecasting situation and he decided to use Winter's and Harrison's forecasting models as well as Census II, Generalized Adaptive Filtering (GAF), and Box-Jenkins techniques. During the week of April 14 in which he carried out the comparison of these forecasting methods, new data on the number of students eating was collected. There were no holidays or exams and the actual numbers of students eating on each day of that week were as follows:

Week of April 14

Mon: 305
Tue: 279
Wed: 268
Thu: 251
Fri: 215
Sat: 132
Sun: 145

Mr. Dahlem desired to determine how each of the models had performed in comparison to the actual values for the past five months and for the past week. To do this, he planned first to measure the accuracy of each model (a) in terms of the MSE (mean squared error) and MAPE (mean absolute percentage error) for all of the fitted data values (Table 1), and (b) in terms of the MSE and MAPE for the past week's actual values.

TABLE 1   Number of Students Eating at INSEAD's Restaurant Daily — November 1, 1974, to April 13, 1975

	1974				1975							
	November		*December*		*January*		*February*		*March*		*April*	
Sat									1	132		
Sun			1	176					2	165		
Mon			2	278					3	273		
Tue			3	300					4	289	1	280
Wed			4	283	1				5	285	2	287
Thu			5	292*	2				6	267	3	294
Fri	1	242*	6	210	3				7	250	4	260
Sat	2	178	7	248	4		1	87	8	176	5	181
Sun	3	233	8	211	5		2	116	9	189*	6	172
Mon	4	294*	9	303	6		3	294	10	282	7	296
Tue	5	329	10	300	7		4	308	11	310	8	281
Wed	6	310	11	289	8	267	5	302	12	285	9	288
Thu	7	312	12	260	9	280	6	299	13	272	10	260
Fri	8	242*	13	271	10	255	7	244	14		11	220
Sat	9	133	14	285	11	281	8	245	15		12	128
Sun	10	131	15	145	12	223	9	239	16		13	146
Mon	11	294*	16	278	13	309	10	294*	17		14	
Tue	12	293	17	313	14	302	11	301	18		15	
Wed	13	290	18		15	294	12	282	19		16	
Thu	14	285	19		16	299	13	271	20		17	
Fri	15	273	20		17	257	14	206	21		18	
Sat	16	168	21		18	185	15	137	22		19	
Sun	17	130	22		19	207	16	158	23		20	
Mon	18	320	23		20	324	17	277	24		21	
Tue	19	293	24		21	304	18	287	25		22	
Wed	20	286	25		22	323	19	279	26		23	
Thu	21	289	26		23	293	20	295	27		24	
Fri	22	285	27		24	242	21	228	28	242*	25	
Sat	23	197	28		25	124	22	152	29	182*	26	
Sun	24	203	29		26	164	23	190	30	189*	27	
Mon	25	286	30		27	315	24	278	31	294*	28	
Tue	26	301	31		28	309	25	280			29	
Wed	27	290			29	310	26	281			30	
Thu	28	284			30	273	27	295				
Fri	29	204			31	243	28	217				
Sat	30	180										

JACKSON HOLE AIRPORT

Like many U.S. airports in the early '70s, the Jackson Hole Airport located at Jackson Hole, Wyoming was faced with the problem in mid-1973 of planning airport facilities for the coming decade. The previous three years had seen an increase in traffic of almost 50% (in terms of total passenger boardings) and the Airport Planning Authority was anxious to provide adequate facilities for future growth. This was particularly brought to mind as they began hearings in the summer of 1973 on a proposed 20% expansion in the facilities at the airport. Several local businessmen had attended those hearings to complain of the fact that they felt considerable business had been lost the previous winter due to overcrowded airport facilities and limitations placed on the number of incoming flights and the size of groups that could be handled at the airport. These businessmen were anxious to avoid any repeat of that in the future and felt that the proposed expansion of the airport was not nearly sufficient to handle anticipated future requirements.

The counter to the business community's argument had been put forward by a couple of local environmental groups. Those groups felt that any expansion of the airport would not only damage the scenic field immediately surrounding the airport location but would also attract an increased number of tourists whose presence would have a more lasting impact on the entire Jackson Hole area. This group argued that the summer of 1972 and the late fall and early winter going into '73 had been a period of unusually high demand and could not be expected to continue. Thus, they were arguing that no substantial increase in airport facilities was really required.

In order to help the Airport Planning Authority to arrive at a factually based recommendation, a local consultant, Roger Schmenner, had been hired to prepare a forecast of total passenger boardings for the Jackson Hole Airport during the next three years. As background for Roger, the Airport Planning Authority had supplied him with the information shown in Figure 1. Unfortunately the Authority did not have available any projections on total visitors to either Grand Teton or Yellowstone National Park during the coming years. Roger had agreed to do the best he could with the available data and had also suggested that he might try and think of other types of data that might readily be available and that could be used as either a double check against the forecast he would prepare or perhaps might provide the information necessary to use a more complex forecasting technique, such as regression analysis, at some future point in time.

TABLE 1   Jackson Hole Airport, 1965-1972
Comparison of Park Visitors and Passenger
Boardings

Year	Total Visitors Grand Teton	Total Visitors Yellowstone	Total Passenger Boardings
1965	2,507,047	2,060,896[a]	8,392
1966	2,673,090	2,129,017[a]	12,282
1967	2,643,708	2,166,397[a]	15,816
1968	2,970,255	2,229,662	17,285
1969	3,134,354	2,193,894	18,849
1970	3,352,464	2,297,290	17,577
1971	3,284,539	2,120,487	19,473
1972	3,002,230	2,246,827	24,405

[a]Includes only visitors by automobile.
Source: Grand Teton National Park Headquarters and Yellowstone National Park Headquarters.

CHAPTER 19

CHOOSING A FORECAST-ING METHOD (SIBYL)

This book has examined a variety of forecasting methods. They range in cost, accuracy, complexity, and in the type of data they can handle. On the one extreme, there are the exponential smoothing methods, which are cheaper, and easier to use, but possibly less accurate. They are intended for specific types of data and are mainly used when large numbers of forecasts are involved. On the other hand, there are the autoregressive-moving average methods which are the most general, under certain conditions more accurate, but at the same time more costly and more difficult to use.

A major characteristic of time series forecasting is that the user has the choice of several methods and can select the one that best fits the data and accuracy requirements at an appropriate cost. SIBYL is an aid in making this choice of a forecasting method for a specific situation.

The choice of a forecasting method depends upon several factors:

1. The type of pattern in the data;
2. The time horizon of forecasting;
3. The cost of using the forecasting method;
4. Its accuracy;
5. The difficulty, or complexity to the user of using it.

The type of data pattern is an important factor in determining the type of method to be used. The data can be seasonal, or have a trend which makes certain methods appropriate or inappropriate. For example, a linear exponential smoothing method cannot be used when the data are seasonal and single exponential smoothing cannot be used when there is a strong trend in the data. Similarly, a seasonal method will add nothing if the data are non-seasonal. SIBYL plots the data (see Figure 19-1) and analyzes them by finding the autocorrelations of the original data and any of their differences. The objective is to discover:

    a. the level at which the data are stationary (this is important in deciding what method to choose and is needed when using several forecasting methods);

    b. whether or not the data are seasonal;

    c. the length of seasonality.

The data analysis can be done automatically if the user desires. However, the user's judgment is a valuable input to effective data analysis. Figure 19-2 presents the results of data analysis on a time series using SIBYL. This data series represents monthly values of the French Index of Industrial Production for over eleven years.

Another part of SIBYL decomposes the data (See Figure 19-3) and gives the average absolute percentage change of each component as well as that of the original data. (The concept of decomposition is described in Chapters 15 and 16 and the percentage changes of components are explained in Chapter 16 in INPUT/OUTPUT 3). The average value of, for example, variations in the original values of the French Index of Industrial Production is 11.93% (see Figure 19-3). This means that the data are reasonably smooth, with no wide month-to-month variations. Most of the variation is caused by seasonality, whose monthly average absolute change is 9.39%. The trend-cycle is not very important since on the average it causes only .54% of the monthly variation. Further, this means that the seasonality causes seventeen times as much variation as the trend-cycle. Another important variation is that of randomness, which is 2.45% in this case. This provides an estimate of the extent to which forecasting can be successful in estimating a time pattern in the data. Thus, one should expect average errors (inaccuracies) higher than 2.45%. In practice, the actual errors are usually between 0.5% and 2% higher than the average percentage change in randomness for more than one period ahead forecasts. A very simple forecasting approach could apply the current period's figure as the forecast for the next period. This would give an average error of 11.93%. (This is called Naive Method Number 1.) Since the best that one can do is to forecast everything but changes due to randomness, the maximum possible improvement is 9.47%. This is a factor which should influence any decision of whether to use a formal forecasting method, and if so, which one.

The second factor, the time horizon of forecasting, is external to the system and generally one over which the user has little control. Depending upon the time interval of forecasting (i.e., one period, three periods or less,

FIGURE 19-1 Plot of the Data (SIBYL)

```
EXE SIBYL
  8K
***SIBYL - INTERACTIVE FORECASTING***

NEED HELP (Y OR N)?N

DO YOU WANT TO USE..
(1) PROGRAM DATA
(2) YOUR OWN DATA, IN A FILE
(3) YOUR OWN DATA, TO BE TYPED IN AT THE TERMINAL
(4) DATA GENERATED BY THE PROGRAM ,WHICH (1,2,3 OR 4)?2
DATA FILENAME?FIIP
HOW MANY OBSERVATIONS DO YOU WANT TO USE?133

DO YOU WANT A GRAPH OF YOUR DATA?Y
HOW MANY OF YOUR 133 OBSERVATIONS DO YOU WANT PLOTTED?24
```

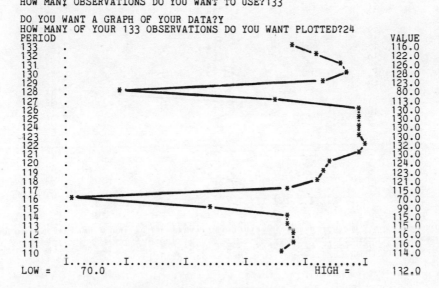

```
PERIOD                                                        VALUE
133                                                           116.0
132                                                           122.0
131                                                           126.0
130                                                           128.0
129                                                           123.0
128                                                            80.0
127                                                           113.0
126                                                           130.0
125                                                           130.0
124                                                           130.0
123                                                           130.0
122                                                           132.0
121                                                           130.0
120                                                           124.0
119                                                           123.0
118                                                           121.0
117                                                           115.0
116                                                            70.0
115                                                            99.0
114                                                           115.0
113                                                           115.0
112                                                           116.0
111                                                           116.0
110                                                           114.0
       I........I........I........I........I........I
LOW =    70.0                                  HIGH =    132.0
```

```
DO YOU WANT THE AUTOCORRELATIONS OF YOUR DATA (Y OR N)?Y


*****   AUTOCORRELATION ANALYSIS    *****


        TABLE    1    AUTOCORRELATIONS    0TH DIFFERENCE
        -----         ----------------    ----------

MEAN AUTOCORRELATION      .458       STANDARD ERROR=     .087
MEAN OF FIRST  12  VALUES=    .552   CHI SQUARE (COMPUTED)=    719.4
MEAN OF LAST   12  VALUES=    .363   CHI SQUARE (FROM TABLE)=   13.1
COEFFICIENT OF VARIATION =    .223

DO YOU WANT A GRAPH OF THE AUTOCORRELATIONS?N

A STUDY OF THE AUTOCORRELATION COEFFICIENTS TELLS ME THAT THERE IS
SOME PATTERN IN YOUR DATA,I.E. THEY ARE NOT RANDOMLY DISTRIBUTED
AROUND THEIR MEAN (THIS IS BECAUSE   24 VALUES LIE OUTSIDE THE
CONTROL LIMITS--DOTTED LINES)   SEE PP11-19
```

FIGURE 19-2 Data Analysis Done by SIBYL

```
TABLE     2   AUTOCORRELATIONS      1ST DIFFERENCE
-----         ------------------        ----------

MEAN AUTOCORRELATION      -.005     STANDARD ERROR=        .087
MEAN OF FIRST  12  VALUES=  -.006   CHI SQUARE (COMPUTED)=     230.0
MEAN OF LAST  12   VALUES=  -.003   CHI SQUARE (FROM TABLE)=    13.1
COEFFICIENT OF VARIATION =   .163

DO YOU WANT A GRAPH OF THE AUTOCORRELATIONS?Y
```

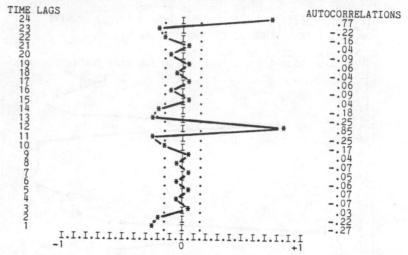

```
TIME LAGS                                          AUTOCORRELATIONS
  24                                                    .77
  23                                                   -.22
  22                                                   -.16
  21                                                    .04
  20                                                   -.09
  19                                                    .06
  18                                                   -.04
  17                                                    .06
  16                                                   -.09
  15                                                    .04
  14                                                   -.18
  13                                                   -.25
  12                                                    .85
  11                                                   -.25
  10                                                   -.17
   9                                                    .04
   8                                                   -.07
   7                                                    .05
   6                                                   -.06
   5                                                    .07
   4                                                   -.07
   3                                                    .03
   2                                                   -.22
   1                                                   -.27
   I.I.I.I.I.I.I.I.I.I.I.I.I.I.I.I.I.I.I.I.I.I.I.I
  -1                      0                   +1
```

A STUDY OF THE AUTOCORRELATION COEFFICIENTS TELLS ME THAT THERE IS
SOME PATTERN IN YOUR DATA,I.E. THEY ARE NOT RANDOMLY DISTRIBUTED
AROUND THEIR MEAN (THIS IS BECAUSE   8 VALUES LIE OUTSIDE THE
CONTROL LIMITS--DOTTED LINES)   SEE PP11-19

```
TABLE     3   AUTOCORRELATIONS      2ND DIFFERENCE
-----         ------------------        ----------

MEAN AUTOCORRELATION      -.004     STANDARD ERROR=        .087
MEAN OF FIRST  12  VALUES=  -.006   CHI SQUARE (COMPUTED)=     314.7
MEAN OF LAST  12   VALUES=  -.003   CHI SQUARE (FROM TABLE)=    13.1
COEFFICIENT OF VARIATION =   .262

DO YOU WANT A GRAPH OF THE AUTOCORRELATIONS?N
```

A STUDY OF THE AUTOCORRELATION COEFFICIENTS TELLS ME THAT THERE IS
SOME PATTERN IN YOUR DATA,I.E. THEY ARE NOT RANDOMLY DISTRIBUTED
AROUND THEIR MEAN (THIS IS BECAUSE   6 VALUES LIE OUTSIDE THE
CONTROL LIMITS--DOTTED LINES)   SEE PP11-19

YOUR DATA ARE STATIONARY AT THEIR   1ST DIFFERENCE

THEREFORE I AM LOOKING AT TABLE   2 TO DETERMINE THE LENGTH
OF SEASONALITY,IF ANY EXISTS

AUTOCORRELATIONS SIGNIFICANTLY DIFFERENT THAN ZERO (ORDERED)

```
TIME LAG          AUTOCORRELATIONS
  12                   .85
  24                   .77
   1                   .27
  11                   .25
  13                   .25
  23                   .22
   2                   .22
  14                   .18
```

I CONSIDER THE LENGTH OF YOUR SEASONALITY IS =   12

DO YOU AGREE WITH ME (Y OR N)?Y

FIGURE 19-3   Decomposition and Average Absolute Percentage Change (SIBYL)

```
SEASONAL INDICES BASED ON MEDIAL AVERAGE(Y OR N)?Y


        *** SUMMARY STATISTICS ***
COMPOSITION OF      AVERAGE PC   APPROXIMATE
PER-TO-PER CHANGE    CHANGE      % EXPLAINED

ORIGINAL DATA       11.93         100.
TREND-CYCLE          .54            4.
 TREND(LINEAR)       .47
 CYCLE               .36
SEASONALITY         9.39           76.
RANDOMNESS(NOISE)*  2.45           20.

*OPTIMAL..LOWEST ERROR POSSIBLE IN PERFECT FORC. MODEL

SEASON MEDIAL AVERAGE NORMALIZED
  2      105.99        105.76
  3      105.59        105.36
  4      105.93        105.71
  5      104.14        103.91
  6      105.26        105.03
  7       92.45         92.25
  8       60.09         59.97
  9      102.36        102.14
 10      105.64        105.41
 11      106.43        106.20
 12      105.86        105.63
  1      102.85        102.63


TOTAL   1202.57       1200.00
DO YOU WANT TO OVERRIDE THESE SEAS. FACTORS(Y OR N)?N

COEFFICIENT OF VARIATION                  .16

WHAT IS THE TIME HORIZON OF FORECASTING(1,2,3 OR 4)?HELP


TYPE 1,2,3 OR 4 DEPENDING ON THE TIME HORIZON OF YOUR FORECAST
1 - IF 1 PERIOD OUT ONLY
2 - IF LESS THAN 3 PERIODS
3 - FROM 4 PERIODS TO LESS THAN 24 PERIODS
4 - 24 PERIODS OR LONGER
WHAT IS THE TIME HORIZON OF FORECASTING(1,2,3 OR 4)?2


     *** HISTORICAL ACCURACY (% ERROR) AND COST INDEX     ***
                  SEASONAL METHODS

METHOD                    FORECAST PERIODS AHEAD
 NAME     MODEL    1      2      3      6      12     COST
EXPOW     6.45%   7.31%  9.12%  10.30% 11.90% 12.31%  3.0
DCOMP     6.61%   8.69%  8.85%   9.04%  9.60% 10.71%  1.2
CENSUS    6.59%   8.49%  8.61%   8.72%  9.29% 10.47%  2.1
HARM      6.62%  10.05% 10.21%  10.31% 10.71% 11.62%   .8
MREG      7.49%  10.20% 11.18%  12.00% 15.84% 18.43%  2.8
GAF       6.50%   7.30%  8.02%   8.30%  9.20%  9.80%  3.8
BOXJEN    6.00%   7.05%  7.56%   8.20%  9.30%  9.75%  6.5
NAIVE2    7.50%   8.70%  9.63%  10.52% 14.60% 18.80%   .4

DO YOU WANT A DESCRIPTION OF ANY OF THE ABOVE METHODS?N


SOME RECOMMENDED METHODS TO START WITH ARE

METHOD    COMPLEXITY   RECOMENDED MIN
 NAME     TO USER      DATA POINTS
EXPOW       .35            24
DCOMP       .5             60
CENSUS      .8             72
HARM        .7             36
MREG        .6             72
GAF         .6             60
BOXJEN     1               72


DO YOU WANT TO RUN ANY FORECASTING PROGRAM(Y OR N)?N
 03600,STOP
```

four periods to less than two years, and two years or longer (see Figure 19-3)), the SIBYL program will identify the specific methods that seem more appropriate, even though this will depend upon the specific requirements of the user and can vary from one situation to another. These methods are listed on the bottom of Figure 19-3 after the accuracies of the seasonal methods have been listed. It is left to the user to evaluate tradeoffs among the remaining factors of cost, accuracy and complexity. SIBYL does provide general information about costs, accuracy and complexity for each of the methods as an aid to making such a choice.

The different factors involved and the various forecasting methods available are shown in Figures 19-4 and 19.5. The decision matrix that is used by SIBYL in selecting the most suitable forecasting methods is shown in Figure 19-6.

References for further study:

> Chambers, et al, (1971)
> Makridakis, et al. (1974)
> Makridakis, S. and S. Wheelwright (1978, Chapter 16)
> Wheelwright, S. and S. Makridakis (1977, Chapter 12)

INPUT/OUTPUT FOR SIBYL

Input

1. *Do You Want to Graph Your Data?*

A graph can be plotted of as many of the data points as the user specifies.

2. *Do You Want Autocorrelations of Your Data?*

If the answer is YES, SIBYL calls ACORN which analyzes the data to determine:

> a. the level at which the data are stationary (original, first difference, second difference)
> b. whether or not the data are seasonal
> c. the length of seasonality, if the data are seasonal.

Once ACORN has been used (a), (b) and (c) can be determined automatically or the user may guide the program by answering 3 and 4 below.

3.   *At What Level Are the Data Stationary?*

The answer should by 0, 1 or 2 depending upon the level at which the data are stationary.

4.   *Length of Seasonality (0 if none)?*

The length of seasonality is the time span, in periods, between successive seasonal peaks (or troughs). For example, in monthly data it is commonly 12.

4a.   *Seasonal Indices Based on the Medial Average (Y or N)?*

A yes (Y) answer calls for seasonal indices to be computed using the medial method which excludes the highest and lowest values for each month. A no (N) indicates that all available values are to be used in computing the seasonal indices.

4b.   *Do You Want to Override These Factors?*

A yes (Y) will cause the program to prompt you for your own estimates of the seasonal indices. Otherwise the values computed by the program will be used.

5.   *Time Horizon of Forecasting?*

Depending upon the time horizon of the data, the answer should be 1, 2, 3 or 4: 1 if the data are taken for forecast intervals of one period, 2 if the interval is between one and three periods, 3 if it is four periods to two years, and 4 if it is two years or more.

6.   *Do You Want a Description of Any Program?*

If the answer is YES, the user should type the appropriate number and SIBYL will provide a short description of each of the programs listed.

7.   *Do You Want to Run Any of the Programs?*

If the answer is YES, control is passed to RUNNER which can run any available program. With an answer of NO, the program stops.

Output

1.   *Time Series Decomposition*

   a. percentage changes of each component.
   b. seasonal indices. (For details see Chapter 16.)

FIGURE 19-4 Dependent Variable is the Mean Absolute Percentage Error (MAPE) of Model Fitting

	Forecasting Method	R^2	Constant	INDEPENDENT VARIABLES				Standard Error of Estimate	F-test
				No of Data	MEAN ABSOLUTE % CHANGE				
					Trend Cycle	Randomness	Seasonality		
1	Naive 1	.98	0			.95 (38.29)	1.02 (51.51)		2051
2	Single Moving Average	.95	1.40 (3.24)			.69 (18.08)	.92 (33.84)	2.48	974
3	Single Exponential Smoothing	.96	1.35 (3.06)			.71 (17.99)	.90 (36.80)	2.58	1081
4	Adaptive Response Rate Exponential Smoothing	.93	2.02 (3.37)			.75 (13.94)	.95 (28.43)	3.49	646
5	Linear Moving Average	.94	1.55 (2.74)			.80 (15.88)	1.04 (28.96)	3.26	724
6	Brown's Linear Exponential Smoothing	.95	1.04 (2.15)			.76 (17.65)	.94 (34.77)	2.83	984
7	Holt's (2 parameters) Linear Exp. Sm.	.97	.79 (2.03)			.80 (23.20)	.97 (39.33)	2.26	1393
8	Brown's Quadratic Exponential Smoothing	.95	1.24 (2.36)			.77 (16.44)	.97 (33.43)	3.05	895
9	Linear Trend (Regression Fit)	.76	0		2.72 (4.83)	.64 (5.4)	.90 (14.03)	7.01	104

Original Data: Non-Seasonal Methods

10	Harrison's Harmonic Smoothing	.76	1.06 (1.75)		1.65 (5.66)	.58 (10.54)	.10 (4.04)	3.27	108
11	Winter's Linear and Seasonal Exp. Sm.	.90	3.61 (5.55)	-.02 (-3.72)		.87 (27.71)	.06 (3.72)	2.00	310
12	Adaptive Filtering	.86	2.09 (2.36)	-.01 (-.82)		.99 (22.49)	.10 (4.60)	2.70	212
13	AutoRegressive Moving Average (Box-Jenkins)	.84	2.30 (2.44)	-.02 (-1.86)		.94 (21.25)	.07 (3.13)	2.89	178
14	Naive 2	.96	.61 (2.94)		.33 (3.49)	.90 (46.22)		1.15	1384
15.	Single Moving Average	.92	.59 (2.29)		.71 (5.82)	.68 (28.22)		1.47	619
16	Single Exp. Sm.	.93	.71 (3.00)		.62 (5.49)	.69 (31.15)		1.36	730
17	Adaptive Response Rate Exponential Smoothing	.92	1.04 (3.35)		.73 (5.76)	.72 (28.45)		1.53	627
18	Linear Moving Average	.91	.53 (1.83)		.30 (2.03)	.82 (27.66)		1.80	518
19	Brown's Linear Exponential Smoothing	.93	.48 (1.83)		.28 (2.31)	.77 (32.04)		1.47	695
20	Holt's (2 parameters) Linear Exp. Sm.	.95	.32 (3.74)			.83 (42.29)		1.34	1788
21	Brown's Quadratic Exponential Smoothing	.93	.55 (2.15)		.31 (2.55)	.78 (32.57)		1.46	722
22	Linear Trend (Regression Fit)	.72	1.72 (2.37)		1.44 (4.72)	.62 (12.29)		3.33	126

Rows 10–14: *Seasonal and Non-Seasonal Methods*

Rows 15–22: *Seasonally Adjusted Data: Non-Seasonal Methods*

FIGURE 19-5   Dependent Variable is the Mean Absolute Percentage Error (MAPE) of Forecasts

	Forecasting Method	R^2	Constant	MEAN ABSOLUTE % CHANGE			Absolute % Change of Trend-Cycle at Period t	Number of Periods Ahead Forecasting	Standard Error of Estimate	F-test
				Trend-Cycle	Random-ness	Season-ality				
1	Naive 1	.47	-5.91 (-4.68)	2.22 (4.81)	1.11 (11.69)	.12 (2.48)	.66 (9.74)	1.44 (9.52)	15.06	145
2	Single Moving Average	.42	-4.02 (-3.47)	1.71 (4.03)	.73 (8.68)	.30 (6.96)	.45 (7.23)	1.51 (10.90)	13.80	121
3	Single Exponential Smoothing	.48	-5.79 (-5.12)	2.51 (6.06)	.81 (9.84)	.25 (5.82)	.54 (8.97)	1.54 (11.36)	13.45	149
4	Adaptive Response Rate Exponential Smoothing	.38	-2.19 (-2.02)	2.27 (5.70)	.83 (10.53)	.20 (4.78)	.13 (2.26)	1.26 (9.68)	12.89	101
5	Linear Moving Average	.37	-5.69 (-3.27)		1.07 (9.05)	.63 (10.25)	.22 (2.18)	2.51 (11.42)	22.00	123
6	Brown's Linear Exponential Smoothing	.41	-5.02 (-4.12)	.99 (2.21)	.79 (8.87)	.29 (6.57)	.42 (6.47)	1.86 (12.73)	14.62	117
7	Holt's (2 parameters) Linear Exp. Sm.	.40	-3.91 (-3.19)	1.43 (3.19)	.85 (9.06)	.18 (4.05)	.43 (6.50)	1.60 (10.84)	14.74	109
8	Brown's Quadratic Exponential Smoothing	.34	-5.46 (-3.68)	1.66 (3.04)	.72 (6.65)	.18 (3.45)	.37 (4.71)	2.49 (13.97)	17.76	84
9	Linear Trend (Regression Fit)	.36	-4.10 (-2.71)	5.30 (9.53)	.39 (3.61)	.46 (8.15)	.40 (5.06)	1.57 (8.70)	18.01	92

Original Data: Non-Seasonal Methods

	Method									
		Seasonal and Non-Seasonal Methods								
10	Harrison's Harmonic Smoothing	.38	0	5.50 (9.88)	.62 (5.48)	.37 (7.02)	.63 (7.39)	.75 (5.09)	18.80	103
11	Winter's Linear and Seasonal Exp. Sm.	.43	−2.45 (−2.69)	1.84 (5.52)	.65 (9.77)	.09 (3.02)	.50 (10.16)	.91 (8.42)	10.77	122
12	Adaptive Filtering	.48	−3.06 (−3.33)	1.52 (4.54)	.82 (11.85)	.17 (5.30)	.49 (9.63)	.91 (8.33)	10.95	155
13	AutoRegressive Moving Average (Box-Jenkins)	.43	−3.85 (−4.09)	2.08 (5.87)	.69 (10.00)	.19 (5.83)	.37 (6.97)	.98 (8.68)	11.14	123
14	Naive 2	.47	−2.33 (−2.93)	2.17 (7.83)	.78 (14.19)		.43 (10.33)	.62 (6.47)	9.48	182
		Seasonally Adjusted Data: Non-Seasonal Methods								
15	Single Moving Average	.43	−1.31 (−1.81)	1.98 (7.46)	.54 (10.27)	.15 (6.06)	.22 (5.68)	.72 (8.25)	8.65	126
16	Single Exp. Sm.	.46	−1.76 (−2.44)	2.18 (8.26)	.64 (12.07)	.12 (4.85)	.22 (5.71)	.65 (7.67)	8.48	141
17	Adaptive Response Rate Exponential Smoothing	.41	0	2.15 (8.18)	.53 (9.44)	.14 (5.74)	.17 (4.24)	.66 (9.36)	8.89	114
18	Linear Moving Average	.39	−2.57 (−2.52)	1.62 (4.32)	.68 (9.24)	.08 (2.38)	.56 (9.92)	1.07 (8.83)	12.00	105
19	Brown's Linear Exponential Smoothing	.46	−2.37 (−2.86)	.95 (3.13)	.77 (12.80)	.09 (3.17)	.44 (9.88)	.92 (9.28)	9.84	142
20	Holt's (2 parameters) Linear Exp. Sm.	.40	−2.75 (−3.21)	1.57 (5.02)	.65 (9.97)	.08 (2.86)	.26 (5.60)	1.07 (10.46)	10.18	111
21	Brown's Quadratic Exponential Smoothing	.41	−3.95 (−3.88)	1.61 (4.46)	.62 (9.04)		.65 (12.38)	1.40 (11.46)	12.12	144
22	Linear Trend (Regression Fit)	.32	0	4.33 (9.61)	.31 (3.41)	.25 (5.99)	.46 (6.77)	.82 (6.86)	15.16	77.30

Numbers in parentheses are the t-tests of regression coefficients

2. Comparison of Accuracies (Percentage Errors) and Costs

This table lists the average errors for each one of the seasonal and non-seasonal methods, as well as their costs based on about 2000 generated data series. However, when 111 series was used in another study the accuracies obtained were somewhat different, with the Naive methods and the exponential smoothing ones doing relatively better than those currently printed. It was found that the accuracies depended upon several factors, the magnitude of which can be seen in Figure 19-4 and 19-5. The errors involved are the ones obtained in fitting the model (MAPE) and those obtained by 1, 2, 3, 6, and 12 period-ahead forecasts. The cost is estimated in French francs using an HP 2000F with charges of 70 FF/hour of connect time. Research is currently underway to estimate accuracies not for the average but for each specific set of data. Once this research is completed, the new specific accuracies will replace the average ones.

3. Some Recommended Methods

By using the decision matrix of Figure 19-6, the most suitable methods are printed together with their complexity (estimated by a survey of 100 users) and the minimum number of data points required (as suggested by literature on that methodology).

FIGURE 19-6   Decision Matrix

#	Method	TYPE (PATTERN) OF DATA — NON-SEASONAL: No Trend / Stationary	First level	Second level	Third level	Seasonal	TIME HORIZON — Immediate	Short	Medium	Long	TYPE OF MODEL — Time Series	Explanatory or "Causal"	Recommended Minimum Data Requirements
1.	Mean	X					X	X			X		30
2.	Single Moving Average	X					X	X			X		5
3.	Single Exponential Smoothing	X					X	X			X		2
4.	Adaptive Response Rate Exponential Smoothing		X				X	X			X		3
5.	Linear Moving Average		X				X	X			X		10
6.	Linear Exponential Smoothing (Brown's)		X				X	X			X		3
7.	Linear Exponential Smoothing (Holt's)			X			X	X			X		3
8.	Quadratic Exponential Smoothing (Brown's)			X	X		X		X		X		4
9.	Simple Regression		X				X				X		20
10.	Naive Method Number 1	X	X				X	X			X		1
11.	Linear and Seasonal Exponential Smoothing			X		X		X			X		2(L)
12.	Classical Decomposition					X		X			X		5(L)
13.	Census II					X		X			X		72
14.	Harrison's Harmonic Smoothing					X		X			X		3(L)
15.	Time Series Multiple Regression			X		X		X			X		6(L)
16.	Generalized Adaptive Filtering		X	X	X	X		X			X		5(L)
17.	Box-Jenkins Approach	X				X		X			X		7(L)
18.	Naive Method Number 2		X					X	X		X		L
19.	S-Curve Fitting		X		X				X	X	X		10
20.	Multiple Regression				X	X			X			X	4(M)
21.	Multivariate Time Series	X	X	X	X	X		X			X	X	9(L)

L is the length of seasonality
M is the number of Independent Variables

CHAPTER 20

Predicting the Cycle

Predicting changes in the level of economic activity is frequently the most difficult part of any forecasting effort, whether formal or informal. At the same time, it is the most crucial task during periods of rapidly changing economic conditions. The oil crisis of 1974-75 and the subsequent worldwide recession made forecasting extremely difficult. The question often raised was "Could forecasting have helped in predicting the downturn?" This chapter focuses on how the cycle can be predicted and what the chances are of a successfull prediction. McLaughlin's concept of paired indices will be used as the main approach.

In Chapter 18, the monthly sales of printing and writing paper in France during the years 1963 through 1972 were used as a sample data series. Figure 20-1 shows the comparison of the sales forecast with the actual values of 1973. In addition to the Box-Jenkins approach, 1973 values were forecast using the method of generalized adaptive filtering.

Both methods did well in forecasting 1973 sales in an ex-post comparison. The randomness of the series (found by running CENSUS, or SIBYL) is about 4.5%, so that the errors of 5.22% and 5.55% of BOX-JENKINS and GAF are as close to the randomness of the series as one might expect to get.

*The authors would like to thank Mr. Bernard Majani of Societe Aussedat-Rey for his many suggestions on chapters of this book and some of the methods described. His advice concerning approaches for predicting the cycle and the use of several of his data series have been extremely helpful.

FIGURE 20-1   Comparison of French Paper Forecasts With Actual Values
(1973)

					GENERALIZED ADAPTIVE FILTERING	
SALES (1973)			*BOX-JENKINS*			
Month	*Period*	*Actual*	*Forecast*	*% Error*	*Forecast*	*% Error*
J	121	946.23	946.11	.01	908.63	3.97
F	122	1013.38	996.70	1.65	1000.99	1.22
M	123	1051.97	1029.98	2.09	973.05	7.5
A	124	1019.86	981.83	3.73	973.51	4.54
M	125	1007.72	929.50	7.76	861.09	14.55
J	126	1020.73	1022.41	- .16	1016.12	.45
J	127	867.26	868.72	- .17	843.84	2.69
A	128	326.12	423.36	-29.82	357.24	-9.55
S	129	911.28	894.53	1.84	859.4	5.69
O	130	960.00	978.55	-1.93	896.7	6.59
N	131	870.00	907.49	-4.31	860.47	1.09
D	132	915.00	999.30	-9.21	994.71	-8.72
			MAPE =	5.22	MAPE =	5.55

1973 was a "normal" year in terms of economic conditions. The French economy continued to grow in an orderly fashion, and so did the paper industry (see Figure 20-2). On the other hand, 1974 was different. The effects of the October 1973 Arab-Israeli war and the oil embargo began to be felt by the middle of 1974 with the quadrupling of oil prices and double digit inflation in the western European economies. The resulting economic slow-down started in the last few months of 1974 and continued through the last available data for February 1975. An ex-post analysis (done in February 1975) of the trend-cycle can be seen in Figure 20-2. This shows the continuation of growth in the trend-cycle during the first months of 1973, and a leveling off between May and December of the same year. This was followed by a slight increase in the first five months of 1974, a leveling off and then a steep drop starting with October 1974. (The trend-cycle values are shown in Figure 20-5 up through period 146-February 1975.)

It is interesting to examine the effect of the sudden drop in the cycle on forecasting accuracy. Assume that one knew only the data up through 1973, i.e., 132 values, and that forecasts for twelve months are obtained using the Box-Jenkins approach and the method of GAF. The results are shown in Figure 20-3.

It can be seen that both models did well up through October, and then the errors increased significantly. Up through October, the MAPE for the Box-Jenkins forecasts is 5.42% while for GAF it is 6.26%, both within the same range as the 1973 errors. The situation quickly deteriorates if one

FIGURE 20-2 Trend-Cycle Analysis (Using CENSUS II) -- Printing and Writing Paper

```
PERIOD                                                              VALUE
  144      .              *                                        678.724
  143      .                      *                                762.809
  142      .                             *                         848.103
  141      .                                     *                 889.646
  140      .                                        *              930.06
  139      .                                          *            952.692    1974
  138      .                                         *             940.183
  137      .                                         *             941.852
  136      .                                           *           956.408
  135      .                                           *           956.681
  134      .                                        *              936.414
  133      .                                         *             940.748
  132      .                                       *               922.038
  131      .                                     *                 911.433
  130      .                               *                       887.829
  129      .                               *                       895.706
  128      .                               *                       890.759
  127      .                               *                       898.173    1973
  126      .                               *                       903.067
  125      .                                 *                     913.078
  124      .                                 *                     913.864
  123      .                                 *                     920.142
  122      .                                *                      922.451
  121      .                                *                      917.916
  120      .                                *                      910.286
  119      .                              *                        897.094
  118      .                              *                        891.687
  117      .                           *                           878.808
  116      .                           *                           873.378
  115      .                          *                            866.424
  114      .                          *                            872.01     1972
  113      .                         *                             860.394
  112      .                         *                             862.424
  111      .                       *                               853.012
  110      .                       *                               855.255
  109      .                      *                                841.096
```

attempts to forecast the last four months of 1974 and the first two months of 1975. The results are shown in Figure 20-4. The MAPE is 41.36% and 34.56% respectively, indicating very inccaccurate forecasting.

It should be obvious that formal forecasting methods attempt to separate randomness and pattern and then extrapolate the latter. Thus, at the end of 1972 (see Figure 20-2), it is only logical that the method will extrapolate the pattern of trend-cycle of the previous years. As long as the trend-cycle continues to behave in about the same way as it has in the past, forecasting will be good. But at the end of 1973, there is no way for the method to know that there will be such a strong downturn in October, November, and December of 1974. This has never happened in the past in a similarly consistent manner, so there is no way for the method to suspect that this will take place. This is the reason for the large errors at the end of 1974 (see Figure 20-3), and the beginning of 1975 (see Figure 20-4). In this sense, one can say that formal forecasting methods do not provide satisfactory results in cases of drastic changes in the economic activity. Does this make formal

FIGURE 20-3   Forecasts for Printing and Writing Paper (1974)

ACTUAL			BOX-JENKINS		GENERALIZED ADAPTIVE FILTERING	
Month	Period	Sales	Forecast	% Error	Forecast	% Error
J	133	975.	961.98	1.33	926.31	4.99
F	134	1039.59	1020.42	1.84	1011.78	2.67
M	135	1174.84	1054.59	10.24	1041.71	11.33
A	136	973.27	1014.22	-4.21	1028.4	-5.67
M	137	1063.44	977.20	8.11	1009.75	5.04
J	138	1110.79	1038.62	6.50	1028.77	7.38
J	139	834.24	884.83	-6.06	872.93	-4.64
A	140	381.42	398.02	-4.35	334.73	12.24
S	141	965.79	917.51	5.00	912.90	5.47
O	142	926.18	986.91	-6.56	955.58	-3.18
N	143	724.34	907.86	-25.34	857.72	18.42
D	144	703.69	981.79	-39.52	909.42	-29.24
			MAPE =	9.92	MAPE =	9.19

FIGURE 20-4   Forecasts for Printing and Writing Paper 1974-1975

ACTUAL				BOX-JENKINS		GENERALIZED ADAPTIVE FILTERING	
Year	Month	Period	Value	Forecast	% Error	Forecast	% Error
1974	S	141	966	934.09	3.3	961.73	.44
1974	O	142	926	999.13	-7.9	968.49	-4.59
1974	N	143	724	919.27	-26.97	862.47	-19.13
1974	D	144	704	993.27	-41.08	907.29	-28.87
1975	J	145	560	1011.12	-80.56	965.54	-72.42
1975	F	146	570	1073.53	-88.39	1036.95	-81.56
				MAPE =	41.36	MAPE =	34.56

forecasting obsolete? For one thing, how often do big recessions take place? Apart from the 1930, the recession that followed the Korean war, the recession of 1969 (in Europe the intensity of this recession was less than in the United States), and the one at the end of 1974-1975, there have only been minor downturns.

Formal forecasting methods should not be judged by the few exceptions but rather by the rule of continuity. If a recession (or a boom) is imminent and one understands that formal forecasting methods will not do well, then one can either suspend formal forecasting for that period or one can adjust the predictions downward (or upward) X = percentage, equal to the extent of the slow down (speed up) of the general economic activity. The problem is that even experts doing forecasting cannot predict changes in the direction of economic activity very well. And even if they can do it for the entire economy, it is even more difficult to tell when the effect of such changes will occur in a particular industry. (The lead-lag relationships are not constant, unfortunately.) Thus, it may be that even in periods of poor forecasting by quantitative methods, the accuracy obtained may be better than straight judgmental forecasting. What is not known well, however, is if naive methods could do better than formal quantitative approaches.

Forecasting requires two types of predictions. The first involves continuation of existing conditions, as shown in Figure 20-1. The other involves the prediction of turning points, as shown in Figures 20-3 and 20-4. In the first case, most time series methods will do well. The user should, therefore, worry little about predicting the cycle. However, with the first signs of changes, additional work should be undertaken to predict changes in the cycle. Some of the different means available to do this will be discussed in the remainder of this chapter. The major purpose is to be able to supplement the quantitative methods and essentially modify their forecasts in order to improve predictions by taking into account the changes in the level of economic activity that are difficult to forecast with traditional quantitative methods.

The means available to predict changes in the cycle (or trend-cycle) are of two kinds. The first involves following economic publications and being aware of leading indicator statistics. Publications such as *The Economist, Business Week, L'Express, The Financial Times, Wall Street Journal* and *L'Echo* provide valuable insights about the economy. No one should attempt to forecast without being informed about the latest economic news. The problem, however, is to assess the impact of this news on the specific industry and organization and to estimate the time delay between an economic or political event and its impact on the organization's performance. There are few generalizations that can be made from one situation to another. Furthermore, it is not uncommon for leading economists to predict one thing and the economy to do something quite different. The 1974-75 recession was predicted by few economists.

The second approach to dealing with the cycle is to look at its behavior in a purely quantitative way. This can be done using a decomposition program or through ORACLE. Figure 20-5 shows the percentage changes of the trend-cycle for the writing paper example during periods 123-146. The trend-cycle has been calculated in the usual way for decomposition methods. The moving average used is a double 3 x 3 moving average. This means that the last two values are lost. The program, however, approximates them by switching to a weighted three-months moving average and then a two-months moving average adjusted for trend. This makes the trend-cycle values of the last two periods less reliable, since they might involve a higher amount of randomness; however, it allows one to obtain trend-cycle values for all periods. The first column of Figure 20-5 assumes that one is at period 134 and computes the values of the trend-cycle during the last thirteen periods and then calculates the percentage change between each successive month. There are twelve such changes in total (see Figure 20-5).

The obvious question is, what will happen after period 146. This question can be answered by looking at the percentage changes in the trend-cycle (Figure 20-5), both vertically and horizontally. Either way shows that the percentage changes decrease in magnitude, which indicates a slow down in the rate of decline and a possible upturn in the data series.

The information shown in Figure 20-5 can be seen graphically in Figures 20-6 through 20-9. These figures show the trend-cycle and its percentage changes for the last twelve months ending at periods 143 (November 1974), 144 (December 1974), 145 (January 1975) and 146 (February 1975).

In Figure 20-6, it can be seen that both the trend-cycle and its percentage change decrease. Starting, however, with Figure 20.7, the percentage change of the trend-cycle slows its decline, a sign that a turning point is approaching. For forecasting purposes, therefore, a turn in the trend-cycle should be envisioned a few periods later. Determining exactly when it will happen requires the additional work of finding an appropriate leading indicator.

Figures 20-10 through 20-13 are equivalent to Figures 20-6 through 20-9, except that they also show the trend-cycle of an additional variable, a

FIGURE 20-5   Percentage Changes in the Trend-Cycle (Periods 123-146)

PERIOD	134	135	136	137	138	139	140	141	142	143	144	145	146
123	0.38												
124	1.43	1.29											
125	0.83	0.76	0.75										
126	-0.76	-0.75	-0.79	-0.77									
127	-0.44	-0.42	-0.46	-0.44	-0.59								
128	-1.02	-0.95	-0.97	-0.96	-1.10	-1.02							
129	0.07	0.17	0.17	0.16	0.04	0.04	0.50						
130	-1.27	-1.15	-1.16	-1.16	-1.16	-1.21	-0.98	-0.86					
131	-0.72	-0.62	-0.63	-0.63	-0.61	-0.63	-0.47	-0.34	-0.20				
132	0.02	0.11	0.10	0.10	0.14	0.07	0.05	0.20	0.33	0.37			
133	1.63	2.20	2.20	2.21	2.25	2.12	2.10	2.12	2.27	2.32	2.10		
134	2.37	3.27	1.87	1.87	1.88	1.77	1.60	1.63	1.64	1.65	1.61	1.65	
135		3.46	1.35	1.87	1.88	1.79	1.53	1.53	1.54	1.56	1.60	1.65	1.66
136			0.21	0.67	0.62	0.60	0.36	0.35	0.34	0.34	0.37	0.40	0.40
137				-0.20	1.33	1.01	0.80	0.77	0.75	0.72	0.72	0.72	0.72
138					3.83	0.64	0.18	0.17	0.14	0.08	0.03	-0.05	-0.05
139						-1.78	-0.35	0.41	0.39	0.34	0.39	0.29	0.30
140							-0.49	0.87	-0.15	-0.19	-0.13	-0.23	-0.23
141								3.42	-0.20	-3.35	-3.25	-3.26	-3.29
142									-2.98	-6.94	-7.40	-7.40	-7.45
143										-11.50	-11.92	11.89	11.95
144											-13.49	14.13	12.79
145												-15.33	13.67
146													-16.44

FIGURE 20-6   Trend-Cycle and Percentage Change in Trend-Cycle for Writing Paper Sales (Periods 132-143)

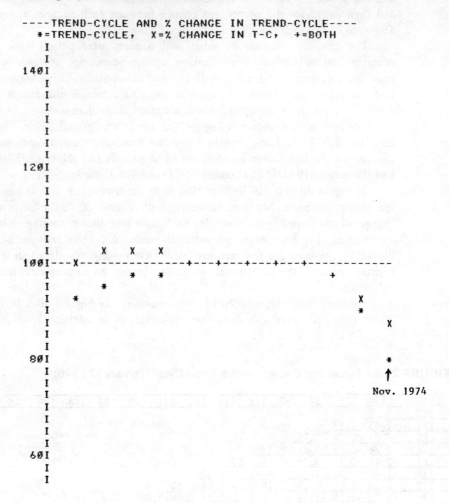

FIGURE 20-7   Trend-Cycle and Percentage Change in Trend-Cycle for Writing Paper Sales (Periods 133-144)

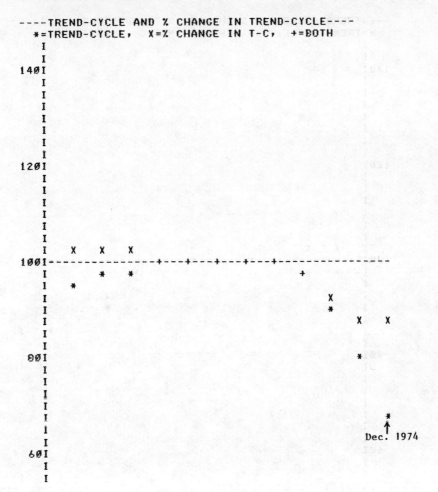

FIGURE 20-8 Trend-Cycle and Percentage Change in Trend-Cycle for Writing Paper Sales (Periods 134-145)

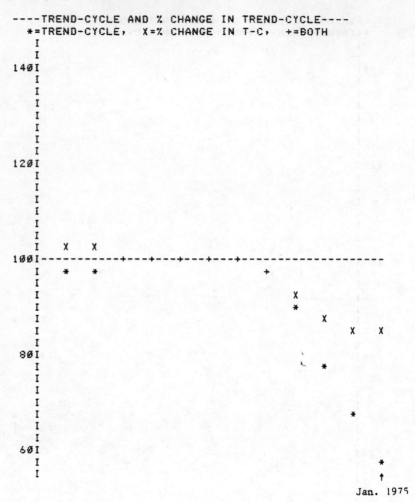

FIGURE 20-9 Trend-Cycle and Percentage Change in Trend-Cycle for Writing Paper Sales (Periods 135-146)

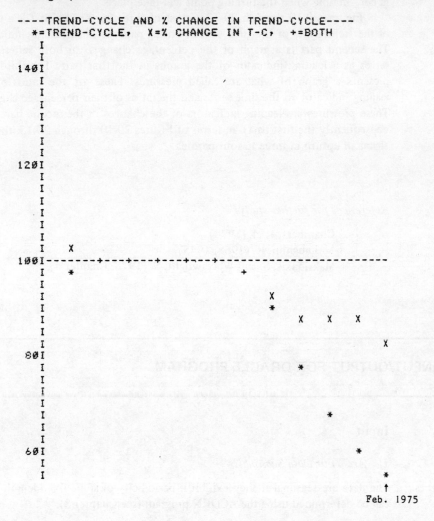

leading indicator of printing and writing paper. Examining the behavior of both the printing and writing paper variable and its leading indicator can help in determining when the turning point will take place.

Figures 20-10 through 20-13 consist of three parts. The first is a graph of the trend-cycle of the time series (writing paper) and its leading indicator. The second part is a graph of the percentage changes for both series. This serves as a leading indicator of the graphs in the first part. The third part presents a graph of what are called pressures. These are the ratios of the leading indicator to the time series and the ratios of their percentage changes. These pressures are leading indicators of the changes of the second part (and consequently the first part). In terms of Figures 20-10 through 20-13, they all signal an upturn in three to four periods.

References for further study:

> Chambers, et al. (1974)
> McLaughlin, R. (1968, 1975)
> Makridakis, S. and Wheelwright, S. (1978, Chapter 4)

INPUT/OUTPUT FOR ORACLE PROGRAM

Input

1. *Are Your Data Seasonal?*

The data are seasonal if they exhibit a periodicity of some fixed length. This can be determined using the ACORN program (see Chapter 3).

2. *What is the Length of Seasonality?*

The length is the time between successive seasonal peaks. It is twelve, for example, when monthly data showing seasonality are considered. The program ACORN can determine the length of seasonality.

3. *Do You Want to Input the Seasonal Indices?*

If the seasonal indices are known, they should be input in the program. This will save computational time. Otherwise, the program will decompose the time series and determine them.

FIGURE 20-10   Trend-Cycle, Percentage Change in Trend-Cycle, and Pressures for the Time Series (Writing Paper Sales) and a Leading Indicator (Last Period - November 1974)

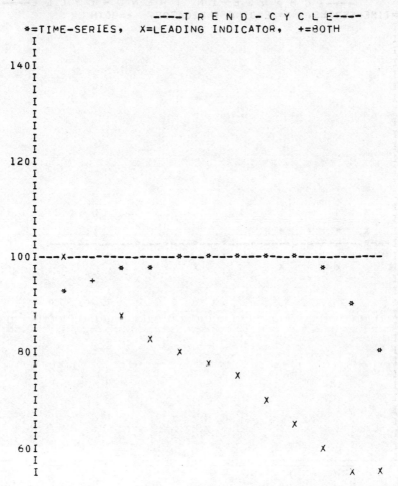

FIGURE 20-10 (continued)

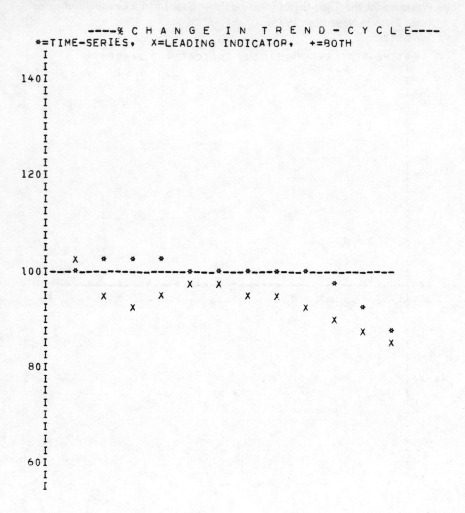

FIGURE 20-10   (continued)

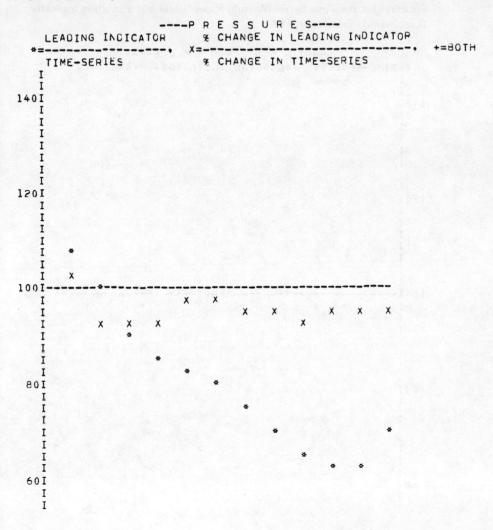

FIGURE 20-11 (Trend-Cycle, Percentage Change in Trend-Cycle, and Pressures for the Time Series (Writing Paper Sales) and a Leading Indicator (Last Period - December, 1974)

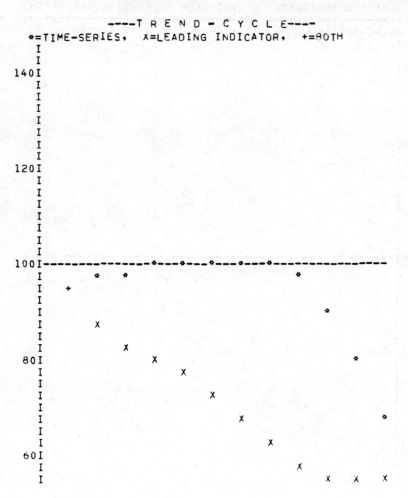

```
                    ----T R E N D - C Y C L E----
         *=TIME-SERIES,  X=LEADING INDICATOR,   +=BOTH
       I
       I
   140 I
       I
       I
       I
       I
       I
       I
       I
   120 I
       I
       I
       I
       I
       I
       I
       I
   100 I-------------------------*-----*-----*-----*-------*------------------
       I           *       *                         *
       I      +
       I
       I                                                    *
       I           X
       I
       I             X
    80 I              X                                          *
       I               X
       I
       I                X
       I                   X                                     *
       I                    X
       I                      X
    60 I                                             X
       I                                               X    X    X
       I
```

FIGURE 20-11   (continued)

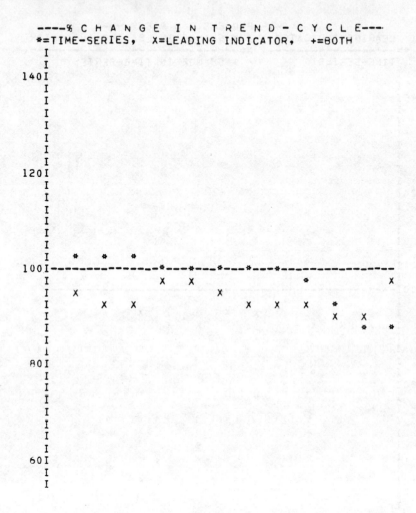

FIGURED 20.11 (continued)

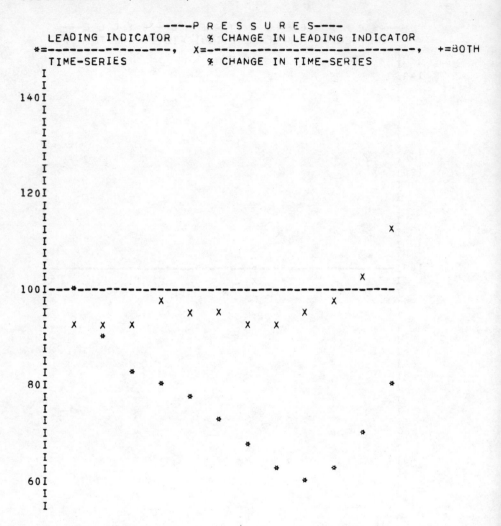

FIGURED 20-12   Trend-Cycle, Percentage Change in Trend-Cycle, and Pressures for the Time Series (Writing Paper Sales) and a Leading Indicator (Last Period - January, 1975)

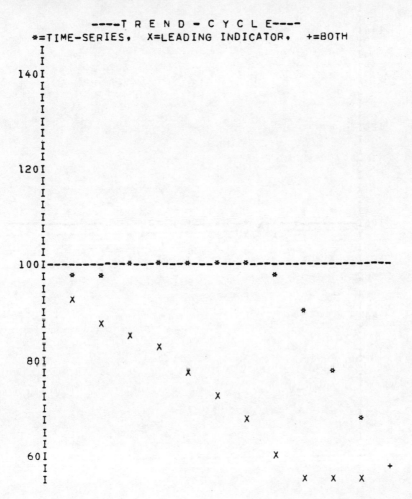

```
                 ----T R E N D - C Y C L E----
         *=TIME-SERIES,   X=LEADING INDICATOR,   +=BOTH
          I
          I
      140 I
          I
          I
          I
          I
          I
          I
          I
      120 I
          I
          I
          I
          I
          I
          I
      100 I-------------*---*----*----*---*------------------
          I    *     *                     *
          I
          I  X
          I
          I     X                      *
          I       X
          I        X
       80 I                 *              *
          I
          I           X
          I
          I              X             *
          I
       60 I              X
          I          X   X   X
          I                          +
          I
          I
```

FIGURE 20-12 (continued)

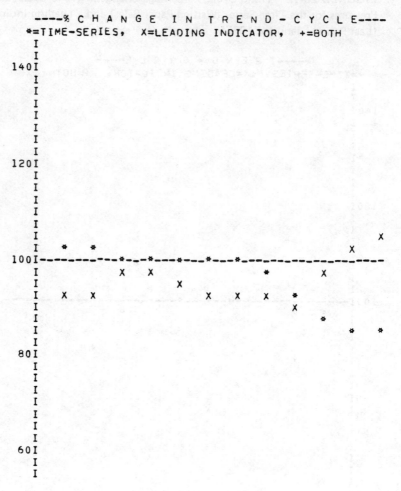

FIGURE 20-12 (continued)

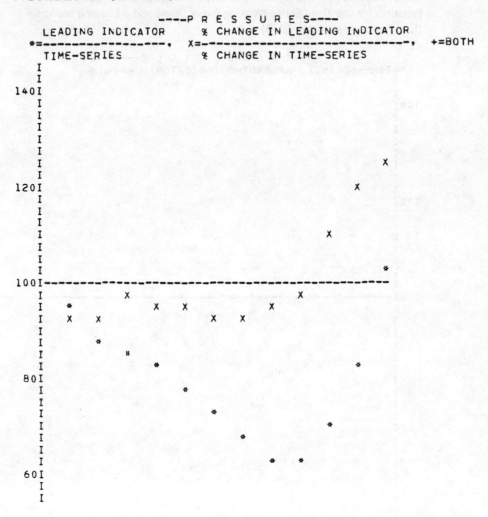

FIGURE 20-13 Trend-Cycle, Percentage Change in Trend-Cycle, and Pressures for the Time Series (Writing Paper Sales) and a Leading Indicator (Last Period - February, 1975)

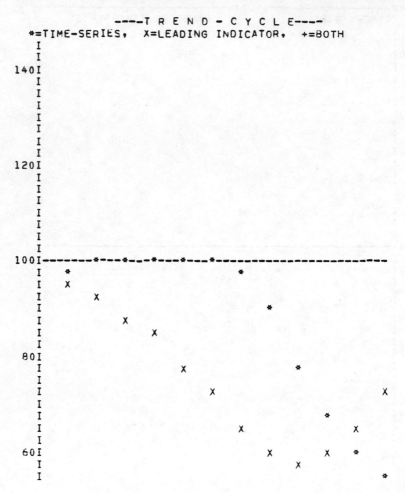

FIGURE 20-13   (continued)

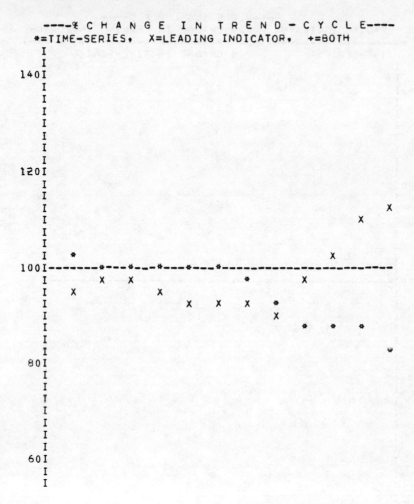

FIGURE 20-13   (continued)

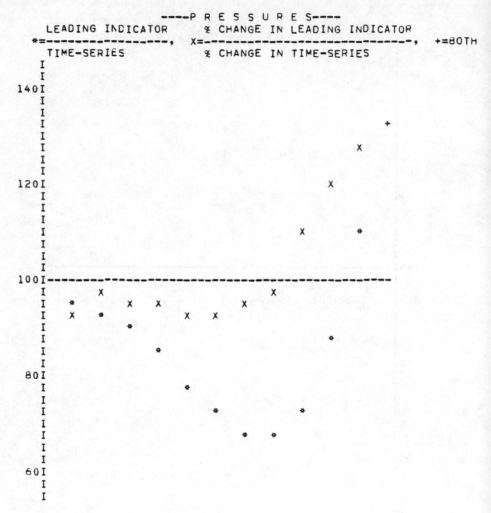

4.   *Do You Only Want the Graph of Trend-Cycle and Its Percentage Change?*

If the answer is yes, the program does not print the intermediate calculations needed to produce the graph; otherwise it does.

5.   *Do You Want to Input a Leading Indicator to Your Series?*

If the answer is yes, input questions 1 through 4 are repeated for the leading indicator series.

Output

1. *Time Series Components*

The program separates each of the components of the time series and prints their values during the last twelve periods. It assumes a linear trend to separate the trend from the cycle. (See Chapters 15 and 16 for details of the decomposition procedure.) The program prints the last twelve periods only because they are the most important in showing the present state of the cycle. If the user wants more information than the last twelve periods, DCOMP or CENSUS should be used.

2. *Percentage Change in the Time Series Components*

This is simply the percentage change between successive months of the values shown under 1 above.

3. *Percentage Error of Time Series Components*

If the actual value is subtracted from the forecast value, an error value is obtained for each period. If the error is divided by the actual value for the period, it will give the percentage error. This is printed for each of the time series components.

4. *Absolute Percentage Error Contribution of Each Time Series Component*

The absolute values of the percentage errors calculated above in 3 are standardized so that their absolute values add up to 100. The standardized errors are then printed. They are an important indication of the contribution of each component in determining the overall error. (The higher the contribution, the more careful one should be in forecasting the corresponding component.) If the highest contribution is under the random component, this means that there is little one can do to improve forecasting accuracy.

5. *Graph of Trend-Cycle (from 1) and Its Percentage Change (from 2) when no leading indicator has been used.*

See Figures 20-6 through 20-9.

6. *Graphs of:*

      a. trend-cycle of time seris and trend-cycle of leading indicator
      b. the percentage changes of (a)
      c. the ratios (pressures) of:

$$\frac{\text{Leading Indicator}}{\text{Time Series}} \quad \text{and} \quad \frac{\%\ \text{Change in Leading Indicator}}{\%\ \text{Change in Time Series}}$$

      See Figures 20-10 through 20-13.

EXERCISE 20.1

The following data show the monthly values of the French Index of Industrial Production from 1964 through 1975. Use the first 120 points and prepare a forecast for 1974. Then use the first 132 points and forecast periods 133 through 144. What can you say about the cycle of the data? Finally, what can you say about the trend-cycle for the year of 1976?

PERIOD		OBSERVATION	PERIOD		OBSERVATION	PERIOD		OBSERVATION
1		69.000	49		83.000	97		105.000
2		70.000	50		85.000	98		109.000
3		62.000	51		85.000	99		110.000
4		71.000	52		84.000	100		109.000
5		74.000	53		83.000	101		104.000
6		75.000	54		87.000	102		108.000
7	1964	66.000	55	1968	75.000	103	1972	96.000
8		44.000	56		49.000	104		61.000
9		74.000	57		85.000	105		109.000
10		76.000	58		87.000	106		112.000
11		77.000	59		89.000	107		115.000
12		77.000	60		90.000	108		115.000
13		77.000	61		87.000	109		113.000
14		78.000	62		90.000	110		114.000
15		78.000	63		91.000	111		116.000
16		79.000	64		91.000	112		116.000
17		79.000	65		91.000	113		115.000
18		78.000	66		93.000	114		115.000
19	1965	68.000	67	1969	82.000	115	1973	99.000
20		43.000	68		54.000	116		70.000
21		76.000	69		89.000	117		115.000
22		79.000	70		96.000	118		121.000
23		79.000	71		99.000	119		123.000
24		77.000	72		99.000	120		124.000
25		75.000	73		90.000	121		130.000
26		78.000	74		97.000	122		132.000
27		77.000	75		97.000	123		130.000
28		78.000	76		100.000	124		130.000
29		77.000	77		100.000	125		130.000
30		80.000	78		100.000	126		130.000
31	1966	69.000	79	1970	39.000	127	1974	113.000
32		46.000	80		56.000	128		80.000
33		78.000	81		97.000	129		123.000
34		82.000	82		103.000	130		128.000
35		83.000	83		101.000	131		126.000
36		83.000	84		102.000	132		122.000
37		79.000	85		102.000	133		115.000
38		83.000	86		105.000	134		115.000
39		83.000	87		105.000	135		112.000
40		83.000	88		107.000	136		112.000
41	1967	83.000	89	1971	103.000	137	1975	108.000
42		85.000	90		104.000	138		113.000
43		75.000	91		92.000	139		111.000
44		47.000	92		59.000	140		111.000
45		84.000	93		102.000	141		111.000
46		85.000	94		106.000	142		113.000
			95		107.000	143		112.000

Partial Solution

(Behavior of Trend-Cycle when n = 144)

ORACLE

   ***** CYCLICAL ANALYSIS OF TIME SERIES *****

NEED HELP ? N

HOW MANY OBSERVATIONS DO YOU HAVE ?   144

WHAT IS THE NAME OF YOUR INPUT FILE   IIPF

IS YOUR DATA SEASONAL   Y

WHAT IS THE LENGTH OF SEASONALITY   12

DO YOU WANT TO INPUT THE SEASONAL INDICES YOURSELF   N

I WILL CALCULATE THE SEASONAL INDICES FOR YOU
SEASON   SEASONAL INDEX
  1   102.509
  2   105.392
  3   104.623
  4   105.462
  5   104.097
  6   105.377
  7   92.5344
  8   60.4819
  9   101.844
 10   105.534
 11   106.382
 12   105.763
DO YOU ONLY WANT THE GRAPH OF TREND-CYCLE AND ITS % CHANGE   N

      *** T I M E   S E R I E S   C O M P O N E N T S ***

PERIOD	ACTUAL	SEAS. ADJ.	TREND-CYCLE	TREND	CYCLE	RANDOM
133	115.0	112.19	112.3	118.5	94.8	115.1
134	115.0	109.12	109.7	117.6	93.3	115.6
135	112.0	107.05	107.5	116.7	92.1	112.5
136	112.0	106.20	106.3	115.9	91.7	112.1
137	108.0	103.75	107.2	115.0	93.2	111.6
138	113.0	107.23	117.6	114.1	103.1	124.0
139	111.0	119.96	128.2	113.3	113.2	118.7
140	111.0	183.53	135.9	112.4	120.9	82.2
141	111.0	108.99	125.9	111.5	112.9	128.3
142	113.0	107.07	115.9	110.7	104.7	122.3
143	112.0	105.28	107.1	109.8	97.6	114.0
144	116.0	109.68	103.1	109.0	94.6	109.0

Explanation of Partial Solution (continued)

```
    *** PERCENTAGE CHANGE IN THE TIME SERIES COMPONENTS ***

PERIOD   ACTUAL   SEAS. ADJ.   TREND-CYCLE   TREND    CYCLE    RANDOM
   133   -5.74     -2.75         -2.58      -0.72    -1.86     -5.57
   134    0.00     -2.74         -2.34      -0.73    -1.62      0.41
   135   -2.61     -1.89         -1.99      -0.73    -1.26     -2.71
   136    0.00     -0.80         -1.15      -0.74    -0.42     -0.36
   137   -3.57     -2.31          0.90      -0.75     1.65     -0.41
   138    4.63      3.36          9.71      -0.75    10.54     11.06
   139   -1.77     11.86          9.00      -0.76     9.83     -4.28
   140    0.00     53.00          5.95      -0.76     6.76    -30.75
   141    0.00    -40.61         -7.31      -0.77    -6.59     56.08
   142    1.80     -1.76         -7.98      -0.77    -7.26     -4.65
   143   -0.88     -1.68         -7.55      -0.78    -6.82     -6.80
   144    3.57      4.18         -3.76      -0.79    -3.00     -4.32
```

```
      *** PERCENTAGE ERROR OF EACH COMPONENT ***

PERIOD   SEAS. ADJ   TREND-CYCLE   TREND     CYCLE     RANDOM
   133      2.45        2.32       -3.01      5.17     -0.13
   134      5.12        4.60       -2.25      6.71     -0.54
   135      4.42        4.00       -4.22      5.42     -0.44
   136      5.18        5.11       -3.45      5.81     -0.08
   137      3.94        0.71       -6.48      0.71     -3.36
   138      5.10       -4.11       -1.01     -4.90     -9.71
   139     -8.07      -15.53       -2.05    -17.29     -6.90
   140    -65.34      -22.40       -1.27    -25.22     25.97
   141      1.81      -13.45       -0.49    -16.97    -15.55
   142      5.24       -2.55        2.05     -6.56     -8.23
   143      6.00        4.34        1.95     -0.18     -1.76
   144      5.45       11.11        6.08      6.18      5.99
```

```
   *** ABSOLUTE PERCENTAGE ERROR CONTRIBUTION OF EACH COMPONENT ***

PERIOD   SEAS. ADJ   TREND-CYCLE   TREND     CYCLE     RANDOM
   133     18.72       17.75      -22.99     39.55     -0.99
   134     26.62       23.95      -11.73     34.89     -2.82
   135     23.89       21.61      -22.82     29.29     -2.39
   136     26.39       26.01      -17.58     29.62     -0.40
   137     25.90        4.66      -42.66      4.66    -22.11
   138     20.55      -16.56       -4.05    -19.74    -39.11
   139    -16.19      -31.16       -4.11    -34.70    -13.85
   140    -46.60      -15.98       -0.90    -17.99     18.52
   141      3.75      -27.87       -1.01    -35.16    -32.21
   142     21.29      -10.36        8.34    -26.61    -33.40
   143     42.15       30.50       13.71     -1.24    -12.40
   144     15.65       31.93       17.46     17.75     17.21
DO YOU WANT TO INPUT A LEADING INDICATOR OF YOUR SERIES
```

Explanation of Partial Solution (continued)

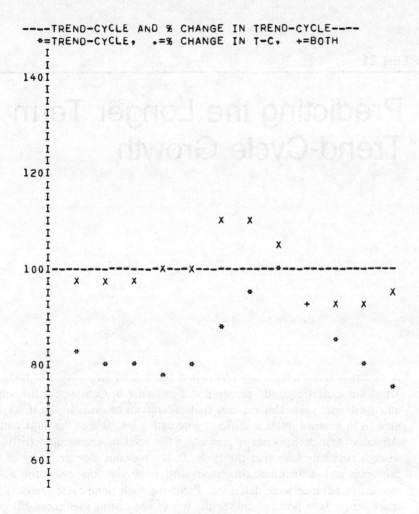

```
----TREND-CYCLE AND % CHANGE IN TREND-CYCLE----
  *=TREND-CYCLE,  .=% CHANGE IN T-C,  +=BOTH
    I
    I
140 I
    I
    I
    I
    I
    I
    I
120 I
    I
    I
    I
    I                        X   X
    I                            X
    I
100 I----------------X---X-------------*--------------
    I     X   X   X
    I                          *                    X
    I                              +   X   X
    I
    I                      *
    I                                *
    I   *
 80 I     .   .     .   .             .
    I           .
    I                                           .
    I
    I
    I
 60 I
    I
```

CHAPTER 21

Predicting the Longer Term Trend-Cycle Growth

A common task facing many planners is the annual preparation of budgets. Thses are most frequently prepared in September or October for the subsequent calendar year. This requires that predictions be made for up to sixteen months in advance, often a difficult forecasting job. One of the most critical elements of such forecasts is predicting the level of economic activity, or what is popularly known as the cycle. In the previous chapter, many of the problems and difficulties associated with predicting the cycle for a few months in advance were described. Preparing such trend-cycle forecasts for much longer time horizons makes the job of forecasting even more difficult.

The challenge of predicting the cycle for budgeting purposes has been further complicated in the past by the fact that in the last two recessions (1969-70 and 1974-75) downtrends in the cycle started in November. While this may simply be due to chance, it has made the job of forecasting for budget purposes even more difficult.

Based on the program CYCLEG, this chapter will describe information that can be used in predicting longer term trend-cycle growth. The next chapter will deal with an important related issue — determining when reality is not as planned so that effective steps can be taken to adjust plans. Both of

*The authors would like to thank Mr. Benard Majani of Societe Aussedat-Rey for his help with the methods described in this and the next chapter.

these chapters are empirically motivated and developed. While what they do is rather simple from a statistical point of view, it has been found to be extremely useful in practice.

The basis of the CYCLEG program is the calculation of trend-cycle values based on the deseasonalized data series and the subsequent fitting of linear, S-shaped, and exponential growth curves to these trend-cycle values. By examining the goodness of fit of each of these three curves, the user can select the best one (or let the program make the selection based on the curve with the largest R^2 value). This curve is then used as the trend that best describes the data and allows the cycle to be separated from the trend-cycle values. The selected curve for the cycle is also extrapolated through the end of the next year. Obviously, if the forecaster has additional information concerning the cycle, it can be used also. However, there may be little information available that will make the longer term growth (the trend) in the trend-cycle a good base for budget forecasts.

References for further study:

> Dauten, C. and L. Valentine (1974)
> McLaughlin, R. (1968)
> Makridakis, S. and Wheelwright, S. (1978, Chapter 4)
> Wheelwright, S. and Makridakis, S. (1977, Chapter 11)

INPUT/OUTPUT FOR THE TREND-CYCLE GROWTH PROGRAM (CYCLEG)

Inputs

1. *What Is the Length of Seasonality?*

The length of seasonality, if not known, can be found using the ACORN program. For monthly data it is usually 12 if a seasonal pattern exists.

2. *Do You Object If I (the computer) Use the Trend to Determine the Least Squares Curve which Best Describes Your Trend-cycle Data?*

The answer should be NO unless the user has some additional information about the longer term growth in the trend of the data.

Outputs

1. a. the R^2 obtained by fitting a straight line.
    b. the R^2 obtained by fitting an exponential growth curve.
    c. the R^2 obtained by fitting an S-curve.

The R^2 values are found by fitting to the trend-cycle of the data of the curve of each of the three forms that gives the least squares results. See chapters 12 and 13.) That is, the data is first deseasonalized and the trend-cycle is computed. The curve is then fitted to the deseasonalized trend-cycle values (see exercise). This procedure eliminates randomness and seasonality without losing any of the data values, as occurs using moving averages. The largest R^2 value indicates the curve form that best fits the data. Unless the user thinks one of the other forms is more appropriate, the program will proceed to use this curve to estimate trend.

2. A graph of the trend-cycle values and the fitted trend values, including forecasts for the trend, up to the end of the next year.

3. Forecast values for the original data and the seasonally adjusted data based on each of the three trend extrapolation curves — linear, exponential growth, and S-curve. The values of the seasonal indexes are also printed.

4. A graph of the cycle (the ratios of the fitted curve values and the trend-cycle values, $T_t C_t / T_t = C_t$).

EXERCISE 21.1

Use the data given in Exercise 19.1 (120 data points) plus nine of the data values shown in Figure 20-1 (i.e., through September 1973) to prepare budget forecasts for 1974. Compare these budget forecasts with the actual data of 1974 shown in Figure 20-3.

Solution: Inputs to the program (see Figure 21-1).

Before fitting the three alternative trend curves, the first step is to calculate seasonal indices. This is done as shown in either Chapter 15 or 16. In the CYCLEG program a centered 12-month (assuming monthly data, otherwise the length is equal to that of seasonality), moving average is used as a first approximation of the trend-cycle and then the seasonal indices are found exactly as in Chapter 15.

FIGURE 21-1   Inputs to CYCLEG Program

```
NEED HELP ?
NO
HOW MANY OBSERVATIONS DO YOU HAVE ?
129
WHAT IS THE NAME OF YOUR INPUT FILE
?$PAPER

IS YOUR DATA SEASONAL ?YES
WHAT IS THE LENGTH OF SEASONALITY ?12
```

The second step is to deseasonalize the data by dividing each actual value by the appropriate seasonal index. Figure 21.2 shows the results obtained for the last thirty data values. The final column of Figure 21.2 is the deseasonalized data ($X'_t = S_t T_t C_t R_t / S_t = T_t C_t R_t$) which include trend, cycle, and randomness. The deseasonalizing is done in such a way that no data values are dropped at the end of the series as is frequently the case when using a centered 12-months moving average.

FIGURE 21-2   Obtaining Seasonally Adjusted Data

Period	Month	$S_t T_t C_t R_t$ Actual Data	S_t Seasonal Index	$T_t C_t R_t$ Seasonally Adjusted Data
100	A	875.000	109.492	799.144
101	M	835.088	105.862	788.845
102	J	934.595	112.952	827.423
103	J	832.500	92.842	896.686
104	A	300.000	37.700	795.761
105	S	791.443	96.236	822.397
106	O	900.000	105.807	850.609
107	N	781.729	99.644	784.526
108	D	880.000	103.983	846.291
109	J	875.024	106.081	824.864
110	F	992.968	110.286	900.359
111	M	976.804	119.115	820.048
112	A	968.697	109.492	884.719
113	M	871.675	105.862	823.406
114	J	1006.852	112.952	891.395
115	J	832.037	92.842	896.188
116	A	345.587	37.700	916.683
117	S	849.528	96.236	882.754
118	O	913.871	105.807	863.719
119	N	868.746	99.644	871.854
120	D	993.733	103.983	955.668
121	J	946.227	106.081	891.986
122	F	1013.375	110.286	918.862
123	M	1051.969	119.115	883.151
124	A	1019.863	109.492	931.448
125	M	1007.722	105.862	951.920
126	J	1020.733	112.952	903.684
127	J	867.258	92.842	934.123
128	A	326.117	37.700	865.039
129	S	911.279	96.236	946.920

The *third* step is to compute a 3 × 3 month moving average (see Chapter 4) of the seasonally adjusted data of Figure 21-2. (A 3 × 3 month moving average is used to minimize the loss of data points.) The results of this step for the last thirty values, are shown in Figure 21-3. The final column of Figure 21-3 contains the trend-cycle values, $C_t T_t$, where R_t has been eliminated by taking the moving average.

FIGURE 21-3 Computing the Trend-Cycle Using a 3 × 3 Moving Average

Period	Month	$S_t T_t C_t R_t$ Actual Data	$T_t C_t R_t$ Seasonally Adjusted Data	$T_t C_t$ Trend Cycle
100	A	875.000	799.144	786.505
101	M	835.088	788.845	807.356
102	J	934.595	827.423	827.574
103	J	832.500	896.686	838.622
104	A	300.000	795.761	833.712
105	S	791.443	822.397	826.786
106	O	900.000	850.609	823.073
107	N	781.729	784.526	821.618
108	D	880.000	846.291	834.283
109	J	875.024	824.864	841.377
110	F	992.968	900.359	857.981
111	M	976.804	820.048	853.166
112	A	968.697	884.719	859.193
113	M	871.675	823.406	859.845
114	J	1006.852	891.395	879.411
115	J	832.037	896.188	890.089
116	A	345.587	916.683	895.885
117	S	849.528	882.754	886.336
118	O	913.871	863.719	885.849
119	N	868.746	871.854	892.111
120	D	993.733	955.668	908.576
121	J	946.227	891.986	908.883
122	F	1013.375	918.862	910.433
123	M	1051.969	883.151	910.433
124	A	1019.863	931.448	920.772
125	M	1007.722	951.920	927.024
126	J	1020.733	903.684	919.949
127	J	867.258	934.123	915.397
128	A	326.117	865.039	910.329
129	S	911.279	946.920	903.445

As an illustration of the moving average computations, the trend-cycle values for periods 125 and 126 are found as:

Period	Seasonally Adjusted Data	3 Months MA	3x3 Months MA (Trend-cycle)
123	883.151		
124	931.448	922.173	
125	951.92	929.017	927.084
126	903.684	929.909	919.949
127	934.123	900.949	915.397
128	865.039	915.361	
129	946.92		

This leaves no trend-cycle values for periods 128 and 129 since two values are lost when a 3×3 months moving average is used. To obtain a value for period 128, a weighted 3-months moving average is used with weights of .3, .4 and .3 (i.e., 3(934.123) + .4(865.039) + .3(946.92) = 910.329). This value of 910.329 is used for period 128. For period 129, a two months moving average is found ((865.039 + 946.92)/2 = 905.98) and adjusted for trend, since it must be centered between periods 128 and 129. This is done by adding half of the trend between periods 127 and 128 ((910.329 − 915.397)/2 = −2.534) to 905.98. The resulting value of 903.445 is used as the trend-cycle value for period 129. Although the last two values may not be as reliable as previous ones (in general they will include more randomness since the length of the moving average is shorter), they provide reasonably good estimates that can form the basis for predicting future values of the trend-cycle.

Once the 3×3 monthly moving averages have been calculated and the last two values estimated, the *fourth* step is to determine the values of the parameters of the trend line using the method of linear least squares. For the linear trend line, equations (12-2) and (12-3) are used. In this instance, the computed parameter values are b = 2.8706 and a = 542.234. Thus, the equation is

$$\text{Trend}_t = 542.234 + 2.8706(t) \qquad (21\text{-}1)$$

where t is the time period (1, 2, 3, . . . , 129).

For the exponential growth curve, the same two equations are used, except the trend-cycle values are first transformed into a logarithmic form (see Chapter 13). The parameter values obtained are b = −.00393 and a = 6.3265. Thus the exponential growth model is

$$\text{Trend}_t = e^{6.3265 + (-.00393)t} \qquad (21\text{-}2)$$

Finally, the S-curve fitting is done as discussed in Chapter 13 and the parameter values obtained for a and b give the model shown in equation (21-3).

$$Trend_t = e^{6.6861+\frac{-5.5305}{t}} \qquad\qquad (21\text{-}3)$$

The R^2 values of each of these three equations – (21-1), (21-2) and (21-3) – are subsequently computed by the CYCLEG program. Since R^2 is a measure of goodness of fit, the highest R^2 will indicate the trend curve which best fits the data. The R^2 values obtained in this instance are shown in Figure 21.4.

In the output in Figure 21.4, the linear trend had the largest R^2 value. However, the R^2 value for the exponential growth curve was very similar, and the user chose to apply the exponential growth curve in forecasting.

A graph of the trend-cycle and trend values also can be obtained. The trend-cycle values for periods 126 through 129 are respectively, 920, 915, 910 and 903. These trend-cycle values are shown on the left hand side of Figure 21.5. The trend values are shown on the right hand side of that figure. For period 126, for example, the trend is

$$Trend_{126} = 2.7183^{6.3265+(-.00393)(126)} = 917$$

For period 127 the trend is

$$Trend_{127} = 2.7183^{6.3265+.00393(127)} = 921$$

etc.

FIGURE 21-4 R^2 Results Obtained from CYCLEG

```
THE R-SQUARE OF THE LINEAR TREND FITTING IS  .952548
THE R-SQUARE OF THE EXPONENTIAL GROWTH FITTING IS  .948835
THE R-SQUARE OF THE S-CURVE FITTING IS  .101209


THE BEST R-SQUARE IS .952548    CORRESPONDING TO LINEAR TREND

DO YOU OBJECT IF I USE THE LINEAR TREND TO DETERMINE THE
LEAST SQUARES CURVE WHICH BEST DESCRIBES YOUR TREND-CYCLE DATA ?YES

DO YOU WANT TO USE :
  1. LINEAR TREND FITTING
  2. EXPONENTIAL TREND FITTING
  3. S-CURVE TREND FITTING
TYPE 1,2 OR 3 ?
2
```

FIGURE 21-5 Plot of Trend-Cycle and Trend Values

```
DO YOU WANT A GRAPH OF THE TREND-CYCLE AND THE FITTED CURVE
(I.E TREND)?YES

PERIOD VALUE                                                VALUE
      OF TREND-CYCLE                                 OF FITTED CURVE
 144                                                        X 984
 143                                                        X 980
 142                                                       X 976
 141                                                       X 973
 140                                                       X 969
 139                                                      X  965
 138                                                      X  961
 137                                                     X   958
 136                                                   X     954
 135                                                   X     950
 134                                                   X     946
 133                                                  X      943
 132                                                  X      939
 131                                                  X      935
 130                                                 X       932
 129  903                                          *  X      928
 128  910                                          *  X      924
 127  915                                           *X       921
 126  920                                          X*        917
 125  927                                          X*        913
 124  921                                          X*        910
 123  910                                          X*        906
 122  910                                          X*        903
 121  909                                         X*         899
 120  909                                         X*         896
 119  892                                          *X        892
 118  886                                         *X         889
 117  886                                         X*         885
 116  896                                         X*         882
 115  890                                        X*          878
 114  879                                        X*          875
 113  860                                      *  X          871
 112  859                                       *X           868
 111  853                                       *X           865
 110  858                                       *X           861
 109  841                                       *X           858
 108  834                                     *  X           854
 107  822                                   *    X           851
 106  823                                    *  X            848
 105  827                                    *  X            844
 104  834                                      *X            841
 103  839                                       X*           838
 102  828                                      *X            835
 101  807                                    *  X            831
 100  787                                  *    X            828
  99  781                                  *    X            825
  98  788                                   *   X            822
  97  797                                    *  X            818
  96  797                                    *  X            815
  95  798                                     *X             812
  94  803                                     *X             809
  93  799                                     *X             806
  92  783                                    *  X            802
  91  754                                 *    X             799
  90  740                             *        X             796
```

FIGURE 21-5 (continued)

```
89   741                                        *      X                    793
88   755                                         *     X                    790
87   764                                          *    X                    787
86   775                                           **X                      784
85   778                                            *X                      781
84   789                                             X*                     778
83   782                                            X*                      775
82   807                                            X      *                 772
81   813                                            X         *              769
80   831                                          X             *            766
79   799                                          X        *                 763
78   781                                          X  *                       760
77   766                                        X*                           757
76   768                                        X  *                         754
75   774                                        X  *                         751
74   775                                       X      *                      748
73   785                                       X        *                    745
72   791                                       X        *                    742
71   797                                      X            *                 739
70   794                                      X          *                   736
69   779                                      X          *                   733
68   762                                     X        *                      730
67   744                                      X  *                           727
66   732                                      X*                             725
65   726                                     X*                              722
64   727                                     X*                              719
63   732                                      X  *                           716
62   734                                     X    *                          713
61   734                                     X    *                          711
60   723                                     X*                              708
59   708                                     X*                              705
58   689                                   *   X                             702
57   686                                   *   X                             699
56   682                                  *  X                               697
55   689                                   *X                                694
54   686                                   *X                                691
53   690                                   X*                                689
52   690                                   X*                                686
51   694                                   X*                                683
50   695                                  X  *                               680
49   696                                  X  *                               678
48   690                                  X*                                 675
47   680                                  X*                                 673
46   660                                 *X                                  670
45   644                               *   X                                 667
44   639                              *    X                                 665
43   650                              *   X                                  662
42   665                                X*                                   659
41   671                               X  *                                  657
40   674                               X  *                                  654
39   670                               X  *                                  652
38   663                              X    *                                 649
37   660                              X    *                                 647
36   667                              X    *                                 644
35   680                              X        *                             642
34   675                             X         *                             639
33   663                             X    *                                  637
32   648                           X  *                                      634
31   652                           X      *                                  632
```

FIGURE 21-5   (continued)

```
30   651                    X   *                              629
29   648                    X   *                              627
28   630                     X*                                624
27   615                    * X                                622
26   602                  *  X                                 619
25   602                  *   X                                617
24   599                 *  X                                  614
23   598                 *  X                                  612
22   589                *   X                                  610
21   580               *   X                                   607
20   574              *    X                                   605
19   573             *     X                                   603
18   578             *   X                                     600
17   581              *  X                                     598
16   583              *  X                                     595
15   586              * X                                      593
14   583              *X                                       591
13   578             *  X                                      588
12   564           *   X                                       586
11   550          *    X                                       584
10   555           *   X                                       582
     I.........I.........I.........I.........I.........I
LOW = 550.424                                    HIGH = 984.194
```

THE ABOVE PLOTS ARE ON A LOGARITHMIC SCALE

The budget forecasts for October 1973 through December 1974 (periods 130 through 144) are obtained by extrapolating each of the three trend curves and reseasonalizing those values to obtain the results in Figure 21.6.

The actual values for periods 133 through 144 (January through December 1974) are shown in Figure 21.7 together with the forecasts for the same period. These comparisons are of an ex-post nature.

From this figure it can be seen that CYCLEG does fairly well in terms of accuracy in this instance. The forecasts from the linear trend extrapolation are a little better than those of the exponential growth extrapolation which suggests that the past behavior of the data as indicated by the R^2 values (the R^2 of the linear trend was .9525 while the R^2 of the exponential was .9488), is a reasonable guide for deciding which trend model to use. Finally, comparing Figure 21.7 with Figure 20.3 indicates that the CYCLEG projections compare very favorably with those of the Box-Jenkins and Generalized Adaptive Filtering approaches. In fact, except for November and December when CYCLEG does worse, it is more accurate than the two sophisticated ARMA methods.

FIGURE 21-6

PERIOD	\multicolumn{7}{c}{F O R E C A S T E D V A L U E S}	
	\multicolumn{2}{c}{L I N E A R}	\multicolumn{2}{c}{E X P O N E N T I A L}	\multicolumn{2}{c}{S-C U R V E}	SEASONAL
	RAW	SEAS. ADJ	RAW	SEAS. ADJ	RAW	SEAS. ADJ	INDEX
130	915.4	968.6	931.6	985.7	767.8	812.4	105.8
131	918.3	915.0	935.2	931.9	768.1	765.3	99.6
132	921.2	957.8	938.9	976.3	768.3	798.9	104.0
133	924.0	980.2	942.6	999.9	768.5	815.3	106.1
134	926.9	1022.2	946.3	1043.6	768.8	847.9	110.3
135	929.8	1107.5	950.0	1131.6	769.0	916.0	119.1
136	932.6	1021.2	953.8	1044.3	769.3	842.3	109.5
137	935.5	990.3	957.5	1013.6	769.5	814.6	105.9
138	938.4	1059.9	961.3	1085.8	769.7	869.4	113.0
139	941.3	873.9	965.1	896.0	769.9	714.8	92.8
140	944.1	355.9	968.9	365.3	770.1	290.3	37.7
141	947.0	911.3	972.7	936.1	770.4	741.4	96.2
142	949.9	1005.0	976.5	1033.2	770.6	815.3	105.8
143	952.7	949.3	980.3	976.8	770.8	768.0	99.6
144	955.6	993.7	984.2	1023.4	771.0	801.7	104.0

FIGURE 21-7   Comparison of Actual and Forecast (1974)

			\multicolumn{4}{c}{Budget Forecasts by CYCLEG}
			\multicolumn{2}{c}{Linear}	\multicolumn{2}{c}{Exponential Growth}
Period	Month	Actual Value	Forecast	% Error	Forecast	% Error
133	J	975.00	980.20	-0.53	999.90	-2.55
134	F	1039.59	1022.20	1.67	1043.60	-0.39
135	M	1174.84	1107.50	5.73	1131.60	3.68
136	A	973.27	1021.20	-4.92	1044.30	-7.30
137	M	1063.44	990.30	6.88	1013.60	4.69
138	J	1110.79	1059.90	4.58	1085.80	2.25
139	J	834.24	873.90	-4.75	986.00	-7.40
140	A	381.42	355.90	6.69	365.30	4.23
141	S	965.79	911.30	5.64	936.10	3.07
142	O	926.18	1005.00	-8.51	1033.20	-11.55
143	N	724.34	949.30	-31.06	976.80	-34.85
144	D	703.69	993.70	-41.21	1023.40	-45.43
				MAPE = 10.18		MAPE = 10.62

The final output of CYCLEG is a graph like that in Figure 21-8 of the cycle at each period. The values used in this graph are the ratios of the trend-cycle divided by the trend. (See Figure 21.5.) Thus, the cycle for period 126 is

$$C_{126} = \frac{T_{126}C_{126}}{T_{126}} = \frac{919.949}{917.49} = 100.36$$

For period 127 it is

$$C_{127} = \frac{T_{127}C_{127}}{T_{127}} = \frac{915.397}{921.138} = 79.49, \text{ etc.}$$

FIGURE 21-8   Graph of the Cycle From CYCLEG

```
DO YOU WANT A PLOT OF THE RATIO OF THE FITTED CURVE AND THE TREND-CYCLE
?YES

PERIOD       RATIO OF FITTED CURVE (TREND) AND TREND-CYCLE) (I.E  CYCLE)
   129                      *******I                          97.41
   128                      ****I                             98.54
   127                      **I                               99.47
   126                      I*                               100.36
   125                      I****                            101.53
   124                      I***                             101.24
   123                      I**                              100.5
   122                      I***                             100.89
   121                      I***                             101.12
   120                      I****                            101.48
   119                     *I                                100.04
   118                     *I                                 99.72
   117                      I*                               100.17
   116                      I****                            101.65
   115                      I****                            101.39
   114                      I**                              100.57
   113                  ****I                                 98.72
   112                   ***I                                 99.03
   111                  ****I                                 98.72
   110                     *I                                 99.67
   109                 *****I                                 98.12
   108                ******I                                 97.68
   107              ********I                                 96.58
   106               *******I                                 97.13
   105                ******I                                 97.95
   104                   ***I                                 99.16
   103                      I*                               100.13
   102                   ***I                                 99.2
   101              ********I                                 97.16
   100          *************I                                95.03
    99          *************I                                94.79
    98            **********I                                 95.96
    97              *******I                                  97.4
    96                ******I                                 97.82
    95                *****I                                  98.33
    94                   **I                                  99.29
    93                   **I                                  99.24
    92                ******I                                 97.66
    91          **************I                               94.37
    90      ******************I                               92.99
    89       *****************I                               93.46
    88           ***********I                                 95.6
    87             ********I                                  97.13
    86                  ***I                                  98.98
   ·85                     *I                                 99.72
    84                      I****                            101.56
    83                      I***                             100.98
    82                      I***********                     104.58
    81                      I**************                  105.89
    80                      I*******************             108.56
    79                      I***********                     104.87
    78                      I********                        102.93
    77                      I***                             101.25
    76                      I****                            101.96
    75                      I********                        103.15
    74                      I*********                       103.74
    73                      I*************                   105.43
    72                      I****************               106.61
    71                      I******************             107.93
```

FIGURE 21-8 (continued)

```
70                        I********************       107.88
69                        I****************           106.28
68                        I***********                104.44
67                        I******                     102.32
66                        I***                        101.08
65                        I**                         100.58
64                        I***                        101.12
63                        I******                     102.24
62                        I********                   102.99
61                        I*********                  103.38
60                        I******                     102.17
59                        I*                          100.45
58                   *****I                            98.12
57                   *****I                            98.06
56                   *****I                            97.88
55                     **I                             99.36
54                     **I                             99.29
53                        I*                          100.24
52                        I**                         100.63
51                        I****                       101.63
50                        I******                     102.24
49                        I*******                    102.68
48                        I******                     102.2
47                        I***                        101.18
46                    ****I                            98.58
45             *********I                              96.61
44             *********I                              96.21
43                   *****I                            98.19
42                        I**                         100.84
41                        I******                     102.2
40                        I########                   102.99
39                        I*******                    102.86
38                        I******                     102.12
37                        I******                     102.12
36                        I#########                  103.53
35                        I****************           106.06
34                        I***************            105.71
33                        I***********                104.21
32                        I******                     102.31
31                        I********                   103.24
30                        I*********                  103.52
29                        I*********                  103.52
28                        I***                        101.01
27                     ***I                            99.04
26                  *******I                           97.25
25                  *******I                           97.58
24                  *******I                           97.51
23                   ******I                           97.74
22                 ********I                           96.59
21            ***********I                             95.63
20           *************I                            94.94
19           *************I                            95.1
18            *********I                               96.32
17               *******I                              97.25
16                 *****I                              98.01
15                   ***I                              98.83
14                  ****I                              98.75
13                  *****I                             98.2
12            **********I                              96.24
11          *************I                             94.31
10           **********I                               95.42
     I.........I.........I.........I.........I.........I
   92.95                100                       108.51
```

CHAPTER 22

Tracking Trend-Cycle and Budget Movements

In the previous chapter, longer term predictions of the trend-cycle were discussed. Such a job is difficult and risky because sudden, major changes can take place in the level of economic activity influencing the trend-cycle. On the other hand, such a job must be done in establishing budgets and other planning targets, keeping in mind that when trend-cycle movements occur, the goal is to detect them as early as possible.

This chapter presents two simple approaches to aid the budgeter/forecaster in these tasks. The first helps the forecaster or planner to determine the actual state of the trend-cycle in comparison to previous trend-cycle values. The second provides a visual summary of the actual and forecast budget. A comparison of actual versus forecast values, simple as it seems, provides substantial information that allows the forecaster to detect early deviations from predicted levels.

Cyclical movements can be compared with each other by standardizing the various trend-cycles in the data and plotting these. With the trend-cycles standardized, the behavior of the most recent one can be seen in relation to the others. Frequently, a trend-cycle will follow a pattern similar to that of a previous standardized trend-cycle, making predictions about its future direction possible. Trend-cycle movement outside of the range of historical movement may indicate that something unusual is occurring. Special measures that can be used to validate such changes are discussed below. However, the most important thing is to quickly become aware of unusual movements.

Figure 22-1 shows for the data of exercise 20-1 (using all 144 data points) five downward trend-cycles standardized around 100. These are the longest five. They can be identified by the numbers 1, 2, 3, 4, 5. When a letter exists, then more than one trend-cycle has a common value at that point (the program provides the numbers corresponding to each letter). The last trend-cycle is the one of concern, and its behavior in comparison to the others must be analyzed. It can be seen in Figure 22-1 that the last trend-cycle started at period 131 (November 1973), reached a plateau around periods 135 to 140, and started declining rapidly after that. The trend-cycle with the most severe downtrend, excluding the most recent one, is trend-cycle No. 3 corresponding to the 1969-1970 recession. Figure 22-2 shows only these two recessions plotted together. (The number 1 of Figure 22-2 corresponds to 3 of 22-1 and the number 2 of 22-2 is the same as number 5.) Both Figures 22-1 and 22-2 show that the 1974-1975 recession was an unusual one. This is clearly indicated in November, 1974, while December confirms it. (It should be noted that the plots refer to standardized values, and thus reflect relative levels of sales rather than absolute levels, which would be larger in the most recent years simply due to sales growth.)

References for further study:

Dauten, C. and L. Valentine (1974)
McLaughlin, R. (1968)
Makridakis, S. and Wheelwright, S. (1978, Chapter 12)

INPUT/OUTPUT OF TCYCLE PROGRAM

Input

1. *Do You Want to Plot Each of the Cycles Individually?*

The answer should be yes or no.

FIGURE 22-1 Graph of the Five Longest Trend-Cycles (Number 5 is the Most Recent)

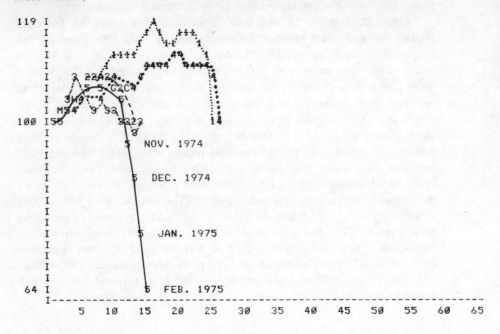

FIGURE 22-2 Comparison of 1969-1970 Recession (1) and the 1974-1975 Recession (2)

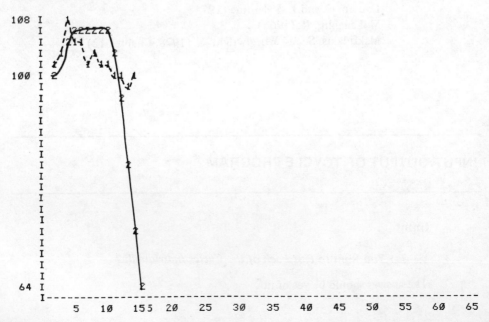

2. *Do You Want to Print the Results?*

If the answer is yes, the program prints the data needed to generate graphs of the trend-cycles.

3. *Do You Want to Plot any Other Cycle?*

If the answer is yes, the user enters (a) the number of trend-cycles to be printed, and (b) three numbers, separated by a comma, representing the cycle number, the period that cycle starts, and the period it finishes.

Output

1. *Upward Turns: Standardized Actual Values and Their Graphs.*

2. *Downward Turns: Standardized Actual Values and Their Graphs.*

EXAMPLE OF CALCULATIONS

In Figure 22-1 the last cycle plotted, 5, has the values 100, 100.3, 102.6, 104.3, 106.1, 106.5, 107.2, 107.2, 107.5, 107.3, 103.7, 96, 84.5, 73.7, 64.3. This is a total of 15 obervations. The starting value of 100 was found by dividing the trend-cycle value for period 131, 898.2, by itself and multiplying the result by 100. For period 132 the trend-cycle value is 900.6. Its standardized value is found by dividing it by the base value, that of period 131, and multiplying by 100:

$$\frac{900.6}{898.2} (100) = 1.003(100) = 100.3$$

The standardized value for period 133 is

$$\frac{919.4}{898.2} (100) = 1.026(100) = 102.6$$

INPUT/OUTPUT OF ZCURVE PROGRAM

The ZCURVE program graphs and prints the actual and budget figures in three forms: (a) seasonally adjusted, (b) cumulative seasonally adjusted, and (c) raw data. These graphs and tables facilitate an easy visual comparison of actual and forecast results.

Input

1. *Most Recent Month of the Year?*

The answer required is a number 1, 2, . . . , 12. If, for example, the most recent month for which actual data is available is April, the response would be 4. If November, the appropriate response would be 11.

2. *Input Forecasts* (budgets)

The forecast (budget) figures must be put into a DATA statement at the end of the program, starting with line 3000.

3. *Input Seasonal Indices*

The twelve seasonal indices must be put into a DATA statement at a line following the budget figures of 2 above.

4. *Input Actual Values*

The *M* actual values (where *M* is the most recent month) should be put into a DATA statement following 3 above.

Output

1. *Graph of Seasonally Adjusted Actual and Forecast Values*

2. *Graph of Seasonally Adjusted Values, Actual and Forecast*

3. *Graph of Raw Data, Actual and Forecast*

4. *Actual, Forecast and Percentage Errors for 1, 2 and 3 above*

EXERCISE 23.1

The forecast values, actual values and seasonal indices for company XYZ are shown below:

	Budget Forecasts	Actual	Seasonal Indices
	YEAR 1975		
Jan	929.27	823.77	1.062
Feb	998.37	795.36	1.098
Mar	1222.12	798.47	1.197
April	1075.33	732.85	1.092
May	1078.52	707.23	1.056
June	1193.84		1.142
July	935.47		.942
Aug	432.42		.382
Sept	985.91		.965
Oct	963.23		1.046
Nov	837.68		.993
Dec	809.63		1.036

Compute the four outputs provided by ZCURVE.

Solution

The program output is as shown in Figure 22-3.

FIGURE 22-3 Application of ZCURVE to Exercise 23

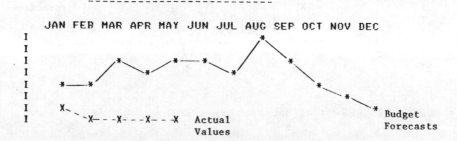

FIGURE 22-3   (continued)

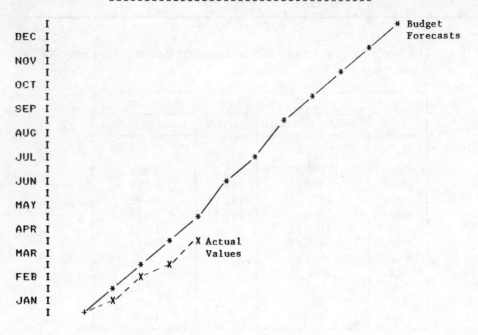

CUMULATIVE SEASONALLY ADJUSTED VALUES

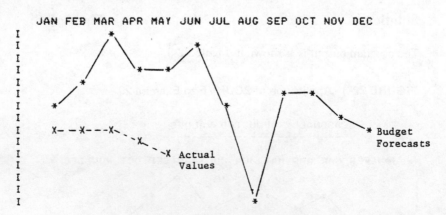

RAW DATA

JAN FEB MAR APR MAY JUN JUL AUG SEP OCT NOV DEC

* BUDGET FIGURES, X ACTUAL VALUES, + BOTH

FIGURE 22-3   (continued)

```
SEASONALLY ADJUSTED VALUES
--------------------------

      I ACTUAL  I  FORECAST I  %ERROR  I
-----I---------I-----------I----------I
 JAN I  775.68 I   874.92 I   -.128  I
 FEB I  724.37 I   909.26 I   -.255  I
 MAR I  667.06 I  1020.99 I   -.531  I
 APR I  671.11 I   984.73 I   -.467  I
 MAY I  669.73 I  1021.33 I   -.525  I



   CUMULATIVE SEASONALLY ADJUSTED VALUES
   -------------------------------------

      I ACTUAL  I  FORECAST I  %ERROR  I
-----I---------I-----------I----------I
 JAN I  775.68 I   874.92 I   -.128  I
 FEB I 1500.05 I  1784.19 I   -.189  I
 MAR I 2167.11 I  2805.17 I   -.294  I
 APR I 2838.22 I  3789.91 I   -.335  I
 MAY I 3507.94 I  4811.23 I   -.372  I



RAW DATA
--------

      I ACTUAL  I  FORECAST I  %ERROR  I
-----I---------I-----------I----------I
 JAN I  823.77 I   929.17 I   -.128  I
 FEB I  795.36 I   998.37 I   -.255  I
 MAR I  798.47 I  1222.12 I   -.531  I
 APR I  732.85 I  1075.33 I   -.467  I
 MAY I  707.23 I  1078.52 I   -.525  I
```

Cases

PERRIN FRERES

Late in the Fall of 1972, the champagne making firm of Perrin Freres located in Epernay, France was preparing its financial plan for the coming calendar year. The impetus behind the preparation of this plan was a request from the company's banker, Societe General, although obviously such a plan would be of general use to the management of Perrin Freres as well. Because the demand for champagne followed a marked seasonal pattern during the course of the year, most champagne companies relied on banks to finance their fluctuating working capital needs. Such was the case at Perrin Freres. Due to the recent financial difficulties of some other companies in the industry, Societe Generale had requested a set of *pro-forma* cash flow, profit and balance sheet statements (for each of the next 12 months) before formally approving Perrin's request for a working capital loan for parts of the coming year.

The Director of Finance at Perrin Freres had asked his assistant, Michel Cunche, to prepare the necessary statements for the Bank. After some initial thought, Michel decided that once he had accurate monthly forecasts for sales, the delay in cash receipts for actual sales, and the various operating costs, it would be primarily an accounting matter to combine these into the proper *pro-forma* statements. From Michel's familiarity with the French champagne industry, and especially Perrin's operations, he felt that the most appropriate method for predicting the company's monthly cash operating expenses and the time delay in cash receipts would be to take the average values for each calendar month over the past three years and use that as the basis of a forecast. Of course, Michel realized that there might be a few instances where he would want to modify one of these forecasts because of new information that was available, but he did not foresee any major problems in doing this.

The task of forecasting the company's champagne sales for each of the next 12 months appeared to be much more difficult. After considerable thought, Michel decided that one approach for doing this would be to obtain a monthly industry forecast of champagne sales and then apply the market share figures of Perrin Freres to this to determine the level of sales his firm could expect to achieve. This approach had an additional benefit besides helping to meet the requirements of the company's bank in that the industry forecast could serve as a basis for some of the other planning activities of the firm, particularly in the marketing area.

As an initial step in developing a forecast of monthly French champagne sales, Michel contacted the Association of French Champagne Firms to which Perrin Freres belonged. While the association did not have an accurate set of monthly forecasts that would meet Perrin's needs, they did have available the historical data in Exhibit 1. In addition, they had computed a 12-month moving average for this data at the request of another member of the association. They were very happy to supply Perrin Freres with this information.

To help him in determining what forecasting technique might be appropriate for this situation, Michel plotted the historical data of Exhibit 1 as shown in Exhibit 2. While he was not surprised to see a strong seasonal pattern in French champagne sales, he was surprised to find what looked like a fairly consistent pattern over each calendar year. Because of this apparent consistency, he felt that some time series approach to forecasting that adequately considered seasonal factors would be most appropriate, and he set out to determine what that technique might be and how he might use it in this situation.

EXHIBIT 1 Monthly Champagne Sales — France (millions of bottles)

(1) Observation	(2) Year	(3) Month	(4) Actual Sales	(5) 12 Month Moving Average*
1	1964	Jan	2.815	—
2		Feb	2.672	—
3		March	2.755	—
4		April	2.721	—
5		May	2.946	—
6		June	3.036	—
7		July	2.282	3.478
8		August	2.212	3.455
9		Sept	2.922	3.439
10		October	4.301	3.462
11		Nov	5.764	3.507
12		Dec	7.312	3.576
13	1965	Jan	2.541	3.593
14		Feb	2.475	3.655
15		March	3.031	3.617
16		April	3.266	3.673
17		May	3.776	3.688
18		June	3.230	3.777
19		July	3.028	3.864
20		August	1.759	3.912
21		Sept	3.595	3.956
22		Oct	4.474	4.041
23		Nov	6.838	4.062
24		Dec	8.357	4.076
25	1966	Jan	3.113	4.139
26		Feb	3.006	4.158
27		March	4.047	4.142
28		April	3.523	4.137
29		May	3.937	4.198
30		June	3.986	4.263
31		July	3.260	4.338
32		August	1.573	4.526
33		Sept	3.528	4.533
34		Oct	5.211	4.506
35		Nov	7.614	4.588
36		Dec	9.254	4.637
37	1967	Jan	5.375	4.683
38		Feb	3.088	4.716
39		March	3.718	4.722
40		April	4.514	4.823
41		May	4.520	4.841
42		June	4.539	4.900
43		July	3.663	5.016
44		August	1.643	4.971
45		Sept	4.739	5.008

EXHIBIT 1   (Continued)

(1) Observation	(2) Year	(3) Month	(4) Actual Sales	(5) 12 Month Moving Average*
46	1967	Oct	5.428	4.975
47		Nov	8.314	4.975
48		Dec	10.651	4.985
49	1968	Jan	3.633	5.003
50		Feb	4.292	5.028
51		March	4.154	5.035
52		April	4.121	5.061
53		May	4.647	5.185
54		June	4.753	5.314
55		July	3.965	5.371
56		August	1.723	5.403
57		Sept	5.048	5.375
58		Oct	6.922	5.404
59		Nov	9.858	5.417
60		Dec	11.331	5.444
61	1969	Jan	4.016	5.438
62		Feb	3.957	5.401
63		March	4.510	5.409
64		April	4.276	5.423
65		May	4.968	5.419
66		June	4.677	5.498
67		July	3.523	5.713
68		August	1.821	5.599
69		Sept	5.222	5.511
70		Oct	6.872	5.416
71		Nov	10.803	5.371
72		Dec	13.916	5.201
73	1970	Jan	2.639	5.143
74		Feb	2.899	5.201
75		March	3.370	5.194
76		April	3.740	5.194
77		May	2.927	5.157
78		June	3.986	5.077
79		July	4.217	5.007
80		August	1.738	5.115
81		Sept	5.221	5.136
82		Oct	6.424	5.213
83		Nov	9.842	5.291
84		Dec	13.076	5.464
85	1971	Jan	3.934	5.538
86		Feb	3.162	5.573
87		March	4.286	5.566
88		April	4.676	5.627
89		May	5.010	5.674
90		June	4.874	5.674
91		July	4.633	5.641

EXHIBIT 1   (Continued)

(1) Observation	(2) Year	(3) Month	(4) Actual Sales	(5) 12 Month Moving Average*
92	1971	August	1.659	5.675
93		Sept	5.951	5.709
94		Oct	6.981	5.733
95		Nov	9.851	5.742
96		Dec	12.670	5.710
97	1972	Jan	4.348	5.746
98		Feb	3.564	5.718
99		March	4.577	5.699
100		April	4.788	5.693
101		May	4.618	—
102		June	5.312	—
103		July	4.298	—
104		August	1.431	—
105		Sept	5.877	—

*These 12 month moving average figures are centered on the 7th observation. For example, the first value of 3.478 which is listed opposite observation 7 is the average for observations 1 through 12. The value opposite observation 8 is average for periods 2 through 13, etc.

EXHIBIT 2 Monthly Champagne Sales — France (millions of bottles)

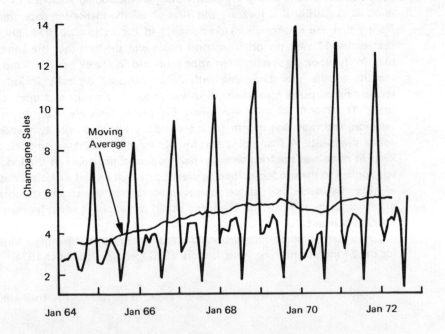

BASIC METALS LTD.

In early 1973 Mr. Ian McDonald, Director of Corporate Planning Research, was contemplating how best to predict the company's future sales of carbon steel alloys. This product was one of many that were bought in bulk form and processed by the company. As with most of the company's products, this was essentially a commodity item and thus the company, while having a significant share of the market, did not determine the world-wide prices for their products. Instead, they were faced with simply accepting whatever the market dictated as the appropriate price at any point in time.

The particular problem that Mr. McDonald had was taking the data shown in Exhibit 1, covering the past 11 years, and trying to apply some forecasting method that would enable him to effectively predict the company's future sales. Among the points that Mr. McDonald wanted to keep in mind in preparing this forecast was first of all the marketing department's feeling that the company's market share had been changing gradually over that entire 11 year period. A second point was the fact that the company used in its reporting system alternating four- and five-week periods. Thus, the months of the year beginning with January would typically include for accounting purposes four weeks, four weeks and five weeks and then repeat again. The month of July was when the plants typically shut down for vacation, and thus that month for the company included only two weeks of actual shipments. A final point that Mr. McDonald thought was important to keep in mind was the long term cyclical pattern that people in the industry felt applied to the product. While he wasn't sure that he had sufficient data to identify that cycle, he did feel it was important to investigate its possible existence since he knew the President would ask about it when he reviewed the final forecast.

As a starting point, Mr. McDonald set as his objective obtaining a forecast for the 24 months running from January 1973 through December 1974.

EXHIBIT 1 Basic Metals Ltd. Weekly Shipments of Carbon Steel Alloys (in 000's of tons)

Year	Jan.	Feb.	Mar.	Apr.	May	June	July	Aug.	Sep.	Oct.	Nov.	Dec.
1961	2079	1952	1941	2311	3035	2276	3514	2398	2734	2650	2847	2495
1962	2592	3190	2741	2806	2993	2704	2999	2640	2360	2413	2501	2391
1963	2328	2691	3045	2918	3071	2847	3141	2730	2546	2255	2536	2495
1964	2373	2734	2985	2889	3109	2638	3940	2986	3178	3122	2559	2625
1965	2584	3271	3287	3536	4042	3165	4119	3178	3449	3687	4187	3112
1966	3571	3755	4303	4651	4128	3953	4646	3622	3929	4183	3936	3758
1967	3605	3919	3526	2997	2883	2346	2773	2539	3052	3311	3836	2942
1968	3356	4070	3657	3348	3562	3335	3660	3426	3530	3263	3471	3212
1969	3600	3408	3921	3462	3408	3315	4463	3213	3195	2976	2759	2659
1970	3419	3348	3311	3488	2941	3142	2935	2440	2591	2314	2346	2252
1971	1990	2354	2639	2915	2977	3105	2994	2454	2372	2709	2833	3173
1972	2250	2831	3351	3275	3271	3218	3732	2813	2845	3036	2996	2670
Number of weeks in month:	4	4	5	4	4	5	2	4	5	4	4	5

Note: Figures are average shipments/week for each month. These were obtained by taking accounting's record of shipments for the month and dividing by the number of weeks over which those shipments were made.

CFE SPECIALTY WRITING PAPERS

During the decade of the sixties, the specialty writing papers division of CFE faced a rapidly expanding market for its product. In response to this increased demand the company, like many of the others in the paper-making industry in Europe, invested heavily in automation equipment and in new facilities in order to provide sufficient capacity to meet demand. While this resulted in annual increases in profitability, it also gradually changed the cost structure of the company resulting in a larger percentage of the costs being fixed costs and thus raising the break-even point of the firm.

When CFE began to encounter the recession of late 1974, management quickly responded by trying to cut some of these fixed costs and yet still maintain sufficient capacity to meet the anticipated upturn in late 1975 or 1976. Because the specialty writing papers division accounted for over 40% of corporate sales, top management was most concerned that the division prepare accurate forecasts that would allow it to plan for the remainder of 1975 and 1976. This appeared to be a particularly difficult task given that in March of 1975 the company was selling at only 55% of the peak rate it had been selling at in 1973. The March 1975 rate was actually very similar to the level of sales the company had faced in 1963.

The person charged with preparing such a forecast for the division was Mr. Bernard Coffinet, planning director. For several years, Mr. Coffinet had collected historical data on the division's sales as shown in Exhibit 1. During the late sixties his forecasting task had simply involved projecting steady but modest increases in demand. Then in the early seventies, he was suddenly faced with recommending to the company appropriate expansions in capacity and/or overtime work that would meet a sharp increase in demand. Finally, at this point in time he was faced with predicting when the bottom of the recession would occur and how fast the recovery would take place for their division. Compounding Mr. Coffinet's concern was the fact that his March 1975 forecasts had been much less accurate than usual. He attributed part of this to the fact that Easter had fallen in March that year and thus reduced the number of working days from what it typically was during March.

As a step to overcome this latter problem, Mr. Coffinet had recently collected data on the number of working days falling in each month since

1963 as shown in Exhibit 2. Using Exhibits 1 and 2 he had adjusted the division's monthly sales so that each month of the year was assumed to have the average number of working days for that month in the years 1963-1974. This was done by simply finding the average number of working days for each month and adjusting the actual sales shown in Exhibit 1 by the appropriate percentage to reflect what actual sales would have been if the month had contained the average number of working days. Thus, if the average for a month had been 20 working days, and that month in a particular year actually had 22 days, then he would reduce the actual sales by 10% so that the month would be comparable to a twenty-day month. The adjusted monthly sales data obtained is shown in Exhibit 3.

In 1974 when Mr. Coffinet had become aware of some changes in the demand for specialty writing papers and the pattern of that demand, he had looked for some other time series that might serve as a leading indicator for the sales of his division. He had identified as such an indicator an index of business activity collected by the government. Unfortunately, this index had not been collected prior to 1968, but in March of 1975 he did have over five years of historical data for that index (Exhibit 4).

In order to determine whether or not the business index was a leading indicator for his division's sales, Mr. Coffinet had applied the Census II technique to both the adjusted data in Exhibit 3 and the business index in Exhibit 4. He had then superimposed the trend-cycle results of the two series on a graph as shown in Exhibit 5. At first glance, it appeared that this business index had some of the properties that he might desire for a leading indicator for the division's sales.

Before proceeding with either the use of the business indicator or some other approach to forecasting division sales, Mr. Coffinet thought it would be important to investigate a couple of key questions. These included determining whether or not the adjustment of the number of working days was worth doing (going from Exhibit 1 representing unadjusted data to Exhibit 3 representing adjusted data) and trying to identify the alternatives that might exist for forecasting the anticipated turning point for the division's sales. The leading indicator approach was simply one methodology available and he felt that others might exist that could be used with even more success. He also wanted to be certain that as he predicted an upturn in the division's sales, he would be able to do so with some degree of confidence so that management could effectively base its decisions on those forecasts.

EXHIBIT 1   CFE Specialty Writing Papers Unadjusted Monthly Sales Date

Jan 1963

PERIOD	OBSERVATION	PERIOD	OBSERVATION	PERIOD	OBSERVATION
1	1355.795	50	1798.615	99	2206.645
2	1278.564	51	1922.900	100	1815.030
3	1508.327	52	1732.955	101	1876.000
4	1419.710	53	1723.575	102	2305.135
5	1440.510	54	1847.860	103	1899.450
6	1424.376	55	1479.695	104	511.210
7	1247.704	56	581.560	105	1906.485
8	455.498	57	1564.115	106	1798.615
9	1278.119	58	1648.535	107	1840.825
10	1422.715	59	1716.540	108	1972.145
11	1221.370	60	1690.745	109	2012.010
12	1314.325	61	1915.865	110	2424.730
13	1528.940	62	1925.245	111	2413.005
14	1486.730	63	2007.320	112	2080.015
15	1606.325	64	1892.415	113	2061.255
16	1545.355	65	1376.515	114	2483.355
17	1489.075	66	1749.370	115	1807.995
18	1667.295	67	1728.265	116	832.475
19	1198.295	68	668.325	117	1953.385
20	506.520	69	1671.985	118	2126.915
21	1350.720	70	2112.945	119	2148.020
22	1402.310	71	1817.375	120	2176.160
23	1475.005	72	1838.480	121	2279.340
24	1458.590	73	2030.770	122	2356.725
25	1416.380	74	1969.800	123	2485.700
26	1531.285	75	2124.570	124	2305.135
27	1761.095	76	1908.830	125	2592.115
28	1632.120	77	1887.725	126	2288.720
29	1711.850	78	2051.875	127	2072.980
30	1664.950	79	1568.805	128	785.575
31	1435.140	80	811.370	129	1995.595
32	572.180	81	1627.430	130	2940.630
33	1477.350	82	2061.255	131	2912.490
34	1622.740	83	1636.810	132	2680.335
35	1730.610	84	2164.435	133	2743.650
36	1641.500	85	1789.235	134	2417.695
37	1554.735	86	1990.905	135	2649.850
38	1739.490	87	2117.535	136	2309.825
39	2002.630	88	2021.390	137	2640.470
40	1648.535	89	1667.295	138	2366.105
41	1657.915	90	2042.495	139	2084.705
42	1948.695	91	1571.150	140	877.030
43	1292.095	92	712.880	141	2220.715
44	555.765	93	1641.500	142	2253.545
45	1486.730	94	1648.535	143	1620.395
46	1535.975	95	1639.155	144	1618.050
47	1650.880	96	1946.350	145	1488.371
48	1700.125	97	1960.420	146	1541.462
49	1800.960	98	1990.905	147	1642.743

(March 1975)

EXHIBIT 2   CFE Specialty Writing Papers

```
            MONTHLY   WORKING DAYS FOR YEARS 1963-1976
            -----------------------------------------------
```

YEAR	JAN	FEB	MAR	APR	MAY	JUN	JUL	AUG	SEP	OCT	NOV	DEC
1963	22	20	21	21	21	19	23	21	21	23	19	21
1964	22	20	21	22	19	22	22	21	22	22	20	22
1965	20	20	23	21	20	21	21	22	22	21	20	23
1966	21	20	23	20	20	22	20	22	22	21	20	22
1967	22	20	22	20	20	22	20	22	21	22	21	20
1968	22	21	21	21	21	19	23	21	21	23	19	21
1969	22	20	21	21	19	21	22	20	22	23	19	22
1970	21	20	21	22	18	22	22	21	22	22	20	22
1971	20	20	23	21	19	22	21	22	22	21	20	23
1972	21	21	23	19	20	22	20	22	21	22	21	20
1973	22	20	22	20	21	20	22	22	20	23	21	20
1974	22	20	21	21	21	19	23	21	21	23	19	21
1975	22	20	20	22	19	21	22	20	22	23	19	22

EXHIBIT 3   CFE Specialty Writing Papers Adjusted Monthly Sales Data*

Jan 1963

PERIOD	OBSERVATION	PERIOD	OBSERVATION	PERIOD	OBSERVATION
1	1323.740	50	1813.603	99	2094.714
2	1289.219	51	1908.332	100	1793.422
3	1568.181	52	1797.941	101	1958.280
4	1402.809	53	1709.211	102	2191.624
5	1360.482	54	1756.867	103	1952.212
6	1568.064	55	1596.837	104	497.655
7	1170.853	56	566.140	105	1855.934
8	505.330	57	1595.149	106	1898.538
9	1303.479	58	1661.024	107	1833.155
10	1376.950	59	1627.988	108	1836.382
11	1280.295	60	1810.506	109	2051.931
12	1340.403	61	1865.066	110	2328.511
13	1488.400	62	1848.846	111	2290.606
14	1499.120	63	2086.975	112	2271.595
15	1670.068	64	1869.886	113	2044.077
16	1457.551	65	1300.042	114	2361.068
17	1640.740	66	1925.842	115	1951.128
18	1585.193	67	1621.814	116	810.402
19	1175.600	68	681.585	117	1992.143
20	516.570	69	1705.159	118	2143.028
21	1314.906	70	2036.293	119	2037.209
22	1412.934	71	1905.055	120	2330.305
23	1468.859	72	1874.958	121	2218.903
24	1419.915	73	1976.924	122	2376.364
25	1516.707	74	1986.215	123	2466.868
26	1544.046	75	2208.878	124	2391.578
27	1671.764	76	1886.105	125	2363.108
28	1612.690	77	1970.520	126	2393.620
29	1697.584	78	2043.733	127	2033.719
30	1658.343	79	1539.093	128	764.745
31	1475.005	80	868.842	129	2136.950
32	557.009	81	1584.279	130	2834.086
33	1438.178	82	1986.572	131	2762.242
34	1712.892	83	1715.779	132	2870.192
35	1723.399	84	2107.044	133	2670.902
36	1528.498	85	1824.736	134	2437.842
37	1585.583	86	2007.496	135	2755.002
38	1754.490	87	2201.563	136	2282.327
39	1901.047	88	1906.538	137	2493.777
40	1710.355	89	1837.112	138	2604.791
41	1644.099	90	1941.917	139	1956.299
42	1852.736	91	1541.393	140	894.431
43	1394.386	92	727.024	141	2264.777
44	541.029	93	1597.975	142	2171.895
45	1447.309	94	1661.024	143	1698.572
46	1621.307	95	1632.325	144	1650.154
47	1644.001	96	1894.742	145	1448.907
48	1655.046	97	2099.283	146	1554.308
49	1753.208	98	2007.496	147	1793.327

(March 1975)

*Adjusted to average working days in each calendar month as computed from Exhibit 2.

EXHIBIT 4 CFE Specialty Writing Papers Business Index Data — Leading Indicator

Jan 1968

PERIOD	OBSERVATION	PERIOD	OBSERVATION	PERIOD	OBSERVATION
1	1233.720	30	2053.529	59	2345.234
2	1160.925	31	1675.544	60	2360.262
3	1124.010	32	474.375	61	2324.610
4	1335.943	33	2122.785	62	2394.341
5	876.300	34	2438.349	63	2751.778
6	1941.170	35	2068.364	64	2159.652
7	1935.795	36	1588.035	65	3016.680
8	556.688	37	1738.441	66	2281.378
9	1840.282	38	1848.244	67	2164.178
10	2154.180	39	2147.556	68	849.514
11	1694.181	40	1891.055	69	2037.566
12	1926.635	41	1781.945	70	2965.551
13	1761.877	42	2289.223	71	2913.014
14	1653.357	43	2012.023	72	3011.850
15	1599.185	44	654.827	73	2339.141
16	1815.276	45	1847.654	74	2360.286
17	1638.602	46	2237.776	75	2283.041
18	1948.580	47	2752.755	76	1973.065
19	1937.282	48	2052.912	77	2172.120
20	446.430	49	2100.988	78	2242.600
21	1750.978	50	2555.070	79	1710.331
22	1918.017	51	2658.225	80	581.670
23	1457.625	52	1869.900	81	1612.526
24	1399.665	53	1891.152	82	1507.722
25	1227.510	54	2143.650	83	1383.529
26	1508.340	55	2035.113	84	1253.464
27	1622.880	56	685.846	85	1614.600
28	1601.088	57	1997.550	86	1712.235
29	1538.355	58	2830.597	87	1755.015

(March 1975)

EXHIBIT 5   CFE Specialty Writing Papers Trend-cycles obtained from Census II

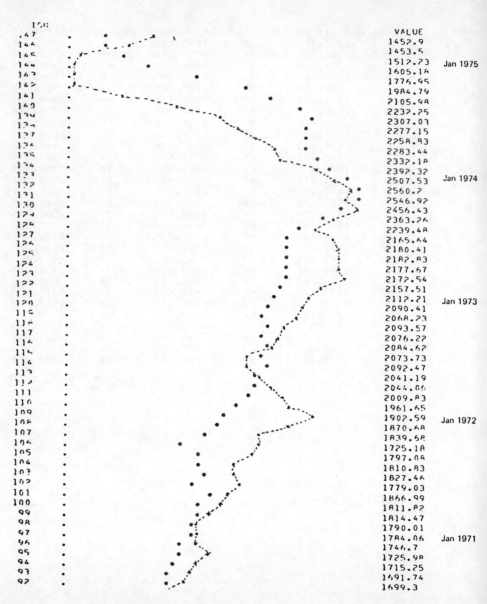

	VALUE	
	1452.9	
	1453.5	
	1512.23	Jan 1975
	1605.18	
	1776.95	
	1984.79	
	2105.98	
	2232.25	
	2307.03	
	2277.15	
	2258.83	
	2283.44	
	2332.18	
	2392.32	
	2507.53	Jan 1974
	2560.2	
	2546.92	
	2456.43	
	2363.26	
	2239.48	
	2165.64	
	2180.41	
	2182.83	
	2177.67	
	2177.54	
	2157.51	
	2112.21	Jan 1973
	2090.41	
	2068.23	
	2093.57	
	2076.22	
	2084.62	
	2073.73	
	2092.47	
	2041.19	
	2044.06	
	2009.83	
	1961.65	
	1902.59	Jan 1972
	1870.68	
	1839.68	
	1725.18	
	1797.08	
	1810.83	
	1827.46	
	1779.03	
	1866.99	
	1811.82	
	1814.47	
	1790.01	
	1784.06	Jan 1971
	1746.7	
	1725.98	
	1715.25	
	1691.74	
	1699.3	

*Adjusted Sales (from Exhibit 3)

x----x Leading Indicator (from Exhibit 4)

EXHIBIT 5 (Continued)

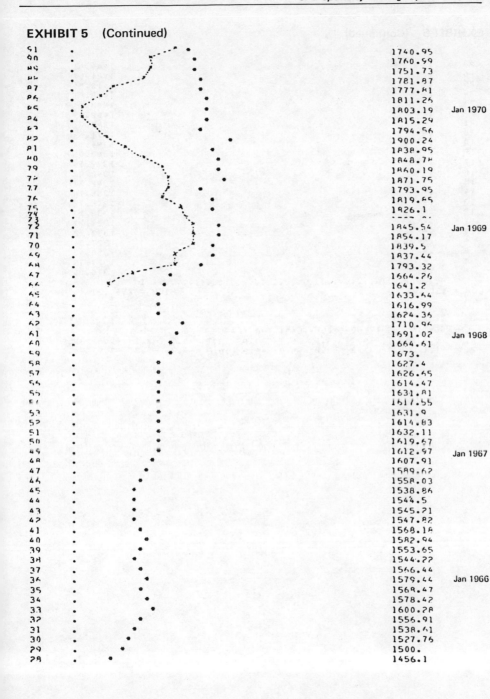

91	1740.95	
90	1760.99	
89	1751.73	
88	1781.87	
87	1777.81	
86	1811.25	
85	1803.19	Jan 1970
84	1815.29	
83	1794.56	
82	1900.24	
81	1838.95	
80	1848.74	
79	1860.19	
78	1871.75	
77	1793.95	
76	1819.65	
75	1926.1	
74		
73		
72	1845.54	Jan 1969
71	1854.17	
70	1839.5	
69	1837.44	
68	1793.32	
67	1664.26	
66	1641.2	
65	1633.64	
64	1616.99	
63	1624.36	
62	1710.96	
61	1691.02	Jan 1968
60	1664.61	
59	1673.	
58	1627.4	
57	1626.65	
56	1614.47	
55	1631.81	
54	1617.55	
53	1631.9	
52	1614.83	
51	1632.11	
50	1619.67	
49	1612.97	Jan 1967
48	1607.91	
47	1589.62	
46	1558.03	
45	1538.86	
44	1544.5	
43	1545.21	
42	1547.82	
41	1568.18	
40	1582.94	
39	1553.65	
38	1544.22	
37	1566.44	
36	1579.44	Jan 1966
35	1568.47	
34	1578.42	
33	1600.28	
32	1556.91	
31	1538.61	
30	1527.76	
29	1500.	
28	1456.1	

EXHIBIT 5 (Continued)

```
27          .      •                                    1449.61
26          .    •                                      1421.57
25          .      •                                    1410.84      Jan 1965
24          .   •                                       1403.61
23          .   •                                       1403.24
22          .   •                                       1402.09
21          . •                                         1380.21
20          .•                                          1371.6
19          .   •                                       1394.17
18          . •                                         1383.03
17          . •                                         1379.14
16          . •                                         1390.02
15          .    •                                      1392.57
14          .•                                          1365.76
13          .•                                          1346.72      Jan 1964
12          .•                                          1333.13
11          .•                                          1337.27
10          .•                                          1333.72
 9          .•                                          1327.24
 8          .•                                          1350.43
 7          .•                                          1327.79
 6          .•                                          1310.04
      I.........I..........I..........I..........I..........I
LOW = 1310.04                                    HIGH = 2560.21
```

% CHANGES IN THE TREND-CYCLE, LAST TWO YEARS
```
 0.2   0.2  -0.1  -0.7   3.4   5.5   3.9   3.7   0.5  -2.1  -4.6  -2.5
-2.1  -1.1   0.8   1.3  -3.2  -5.7  -5.8 -10.5  -9.7  -5.8  -3.9  -0.0
```

Multiple Regression

So far this book has been concerned with a wide range of time series methods and how they can be used for forecasting purposes. As stated in Chapter 2, time series methods assume that history repeats itself and aim at discovering the past (historical) pattern of events so they can be extrapolated to forecast the future. These methods are mechanistic in nature, and they require few judgmental inputs from the user in order to forecast. The exception is when changes in the trend cycle must be predicted, a task that is examined in detail in Chapters 27 through 29.

In addition to the time series approach to forecasting, there is an alternative procedure of "explanatory" methods. This latter category includes multiple regression and econometric models. Explanatory methods assume that there exist some basic casual factors which influence the course of events. The purpose of explanatory methods is to discover these factors and the form and extent of their influence. Once they are known, they can be used to forecast.

In this chapter, multiple regression will be examined as an explanatory method of forecasting. In addition, its use in a time series mode will be described. For single time series, multiple regression is not as efficient a forecasting procedure as the autoregressive-moving average schemes. However, it has the advantage that it can be augmented to include other factors in addition to time and seasonality, thus combining some of the advantages of both time series and explanatory approaches to forecasting.

THE MULTIPLE REGRESSION APPROACH

The general form of a multiple regression equation is:

$$Y = a + b_1 X_1 + b_2 X_2 + b_3 X_3 + \ldots + b_m X_m + u \qquad (23\text{-}1)$$

where $a, b_1\ b_2, b_3, \ldots, b_m$ are the regression coefficients.

Y is the dependent variable, such as Sales, which is to be forecast.

$X_1, X_2, X_3, \ldots, X_m$ can be any factors which affect the dependent variable, such as GNP, advertising, prices, etc. These are the independent variables. The residual term, u is that portion of Y that cannot be explained by the independent variables.

Once $a, b_1, b_2, b_3, \ldots, b_m$ are estimated, the values of Y can be predicted for specific values of $X_1, X_2, X_3, \ldots, X_m$. The difference between multiple regression and time series forecasting is that in the former the values of $X_1, X_2, X_3, \ldots, X_m$ must be known before any forecasting can be attempted. In time series, the corresponding X_i are known past values of the dependent variable.

Multiple regression assumes a linear relationship (or any relationship that can be transformed to linear) and then determines values for $a, b_1, b_2, b_3, \ldots, b_m$ in such a way that the mean squared error between the actual and the forecast figures is as small as possible. This method of estimation is called least squares and, because (23-1) is linear, it is known as the linear least squares method of estimation.

Estimating the values of $a, b_1, b_2, b_3, \ldots, b_m$ is a statistical matter of little interest to a user. (See Exercise 23-1 for the computational procedures.) There are many standard computer programs that can do this. What is important is to decide on the most appropriate independent variables to include in (23-1). The problem is similar to that faced in dealing with ARMA processes where the degree of the model has to be specified.

In order to identify the most appropriate factors to include in a multiple regression equation, one must use a theoretical base, empirical reasoning, and statistical means. The theory usually comes from economics, while the empirical part involves talking to those people in the organization who are doing the forecasting and asking them about the factors they consider when they make, for instance, sales predictions. With some persistence (since often the first answer is that there are no such factors, that the predictions are made only through experience) and by interviewing several people in the organization, one can identify most of the real factors which influence the dependent variable to be forecast. There may be some missing factors which can be introduced based on theoretical considerations and economic theory.

For example, the relationship between price and quantity, or the factors affecting sales and costs have been examined by econimists, and a great deal is known which can be used to develop multiple regression equations.

Once a set of independent variables have been identified, one must find a way of quantifying them so that they can be used as inputs for multiple regression analysis. Next one must decide which of the variables is most appropriate for inclusion in the regression equation.

References for further study:

> Johnston, J. (1976)
> Makridakis, S. and S. Wheelwright (1978, Chapter 6)
> Wheelwright, S. and S. Makridakis (1977, Chapter 7)

INPUT/OUTPUT FOR THE MULTIPLE REGRESSION PROGRAM (MREG)

Input

1. *Data for the Dependent and Independent Variables*

The data must be input in matrix form where the columns are the variables and the rows are the observations. All variables — dependent and independent alike — must be input for each observation. Thus, with k variables ($k = m + 1$) and n observations, the data must be input as follows (variables must be separated by a comma):

Observation	Var. 1	Var. 2	Var. 3	Var. k
1	X_{11}	X_{12}	$X_{13} \cdots X_{1k}$	
2	X_{21}	X_{22}	$X_{23} \cdots X_{2k}$	
3	X_{31}	X_{32}	$X_{33} \cdots X_{3k}$	
4	X_{41}	X_{42}	$X_{43} \cdots X_{4k}$	
.	.	.	.	.
.	.	.	.	.
.	.	.	.	.
n	X_{n1}	X_{n2}	$X_{n3} \cdots X_{nk}$	

2.   *How Many of the k – 1 Independent Variables Do You Want to Use in This Run?*

Since there are k variables in total and one has to be the dependent one, there are k – 1 (or m, see 23-1) remaining independent variables. Up to k –1 can be used for each regression equation. The first time the regression program is run, it is advisable to ask for all k – 1 independent variables: this provides fundamental information for determining the most appropriate variables to be included in the regression equation.

3.   *Which Independent Variables Are They?*

If the dependent variable is known (say it is number 2) one can type 1, 3, 4, . . . , k as the k – 1 independent variables. In subsequent runs, one can use a subset of the independent variables as desired.

4.   *Which Is Your Dependent Variable?*

The dependent variable can be any one, say 2.

5.   *Do You Want the Correlation Matrix?*

For the first run of the program, the answer should be YES. For subsequent runs, it makes little difference whether or not the correlation matrix is obtained.

6.   *Do You Want the Residuals to be Printed?*

On a slow terminal the answer should be NO, unless one has pretty well decided on what the final equation should be. From a statistical and logical point of view the residuals are useful in determining whether or not the equation under consideration is a "correct" one.

7.   *Do You Want to Stop?*

The user can experiment with different regression equations (by varying 2 and 3 above) until one is identified that satisfactorily describes the data. Regression analysis, as will be seen in examining the outputs, gives the user information that can be a guide in discovering a satisfactory regression equation of the type of (23-1).

If the answer to the question, "Do you want to stop?" is NO, the program branches to input 2 and the steps 2 to 7 are repeated.

Output

1. *The Mean and Standard Deviation of Each Variable*

This information can be useful in understanding the sampling distribution of the mean of each variable and can be used in constructing confidence intervals.

2. *Correlation Coefficients*

The correlation coefficients (or alternatively, the simple correlation matrix, which is the aggregation of all correlation coefficients) show how all of the independent variables are correlated between themselves and with the dependent variable. As a general rule, one does not want to include in a regression equation two or more independent variables that are highly correlated between themselves (i.e., more than .7), nor does one wish to include independent variables with low (i.e., less than .05) correlation with the dependent variable. The first case introduces multi-collinearity, which is a computational problem causing the results to become unreliable. (It is like trying to divide 1 by .00000000001: the computer may not have enough significant digits to handle the division.) If there are two independent variables with a high correlation between them, no information is lost by removing one of them. Low correlations imply that there is no relationship between the dependent and the corresponding independent variables. Thus, if such an independent variable is included, it generally will add little in explaining variations in the dependent variable.

Looking at the correlation matrix of all variables, one can immediately exclude those independent variables having a low correlation with the dependent variable, and then add independent variables that do not have a high intercorrelation between them. This can be achieved by starting with the independent variable with the highest correlation with the dependent variable, and then adding the independent variable with the next highest correlation with the dependent variable but with a low correlation with the previously included independent variable(s). (This information can be found by examining the correlation coefficients.) The correlation matrix allows one to exclude many variables (either high or low correlated), which makes the job of deciding upon a final equation much easier.

3. *Regression Coefficients and Related Statistics*

(a) *Regression Coefficients:*   These are the values a, b_1, b_2, ... b_k in (23-1). The value of a is indicated as "constant," while the coefficients, b_i, are indicated by "i". In the example below, the regression equation is:

$$Y = 23.1871 + 35.3895(X_2) + 10.3405(X_3) + 1.33445(X_4)$$

The regression coefficient b_2 indicates that if the value of the corresponding variable, X_2, increases by 1 unit, then the value of the dependent variable will increase by 35.3895 units.

(b) *Standard Error:* These standard errors of the regression coefficients allow one to specify the sampling distribution of the regression coefficients because the latter are assumed to be normally distributed. Dividing the regression coefficients by their standard errors yields the t-test for each coefficient.

(c) *t-tests:* The t-test is a statistical test used to check the hypothesis that a regression coefficient is equal to zero, versus the alternative hypothesis that it is not. If the value of the t-test is less than about 2 (see t-test table for exact value) then the value of the corresponding regression coefficient is not statistically different from zero. (That is, no matter what its value is, that value may have happened by chance.) Obviously, if this is true, one should re-run the regression program and exclude the variables whose t-tests are less than 2. If Figure 23-1 there are three t-tests less than 2: the constant and variables 3 and 4. As a rule, one should not drop all variables together; but those with the smallest t-test. Doing so often increases the t-test of the remaining variables. Dropping variable 4, the t-test of 3 increases to 11.17 (see exercise at the end of the chapter). This is because variables 3 and 4 are highly correlated (see Figure 23-1); their correlation coefficient is .987) which results in multi-collinearity and makes the results of Figure 23-1 unreliable.

FIGURE 23-1 Results of Regression Run Number 1

**REGRESSION NUMBER 1   *:DEPENDENT VARIABLE IS 1*

VARIABLE	MEAN	STANDARD DEVIATION
2	5.95694	1.20541
3	27.7118	12.4542
4	9	5.04975
1	532.565	166.673

CORRELATION COEFFICIENTS

VAR	2	3	4	1	
2	1.000	0.530	0.559	0.688	
3	0.530	1.000	0.987	0.948	
4	0.559	0.987	1.000	0.946	
1	0.688	0.948	0.946	1.000	DEPENDENT VARIABLE

VARIABLE	B	STD. ERROR	T-TEST	PCT. VARIATION EXPLAINED
CONSTANT	23.1871=a	68.1692	.340141	
2	35.3895=b_2	10.81	3.27377	.176033
3	10.3405=b_3	5.438	1.90154	.732597
4	1.33445=b_4	13.7188	9.72717E-02	3.82569E-02

(d) *Percentage Variation Explained:* The percentage variation explained indicates what percentage of the overall variation in the dependent variable is explained by each one of the independent variables. The higher the percentage, the more influential is that variable. The sum of all percentage variations explained is R^2.

(e) *R-Squared (R^2):* R^2 is a measure of the adequacy of the fit. It indicates the percentage of the total variations of the dependent variable explained by variations in the independent variables. R^2 varies from zero to one. If it is zero, it means that the regression equation does not explain the variations in the dependent variable. If it is one, it means that the regression fit is perfect and that any changes in the dependent variable are accounted for by changes in the independent variables. Such a relationship is deterministic and forecasting will be perfect. Between these two extremes of zero and one, R^2 can serve as one indicator of how "good" the regression equation is. (However, an R^2 value of close to 1.0 does not necessarily indicate a good forecasting equation.)

The question is often asked as to how high R^2 should be before the results are valid. The answer to this question, unfortunately, is: "it depends." For some applications, an R^2 of .95 is not good enough, while for others, .50 would be considered adequate. In medicine, for example, regression equations often are not accepted unless $R^2 \geqslant .99$, while in behavioral or marketing studies where human behavior is involved, R^2 values of about .15 or .20 are considered satisfactory.

(f) F-test: The F-test is similar to the t-test except that it tests the simultaneous significance of the variables in the regression equation It is a statistical means of checking the hypothesis that the overall relationship between the dependent variable and all the independent ones is statistically significant. As a general rule, if $F > 5$ it means that there is a statistically significant relationship. The F-test is the first statistic to examine in regression analysis. If it is greater than 5, then one can continue with the other tests. If it is less than 5, one must examine alternative models. If the F-test is significant, one can examine each of the t-tests and the R^2. If the F-test is > 5, and each of the t-tests > 2, then one can decide whether or not the R^2 is satisfactory. If it is, the remaining task is to check the regression assumptions. If R^2 is not satisfactory, it implies that there are other important factors that influence the dependent variable and that have not been included in the regression equation. It may be possible to identify these factors and increase the value of R^2 to some acceptable level.

4.   *The Durbin-Watson Test and Homoscedasticity*

There are four assumptions made in any application of regression: linearity, normality, independence of residuals, and constant variance. The first two are of little concern because they are almost always satisfied. For the last two, however, one must check to be sure that they are not violated.

(a) *Durbin-Watson Test:* The Durbin-Watson (D-W) test checks for the presence of a pattern in the residuals. If the computed value of the test is between 1.5 and 2.5 (one can look at a D-W table for exact values), the residuals are independent and no assumption is violated. If this is not the case, one must find ways to correct the violation of this basic assumption. The reasons for the violation may vary. It may be that there is some important independent variable missing from the regression equation or that the functional form being used is wrong. The further the value is from the range 1.5 and 2.5, the more serious the problem.

(b) *Visual Examination of the Table of Residuals:* Printing the actual, predicted, residual (error) and percentage error values for all data allows one to examine for a constancy in the variance. Most important for this are the percentage errors of the residuals. If these percentages are about the same throughout the entire range of observations, it implies a constant variance. If not, it means that the assumption of homoscedasticity is violated and that steps must be taken to correct it. The main causes of such problems are similar to those for the previous assumption — some important variable may have been omitted or the functional form is incorrect.

5. *Standard Deviation of Regression*

The standard deviation (or error) of regression can be used as a measure of anticipated deviations in the forecasts provided by the regression equation. It is the standard deviation of regression on which estimation of future errors is based.

TIME SERIES MULTIPLE REGRESSION (TMREG)

The time series multiple regression is completely analogous to the general multiple regression approach, except that it uses time and seasonality as the independent factors. Thus, its form is:

$$Y = a + bT + c_1 D_1 + c_2 D_2 + \ldots + c_{L-1} D_{L-1} \tag{23-2}$$

where $a, b, c_1, c_2, \ldots, c_{L-1}$ are the regression coefficients.

T is time, i.e., $1, 2, 3, \ldots, n$.

$D_1, D_2, \ldots, D_{L-1}$ are dummy variables used for seasonality, and L is the length of seasonality.

One can combine (23-1) and (23-2) to have mixed general and time series multiple regression.

Equation (23-2) says that the dependent variable is a function of time and seasonality. The purpose of the time series (TMREG) program is to identify the significant factors and measure their coefficients.

The inputs and outputs for the MREG program are exactly those of the general multiple regression program (MREG) with only two differences. First, the time series multiple regression program requires the length of seasonality as an input. Second, no values for independent variables are needed because in (23-2) they consist of time and dummy variables provided by the program. The user should, therefore, input the values of the dependent variable only.

EXERCISE 23-1

The following data are the yearly sales and some related independent variables of California Plate Glass. They are the program data in MREG. They can be easily accessed by running the multiple regression program. Use them to develop a multiple regression equation.

FIGURE 23-1 Results of Regression Run Number 1

NET SALES CALIFORNIA PLATE GLASS

SALES	AUTO PRODUCTION	BUILDING CONTRACTS	TIME
280	3.909	9.43	1
281.5	5.119	10.36	2
337.4	6.666	14.5	3
404.2	5.338	15.75	4
402.1	4.321	16.78	5
452	6.117	17.44	6
431.7	5.559	19.77	7
582.3	7.92	23.76	8
596.6	5.816	31.61	9
620.8	6.113	32.17	10
513.6	4.258	35.09	11
606.9	5.591	36.42	12
629	6.675	36.58	13
602.7	5.543	37.14	14
656.7	6.933	41.3	15
778.5	7.638	45.62	16
877.6	7.752	47.38	17

VARIABLES   1   2   3   4

Solution

The first step is to obtain the correlation matrix and see how the different variables relate with each other.

FIGURE 23-2 Correlation Matrix

CORRELATION COEFFICIENTS

VAR	2	3	4	1
2	1.000	0.530	0.559	0.688
3	0.530	1.000	0.987	0.948
4	0.559	0.987	1.000	0.946
1	0.688	0.948	0.946	1.000

Variable 1 is the dependent variable, while 2, 3 and 4 are the independent variables.

Starting at the last row of Figure 23-2, the highest correlation with variable 1 is with 3. Therefore, one would like to include 3 in the regression equation. The next highest correlation is with 4, but before including 4, one should check its correlation with 3. This correlation is .987 which implies that 3 (building contracts) and 4 (time) are highly correlated and, therefore, should not be included in the same regression equation. If both 3 and 4 were included, the result would be multi-collinearity, which could be detected by identifying one or more small t-tests (less than 2) and high values for the F-test and the R^2. Figure 23-3 shows the outcome for the regression equation which includes both 3 and 4 as independent variables.

FIGURE 23-3 Outcome of Regression Equation

VARIABLE	B	STD. ERROR	T-TEST	PCT. VARIATION EXPLAINED
CONSTANT	23.1871	68.1692	.340141	
2	35.3895	10.81	3.27377	.176033
3	10.3405	5.438	1.90154	.732597
4	1.33445	13.7188	9.72717E-02	3.82569E-02

R-SQUARED=0.947 R=0.973
F-TEST=124.69   STD. OF REGR.=41.08
DEGREES OF FREEDOM FOR NUMER.=2 FOR DENUMER.=14

The other remaining variable is 2, whose correlation with 3 is .53. Thus it can be incorporated in the regression equation with no problem. This gives the equation:

$$\text{Sales} = a + b_2 \ (\text{Auto Production} = X_2)$$

$$+ \ b_3 \ (\text{Building contracts} = X_3)$$

The regression run of this equation is shown in Figure 23-4. The parameter values are:

$$\text{Sales} = 19.125 + 35.6695(X_2) + 10.8603 \tag{23-3}$$

The F-test value is 124.69, which is very high and means that the relationship in equation (23-3), is statistically significant. The t-tests for variables 2 and 3 are greater than 2, and thus the effects of both variables are significantly different from zero. However, the t-test of a, the constant, is .36822, which means that it is not significantly different from zero. Thus, it might be desireable to force the constant term to zero. (If not, then equation (23-3) will be used for forecasting.) If the constant is forced to zero (see Figure 23-5 the equation becomes:

$$\text{Sales} = 38.914(X_2) + 10.8678(X_3)$$

It can be seen that (23-4) and (23-3) are very similar. The fact that R^2 = .946 means that variations (or changes) in X_2 and X_3 explain 94.6% of the variation (or changes) in sales in (23-4). From this 94.7%, 19.36% is explained (or caused) by variations in X_2 and 76.71% by X_3. (19.36 + 76.74 = 94.6 = R^2/100.)

FIGURE 23-4   Regression Run of New Equation

```
::REGRESSION NUMBER 2       :DEPENDENT VARIABLE IS 1


VARIABLE        B        STD. ERROR   T-TEST   PCT. VARIATION
                                                 EXPLAINED


CONSTANT      19.1248     51.9385     .36822
   2          35.6695     10.0444    3.55118      .177426
   3          10.8603      .972174  11.1712       .769423

R-SQUARED=0.947 R=0.973
F-TEST=124.69 STD. OF REGR.=41.08
DEGREES OF FREEDOM FOR NUMER.=2 FOR DENUMER=14
```

FIGURE 23-5 Regression Run with Constant Term of Zero

VARIABLE	B	STD. ERROR	T-TEST	PCT. VARIATION
CONSTANT	0	41.2782	0	
2	38.914	4.84435	8.03287	.193564
3	10.8276	.972781	11.1306	.767105

R-SQUARED=0.946   R=0.973
F-TEST=123.43   STD. OF REGR.=41.28

FIGURE 23-6 Durbin-Watson Test

DO YOU WANT THE RESIDUALS TO BE PRINTED

ACTUAL	PREDICTED	RESIDUALS	% ERROR
280	260.97	19.0302	6.7965
281.5	314.23	-32.73	-11.627
337.4	414.372	-76.9725	-22.8134
404.2	380.579	23.6212	5.84394
402.1	355.489	46.6109	11.5919
452	426.719	25.2807	5.59308
431.7	432.12	-.420288	-9.73565E-02
582.3	559.669	22.6315	3.88656
596.6	569.874	26.7263	4.47977
620.8	586.549	34.2508	5.5172
513.6	552.095	-38.4946	-7.49506
606.9	614.086	-7.18622	-1.18409
629	654.49	-25.4896	-4.0524
602.7	620.193	-17.4935	-2.90253
656.7	714.953	-58.2532	-8.87059
778.5	787.017	-8.51678	-1.094
877.6	810.197	67.4027	7.68035

DURBIN-WATSON STAT.=1.4211

Once the F-test and the t-tests are significant and the R^2 is satisfactory, one must check to be sure that no assumptions are violated. While there are not enough data points to be absolutely sure, the Durbin-Watson test is 1.42 (see Figure 23-6), which is almost in the needed range, 1.5 and 2.5. Thus, one might conclude that there is no pattern (autocorrelation) in the residuals; even though more careful consideration should be given if the results are to be used for forecasters' purposes, since 1.42 is outside the range normally used. Futhermore, the variance (see % Error) is fairly constant throughout the seventeen observations, meaning that it is homoscedastic.

If one had decided to include only variable 2 in the regression equation, the results would have been as shown in Figure 23-7. The F-test (13.47) is significant, the t-test for a is not significant, and for b_2 it is significant. Thus the equation would have been approximately:

$$\text{Sales} = 95.1(X_2) \tag{23-5}$$

If the constant term, a, is to be omitted because of a lack of statistical significance — as indicated by the t-test — one would generally rerun the regression program while forcing the constant term to be 0.0 in order to obtain a better estimate of b, as was done for equation (23-4). The difference between the remaining coefficients is small, however, and in the case of (23-5), the program was not re-run.

FIGURE 23-7 Regression Run for Variable 2 Only

```
:::REGRESSION NUMBER 3                :DEPENDENT VARIABLE IS 1
```

VARIABLE	B	STD. ERROR	T-TEST	PCT. VARIATION EXPLAINED
CONSTANT	-33.9407	157.328	-.215731	
2	95.1001	25.9163	3.66951	.473043

```
R-SQUARED=0.473  R=0.688
F-TEST=13.47   STD. OF REGR.=124.96
DEGREES OF FREEDOM FOR NUMER.=1  FOR DENUMER.=15
```

DO YOU WANT THE RESIDUALS TO BE PRINTED

ACTUAL	PREDICTED	RESIDUALS	% ERROR
280	337.805	-57.8055	-20.6448
281.5	452.877	-171.377	-60.8798
337.4	599.996	-262.596	-77.8294
404.2	473.703	-69.5035	-17.1953
402.1	376.987	25.1132	6.24552
452	547.786	-95.7864	-21.1917
431.7	494.721	-63.0206	-14.5982
582.3	719.252	-136.952	-23.5191
596.6	519.161	77.4387	12.98
620.8	547.406	73.394	11.8225
513.6	370.995	142.605	27.7657
606.9	497.764	109.136	17.9826
629	600.852	28.1477	4.47499
602.7	493.199	109.501	18.1684
656.7	625.388	31.3119	4.76807
778.5	692.434	86.0664	11.0554
877.6	703.275	174.325	19.8638

DURBIN-WATSON STAT.=.728289

The R^2 value of equation (19-5) is .473 which, although not as high as that of (23-4), is large enough to make (23-5) an acceptable equation. Looking at the D-W test, one sees that it is only .7283, which indicates the existence of some pattern in the residuals. Furthermore, the percentage error column shows non-constancy, going mainly from negative % errors to positive ones and from larger to smaller ones. Both facts suggest the wrong functional form or the omission of an important variable. In this case, it is the omission of variable 3.

Before continuing with an additional exercise illustrating time series multiple regression, it may be useful to illustrate the estimation of the parameters of Figure 23-3. This involves the use of matrix algebra, and some readers not familiar with that methodology may want to skip to the next exercise. (It is extremely difficult to show the computations involved without using matrix algebra because of the amount of space required.)

Suppose one calls the a, b_2, b_3, b_4 the B vector and denotes the elements as b_0, b_1, b_2, b_3, where b_0 refers to the intercept, b_1 to the auto production, b_2 to building contracts, and b_3 to time. The constant can be presented by a series of 1's. (If the constant is to be forced to zero the value of b_0, the intercept, takes the value of 0 instead of 1.) This gives the data matrix for X as

$$X = \begin{bmatrix} 1 & 3.909 & 9.43 & 1 \\ 1 & 5.119 & 10.36 & 2 \\ 1 & 6.666 & 14.5 & 3 \\ \cdot & \cdot & \cdot & \cdot \\ \cdot & \cdot & \cdot & \cdot \\ \cdot & \cdot & \cdot & \cdot \\ 1 & 7.752 & 47.38 & 17 \end{bmatrix} \qquad (23\text{-}6)$$

Note: This matrix has 17 rows and 4 columns.

The transpose of (23-6) is:

$$X' = \begin{bmatrix} 1 & 1 & 1 & \cdot \cdot \cdot & 1 \\ 3.909 & 5.119 & 6.666 & \cdot \cdot \cdot & 7.752 \\ 9.43 & 10.36 & 14.5 & \cdot \cdot \cdot & 47.38 \\ 1 & 2 & 3 & \cdot \cdot \cdot & 17 \end{bmatrix} \qquad (23\text{-}7)$$

Note: This matrix has 4 rows and 17 columns.

The sales vector, Y is:

$$Y = \begin{bmatrix} 280 \\ 281.5 \\ 337.4 \\ \cdot \\ \cdot \\ \cdot \\ 877.6 \end{bmatrix} \qquad (23\text{-}8)$$

The regression equation in matrix form is:

$$Y = BX$$

In order to find the regression coefficient vector, B, one must solve for B, which is done by applying the formula:

$$B = (X'X)^{-1} \, X'Y \qquad (23\text{-}9)$$

whoro X' is the transpose of X, and $(X'X)^{-1}$ is the inverse of the cross product of $X'X$.

Next one must multiply (23-7) times (23-6). The result is the 4 × 4 matrix:

$$(X'X) = \begin{bmatrix} 17 & 101.268 & 471.1 & 153 \\ 101.268 & 626.496 & 2933.54 & 965.845 \\ 471.1 & 2933.54 & 15536.7 & 5233.26 \\ 153 & 965.845 & 5233.26 & 1785 \end{bmatrix}$$

$$(23\text{-}10)$$

The matrix (23-10) is known as the cross product matrix of the X's, and it can be seen that it is symmetric.

Finally, to obtain $X'Y$, one multiplies (23-7) times (23-8). The result is the 4 × 1 matrix.

$$X'Y = \begin{bmatrix} 9053.6 \\ 562381 \\ 282381 \\ 94255.1 \end{bmatrix} \qquad (23\text{-}11)$$

In order to solve (23-9) one takes the inverse of (23-10), which is:

$$(X'X)^{-1} = \begin{bmatrix} 2.559 & -.3374 & -.1202 & .3155 \\ -.3374 & .0643 & .0054 & -.0217 \\ -.1202 & .0054 & .0163 & -.0404 \\ -.3155 & -.0217 & -.0404 & .1036 \end{bmatrix} \quad (23\text{-}12)$$

Finally, multiplying (23-12) times (23-11) yields:

$$B = (X'X)^{-1} (X'Y)$$

$$B = \begin{bmatrix} 23.17 \\ 35.39 \\ 10.34 \\ 1.33 \end{bmatrix} \quad (23\text{-}13)$$

$$\left. \begin{array}{l} \text{or } a = 23.17 \\ b_2 = 35.39 \\ b_3 = 10.34 \\ b_4 = 1.33 \end{array} \right\}$$ These are the values in Figure 23-3.

EXERCISE 23.2 (TIMES SERIES MULTIPLE REGRESSION)

The following data show the monthly number of international airline passengers (in thousands) between January 1949 and December 1955. Use them in the multiple regression program, and forecast the number of passengers for the year 1956. Since the actual numbers for 1956 were 284, 277, 317, 313, 318, 374, 413, 405, 355, 306, 271, 306, compare them with those forecast, and calculate the MSE (mean squared error) and the MAPE (mean absolute percentage error).

```
NAME OF VARIABLE FROM DATA FILE?AIRLINE


***** LISTS THE DATA OF A FILE *****
PERIOD   OBSERVATION   PERIOD   OBSERVATION   PERIOD   OBSERVATION
   1        112.00       29       172.00       57       237.00
   2        118.00       30       178.00       58       211.00
   3        132.00       31       199.00       59       180.00
   4        129.00       32       199.00       60       201.00
   5        121.00       33       184.00       61       204.00
   6        135.00       34       162.00       62       188.00
   7        148.00       35       146.00       63       235.00
   8        148.00       36       166.00       64       227.00
   9        136.00       37       171.00       65       234.00
  10        119.00       38       180.00       66       264.00
  11        104.00       39       193.00       67       302.00
  12        118.00       40       181.00       68       293.00
  13        115.00       41       183.00       69       259.00
  14        126.00       42       218.00       70       229.00
  15        141.00       43       230.00       71       203.00
  16        135.00       44       242.00       72       229.00
  17        125.00       45       209.00       73       242.00
  18        149.00       46       191.00       74       233.00
  19        170.00       47       172.00       75       267.00
  20        170.00       48       194.00       76       269.00
  21        158.00       49       196.00       77       270.00
  22        133.00       50       196.00       78       315.00
  23        114.00       51       236.00       79       364.00
  24        140.00       52       235.00       80       347.00
  25        145.00       53       229.00       81       312.00
  26        150.00       54       243.00       82       274.00
  27        178.00       55       264.00       83       237.00
  28        163.00       56       272.00       84       278.00

DO YOU WANT TO RUN ANOTHER PROGRAM (Y OR N)?N
```

Solution

Figure 23-8 presents the results of a regression run with passengers as the dependent variable, and time and dummy variables for seasonality as the independent variables. The F-test is significant, while the t-tests for variables 3, 4 and 12 are less than 2. One should therefore eliminate variables 3, 4 and 12. It is advisable, however, to eliminate one or two independent variables at a time, because doing so may increase the t-tests of the others. Figure 23-9 shows the regression results after 4 and 12 (the variables with the lowest t-tests) have been excluded.

As can be seen from Figure 23-9, the t-test corresponding to variable 3 is still low. Variable 3 must, therefore, be eliminated. The result is shown in Figure 23-10.

The F-test as well as all t-test of Figure 23-10 are significant. However, the regression equation represented in Figure 23-10 is not without problems. For one, thing, the Durbin-Watson Statistic is only .63. For another, there are two negative values under the column headed "percentages variation explained." This shows that variables 6 and 7 not only fail to add anything, but the overall result will be improved if they are removed. This is done, and the results shown in Figure 23-11.

FIGURE 23-8   Regression Run With Passengers as Dependent Variable

VARIABLE	B	STD.ERROR	T-TEST	PCT VARIATION EXPLAINED
CONSTANT	85.810	5.8725	14.6120	
2	2.159	.0612	35.3	.7981
3	3.603	7.2238	.5	-.0025
4	2.302	7.2183	.3	-.0015
5	27.429	7.2134	3.8	.0001
6	19.127	7.2090	2.7	-.0028
7	16.254	7.2051	2.3	-.0027
8	38.095	7.2017	5.3	.0160
9	60.937	7.1989	8.5	.0629
10	57.921	7.1965	8.0	.0586
11	30.619	7.1947	4.3	.0122
12	3.317	7.1934	.5	-.0007
13	-22.127	7.1926	-3.1	.0174

```
R-SQUARED =  .955    R =  .977
F-TEST =  125.827    STD OF REGR. =   13.456
DEGREES OF FREEDOM OF NUMER. =   12    OF DENUMER. =   71
DO YOU WANT A TABLE OF ACTUAL AND PREDICTED (Y OR N)?N


DURBIN-WATSON TEST =    .5573
MEAN SQUARED ERROR (MSE)          =   153.0
MEAN ABSOLUTE PC ERROR   (MAPE)   =    5.0%
MEAN PC ERROR (MPE) OR BIAS       =   -.00%
DO YOU WISH TO STOP?N
```

FIGURE 23-9   Regression Run After Variables 4 and 12 Were Deleted

```
HOW MANY OF YOUR    12 INDEPENDENT VARIABLES DO YOU WANT TO USE?10

WHICH INDEPENDENT VARIABLES ARE THEY ?2 3 5 6 7 8 9 10 11 13

DO YOU WISH THE CORRELATION MATRIX ?N

REGRESSION NUMBER    2         DEPENDENT VARIABLE IS    1
```

VARIABLE	B	STD.ERROR	T-TEST	PCT VARIATION EXPLAINED
CONSTANT	87.728	3.9273	22.3382	
2	2.158	.0602	35.9	.7978
3	1.723	5.8159	.3	-.0012
5	25.550	5.8084	4.4	.0001
6	17.250	5.8056	3.0	-.0026
7	14.378	5.8034	2.5	-.0024
8	36.220	5.8019	6.2	.0153
9	59.062	5.8009	10.2	.0610
10	56.048	5.8006	9.7	.0567
11	28.747	5.8009	5.0	.0114
13	-23.997	5.8034	-4.1	.0189

```
R-SQUARED =  .955    R =  .977
F-TEST =  154.736    STD OF REGR. =   13.291
DEGREES OF FREEDOM OF NUMER. =   10    OF DENUMER. =   73
DO YOU WANT A TABLE OF ACTUAL AND PREDICTED (Y OR N)?N


DURBIN-WATSON TEST =    .5588
MEAN SQUARED ERROR (MSE)          =   153.5%
MEAN ABSOLUTE PC ERROR   (MAPE)   =    5.0%
MEAN PC ERROR (MPE) OR BIAS       =   -.00%
DO YOU WISH TO STOP?N
```

FIGURE 23-10 Regression Run After Variable 3 Was Eliminated

```
HOW MANY OF YOUR    12 INDEPENDENT VARIABLES DO YOU WANT TO USE?9
WHICH INDEPENDENT VARIABLES ARE THEY ?2 5 6 7 8 9 10 11 13
DO YOU WISH THE CORRELATION MATRIX ?N
REGRESSION NUMBER    3          DEPENDENT VARIABLE IS    1
```

VARIABLE	B	STD.ERROR	T-TEST	PCT VARIATION EXPLAINED
CONSTANT	88.213	3.5472	24.8683	
2	2.156	.0597	36.2	.7973
5	25.115	5.5851	4.5	.0001
6	16.816	5.5833	3.0	-.0025
7	13.946	5.5822	2.5	-.0023
8	35.789	5.5817	6.4	.0151
9	58.633	5.5819	10.5	.0605
10	55.619	5.5827	10.0	.0563
11	28.320	5.5841	5.1	.0112
13	-24.422	5.5889	-4.4	.0192

```
R-SQUARED =  .955    R =  .977
F-TEST =  174.065    STD OF REGR. =   13.209
DEGREES OF FREEDOM OF NUMER. =    9    OF DENUMER. =    74
DO YOU WANT A TABLE OF ACTUAL AND PREDICTED (Y OR N)?N


DURBIN-WATSON TEST =    .5605
MEAN SQUARED ERROR (MSE)         =   153.7
MEAN ABSOLUTE PC ERROR   (MAPE)  =     5.0%
MEAN PC ERROR (MPE) OR BIAS      =    -.00%
DO YOU WISH TO STOP?N
```

FIGURE 23-11 Regression Results After Variables 6 and 7 Were Eliminated

```
HOW MANY OF YOUR    12 INDEPENDENT VARIABLES DO YOU WANT TO USE?7
WHICH INDEPENDENT VARIABLES ARE THEY ?2 5 8 9 10 11 13
DO YOU WISH THE CORRELATION MATRIX ?N
REGRESSION NUMBER    4          DEPENDENT VARIABLE IS    1
```

VARIABLE	B	STD.ERROR	T-TEST	PCT VARIATION EXPLAINED
CONSTANT	93.562	3.4346	27.2406	
2	2.151	.0637	33.8	.7953
5	19.974	5.7658	3.5	.0000
8	30.664	5.7633	5.3	.0129
9	53.513	5.7639	9.3	.0553
10	50.505	5.7652	8.8	.0511
11	23.211	5.7672	4.0	.0092
13	-29.520	5.7733	-5.1	.0232

```
R-SQUARED =  .947    R =  .973
F-TEST =  194.321    STD OF REGR. =   14.117
DEGREES OF FREEDOM OF NUMER. =    7    OF DENUMER. =    76
DO YOU WANT A TABLE OF ACTUAL AND PREDICTED (Y OR N)?N


DURBIN-WATSON TEST =    .6301
MEAN SQUARED ERROR (MSE)         =   180.3
MEAN ABSOLUTE PC ERROR   (MAPE)  =     5.3%
MEAN PC ERROR (MPE) OR BIAS      =    -.08%
```

Figure 23-11 shows no negative percentage variation. However, the D-W test is still very low, suggesting that there is some variable missing or a wrong functional form. Therefore, the forecasts given should be treated cautiously.

Figure 23-12 compares the forecasts with the actual values and shows the ex-post errors. The MSE, for the year of 1956, is 1331.80 while the MAPE = 7.74, both indicating acceptable levels of accuracy.

To obtain the forecasts, the program uses the regression equation of Figure 23-11.

Passengers = 93.56 + 2.151 (Time) + 19.97 (March) + 30.66 (June) + 53.51 (July) + 50.5 (Aug.) + 23.211 (Sept.) - 29.52 (Nov.)

Thus, if one wants to forecast for March, it will be: Passengers $_{1956, MH}$ = 93.56 + 2.151(87) + 19.97 = 300.7. For May, 1956 it will be: Passengers$_{1956, M}$ = 93.56 + 2.151(89) + 0 = 285.0: while for November 1956: Passengers $_{1956, N}$ = 93.56 + 2.151(95) - 29.52 = 268.4.

In order to obtain a forecast, one adds the constant to the trend at the particular period to be forecast, and adds (or subtracts) the value of the dummy variable corresponding to the month to be predicted. (See Figure 23-12.)

FIGURE 23-12 Comparison of Forecast with Actual Values

	Period		Actual	Forecast	Error	% Error
	85	J	284.0	276.4	7.6	2.7
	86	F	277.0	278.5	-1.6	-0.6
	87	M	317.0	300.7	16.3	5.1
	88	A	313.0	282.8	30.1	9.6
	89	M	318.0	285.0	33.0	10.4
	90	J	374.0	317.8	56.2	15.0
Year	91	J	413.0	342.8	70.2	17.0
1956	92	A	405.0	342.0	63.0	15.6
	93	S	355.0	316.8	38.2	10.7
	94	O	306.0	295.8	10.2	3.3
	95	N	271.0	268.4	2.6	1.0
	96	D	306.0	300.1	5.9	1.9

MSE = 1331.8   MAPE = 7.74

CHAPTER 24

Introduction: Multivariate Time Series Analysis and Forecasting

The time series methods described in the previous chapters have all been univariate in nature: the forecast is made as a combination of either past values or past errors of a single (dependent) variable or both. Univariate time series forecasting requires first that the characteristics of the series be discovered through time series analysis (using mainly autocorrelation and possibly some other techniques). Based on such analysis, the most appropriate model(s) for the series can be determined and subsequently used for predicting future values.

These univariate time series approaches to forecasting are sometimes viewed as unrealistic by those who feel that no forecasting can be done without taking into consideration external factors that affect the variable being forecast. These critics argue, for example, that an increase in the level of advertising will certainly affect sales and that time series approaches like those presented thus far cannot account for such outside influences.

One step in applying univariate time series methods that does attempt to use another variable as an aid to forecasting has been described in chapter 20. This involves the concept of paired indices. The additional variable in this case was a leading indicator of the dependent variable to be forecast. The objective was to find a visual lead/lag relationship between the dependent variable and a leading (independent) variable that could subsequently be used in predicting cyclical changes. From a practical point of view the approach of paired indices can be extremely useful because it provides causal information.

In this sense it blends time series and regression approaches to forecasting, even though this is done only in a visual form.

The methodologies of multiple regression and econometric modelling can be employed to estimate the full explanatory or causal relationships involving the variable(s) being forecast. Such an approach is the complement of time series forecasting and can be of importance if used properly. Unfortunately, there are problems from a forecasting point of view in applying such explanatory or causal approaches. For example, to predict sales one may decide that GNP and advertising are two factors affecting sales. However, future levels of GNP and advertising must be predicted before sales can be forecast. Thus, even though a perfect relationship may exist between sales and GNP/advertising, it may be of limited forecasting value. The forecasting problem is not necessarily solved by discovering cause and effect relationship.

An alternative to multiple regression may be to use the principle of paired indices but in a more mathematical form. That is, it may be possible to lead or lag certain variables or to use others that would be leaders in a regression equation in such a way that it is not necessary to know their future values in order to prepare a forecast. This is possible in some cases and has been used by economists and some businesses. This approach to forecasting is also not problem-free. For one thing, it requires much more sophistication on the part of the user than univariate time series analysis. However, if the alternative is multiple regression or econometric modelling that require independent variables, these may have their own difficulties, ranging from autocorrelation of the residuals to multicollinearity. The solution to these problems is not always easy and may complicate the use of these methods when an attempt is made to introduce leading or lagging indicators.

The methodology presented in the next few chapters combines the time series and explanatory or causal approaches to forecasting. It solves many of the problems faced by those who attempt to use regression with leading or lagging variables by transforming the data and reducing the residuals to white noise. Even though the nature of the approach is related closely to the methodology and concepts of the ARMA models examined in chapters 17 and 18, it incorporates the advantages of regression methods by allowing introduction of explanatory (independent) variables in order to forecast a dependent variable.

The steps in applying the multivariate approach, as it is called, will be explained in the next three chapters. Chapter 25 will be for multiple time series what chapter 3 is for univariate time series analysis. Use of the cross autocorrelations between any two variables to discover the degree and type of relationship between them and their application in identifying an appropriate model will be described. Then chapter 26 will discuss generalized adaptive filtering and how it can be used in a multivariate form. Finally, chapter 27 will look at multivariate ARMA models from the Box-Jenkins point of view — what Box and Jenkins refer to as Transfer Function models.

The remainder of this chapter will describe the rationale and major steps involved in using multivariate ARMA models. As a starting point, it is useful to review the steps applied in model specification when using multiple regression and some of the difficulties encountered when the assumption of the independence of residuals is violated.

An important step in multiple regression and econometric models is specification of an appropriate model. This is done using one of three approaches. The first, and often thought to be theoretically the most correct, advocates the specification of a model at the outset and then collection of the data needed to test whether that model should be accepted or rejected. Minor modifications of this model are acceptable, and they take the form of dropping some of the independent (or exogenous) variables that are not significantly different from zero.

The second and third approaches start with only a vague idea of what the final model should be. The data collection, therefore, is not limited to a specific set of variables; instead, data are collected for as many variables as "possible." The idea is that some of these variables may be useful later on. The second approach starts with one or two variables and then proceeds to include additional variables in the model. This is done until the R^2 value cannot be improved further, or until a satisfactory model is specified. (The manifistation of this approach is step-wise multiple regression.) The third approach starts with all of the variables (and their transformations) in the model and proceeds by eliminating those variables whose influence on the dependent (endogenous) variable is not statistically significant.

No matter what approach is used, a procedure for specification is required. All three approaches seek to identify a model in which all the variables are statistically significant in their influence. This involves repeated applications of regression analysis and the testing of a series of hypotheses to determine whether a specific independent (exogenous) variable is significantly related with the dependent (engogenous) variable. If it is, it is kept in the regression model (equation). If it is not, it is dropped from the model. The testing of hypotheses utilizes the variances of the individual regression coefficients (see chapter 23) to compute the required t-tests. It is, therefore, imperative in regression analysis that the variances be correct. If they are not, the model specification may be erroneous because the t-tests on which the specification is based will not be reliable. The same is true for the F-test or the R^2 values, which can be smaller or larger than those given in the program when the variance computer is not correct.

Unfortunately, the computed variances will not be correct if the residuals of the regression model are not random. This is so because in regression a major assumption is that the residual errors are independent. The variance is, therefore, estimated as if this assumption of independence of residuals is not violated. If it is violated, this will affect the magnitude of the computed variances and the resulting t-tests, F-test and R^2. The extent of the impact of a non-zero covariance is not easy to determine. The fact is, however, that the variances may be larger or smaller than those calculated by regression. The end effect is that the user may arrive at highly erroneous conclusions. It is possible to estimate the extent of over estimation or underestimation of the variance. If only the first autocorrelation is significant, the correct variance is approximately equal to:

$$\sigma_{uc}^2 = \sigma_u^2 \left[n - \frac{1 + \rho_1^2}{1 - \rho_1^2} \right]$$

where σ_{uc}^2 is the correct variance

σ_u^2 is the variance of regression

n is the sample size

and ρ^2 is the square of the first order autocorrelation coefficients.

The problem, however, is that non-random disturbances in the residuals are not restricted to first order, and it is extremely difficult to estimate the extent of over or under (usually it is under) estimation of the variance for such higher order autocorrelations. Furthermore, the standard statistical tests used in regression for testing autocorrelation do not apply where lagged independent variables are concerned, except when several additional assumptions are made. Finally, some of these tests are even useless when lagged values of the dependent variable are used in the equation. For an excellent discussion of these problems, beyond the scope of this text, the interested reader is encouraged to look at Johnston, chapters 8 and 10.

It should be clear that in regression models where some of the independent variables (or the dependent variable) are lagged or lead within the same equation there may be serious problems in specifying an appropriate model and estimating its parameters. Unfortunately, these problems have been largely ignored by both academicians and practitioners. The result has been the utilization of regression equations that may be useless and misleading. (The case, Forecasting IBM-Paris, illustrates many of the dangers involved.) Because this influences the variances of the parameters, it makes the t-tests unreliable, and it becomes impossible to be sure that the right model has been specified.

Other problems also arise when regression is used for multivariate time series forecasting. If two variables, each with a strong trend, are included in the same equation, the trend tends to dominate the relationship and does not allow the remaining pattern to be fully utilized. Trend can be eliminated by differencing, but this still does not solve the problem satisfactorily because strong seasonality, for example, can produce similar problems. In other words, if there is more than one component in the pattern, as in most series, these component patterns will tend to obscure each other and make identification and estimation difficult if not impossible. This is where regression methods seem to be at a serious disadvantage. As presently used, regression creates relationships that are the result of spurious correlations, rather than real cause and effect relationships. It has been shown, for example, that re-estimating a relationship after trends and seasonality have been removed yields completely different relationships and an R^2 which can be very close to zero, or regression equations that are not statistically significant. (The interested reader is encouraged to look at Pierce, Granger, and Granger and Newbold for an in depth coverage of this topic. Also the case "Forecasting IBM Paris" illustrates the points mentioned.)

The existence of such problems can be easily seen by removing a single variable from the model and re-estimating the parameters to see if the results may become quite different. Also, using the two halves of the data to obtain two estimations of the model may produce completely different results. In summary, multiple regression and econometric models may encounter considerable difficulties when attempts are made to combine them with time series principles. These difficulties become even more serious when lagging or leading of variables is involved. This is so because the assumption of independence of residuals is usually violated, and multicollinearity becomes a computational problem.

The multivariate ARMA (MARMA) approach to apply time series methods where additional independent variables are to be included does not suffer from the difficulties described above for regression methods. This is because model specification (referred to as identification in the ARMA methodology) is not based on the t-tests but rather on the autocorrelation and cross autocorrelation coefficients. This can be seen by examining the mathematics of MARMA models.

The general form of a MARMA model is:

$$Y_t = \delta_1 Y_{t-1} + \delta_2 Y_{t-2} + \cdots + \delta_r Y_{t-r} + \omega_o X_{t-b} - \omega_1 X_{t-b-1} - \cdots - \omega_s X_{t-b-s}$$

$$+ \varsigma_0 Z_{t-c} - \varsigma_1 Z_{t-c-1} - \cdots - \varsigma_m Z_{t-b-m}$$

$$+ \ .$$

$$+ \ .$$

$$+ \ .$$

$$+ \xi_0 W_{t-d} - \xi_1 W_{t-d-1} - \cdots - \xi_v W_{t-d-v}$$

$$+ e_t \qquad\qquad (24\text{-}1)$$

where Y_t is the dependent variable,

$Y_{t-1}, Y_{t-2}, \ldots Y_{t-r}$ are past values of the dependent variable,

X, Z and W are independent variables, and,

X_{t-b-i}, Z_{t-c-i}, and W_{t-d-i} (i=0,1, . . .) are past values of these independent variables.

If b, c and d are greater than zero, the corresponding independent variables (X, Z or W) will be leading indicators of Y_t. The purpose of the MARMA methodology is to facilitate determination of r, s, b, c, m, d and v and estimation of the parameter values, δ_i, ω_i, ζ_i, and ξ_i.

Once this is done, e_t (the residual) may be random or it may still include some pattern. If it is not random, additional noise terms may be added to (24-1) so that the subsequent residual, e_t', will be white noise (completely random). This is done using the model,

$$e_t = \phi_1 e_{t-1} + \phi_2 e_{t-2} + \ldots + \phi_p e_{t-p} + \theta_1 e_{t-1}'$$

$$- \theta_2 e_{t-2}' - \ldots - \theta_q e_{t-q}' \qquad (24\text{-}2)$$

Application of this model requires that the order of p and q be specified and the noise parameters, ϕ_i and θ_i, be estimated. The resulting errors, e_t', will be white noise, assuring zero covariance. Achieving white noise residuals can be done routinely by following the MARMA methodology, which involves several steps and requires some additional sophistication on the part of the user over that required in using univariate ARMA models.

At the present time, the MARMA models practically available are restricted to two variables (bivariate), and only this form will be examined in the following three chapters. Aside from complexity, there is no reason for such a restriction in the long term, and MARMA models involving more than one independent variable should become available in the near future.

Figure 24-1 shows the different stages required in using MARMA models. These four phases are: prewhitening, identification, estimation and diagnostic checking, and forecasting. Each will be briefly examined to give a more complete overview of this methodology.

FIGURE 24-1 A Schematic Presentation of the Basic Steps in Developing a MARMA Model

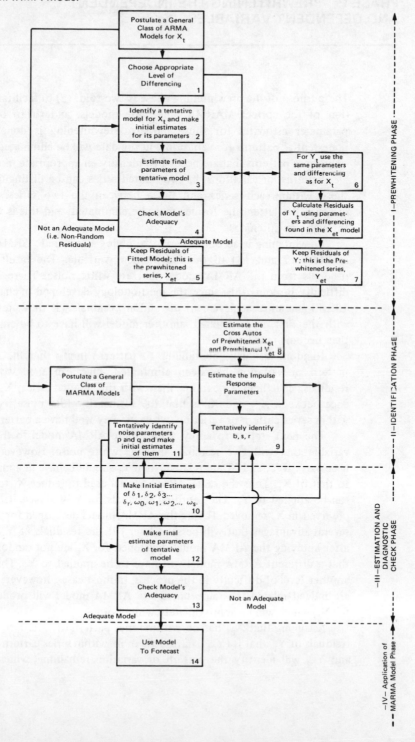

PHASE I: PREWHITENING THE INDEPENDENT AND DEPENDENT VARIABLES.

The purpose of the prewhitening phase is two fold: (a) to facilitate identification of the correct MARMA and noise models, and (b) to obtain initial parameter estimates for those models. Prewhitening is done so that all nonessential patterns (or variations) in the data will be eliminated. This leaves only those patterns that are needed to identify an appropriate model.

Patterns, or variations, in two related series can be distinguished, as can those within each series and those between the two series. In order to discover the latter, the former must be eliminated, and this is what is done through prewhitening.

Prewhitening is accomplished by applying a univariate ARMA model (see chapters 17 and 18) to the independent variable. The result is a set of residuals from the ARMA model that are white noise. There is no special difficulty in doing this, since the methodology developed in chapters 17 and 18 will guarantee achievement of random residuals. (If they are not random with the first model applied, another model will have to be employed until they become random.)

Random residuals (containing no pattern) imply that the within series pattern or variation has been eliminated. One can then work with the residuals, call them X_{et}, rather than with the original series, X_t. The difference between X_{et} and X_t is that the former is white noise, containing no within series pattern or variation, while X_t may well have a pattern.

The next step is to apply a univariate ARMA model to the dependent variable that will give residuals that are white noise. However, this is not necessary. If Y_t and X_t are related, then Y_t should behave in a manner similar to that of X_t. Thus the same ARMA model used to reduce X_t to white noise can be applied to Y_t. The resulting residuals for Y_t will have had the pattern observed in X_t removed. (Using the ARMA model developed for X_t on Y_t has several advantages that will be seen later.) If the residuals of Y_t that remain after applying the ARMA model developed for X_t are not random, it may be that a different ARMA model will have to be applied to Y_t. This introduces another level of difficulty in the process. In most cases, however, if X_t and Y_t are indeed related the same univariate ARMA model will produce residuals for Y_t, which will be approximately white noise.

Assuming that X_{et} has no within series pattern or variation, and that the residuals of Y_t, that is (Y_{et}), have little or no within series pattern, relating X_{et} and Y_{et} will identify the pattern or variation remaining, which will be the between X_t and Y_t pattern or variation only. In other words, by prewhitening X_t and Y_t to obtain X_{et} and Y_{et}, the within series patterns or variations are eliminated. This allows the between series pattern to be used in phase II as the basis for model identification and in phase III as the basis for parameter estimation.

PHASE II:　APPROPRIATE MODEL IDENTIFICATION

A complete MARMA model including two variables is given by:

$$Y_t = \delta_1 Y_{t-1} + \delta_2 Y_{t-2} + \ldots + \delta_r Y_{t-r} + \omega_0 X_{t-b}$$

$$- \omega_1 X_{t-b-1} - \ldots - \omega_s X_{t-b-s} + e_t \qquad (24\text{-}3)$$

where

$$e_t = \phi_1 e_{t-1} + \phi_2 e_{t-2} + \ldots + \phi_p e_{t-p} + e_t' - \theta_1 e_{t-1}'$$

$$- \theta_2 e_{t-2}' - \ldots - \theta_q e_{t-q}' \qquad (24\text{-}4)$$

Substituting (24-4) into (24-3) yields the full MARMA model,

$$Y_t = \delta_1 Y_{t-1} + \delta_2 Y_{t-2} + \ldots + \delta_r Y_{t-r} + \omega_0 X_{t-b} - \omega_i X_{t-b-1}$$

$$- \ldots - \omega_s X_{t-b-s} + \phi_i e_{t-1} + \phi_2 e_{t-2} + \ldots + \phi_p e_{t-p}$$

$$+ e_t' - \theta_1 e_{t-1}' - \theta_2 e_{t-2}' - \ldots - \theta_q e_{t-q}' \qquad (25\text{-}5)$$

Identification of (24-5) requires that values be detemined for

- r, the number of terms of the dependent variable included;
- s, the number of terms, plus one, of the independent variable included;
- b, the lag between when something happens in the independent variable and when its influence on the dependent variable occurs (the number of time periods the independent variable leads the dependent one);

- p, the number of autoregressive terms in the noise model (24-4); and
- q, the number of moving average terms in the noise model (24-4).

The order of r, s, and b can be identified from the cross autocorrelations between Xe_t and Ye_t, and the noise parameters, p and q, can be identified from the autocorrelations and the partial autocorrelations of the residuals of (24-4). The detailed steps required for identification will be described in the next chapter.

PHASE III: INITIAL AND FINAL PARAMETER ESTIMATION AND DIAGNOSTIC CHECKING

The estimation of initial and final parameters is similar to that for univariate ARMA models. The initial MARMA estimates of (24-3) are based on the cross autocorrelations, while the initial noise parameter estimates are based on the autocorrelations of (24-4). Final parameter estimates are obtained using a non-linear estimation algorithm proposed by Marquardt (1963). The parameter estimation phase and phase IV are the most routine steps in MARMA model applications.

Once final MARMA and noise parameters have been estimated, the residuals (e_t') are tested for randomness. If they are random, the next phase can be pursued. If they are not random, a new model must be tested. In addition to testing the residuals of e_t' for randomness, the cross autocorrelations between e_t' and Xe_t are also checked for randomness.

PHASE IV: FORECASTING

Once an adequate model has been found, it can be used for forecasting purposes. This phase presents no special difficulties. A detailed description of the four phases outlined above as well as examples illustrating their use will be given in the next three chapters.

References for Further Study:

Box, G. and G. Jenkins (1976)
Granger, C. W. J. (1975)
Granger, C. W. J. and P. Newbold (1974)
Makridakis, S. and S. Wheelwright (1978, Chapter 11)
Marquardt, D. (1963)
Pierce, D. A. (1977)

CHAPTER 25

Cross Autocorrelation Analysis

The cross autocorrelation coefficients,* or cross autos, describe the degree of association between two variables for various time lags. They can be visualized as combining the characteristics of correlation coefficients and autocorrelation coefficients into a single measure of association.

As with correlations and autocorrelations, the cross autocorrelations are relative measures of association that vary from -1 to $+1$. A value close to -1 indicates a strong negative relationship between X_t and Y_{t+k}, i.e., k time lags apart; a value close to $+1$ indicates a strong positive relationship, while a value close to zero indicates the absence of an association between X_t and Y_{t+k}.

Estimated cross autocorrelations based on sample data are usually expressed as

$$r_{XY}(k) \quad k = 0, \pm 1, \pm 2, \pm 3, \ldots, \pm m$$

where

X is the independent variable,

*The authors prefer the term cross autocorrelation coefficients to the more common phrase cross correlation coefficients, to recognize the similarities of the coefficients calculated in (25-1) to both correlations and autocorrelations. The authors realize that if taken literally, the words "cross" and "auto" contradict, but feel that cross autocorrelations is a name which more fully describes what $r_{xy}(k)$ represents and implies.

Y is the dependent variable,

k is the number of time lags separating X_t and Y_{t+k}, and

m is the number of time lags computed.

The formula for computing the cross autos for positive values of k is:

$$r_{XY}(k) = \frac{\sum_{t=1}^{n-k} (X_t - \bar{X})(Y_{t+k} - \bar{Y})}{\sqrt{\left[\sum_{t=1}^{n} (X_t - \bar{X})^2\right]\left[\sum_{t=1}^{n} (Y_t - \bar{Y})^2\right]}} \qquad (25\text{-}1)$$

The term in the numerator of (25-1) when $k = 0$ is similar to the covariance between Y and X but is not divided by n. If $k \neq 0$ the numerator is similar to the auto covariance between X_t and Y_{t+k}, for each k, except that it is not divided by n. The denominator of (25-1) is the standard deviation of X multiplied by the standard deviation of Y (except that neither is divided by n). While $r_{XY}(k)$ indicates the cross auto between X_t and Y_{t+k} of k time lags, $r_{XY}(-k)$, by definition, indicates the cross auto between Y_t and X_{t+k} of k time lags. Therefore,

$$r_{YX}(k) = r_{XY}(-k)$$

This is why the range of cross auto values is stated for $- m$ to $+ m$.

As an illustration of the application of (25-1), the ten values of X_t and Y_t shown in Table 25-1 can be used and the values of $r_{XY}(0)$, $r_{XY}(1)$, $r_{XY}(3)$, and $r_{XY}(-3)$ can be calculated.

For $k = 0$, expression (25-1) becomes

$$r_{XY}(0) = \frac{\sum_{t=1}^{10} (X_t - \bar{X})(Y_t - \bar{Y})}{\sqrt{\left[\sum_{t=1}^{10} (X_t - \bar{X})^2\right]\left[\sum_{t=1}^{10} (Y_t - \bar{Y})^2\right]}}$$

Thus, the values needed are the sum of the cross product of column (4) and (7), and the sum of the squares of columns (4) and (7). These values can

TABLE 25-1   Data Values for Calculating the Cross Autos Between X_t and Y_{t+k}

(1) Time	(2) X_t	(3) $\bar{X}$	(4) $(X_t - \bar{X})$	(5) Y_t	(6) $\bar{Y}$	(7) $(Y_t - \bar{Y})$	(8) $(Y_{t+1} - \bar{Y})$	(9) $(Y_{t+2} - \bar{Y})$	(10) $(Y_{t+3} - \bar{Y})$
1	6.0	7.1	-1.1	16.0	9.6	6.4	-9.6	5.4	-5.6
2	5.0	7.1	-2.1	0.0	9.6	-9.6	5.4	-5.6	-9.6
3	3.0	7.1	-4.1	15.0	9.6	5.4	-5.6	-9.6	-6.6
4	11.0	7.1	3.9	4.0	9.6	-5.6	-9.6	-6.6	1.4
5	0.0	7.1	-7.1	0.0	9.6	-9.6	-6.6	1.4	13.4
6	3.0	7.1	-4.1	3.0	9.6	-6.6	1.4	13.4	0.4
7	13.0	7.1	5.9	11.0	9.6	1.4	13.4	0.4	4.4
8	13.0	7.1	5.9	23.0	9.6	13.4	0.4	4.4	0.0
9	10.0	7.1	2.9	10.0	9.6	0.4	4.4	0.0	0.0
10	7.0	7.1	-0.1	14.0	9.6	4.4	0.0	0.0	0.0
Sum	71		0	96		0			
Average	7.1			9.6					

be seen in columns (4)–(6) of Table 25-2. Substituting these values into the above equation gives

$$r_{XY}(0) = \frac{152.4}{\sqrt{(182.9)\,(530.4)}} = .489$$

The cross auto of 1 time lag, i.e., $k = 1$, is given by

$$r_{XY}(1) = \frac{\displaystyle\sum_{t=1}^{9} (X_t - \bar{X})(Y_{t+1} - \bar{Y})}{\sqrt{\left[\displaystyle\sum_{t=1}^{10}(X_t - \bar{X})^2\right]\left[\displaystyle\sum_{t=1}^{10}(Y_t - \bar{Y})^2\right]}}$$

To calculate $r_{XY}(1)$, the sum of the cross products of columns (4) and (8) (see Table 25-1) is needed as well as the sum of the squares of columns (4) and (7) of Table 25-1. These values are 182.9 and 530.4, respectively, as can be seen from Table 25-2. The relevant calculations are shown in Table 25-3.

TABLE 25-2 Relevant Values for Calculating the Cross Autos Between X_t and Y_{t+k} when $k=0$

(1)	(2)	(3)	(4)	(5)	(6)
Time	$(X_t-\bar{X})$	$(Y_t-\bar{X})$	$(X_t-\bar{X})(Y_t-\bar{Y})$	$(X_t-\bar{X})^2$	$(Y_t-\bar{Y})^2$
1	-1.1	6.4	- 7.04	1.21	40.96
2	-2.1	-9.6	+20.16	4.41	92.16
3	-4.1	5.4	-22.14	16.81	29.16
4	3.9	-5.6	-21.84	15.21	31.36
5	-7.1	-9.6	+68.16	50.41	92.16
6	-4.1	-6.6	+27.06	16.81	43.56
7	5.9	1.4	+ 8.26	34.81	1.96
8	5.9	13.4	+79.06	34.81	179.56
9	2.9	.4	+ 1.16	8.41	.16
10	- .1	4.4	- .44	.01	19.36
Sum	0	0	152.4	182.9	530.4

TABLE 25-3 Data Values for Calculating the
Cross Autos Between X_t and Y_{t+k} when k=1

Time	$(X_t-\bar{X})$	$(Y_{t+1}-\bar{Y})$	$(X_t-\bar{X})(Y_{t+1}-\bar{Y})$
1	−1.1	−9.6	10.56
2	−2.1	5.4	−11.34
3	−4.1	−5.6	22.96
4	3.9	−9.6	−37.44
5	−7.1	−6.6	46.86
6	−4.1	1.4	− 5.74
7	5.9	13.4	79.06
8	5.9	0.4	2.36
9	2.9	4.4	12.76
10	− .1	0	0
Sum			120.04

Thus,

$$r_{XY}(1) = \frac{(X_1-\bar{X})(Y_2-\bar{Y}) + (X_2-\bar{X})(Y_3-\bar{Y}) + ... + (X_8-\bar{X})(Y_9-\bar{Y}) + (X_9-\bar{X})(Y_{10}-\bar{Y})}{\sqrt{\left[(X_1-\bar{X})^2+(X_2-\bar{X})^2+ ... +(X_{10}-\bar{X})^2\right]\left[(Y_1-\bar{Y})^2+(Y_2-\bar{Y})^2+ ... +(Y_{10}-\bar{Y})^2\right]}}$$

$$= \frac{(-1.1)(-9.6) + (-2.1)(5.4) + ... + (5.9)(.4) + (2.9)(4.4)}{\sqrt{\left[-1.1^2 + (-2.1)^2 + ... + (-.1)^2\right]\left[6.4^2 + (-9.6)^2 + ... + 4.4^2\right]}}$$

$$= \frac{120.04}{\sqrt{(182.9)(530.4)}} = .385$$

To calculate $r_{XY}(3)$, the sum of cross products of columns (4) and (10) of Table 25-1 is needed as well as the sum of the squares of columns (5) and (6) of Table 25-2. The relevant values are shown in Table 25-4. Thus,

$$r_{XY}(3) = \frac{-11.98}{\sqrt{(182.9)(530.4)}} = .038$$

Finally, one can verify that $r_{XY}(-3) = r_{YX}(3)$. This will require lagging X_t. The calculations required are shown in Table 25-5

TABLE 25-4 Data Values for Calculating the Cross Autos Between X_t and Y_{t+k} when k=3

Time	$(X_t - \bar{X})$	$(Y_{t+3} - \bar{Y})$	$(X_t - \bar{X})(Y_{t+3} - \bar{Y})$
1	-1.1	-5.6	6.16
2	-2.1	-9.6	20.16
3	-4.1	-6.6	27.06
4	3.9	1.4	5.46
5	-7.1	13.4	-95.14
6	-4.1	0.4	- 1.64
7	5.9	4.4	25.96
8	5.9	0	
9	2.9	0	
10	.1	0	
Sum			-11.98

TABLE 25-5 Data Values for Calculating the Cross Autos Between X_{t+k} and Y_t when k=3

Time	$Y_t - \bar{Y}$	$X_t - \bar{X}$	$X_{t+3} - \bar{X}$	$(Y_t - \bar{Y})(X_{t+3} - \bar{X})$
1	6.4	-1.1	3.9	24.96
2	-9.6	-2.1	-7.1	68.16
3	5.4	-4.1	-4.1	-22.14
4	-5.6	3.9	5.9	-33.04
5	-9.6	-7.1	5.9	-56.64
6	-6.6	-4.1	2.9	-19.14
7	1.4	5.9	- .1	- .14
8	13.4	5.9	0	
9	0.4	2.9	0	
10	4.4	- .1	0	
				-37.98

Thus,

$$r_{XY}(-3) = \frac{\sum\limits_{t=1}^{7} (Y_t - \bar{Y})(X_{t+3} - \bar{X})}{\sqrt{\left[\sum\limits_{t=1}^{10} (Y_t - \bar{Y})^2\right]\left[\sum\limits_{t=1}^{10} (X_t - \bar{X})^2\right]}}$$

$$= \frac{-37.98}{\sqrt{(182.9)(530.4)}} = -.122$$

(See Exercise 25-3 for the cross autos of other time lags.)

References for further study:

> Box, G. and G. Jenkins (1976)
> Makridakis, S. and S. Wheelwright (1978, Chapter 11)

INPUT/OUTPUT FOR THE CROSS AUTOCORRELATION PROGRAM (CROSS)

Input

1. The file containing the data values for the independent variable (the time series, X). It is recommended that X_t be prewhitened before being used in this program.
2. The file containing the data values for the dependent variable (the time series, Y_t). Y_t must also be prewhitened, otherwise the cross autos may have little meaning.
3. Number of cross autocorrelations.
   The user can specify as many as desired, although no more than about n/3 should be needed.
4. The user can continue by finding the cross autos of the differenced Y_t series, or of the moving average of the Y_t series. This is helpful when the order of r, s and b is not obvious. (See Priestley (1961) and Makridakis and Wheelwright (1978)). The order of r is then specified as the level of differencing or the length of the moving average at which no more contraction in the cross autos is possible. The order of s is equal to the number of cross autos significantly different from zero at the given r, while b is specified as the time lag at which the cross autos peak.

Output

A graph of the cross autos and the parameters in the equation (the transfer function) that relates an input series and an output series is provided by the program. (These parameters are frequently called inpulse response parameters.)

Cross autos, like regular autocorrelations, facilitate determination of:

a. whether the data, both X_t and Y_t, are stationary.
b. whether the data, X_t or Y_t, show any repetitive pattern in their interrelationship.
c. the length of this pattern.
d. whether the two variables, X_t and Y_t, lack any relationship between them.
e. the type of relationship between X_t and Y_t. (It therefore enables identification of an appropriate MARMA model.)

Usually (a) is of little practical value since the level of stationarity of X_t and Y_t can be determined individually through the regular autocorrelations. (The cross autos are more often used to confirm stationarity.) Points (b) through (e) are unique features of the cross autocorrelations. The use of cross autos will be illustrated in Figures 25-1 through 25-9.

Illustrations of the use of Cross Autocorrelations

Figure 25-1 shows the cross autos of two series, X_t and Y_t, consisting of 100 observations. X_t was generated from a normal distribution with a mean of zero and a variance of one. Y_t is simply $2X_t$. It *must* be noted that X_t is automatically in a prewhitened form since it is generated as such. The same thing is true with Y_t which is simply $2X_t$. Thus, no prewhitening is necessary to determine the relationship between X_t and Y_t since both series are generated to be "prewhitened" (or are white noise). This is not true in general, in which case one *must* prewhiten X_t and Y_t to discover their between relationship. Furthermore, all series used in this chapter (except for Figure 25-8) are in prewhitened form because the main purpose is to illustrate the behavior of the cross autos rather than to explain the process of prewhitening. (This will be done in the next chapter.) The case "Forecasting IBM Paris" can be used to illustrate the process of prewhitening and the almost impossible task of discovering a relationship between two variables if they have not been prewhitened.

If nothing is known about X_t and Y_t and their relationship to each other, this information can be inferred by examining the cross autos shown in Figure 25-1. Since $r_{XY}(0) = 1$, it indicates that for a time lag of zero periods, X_t and Y_t are perfectly correlated. For all other time lags, the cross autos are small and fall within the dotted lines which denote the 95% confidence limits

FIGURE 25-1   Cross Autos of Two Random Series, X_t and $Y_t = 2X_t$

(Note that all cross autos, except for k = 0, are within the dotted lines)

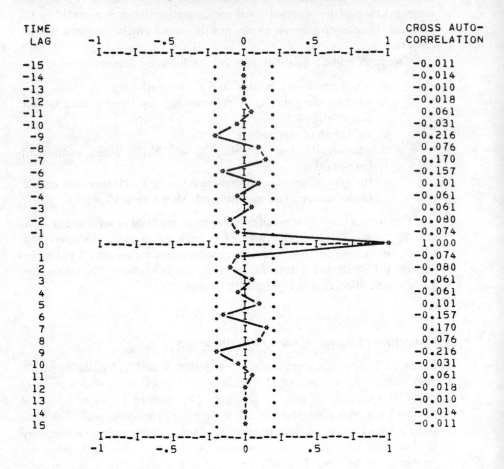

TIME LAG								CROSS AUTO-CORRELATION
	-1	-.5		0	.5		1	
-15								-0.011
-14								-0.014
-13								-0.010
-12								-0.018
-11								0.061
-10								-0.031
-9								-0.216
-8								0.076
-7								0.170
-6								-0.157
-5								0.101
-4								-0.061
-3								0.061
-2								-0.080
-1								-0.074
0								1.000
1								-0.074
2								-0.080
3								0.061
4								-0.061
5								0.101
6								-0.157
7								0.170
8								0.076
9								-0.216
10								-0.031
11								0.061
12								-0.018
13								-0.010
14								-0.014
15								-0.011
	-1	-.5		0	.5		1	

for a random series. These cross autos for time lags other than zero indicate that X_t and Y_{t+k} are unrelated.

In Figure 25-2, X_t has been generated as before, but $Y_t = X_{t-4}$. In this case, Y_t and X_t are related, even though there is no pattern within either series individually.

Figure 25-2 clearly shows the interrelationship between Y_t and X_t for a time lag of 4 periods ($r_{XY}(4) = .934$). All other cross autos are not significantly different from zero, indicating no relationship for other time lags.

The data used in Figure 25-3 for X_t are similar to those of the previous two figures. However, $Y_t = .7Y_{t-1} + X_{t-4}$. ($Y_t$ is generated with an additional lagged term of its own.)

Figure 25-3 has more than one cross auto that is significantly different from zero. The first is when $k = 4$ and for $k > 4$, the cross autos decline to zero exponentially, as would be expected of an AR-type model. As a matter of fact, the equation by which Y_t was generated is an AR(1). This can be seen from

$$Y_t = .7Y_{t-1} + X_{t-4} \tag{25-2}$$

Since X_{t-4} is random noise (it was generated as such), equation (25-1) can be rewritten as

$$Y_t = .7Y_{t-1} + e_t \tag{25-3}$$

The autocorrelations of (25-2), or its equivalent (25-3), could be computed easily. They would be found to behave as an AR(1) model. This behavior can be seen in Figure 25-3, starting at $k = 4$ since Y_t is related to X_t by (25-2). An interesting comparison can be made by calculating the regular autocorrelations of (25-3) and comparing them to the cross autos of (25-2). Except for the difference in time lags, the autos and cross autos behave almost exactly the same. (See Figure 25-4.) This is a visual confirmation of the behavior of equations (25-2) and (25-3) and their relationship.

Figure 25-5 represents the cross autos for a Y_t which is similar to that of Figure 25-3, except that it includes one additional term,

$$Y_t = .7Y_{t-1} - .5Y_{t-2} + X_{t-4} \tag{25-4}$$

Since X_{t-4} is white noise, Y_t will behave as an AR(2) model except that its cross autos will start being significant at a time lag of 4 periods. This can be clearly seen in Figure 25-5, which shows the cross autos between X_t and Y_t. The cross autos oscillate as a sine wave around zero, while declining in magnitude. This is typical of an AR(2) model.

FIGURE 25-2   Cross Autos of Two Random Series Related to Each Other
by $Y_t = X_{t-4}$

(Note that all cross autos, except for k = 4, are within the dotted lines)

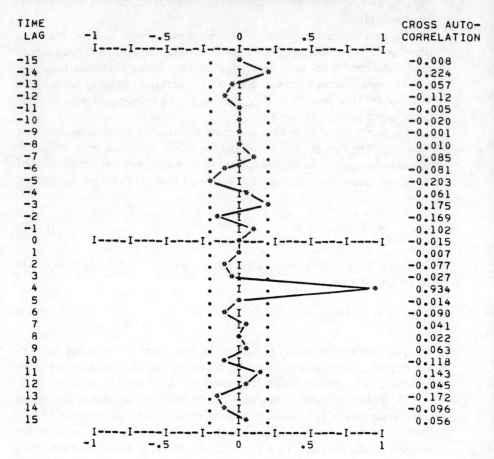

TIME LAG	CROSS AUTO-CORRELATION
-15	-0.008
-14	0.224
-13	-0.057
-12	-0.112
-11	-0.005
-10	-0.020
-9	-0.001
-8	0.010
-7	0.085
-6	-0.081
-5	-0.203
-4	0.061
-3	0.175
-2	-0.169
-1	0.102
0	-0.015
1	0.007
2	-0.077
3	-0.027
4	0.934
5	-0.014
6	-0.090
7	0.041
8	0.022
9	0.063
10	-0.118
11	0.143
12	0.045
13	-0.172
14	-0.096
15	0.056

FIGURE 25-3   Cross Autos of Two Random Series Related to Each Other
by $Y_t = .7Y_{t-1} + X_{t-4}$

(Note that the first significant cross auto corresponds to a time lag of 4, and that from
there on the cross autos decline to zero exponentially)

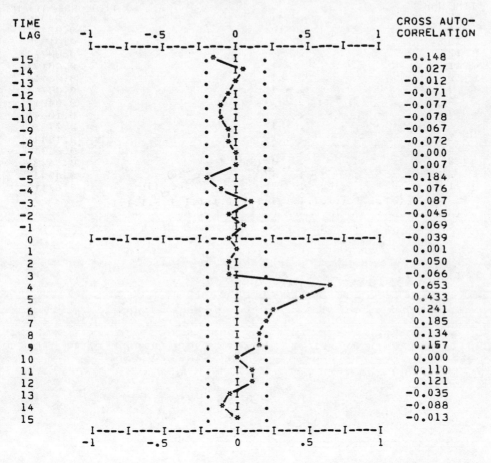

TIME LAG		CROSS AUTO-CORRELATION
-15		-0.148
-14		0.027
-13		-0.012
-12		-0.071
-11		-0.077
-10		-0.078
-9		-0.067
-8		-0.072
-7		0.000
-6		0.007
-5		-0.184
-4		-0.076
-3		0.087
-2		-0.045
-1		0.069
0		-0.039
1		0.001
2		-0.050
3		-0.066
4		0.653
5		0.433
6		0.241
7		0.185
8		0.134
9		0.157
10		0.000
11		0.110
12		0.121
13		-0.035
14		-0.088
15		-0.013

FIGURE 25-4   Autocorrelations (*) of Equation (25-3) and Cross
Autocorrelations (X) of (25-2)

(Where both the autos and cross autos overlap, a (o) is used) (See also Figure 25.3)

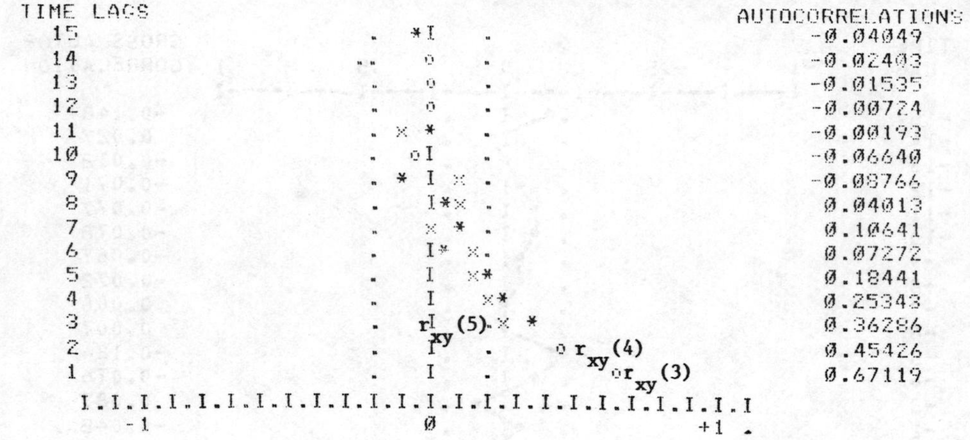

```
TIME LAGS                                              AUTOCORRELATIONS
   15              .   *I   .                               -0.04049
   14              .. .  o   .                              -0.02403
   13              .       o   .                            -0.01535
   12              .       o   .                            -0.00724
   11              . x  *   .                               -0.00193
   10              .  oI    .                               -0.06640
    9              . *  I x .                               -0.08766
    8              .    I*x .                                0.04013
    7              .     x *  .                              0.10641
    6              .    I* x.                                0.07272
    5              .    I   x*                               0.18441
    4              .    I     x*                             0.25343
    3              .   rI (5). x  *                          0.36286
    2              .   xy .       o r   (4)                   0.45426
    1              .    I   .      xy o r   (3)               0.67119
                                           xy
    I.I.I.I.I.I.I.I.I.I.I.I.I.I.I.I.I.I.I.I.I.I.I
      -1                   0                 +1  .
```

FIGURE 25-5   Cross Autos of Two Random Series Related Through
$Y_t = .7Y_{t-1} - .5Y_{t-2} + X_{t-4}$

(The first significant cross auto is at k = 4 and subsequent values decline to zero
while oscillating around zero)

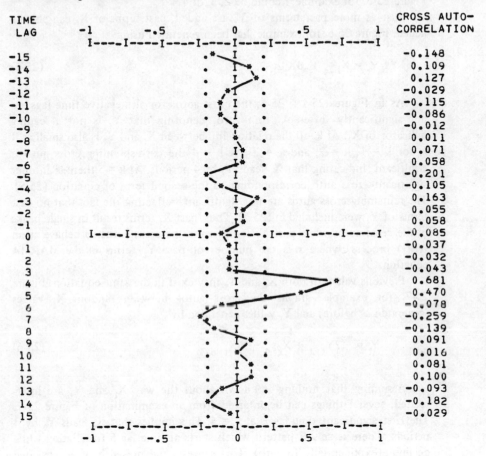

TIME LAG	-1	-.5	0	.5	1	CROSS AUTO-CORRELATION
-15						-0.148
-14						0.109
-13						0.127
-12						-0.029
-11						-0.115
-10						-0.086
-9						-0.012
-8						0.011
-7						0.071
-6						0.058
-5						-0.201
-4						-0.105
-3						0.163
-2						0.055
-1						0.058
0						-0.085
1						-0.037
2						-0.032
3						-0.043
4						0.681
5						0.470
6						-0.078
7						-0.259
8						-0.139
9						0.091
10						0.016
11						0.081
12						0.100
13						-0.093
14						-0.182
15						-0.029

If X_{t-4} in equation (25-4) is left unchanged but additional Y_{t-i} terms are added, the behavior of the cross autos will be like that of an AR model of order r, where r is equal to the number of past terms of Y_t that are included. The exact order of r can only be inferred, however, from the cross autos. In Figure 25-5, for example, r could be 2, 3, or 4.

Just as more past terms of Y_t are added, past terms of X_t can also be added. Figure 25-6, for example, has been generated using

$$Y_t = X_{t-4} + .6X_{t-5} \qquad (25\text{-}5)$$

As in Figures 25-1 – 25-5, the cross autos for all negative time lags are not significantly different from zero, denoting that Y_t is not a leading indicator of X_t. At k = 0 the relationship between X_t and Y_t is also small as it is for k = 1, k = 2, and k = 3. At k = 4 the corresponding cross auto is significant, indicating that X_t leads Y_t by 4 periods. At k = 5 there is another significant cross auto corresponding to the second term of equation (25-5). The remaining cross autos are not significant, reflecting the fact that no past terms of Y_t were included in (25-5). Thus, past X_t terms result in single peaks in the cross autos while past Y_t terms cause the cross autos to behave as an AR(r) process, where r is the number of past Y_t terms in the MARMA equation.

Previous values of both X_t and Y_t may exist in the same equation. Figure 25-7, for example, graphs the cross autos between random X_t values (generated as before) and Y_t values generated by

$$Y_t = -.7Y_{t-1} + X_{t-4} - .5X_{t-5} \qquad (25\text{-}6)$$

Assuming that nothing is known about the way X_t and Y_t are inter-related, several things can be inferred from an examination of Figure 25-7. The first significant time lag is at k = 4. This implies that X_t leads Y_t by 4 periods. There is an AR pattern which starts at time lag 5 (oscillating while declining exponentially to zero). This suggests the presence of at least one past Y_t term. Finally, with the pattern starting at 5 and the first significant auto at 4, it indicates two X_t terms (X_{t-4} and X_{t-5}).

Figures 25-1 through 25-7 summarize the behavior of a fairly wide range of cross autos, although all are from specially generated series. Obviously, with real data the exact underlying model for each series and the way in which the series are interrelated with each other are not known. In addition, randomness may be much more substantial, further obscuring the underlying patterns. However, the cross autos are still the only available approach for identifying the pattern of interrelationship between two series, and they must be examined carefully. It is hoped that more reliable identification method-ologies will become available in the future.

FIGURE 25-6   Cross Autos of Two Random Series Related by
Equation (25-5)

(Two are significant, k = 4 and k = 5, and all others are not significant)

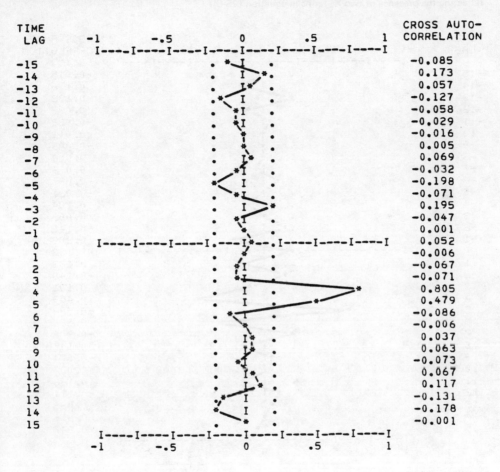

TIME LAG		CROSS AUTO-CORRELATION
-15		-0.085
-14		0.173
-13		0.057
-12		-0.127
-11		-0.058
-10		-0.029
-9		-0.016
-8		0.005
-7		0.069
-6		-0.032
-5		-0.198
-4		-0.071
-3		0.195
-2		-0.047
-1		0.001
0		0.052
1		-0.006
2		-0.067
3		-0.071
4		0.805
5		0.479
6		-0.086
7		-0.006
8		0.037
9		0.063
10		-0.073
11		0.067
12		0.117
13		-0.131
14		-0.178
15		-0.001

FIGURE 25-7 Cross Autos of Two Random Series Related by Equation (25-6)

(The first significant cross auto is at k = 4, while the AR pattern starts at k = 5, reflecting the presence of two X_t terms in equation (25-6))

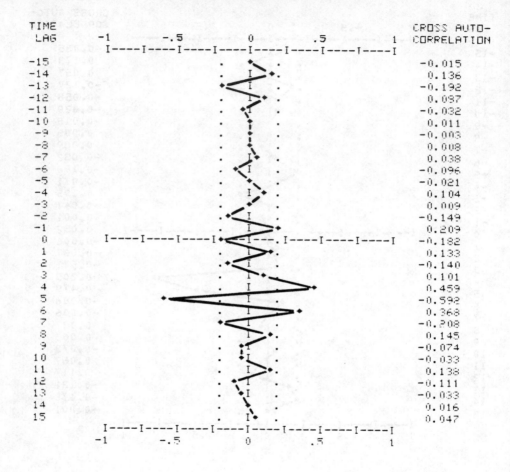

```
TIME                                                        CROSS AUTO-
 LAG      -1        -.5         0         .5          1     CORRELATION
        I----I----I----I----I----I----I----I----I
-15                            .  ◆  .                       -0.015
-14                            .  I  ◆.                        0.136
-13                          ◆  .I  ◆.                        -0.192
-12                            .  I◆ .                         0.087
-11                            .◆ I  .                        -0.032
-10                            .  I  .                         0.011
 -9                            .  I  .                        -0.003
 -8                            .  I  .                         0.008
 -7                            .  I◆ .                         0.038
 -6                          . ◆I  .                          -0.096
 -5                            .◆ I  .                        -0.021
 -4                            .  I ◆.                         0.104
 -3                            .  I  .                         0.009
 -2                          . ◆ I  .                         -0.149
 -1                            .  I   ◆                        0.209
  0   I----I----I----I◆---I----.I----.I----I----I----I       -0.182
  1                            .  I ◆.                         0.133
  2                          . ◆ I  .                         -0.140
  3                            .  I ◆.                         0.101
  4                            .  I      ◆                     0.459
  5              ◆              .  I  .                       -0.592
  6                            .  I    ◆.                      0.368
  7                          . ◆ I  .                        -0.208
  8                            .  I ◆.                         0.145
  9                            .◆ I  .                        -0.074
 10                            .  I  .                        -0.033
 11                            .  I ◆.                         0.138
 12                          .◆  I  .                         -0.111
 13                            . ◆I  .                        -0.033
 14                            .  I  .                         0.016
 15                            .  I◆ .                         0.047
        I----I----I----I----I----I----I----I----I
        -1        -.5         0         .5          1
```

As shown in the previous chapter, the general form of a MARMA model involving two variables can be written as

$$Y_t = \delta_1 Y_{t-1} + \delta_2 Y_{t-2} + \ldots + \delta_r Y_{t-r} + \omega_0 X_{t-b}$$

$$- \omega_1 X_{t-b-1} - \ldots - \omega_s X_{t-b-s} + e_t \tag{25-7}$$

Application of this model requires that its order be specified and the value of b be known. Specifically, the values of r (the number of past Y_t terms), s (the number of past X_t terms minus 1) and b (the number of periods by which X_t leads Y_t) must be determined. These values of r, s and b can be identified using the cross autos as follows:

- b. is equal to the time lag corresponding to the first significant cross auto. In the examples examined previously in this chapter, the first significant cross auto was at k = 4, thus b = 4.
- r. is found by examining the cross autos for any pattern in them. This pattern, if it exists, will be of AR(r) nature. If it does not exist (as with Figures 25-2 and 25-6), it implies that r = 0, i.e., that the model contains no past terms of Y_t.
- s. is the number of time lags that the start of a pattern* in the cross autos is delayed. For example, in Figure 25-7, $r_{XY}(4)$ is significant but the AR(1) pattern doesn't start until k = 5. Thus, s = 5 – 4 = 1 in this case.

With the above guidelines, values for r, s and b can be tentatively identified and a model postulated. This can be done as long as X_t is white noise (i.e., it includes no within series pattern) so that the interrrelationship between X_t and Y_t can be isolated and subsequently measured using the cross autos. When X_t is not white noise, the cross autos behave as shown in Figure 25-8.

Figure 25-8 does not indicate much about the underlying MARMA model. It seems that Y_t leads X_t (there are significant cross autos for negative time lags) by about the same magnitude as X_t leads Y_t. This is not of much use for forecasting purposes. The fact that the cross autos do not decline to zero rapidly (after three or four time lags) suggests nonstationarity in X_t or Y_t, or probably in both. But even removing nonstationarity does not improve the behavior of the cross autos much because of the existence of a pattern within X_t and within Y_t. If this pattern is removed, the cross autos (shown in Figure 25-9) are much more useful in identifying the relationship between X_t and Y_t and Y_t and the appropriate MARMA model.

Examining Figure 25-9, it can be seen that b = 5. However, it is not clear what the order of r and s should be. It can be noted that there are two

---

*This use of the concept of a pattern differs from the use of pattern described by Box and Jenkins (1969).

FIGURE 25-8   Cross Autos Between X_t and Y_t When X_t and Y_t are not Prewhitened

(The cross autos reveal little about the underlying model because of the presence of within series patterns in X_t and Y_t)

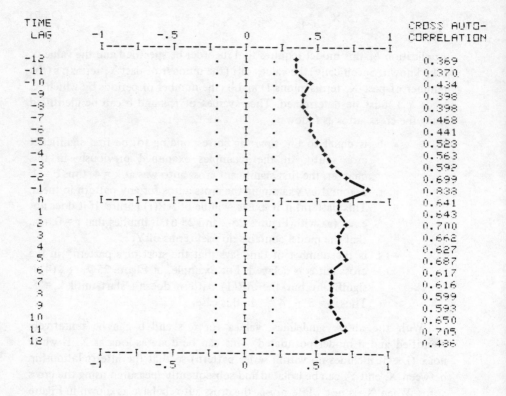

TIME LAG		CROSS AUTO-CORRELATION
-12		0.369
-11		0.370
-10		0.434
-9		0.398
-8		0.398
-7		0.468
-6		0.441
-5		0.523
-4		0.563
-3		0.592
-2		0.699
-1		0.833
0		0.641
1		0.643
2		0.700
3		0.662
4		0.627
5		0.687
6		0.617
7		0.616
8		0.599
9		0.593
10		0.650
11		0.705
12		0.436

FIGURE 25-9   Cross Autos Between X_t and Y_t (see Figure 25-8) When X_t and Y_t are Prewhitened

(The interrelationship between X_t and Y_t is now clear)

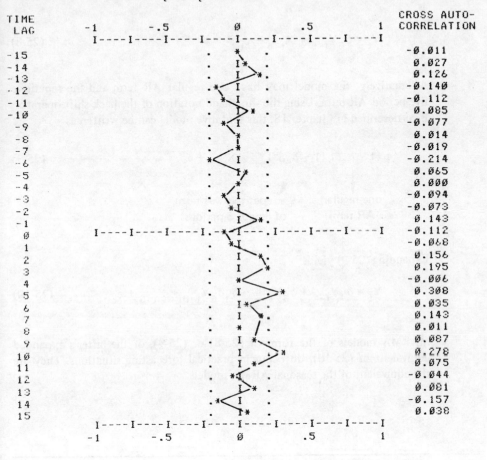

significant cross autos, at k = 5 and k = 10. This suggests a repetitive pattern (like seasonality), whose length is equal to 5. This pattern is similar to that of an AR and is not delayed. It can be inferred, therefore that s = 0 and that r has a reptitive term of order. 5. The tentative model might therefore be stated as

$$Y_t = \delta_5 Y_{t-5} + X_{t-5} + e_t \qquad (25\text{-}8)$$

Alternatively, the model may have one regular AR term and the repetitive five-period AR term. Using the shorthand notation of the back shift operator, B, as described in Chapter 18, this tentative model can be written as

$$(1 - \delta_1\beta)\ (1 - \delta_5\beta^5)Y_t = X_{t-5} + e_t \qquad (25\text{-}9)$$

one regular   a repetitive AR term
AR term   of length 5 periods

Expanding (25-9) yields

$$Y_t = \delta_1 Y_{t-1} + \delta_5 Y_{t-5} - \delta_1\delta_6 Y_{t-6} + X_{t-5} + e_t \qquad (25\text{-}10)$$

MARMA models of the form of (25-8) or (25-9), or the latter's expanded equivalent of (25-10), do occur in practical forecasting situations. They are the equivalent of the seasonal ARMA models.

EXERCISES

25-1 Identify the values of r, s and b using the cross autos shown in Figure 25-10.

25-2 Identify the order of r, s and b by examining the cross autos between X_t and Y_t shown in Figure 25-11. What is the relationship between Figure 25-11 and the autocorrelations of Y_t shown in Figure 25-12?

FIGURE 25-10 Cross Autos Between X_t and Y_t (exercise 25.1)

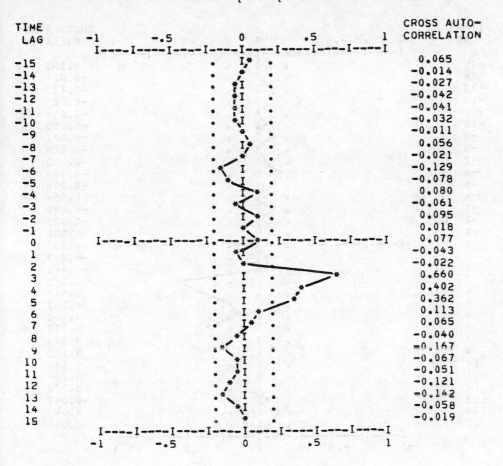

TIME LAG	CROSS AUTO-CORRELATION
-15	0.065
-14	-0.014
-13	-0.027
-12	-0.042
-11	-0.041
-10	-0.032
-9	-0.011
-8	0.056
-7	-0.021
-6	-0.129
-5	-0.078
-4	0.080
-3	-0.061
-2	0.095
-1	0.018
0	0.077
1	-0.043
2	-0.022
3	0.660
4	0.402
5	0.362
6	0.113
7	0.065
8	-0.040
9	-0.167
10	-0.067
11	-0.051
12	-0.121
13	-0.142
14	-0.058
15	-0.019

FIGURE 25-11 Cross Autos Between X_t and Y_t (exercise 25.2)

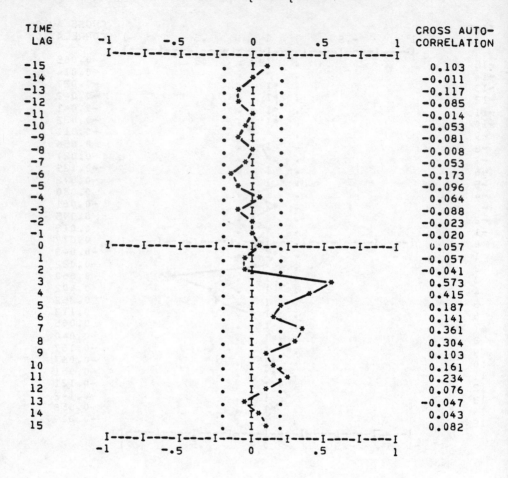

```
TIME                                                          CROSS AUTO-
 LAG    -1        -.5         0          .5          1        CORRELATION
        I----I----I----I----I----I----I----I----I
-15                        .   I *  .                            0.103
-14                        . *  I   .                           -0.011
-13                        . * I    .                           -0.117
-12                        . * I    .                           -0.085
-11                        .  *I    .                           -0.014
-10                        . * I    .                           -0.053
 -9                        . * I    .                           -0.081
 -8                        .   I*   .                           -0.008
 -7                        .  *I    .                           -0.053
 -6                        .* I     .                           -0.173
 -5                        . *I     .                           -0.096
 -4                        .   I *  .                            0.064
 -3                        . * I    .                           -0.088
 -2                        . * I    .                           -0.023
 -1                        .   *I   .                           -0.020
  0    I----I----I----I----I---*I---.I----I----I----I           0.057
  1                        . *I     .                           -0.057
  2                        . *I     .                           -0.041
  3                        .   I         *                       0.573
  4                        .   I      *  .                       0.415
  5                        .   I  *     .                        0.187
  6                        .   I *      .                        0.141
  7                        .   I    *   .                        0.361
  8                        .   I    *   .                        0.304
  9                        .   I*   .                            0.103
 10                        .   I *  .                            0.161
 11                        .   I  * .                            0.234
 12                        .   I*   .                            0.076
 13                        . *I     .                           -0.047
 14                        .   I*   .                            0.043
 15                        .   I *  .                            0.082
        I----I----I----I----I----I----I----I----I
       -1        -.5         0          .5          1
```

FIGURE 25-12   Autocorrelations of Y_t (exercise 25.2)

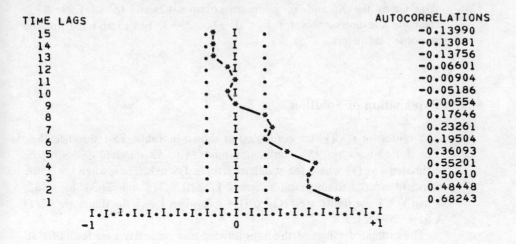

```
TIME LAGS                                               AUTOCORRELATIONS
   15              . *   I     .                            -0.13990
   14              .*•   I     .                            -0.13081
   13              .*•   I     .                            -0.13756
   12              .    •I     .                            -0.06601
   11              .     •I    .                            -0.00904
   10              .    •I     .                            -0.05186
    9              .     I•    .                             0.00554
    8              .     I    •.                             0.17646
    7              .     I     •.                            0.23261
    6              .     I•    .                             0.19504
    5              .     I     . •                           0.36093
    4              .     I     .  •                          0.55201
    3              .     I     .  •                          0.50610
    2              .     I     . •                           0.48448
    1              .     I     .     •                       0.68243
        I.I.I.I.I.I.I.I.I.I.I.I.I.I.I.I.I.I.I.I.I.I.I.I
       -1                    0                    +1
```

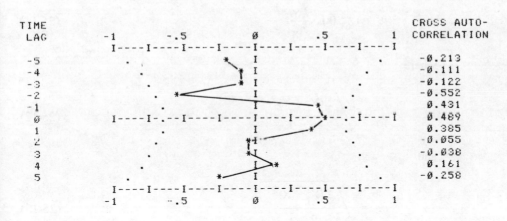

```
TIME                                                      CROSS AUTO-
LAG      -1       -.5       Ø        .5        1           CORRELATION
         I----I----I----I----I----I----I----I----I
  -5      .    ,     *      I               .             -0.213
  -4      .         *•      I               .             -0.111
  -3      .         *•      I               .             -0.122
  -2      .    *           I               .             -0.552
  -1      .               I          *    .               0.431
   Ø      I----I-.--I----I----I----I----*--.-I----I        0.489
   1      .               I        *      .               0.385
   2      .         **•I          .                      -0.055
   3      .         *•I           .                      -0.038
   4      .               I    *         .                0.161
   5      .         *      I               .             -0.258
         I----I----I----I----I----I----I----I----I
         -1       -.5       Ø        .5        1
```

```
ESTIMATED IMPULSE RESPONSE WEIGHTS (PARAMETERS)
 K    --------------------------------Ø--------------------------------VALUE
 Ø                                    I**********************            .833
 1                                    I*****************                 .656
 2                        ***I                                          -.094
 3                         **I                                          -.066
 4                                    I*******                           .274
 5              *************I                                          -.439
```

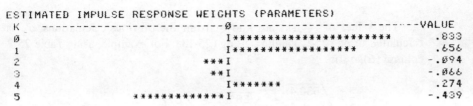

Solved Exercise

25.3 Given the X_t and Y_t values in columns (2) and (5) of Table 25.1, calculate the cross autos for k = 0, +1, +2, +3, +4, +5 and the impulse response parameters.

Explanation of Solution

The values for $r_{XY}(k)$ are computed as shown in Tables 25-1 through 25-5. The dotted lines are 95% confidence limits, i.e., ±2 standard errors of the estimated $r_{XY}(k)$ where the standard error is $1/\sqrt{n-k}$. Thus, when k = 0, the standard error, assuming randomness, is $1/\sqrt{10} = .316$ and $\pm2(.316) = \pm.632$. When k = 1 the limits are $\pm2(1/\sqrt{9}) = \pm.6$, when k = 5 the limits are $\pm2(1/\sqrt{5}) = \pm.894$, etc.

The estimated values of the impulse response parameters are found for all positive time lags (i.e., k = 0, 1, 2, . . ., 5) and are equal to:

$$\nu_k = r_{XY}(k) \cdot \frac{\sigma_Y}{\sigma_X} \tag{25-11}$$

where

$$\sigma_Y = \sqrt{\frac{\Sigma(Y_i - \bar{Y})^2}{n-1}} \tag{25-12}$$

and

$$\sigma_X = \sqrt{\frac{\Sigma(X_i - \bar{X})^2}{n-1}} \tag{25-13}$$

The values of $r_{XY}(k)$ are known from the previous calculations. σ_Y is calculated using (25-12) and σ_X using (25-13). For example, using Table 25.2 column (6), σ_Y is

$$\sigma_Y = \sqrt{\frac{530.4}{9}} = \sqrt{58.93} = 7.677$$

and using table 25.2 column (5), σ_X is

$$\sigma_X = \sqrt{\frac{182.9}{9}} = \sqrt{20.322} = 4.508$$

Thus,

$$\nu_0 = r_{XY}(0) = .489 \frac{7.677}{4.508} = .833$$

$$\nu_1 = r_{XY}(1) = .385 \frac{7.677}{4.508} = .656$$

$$\nu_2 = r_{XY}(2) = -.055 \frac{7.677}{4.508} = -.094$$

and

$$\nu_3 = r_{XY}(5) = -.258 \frac{7.677}{4.508} = -.439$$

The impulse response parameters can be used to determine initial estimates for the parameters of the MARMA model as will be discussed in the next chapter.

CHAPTER 26

Multivariate Box-Jenkins Methodology-Transfer Functions

Chapter 24 provided an overall description of Multivariate Autoregressive/ Moving Average (MARMA) models and discussed the procedures involved in using them. Chapter 25 dealt with time series analysis as it applies to MARMA models and illustrated how cross autos can be used to identify appropriate MARMA and noise models (see chapter 24). In this chapter a specific example will be used to illustrate the steps of identification, estimation, and forecasting for MARMA models using the Box-Jenkins approach. (The reader who has not read chapters 24 and 25 is encouraged to do so before continuing.) The programs needed to complete the analysis and forecasts of MARMA models are:

I) ID – Identification of a Univariate ARMA Model and initial estimation of its parameters. This program is used to identify the process generating the independent variable and to obtain initial estimates for the ARMA model identified.

II) UNIVBJ – Univariate ARMA Model Estimation. This program is used to compute final parameter estimates, to test the residuals for randomness and, once random, to place those residuals in a file so they can be used later as the prewhitened series, X_{et} and Y_{et}.

III) CROSS – Computation of Cross Autocorrelations and Impulse Response Parameters. This program (see chapter 25) can be used to identify an appropriate MARMA model.

IV) INIREG – Initial MARMA Estimates using Ordinary Linear Regression. This program uses the prewhitened series, X_{et} and Y_{et}, as its inputs and calculates initial estimates for the MARMA parameters. The residuals of the MARMA model (i.e., the noise model) are subsequently used to identify an appropriate noise model and to make initial estimates of its parameters.

V) NLEST — Non-Linear Estimation. This program makes final estimates for the ARMA, MARMA and noise model parameters using Marquardt's non-linear least-squares algorithm.

VI) FORCAS — Forecasting. This program prepares forecasts based on either univariate or Multivariate ARMA models.

VII) MULTBJ — Multivariate ARMA, Box-Jenkins Procedure. This is the main control program. It can be called when the user does not wish to use each of the programs individually. The user is encouraged to try MULTBJ at least once before running the programs individually. Once some experience is gained, it is much faster to use the programs individually.

The program UNIVBJ can be run by itself and used in place of BOXJEN when the user desires a wider choice of models than that given by BOXJEN. UNIVBJ is also faster than BOXJEN because it uses a somewhat different optimization procedure. (The same is true with ID which can be used to identify a univariate ARMA model.)

References for further study:

> Box, G. and G. Jenkins (1976)
> Makridakis, S. and S. Wheelwright (1978, Chapter 11)

INPUT/OUTPUT FOR PROGRAMS

I) ID

Input

1. What is the number of short term differences?
   Short (regular) or period to period differencing, i.e., $X_t - X_{t-1}$, is taken to make the data stationary.
2. What is the number of long term differences?
   Long (seasonal) or yearly differencing, i.e., $X_t - X_{t-L}$ where L is the length of seasonality, is taken to achieve long term stationarity. Some combination of short and long differences normally is required.
3. Order of p and q in ARMA (p,q)? (also the seasonal parameters if a seasonal pattern exists)
   This is found by examining item 2 in *Output*.

Output

1. Graph of the autocorrelations of the original data and a graph of the autocorrealtions of the first differenced data. This is given to allow the user to determine the length of seasonality.
2. Graph of the autos and partial autos at the specified level of differencing. This graph allows the user to identify an appropriate ARMA model when running UNIVBJ or an appropriate noise model when running MULTBJ.
3. Initial estimates of the ARMA (or noise model) parameters.

II) UNIVBJ

Inputs (In addition to those of ID)

1. Order of the AR Model (enter p)?
   This is the order of the AR process desired (see example below).
2. Order of the MA Model (enter q)?
   See example below.
If the data are seasonal, there are two additional inputs.
3. Is there a seasonal parameter in the AR?
   The answer required is Yes or No. (See example below.)
4. Is there a seasonal parameter in the MA?
   The answer required is Yes or No.
   In the ARMA (2,1) model,

$$X_t = \phi_1 X_{t-1} + \phi_2 X_{t-2} + e_t - \theta_1 e_{t-1}$$

the inputs are:

1. 2
2. 1

Questions 3 and 4 would not be asked because this model is not seasonal.
   In the ARMA(1,1) seasonal model (seasonality in the moving average portion),

$$(1 - \phi_1 B)X_t = (1 - \phi_1 B) (1 - \phi_{12} B^{12})e_t$$

the inputs are:

1. 1
2. 1
3. No
4. Yes

Finally, for the AR(2) seasonal model (seasonality in the AR and MA portions),

$$(1 - \phi_1 B - \phi_2 B^2)(1 - \phi_{12} B^{12})X_t = (1 - \phi_{12} B^{12})e_t$$

the inputs are:

1. 2
2. 0
3. Yes
4. Yes
5. Can you enter initial estimates for the parameters?
   These estimates can be found through ID and their entry will reduce the computational time required at this stage.
6. Number of iterations (0 if none)?
   This specifies the maximum number of iterations to be used to optimize the parameters of the model. If the answer is 0, the initial estimates will be used as final estimates. This can be useful, when one wants to prewhiten X_t or Y_t and the final parameters for doing so are known.
7. Do you want to put the residuals into a file?
   If the MULTBJ is to be used, the answer should be yes since the residuals will be used as the prewhitened series.

Output

1. Final parameter estimates.
2. Residual autocorrelations.
3. Residual errors.

III) CROSS

(See Chapter 25 for a complete description.)

IV) INIREG

Input

1. What is the name of the file containing the prewhitened independent (input) variable?
   This is the file containing X_{et}, frequently called the independent, exogenous, or input variable. (See 24-3.)

2. What is the name of the file containing the prewhitened dependent (output) variable?

    The output variable, Y_{et}, often called the endogenous variable, should be in prewhitened form, as should the input variable.

3. Enter values for r, s and b?

    This requires the user to specify the order of the MARMA model,

$$Y_t = \delta_1 Y_{t-1} + \delta_2 Y_{t-2} + \ldots + \delta_r Y_{t-r} + \omega_0 X_{t-b}$$

$$- \omega_1 X_{t-b-1} - \ldots - \omega_s X_{t-b-s} + e_t$$

where r is the number of past Y_t values

b is the delay lag (the number of periods until the effect of a change in X_t affects Y_t), and

s is the number of past X_t terms, minus one.

4. Do you want to identify the order of the noise model?

    This will allow the user to see possible patterns in e_t.

Output

1. Initial estimates of $\delta_1, \delta_2, \ldots, \delta_r, \omega_0, \omega_1, \ldots \omega_s$.
2. Autos and partials for the residual errors, e_t. These are used to identify an appropriate noise model.

V) NLEST

Input

1. Do you want to estimate for a univariate ARMA (Box-Jenkins) model?

    If the answer is yes, the program estimates $\phi_1, \phi_2, \ldots, \theta_1, \theta_2, \ldots$ for a univariate ARMA model. If the answer is No, the program estimates $\delta_1, \delta_2, \ldots \delta_r, \omega_0, \omega_1, \ldots, \omega_s$, (the MARMA parameters), and $\phi_1, \phi_2, \ldots, \theta_1, \theta_2, \ldots$ (the noise parameter).

2. Enter initial parameter estimates in the sequence specified for the model being investigated.

    If the model is ARMA(1,1) with the seasonality in AR, three parameters must be input – $\phi_1, \theta_1, \phi_{12}$. If the program being used is MULTBJ and r = 2, s = 1, b = 3, and p = 1, q = 1 and the seasonal is in the AR portion, seven parameters must be input:

$$\delta_1, \delta_2, \omega_0, \omega_1, \phi_1, \theta_1, \phi_{12}$$

Output

1. The iteration number, mean squared error (MSE), and parameter values are printed for each iteration.
2. Final MSE and final parameter estimates.
3. The correlation matrix of the parameters. The smaller the values in the correlation matrix, the better, since it shows independence among the variables. If they are not close to zero, it is better they have different signs than if all of the values of the correlation matrix have the same sign.
4. Lower and upper 95% confidence limits for each of the final parameter estimates.

VI) FORCAS

(Inputs are the same as in other programs, the outputs are m forecasts.)

VII) MULTBJ

(Inputs and outputs are the same as for other programs.)

Solved Exercise

Figure 26.1 shows 40 values of the independent variable, X_t, while Figure 26.2 contains 40 values of the dependent variable, Y_t. The values of X_t were generated by the process,

$$(1 - B)(1 - .9B)X_t = (1 - .4B)e_t \qquad (26\text{-}1)$$

where e_t is unit normally distributed.

Equation (26-1) is an ARMA (1,1) process with one level of non-stationarity. The values of Y_t, on the other hand, were generated using

$$Y_t = .5Y_{t-1} + 1.7X_{t-2} \qquad (26\text{-}2)$$

Equation (26-2) is a MARMA model with $r = 1$, $s = 0$ and $b = 2$.

FIGURE 26-1   Values of the Independent Variable, X_t, Generated by (26-1)

PERIOD	OBSERVATION	PERIOD	OBSERVATION	PERIOD	OBSERVATION
1	5.870	14	24.260	27	27.050
2	7.730	15	26.660	28	26.250
3	10.040	16	27.720	29	26.760
4	12.060	17	29.130	30	26.570
5	13.490	18	30.630	31	24.270
6	14.010	19	32.970	32	23.860
7	13.380	20	32.950	33	22.200
8	15.300	21	33.120	34	20.630
9	16.480	22	32.660	35	20.680
10	19.310	23	30.390	36	20.060
11	21.250	24	28.900	37	18.060
12	23.000	25	27.550	38	17.220
13	23.690	26	27.330	39	18.210
				40	18.150

FIGURE 26-2   Values of the Independent Variable, X_t, Generated by (26-2)

PERIOD	OBSERVATION	PERIOD	OBSERVATION	PERIOD	OBSERVATION
1	9.062	14	71.893	27	98.177
2	12.725	15	76.220	28	95.549
3	16.341	16	79.352	29	93.760
4	21.312	17	84.998	30	91.505
5	27.724	18	89.623	31	91.244
6	34.364	19	94.332	32	90.791
7	40.115	20	99.237	33	86.655
8	43.874	21	105.668	34	83.889
9	44.683	22	108.849	35	79.685
10	48.352	23	110.728	36	74.913
11	52.192	24	110.886	37	72.613
12	58.923	25	107.106	38	70.408
13	65.586	26	102.683	39	65.906
				40	62.227

Use the data of Figures 26-1 and 26-2 to estimate the relevant parameters, assuming the generating process is unknown. What can you say about the difference in estimated and actual values? Prepare forecasts for the next 5 periods.

Solution

Figure 24-1 will be followed to achieve the final objective of forecasting with a MARMA model. Once this is done, the same forecasts will be obtained using MULTBJ, the control program. The basic logic of this control program follows that of Figure 24-1. The numbered phases in the sequence below correspond to those in Figure 24-1. (See chapter 24.)

The objective of phase I for a MARMA model is to reduce X_t and Y_t to white noise. This requires that the following be known:

1. The level at which X_t is stationary.
2. The univariate ARMA process generating X_t and initial estimates for the parameters of the tentative model identified to approximate that process.
3. Final estimates for the parameters.
4. A residual autocorrelation test to make sure that the residuals are indeed white noise (i.e., they are random).

1. *Differencing*

A set of data is stationary if its autocorrelations drop rapidly to zero. Figure 26-3 indicates that this is not the case for X_t. Thus the first differences must be taken.

2. *Tentative ARMA Model*

Figure 26-4 shows the autos and partials of the first differenced data. Both the autos and partials tail off to zero exponentially, though this is not completely obvious with the partials. This suggests a mixed ARMA process. There is one significant partial, indicating AR(1), there are 3 significant autos indicating a MA of up to 3. However, this may also be caused by nonstationarity. (The fact that $\phi_1 = .9$ actually causes this.) As a starting point, one can settle for the smallest possible model, MA(1), and assume that the total process is ARMA(1,1). The diagnostic step can then be used to validate the appropriateness of this model.

The model identified is:

$$(X_t - X_{t-1}) = \phi_1 (X_{t-1} - X_{t-2}) + e_t - \theta_1 e_{t-1} \qquad (26\text{-}3)$$

or if parentheses are removed, this becomes

$$X_t = (1 + \phi_1)X_{t-1} + e_t - \theta_1 e_{t-1} \tag{26-4}$$

Equation (26-4) is an algebraic form more convenient to work with than is (26-3) from a programming point of view, even though it is not as easily recognized as an ARMA(1,1) model with one level of short term differencing.

Once (26-3) has been tentatively identified as an appropriate ARMA model, initial estimates for ϕ_1 and θ_1 must be obtained. These initial estimates can be based on the autocorrelation coefficients as follows:

$$\phi_1 = \frac{r_2}{r_1} = \frac{.497}{.626} = .794 \tag{26-5}$$

and,

$$r_1 = \frac{(1 - \phi_1\theta_1)(\phi_1 - \theta_1)}{1 + \phi_1^2 - 2\phi_1\theta_1} \tag{26-6}$$

If $r_1 = .626$ and $\phi_1 = .794$ are substituted into (26-5) and that is then solved for θ_1, one obtains:

$$-.1680\theta_1^2 + .6365\theta_1 - .1683 = 0 \tag{25-7}$$

Solving (26-7) yields two values, $\theta_1 = 3.5$ and $\theta_1 = .286$. The value of 3.5 is outside the allowable limits. Thus, the value of $\theta_1 = .286$ is used as an initial estimate.

Solving (26-6) is not trivial and cannot always be done routinely using analytical methods. However, there are adequate iterative procedures for determining the value of θ_1 without having to solve (26-6) analytically.

Expressions (26-5) and (26-6) can be generalized so that initial estimates of any ARMA model can be obtained. These estimates are based on solving the Yule-Walker equations for AR models and systems of nonlinear equations for MA models.

3. *Estimation of Final Parameters*

Final parameters are found by using Marquardt's (1963) algorithm for least-squares estimation of nonlinear parameters. The algorithm and the computations involved are based on the first differenced data and are as follows:

(a) Defining $Z_{t-1} = X_t - X_{t-1}$ and substituting the intial estimates of ϕ_1 = .794 and θ_1 = .286 into (26-3) gives

$$\hat{Z}_t = \phi_1 Z_{t-1} - \theta_1 e_{t-1} \tag{26-8}$$

FIGURE 26-3 Autocorrelations of Original, X_t, Data (see Figure 26.1)

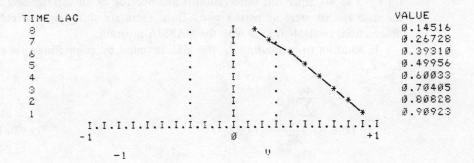

```
                    AUTOCORRELATIONS

TIME LAG                                            VALUE
    8                         I    *                0.14516
    7                         I    .*               0.26728
    6                         I    .  *             0.39310
    5                         I    .    *           0.49956
    4                    .    I    .      *         0.60033
    3                    .    I    .        *       0.70405
    2                    .    I    .          *     0.80828
    1                    .    I    .            *   0.90923
         I.I.I.I.I.I.I.I.I.I.I.I.I.I.I.I.I.I.I.I.I
        -1                    0                   +1
           -1                      0
```

FIGURE 26-4 Autos and Partials of First Differenced X_t Data

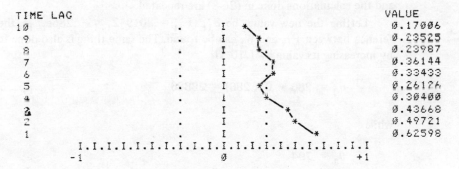

```
TIME LAG                                            VALUE
   10                    .    I    *                0.17006
    9                    .    I    .*               0.23525
    8                    .    I    .*               0.23987
    7                    .    I    .   *            0.36144
    6                    .    I    .  *             0.33433
    5                    .    I    .*               0.26176
    4                    .    I    . *              0.30400
    3                    .    I    .    *           0.43668
    2                    .    I    .     *          0.49721
    1                    .    I    .        *       0.62598
         I.I.I.I.I.I.I.I.I.I.I.I.I.I.I.I.I.I.I.I.I
        -1                    0                   +1
```

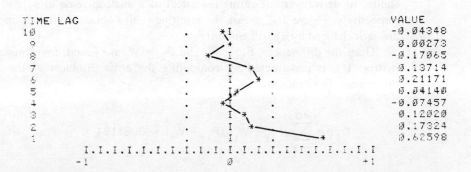

```
                 PARTIAL AUTOCORRELATIONS

TIME LAG                                            VALUE
   10                    .    *I   .               -0.04348
    9                    .    *    .                0.00273
    8                    .  *      I    .          -0.17065
    7                    .    I  *      .           0.21171
    6                    .    I  *      .           0.13714
    5                    .    I *       .           0.04140
    4                    .   *I        .           -0.07457
    3                    .    I  *      .           0.12020
    2                    .    I  *      .           0.17324
    1                    .    I    .        *       0.62598
         I.I.I.I.I.I.I.I.I.I.I.I.I.I.I.I.I.I.I.I.I
        -1                    0                   +1
```

The estimated values for (26-8) are then

$$\hat{Z}_t = .794Z_{t-1} - .286e_t - 1 \qquad\qquad (26\text{-}9)$$

for t = 3 to 40, since one observation is lost because of the differencing and the calculations start at period two. Thus, there are thirty-eight effective observations available for use with the MARMA program.

In addition to computing $\hat{Z}_t$, the MSE is found by computing e_t as $e_t = Z_t - \hat{Z}_t$.

$$\text{MSE} = \frac{\displaystyle\sum_{t=3}^{40} e_t^2}{38} = 1.082 \qquad\qquad (26\text{-}10)$$

(b) The value of $\phi_1 = .794$ is increased by 1/10 and the calculations done in (a) are repeated. Thus

$$\phi_1 = .794 + .01(.794) = .80194$$

and the calculations done in (26-8) are repeated.

Letting the new values be F_{1t} ($F_{1t} = .80194Z_{t-1} - .286e_{t-1}$), the difference between F_{1t} and $\hat{W}_t$ can be found. The same thing is also done for θ_1 by increasing its value by 1/10. Or

$$\theta_1 = .286 + .01(.286) = .28886$$

while

$$\theta_1 = .794$$

A new value, F_{2t}($F_{2t} = .794Z_{2t-1} - .28886e_{t-1}$), can be found now and the difference between F_{2t} and W_t can be computed. These differences are similar to derivatives, indicating the effect of a small increase in ϕ_1 and θ_1, respectively. Figure 26-5 shows the resulting values obtained after ϕ_1 and θ_1 are each changed by a small amount.

Once the differences, $F_{1t} - \hat{W}_t$ and $F_{2t} - \hat{W}_t$ are found, the symmetric matrix, D_{ij}, is constructed by computing the cross products of the two differences. For example:

$$D_{11} = \sum_{t=3}^{40} (F_{1t} - \hat{W}_t)(F_{1t} - \hat{W}_t) = .0081194$$

FIGURE 26-5   A - Resulting Differences After a Small Change in ϕ_1

Period	$\phi_1 + \delta$	θ_1	F_{1t}	$\hat{W}_t$	$F_{1t} - \hat{W}_t$
3	0.80194	0.28600	1.492	1.477	0.015
4	0.80194	0.28600	1.618	1.596	0.023
5	0.80194	0.28600	1.505	1.483	0.022
6	0.80194	0.28600	1.168	1.150	0.018
7	0.80194	0.28600	0.602	0.593	0.009
8	0.80194	0.28600	-0.153	-0.150	-0.002
9	0.80194	0.28600	0.947	0.932	0.015
10	0.80194	0.28600	0.880	0.866	0.014
11	0.80194	0.28600	1.712	1.685	0.026
12	0.80194	0.28600	1.490	1.468	0.023
13	0.80194	0.28600	1.329	1.309	0.020
14	0.80194	0.28600	0.736	0.725	0.011
15	0.80194	0.28600	0.505	0.497	0.008
16	0.80194	0.28600	1.383	1.361	0.021
17	0.80194	0.28600	0.942	0.928	0.015
18	0.80194	0.28600	0.997	0.982	0.015
19	0.80194	0.28600	1.059	1.043	0.016
20	0.80194	0.28600	1.510	1.487	0.023
21	0.80194	0.28600	0.422	0.415	0.006
22	0.80194	0.28600	0.208	0.205	0.003
23	0.80194	0.28600	-0.178	-0.175	-0.003
24	0.80194	0.28600	-1.222	-1.203	-0.019
25	0.80194	0.28600	-1.118	-1.101	-0.017
26	0.80194	0.28600	-1.016	-1.001	-0.016
27	0.80194	0.28600	-0.404	-0.398	-0.006
28	0.80194	0.28600	-0.260	-0.256	-0.004
29	0.80194	0.28600	-0.487	-0.480	-0.007
30	0.80194	0.28600	0.124	0.122	0.002
31	0.80194	0.28600	-0.063	-0.062	-0.001
32	0.80194	0.28600	-1.205	-1.186	-0.019
33	0.80194	0.28600	-0.556	-0.547	-0.009
34	0.80194	0.28600	-1.015	-1.000	-0.016
35	0.80194	0.28600	-1.100	-1.084	-0.017
36	0.80194	0.28600	-0.289	-0.284	-0.004
37	0.80194	0.28600	-0.403	-0.396	-0.006
38	0.80194	0.28600	-1.147	-1.129	-0.018
39	0.80194	0.28600	-0.761	-0.750	-0.012
40	0.80194	0.28600	0.293	0.289	0.005

FIGURE 26-5   (continued)

B —   Resulting Differences After A Small Change in Both ϕ_1 (as above) and θ_1

Period	ϕ_1	$\theta_1 + \delta$	F_{2t}	$\hat{W}_t$	$F_{2t} - \hat{W}_t$
3	0.79400	0.28886	1.477	1.477	0.000
4	0.79400	0.28886	1.593	1.596	-0.002
5	0.79400	0.28886	1.481	1.483	-0.002
6	0.79400	0.28886	1.150	1.150	-0.000
7	0.79400	0.28886	0.595	0.593	0.002
8	0.79400	0.28886	-0.146	-0.150	0.004
9	0.79400	0.28886	0.928	0.932	-0.005
10	0.79400	0.28886	0.864	0.866	-0.002
11	0.79400	0.28886	1.679	1.685	-0.006
.	.	.	.	.	.
.	.	.	.	.	.
.	.	.	.	.	.
29	0.79400	0.28886	-0.478	-0.480	0.001
30	0.79400	0.28886	0.119	0.122	-0.002
31	0.79400	0.28886	-0.062	-0.062	0.000
32	0.79400	0.28886	-1.180	-1.186	0.006
33	0.79400	0.28886	-0.548	-0.547	-0.000
34	0.79400	0.28886	-0.997	-1.000	0.003
35	0.79400	0.28886	-1.081	-1.084	0.003
36	0.79400	0.28886	-0.287	-0.284	-0.003
37	0.79400	0.28886	-0.396	-0.284	0.000
38	0.79400	0.28886	-1.125	-1.129	0.005
39	0.79400	0.28886	-0.749	-0.750	0.001
40	0.79400	0.28886	0.284	0.289	-0.005

$$D_{12} = D_{21} = \sum_{t=3}^{40} (F_{1t} - \hat{W}_t)(F_{2t} - \hat{W}_t') = -.00118724$$

$$D_{22} = \sum_{t=3}^{40} (F_{2t} - \hat{W}_t)(F_{2t} - \hat{W}_t) = .000349934$$

The 2×2 D_{ij} matrix is, therefore

$$D_{ij} = \begin{bmatrix} .008119 & -.001187 \\ -.001187 & .000350 \end{bmatrix}$$

Subsequently, D_{ij} is scaled and standardized so that it can form the basis for applying the steepest descent algorithm. The scaling is done by dividing D_{ij} by the product of P_i times P_j where

$$P_1 = \delta\phi_1 = .01(.794) = .00794$$

$$P_2 = \delta\theta_1 = .01(.286) = .00286$$

Thus, D_{12} becomes

$$D_{12} = \frac{-.00187}{(.00794)(.00286)} = -52.1496$$

and D_{ij} becomes

$$D_{ij} = \begin{bmatrix} 128.672 & -52.2819 \\ -52.2819 & 42.7813 \end{bmatrix}$$

D_{ij} is then constrained by being divided by o_i and o_j where

$$o_1 = \sqrt{D_{11}} = \sqrt{128.672} = 11.3434$$

$$o_2 = \sqrt{D_{22}} = \sqrt{42.7813} = 6.5407$$

The resulting matrix, J_{ij}, is

$$J_{ij} = \begin{bmatrix} 128.672/(11.3433)(11.3433) & -.52.28/(11.343)(6.5407) \\ -52.28/(6.5407)(11.3433) & 42.7813/(6.5407)(6.5407) \end{bmatrix}$$

$$J_{ij} = \begin{bmatrix} 1 & -.704665 \\ -.704665 & 1 \end{bmatrix}$$

The importance of starting with the original D_{ij} matrix and ending up with J_{ij} is that the former could take any values (as it can be seen from the original D_{ij} whose values are extremely small), while the latter is scaled between -1 to $+1$ and is within the desired bounds of the unit circle constraint.

J_{ij} is a scaled and constrained matrix of the differences in errors between the parameters ϕ_1 and θ_1, ϕ_1 and δ and θ_1, and $\phi_1 + \delta$. The signs and relative magnitudes of these errors are important in providing information regarding in what direction the parameters should be changed. In order to proceed, some additional quantities need to be calculated.

$$q_1 = \sum_{t=3}^{40} [(Z_t - \hat{W}_t)(F_{1t} - \hat{W}_t)]/\delta\phi_1 = -.0040193/.01(.794)$$

$$= -.5062$$

$$q_2 = \sum_{t=3}^{40} [(Z_t - \hat{W}_t)(F_{2t} - \hat{W}_t)]/\delta\theta_1 = .009265/.01(.286)$$

$$= 3.2394$$

Next, q_1 and q_2 are scaled as was the D_{ij} matrix (divided by $\sqrt{D_{ij}}$, o_i)

$$h_1 = q_1/o_1 = -.5062/11.343 = -.04463$$

$$h_2 = q_2/o_2 = 3.2394/6.5407 = .49527$$

Note that

$$\sum_{t=3}^{40} (Z_t - \hat{W}_t)(F_{1t} - \hat{W}_t)$$

is the cross product of errors when $\phi_1 = .794$, and when $\phi_1 = \phi_1 + \delta = .80194$ (θ_1 is the same in both). Also

$$\sum_{t=3}^{40} (Z_t - \hat{W}_t)(F_{2t} - \hat{W}_t)$$

is the cross product of errors when $\theta_1 = .286$ and $\theta_1 = \theta_1 + \delta = .28886$ (ϕ_1 is the same in both).

Dividing this cross product by $\delta\phi_1$ gives another relative measure of how a change in the parameter affects the errors. The h_1 and h_2 are the same as q_1 and q_2 except they are scaled. This is the equivalent of the original matrix, D_{ij}, and the one that was divided by the product of o_i and o_j. In other words, D_{ij} and q_1 and q_2 are related and denote how a change in the parameters affects the errors.

The relationship, $J_{ij}p_i = h_i$, must hold in such a way that the scaled and constrained J_{ij} matrix (including all possible changes in the parameters) will be equal to h_i. (This includes changes in the errors from changing one parameter at a time.) Instead of using the J_{ij} matrix, however, a constant quantity, GO, could be added to its diagonal. This gives

$$J_{ij} = J_{ij} + GO = \begin{bmatrix} 1.01 & -.704665 \\ -.704665 & 1.01 \end{bmatrix}$$

when $GO = .01$.

The new relationship, $J_{ij}p_i = h_i$, can be subsequently solved for p_i. The result in matrix form is

$$p = J^{-1}h$$

For this example,

$$p = \begin{bmatrix} 1.92915 & 1.34595 \\ 1.34595 & 1.92915 \end{bmatrix} \begin{bmatrix} -.04463 \\ .49529 \end{bmatrix} = \begin{bmatrix} .5805 \\ .8954 \end{bmatrix}$$

Therefore, $p_1 = .5805$ and $p_2 = .8954$ are the correction factors to be applied to ϕ_1 and θ_1. Before being used, these correction factors have to be scaled by dividing them by o_1 and o_2.

$$P_1 = p_1/o_1 = .5805/11.343 = .05118$$

$$P_2 = p_2/o_2 = .8954/6.5407 = .13689$$

From these, new values for ϕ_1 and θ_1 are found.

$$\phi_1^* = \phi_1 + P_1 = .794 + .0512 = .845$$

$$\theta_1^* = \theta_1 + P_2 = .286 + .1369 = .423$$

Once ϕ_1^* and θ_1^* are computed, a new MSE_1 can be found.

$$MSE_1 = \frac{\sum_{t=3}^{40} e_{1t}^2}{38} = 1.0684 \tag{26-10}$$

Since $MSE_1 < MSE$ (see 26-10), the corrections from ρ_1 and ρ_2 are tested to determine whether they are too small to be worth continuing. If they are not too small, the program returns to (a) but with $\phi_1 = .845$ and $\theta_2 = .423$, and all computations are repeated. In this example, the corrections are very small by the second iteration and the program stops with final parameter estimates of $\phi_1 = .845$ and $\theta_1 = .423$.

Both the univariate and multivariate programs use the same optimization routine. As shown previously, the results are independent of the number of parameters. What is most important is the order (sequence) of the parameters — it must always be the same — and the way the errors in each model are calculated — it must be done consistently. (This is why so much space has been devoted to explaining this nonlinear optimization algorithm.)

4. *Autocorrelations of Residuals*

Using the final parameter estimates of $\phi_1 = .838$ and $\theta_1 = .406$, the residuals of the model are

$$e_t = Z_t - .838Z_{t-1} + .406e_{t-1} \tag{26-11}$$

The autocorrelations of (26-11) are shown in Figure 26.6. Since no pattern remains, the e_t can be used as the prewhitened series, X_{et}.

FIGURE 26-6   Autocorrelation of Residuals of (26-11)

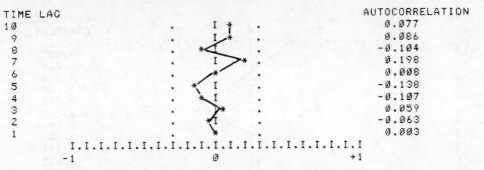

TIME LAG		AUTOCORRELATION
10		0.077
9		0.086
8		-0.104
7		0.198
6		0.008
5		-0.138
4		-0.107
3		0.059
2		-0.063
1		0.003

5. *Keeping the Residuals*

If MULTBJ is run, the residuals are kept in the file WHITEX and used later in CROSS and INIREG. If MULTBJ is not used, any file name can be specified for storing the residuals of (26-11) once they are white noise. If the residuals are not white noise, a new ARMA model must be specified by returning to 1 or 2.

6. *Prewhitening Y_t*

Using the same model and parameter estimates found for the X_t series, the residuals of Y_t are found as:

$$c_t^* = (Y_t - Y_{t-1}) - .838(Y_{t-1} - Y_{t-2}) + .406e_{t-1}^* \qquad (26\text{-}12)$$

7. *Keeping the Residuals of Y_t (26-12)*

This is done automatically in the file WHITE if MULTBJ is run; otherwise the user must specify a file name.

8. *Estimating Cross Autos*

This is done as in Chapter 25. The results are shown in Figure 26-7.

9. *Identifying r, s and b*

The cross autos for negative time lags are not significantly different than zero, neither are $r_{xy}(0) = -.055$ and $r_{xy}(1) = 0$. The first significant cross auto is at time lag 2. This suggests that $b = 2$. Starting immediately from $k = 2$, there is an almost perfect exponential decay pattern. This suggests that $s = 0$ and $r = 1$. Thus the MARMA function model, using $W_t = Y_t - Y_{t-1}$ and $Z_t = X_t - X_{t-1}$, is

$$W_t = \delta_1 W_{t-1} + \omega_0 Z_{t-2} + e_t \qquad (26\text{-}13)$$

FIGURE 26-7   Cross Autos of the Prewhitened X_t and Y_t Series

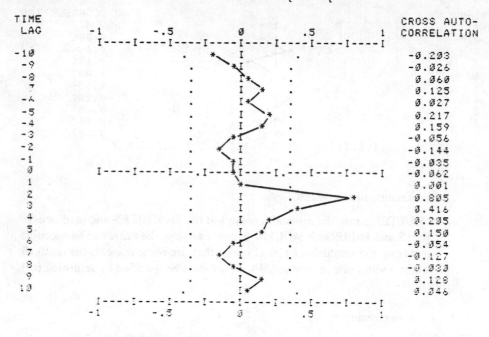

10. *Initial Estimates of the MARMA Model*

A multiple regression analysis could be run using W_t as the dependent variable and W_{t-1} and Z_{t-2} as the independent variables (see chapter 19). However, it is known that W_t and W_{t-1} are autocorrelated and that W_{t-1} and Z_{t-2} are dependent. This can cause two kinds of problems — multicollinearity, which can give erroneous results, non-zero covariances between the independent variables and other forms of bias (see Johnston, Chapter 10), which violate a major assumption of regression. These problems can be avoided to a great extent by using the prewhitened X_t and Y_t instead of their differenced values. Thus, (26-13) can take the form

$$Y_{e_t} = \delta_1 Y_{e_{t-1}} + \omega_0 X_{e_{t-1}} + e_t \tag{26-14}$$

The results from regression analysis for this model are

```
DELTA PARAMETERS (DEPENDENT VARIABLE)
  1            .518563

OMEGA PARAMETERS (INDEPENDENT VARIABLE)
  0           1.71774

R-SQUARE OF INITIAL ESTIMATES = .996216
```

The tentative MARMA model is, therefore:

$$W_t = .519 \, W_{t-1} + 1.718 \, Z_{t-2} + e_t \qquad (26\text{-}15)$$

The residuals of this model are

$$e_t = W_t - .519 \, W_{t-1} - 1.718 \, Z_{t-2} \qquad (26\text{-}16)$$

11. *Identify the Noise Model and Initial Estimates of its Parameters*

The autocorrelations of the residuals given by (26-16) can be found as shown in Figure 26-8.

The autos drop off exponentially and a single partial is significantly different from zero. This suggests an AR(1) model. The total tentative model (i.e. MARMA and noise) is therefore:

$$W_t = \delta_1 W_{t-1} + \omega_0 Z_{t-2} + \phi_1 e_{t-1} + e_t' \qquad (26\text{-}17)$$

where e_t' is white noise.

Initial estimates for ϕ_1 can be found as before (i.e., $\phi_1 = r_1 = .466$) giving $\phi_1 = .466$. Equation (26-17) therefore becomes

$$\hat{W}_t = .519 \, \hat{W}_{t-1} + 1.718 \, Z_{t-2} + .467 \, e_{t-1} \qquad (26\text{-}18)$$

where $\hat{W}_t$ is the estimate of W_t and

$$e_t = W_t - \hat{W}_t \qquad (26\text{-}19)$$

12. *Final Parameter Estimates*

Final parameter estimates can be found by again using the program NLEST. The W_t values are found by computing e_t in (26-15) for all periods and then estimating its pattern using (26-18) in a second estimation process, since e_t is required in (26-18). The difference between e_t and $.467e_{t-1}$ is the total error, e_t'. (See Figure 26-9 for calculations.)

The estimated value for period 21, for example, during the last iteration when $\delta_1 = .5$, $\omega_0 = 1.2$ and $\phi_1 = .5$ is:

$$\hat{W}_{21} = .5 \, \hat{W}_{20} + 1.7 \, Z_{19} = .5(4.9) + 1.7(2.34) = 6.43$$

For period 22:

$$\hat{W}_{22} = .5 \, \hat{W}_{21} + 1.7 \, Z_{20} = .5(6.43) + 1.7(-.02) = 3.18$$

FIGURE 26-8 Autocorrelations and Partial Autocorrelations of Residuals of (26-16)

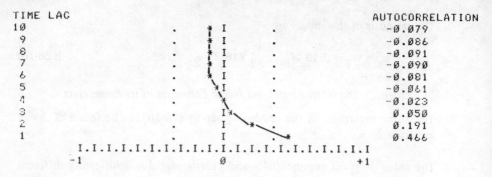

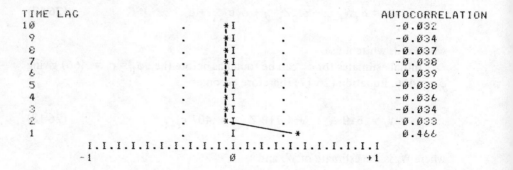

The corresponding errors are:

$$e_{21} = W_{21} - \hat{W}_{21} = 6.43 - 6.43 = 0$$

$$e_{22} = W_{22} - \hat{W}_{22} = 3.18 - 3.18 = 0$$

In the second iteration the error values, e_t, are estimated. Thus, for period 7, for example:

$$\hat{e}_7 = \phi_1 e_6 = .5(.9) = .45$$

and

$$e_7' = e_7 - \hat{e}_7 = .45 - .45 = 0$$

FIGURE 26-9   Relevant Values for Estimating (a) MARMA and (b) Noise Model Forecasts and Errors

(a) Period	δ_1	ω_0	$W_t = Y_t - Y_{t-1}$	$Z_t = X_t - X_{t-1}$	$\hat{W}_t$	e_t
4	0.50000	1.70000	4.970	2.02	3.162	1.808
5	0.50000	1.70000	6.412	1.43	5.508	0.904
6	0.50000	1.70000	6.640	0.52	6.188	0.452
7	0.50000	1.70000	7.751	-0.63	5.525	0.226
8	0.50000	1.70000	3.760	1.92	3.647	0.113
9	0.50000	1.70000	0.809	1.18	0.752	0.057
10	0.50000	1.70000	3.668	2.83	3.640	0.028
11	0.50000	1.70000	3.840	1.94	3.826	0.014
12	0.50000	1.70000	6.731	1.75	6.724	0.007
13	0.50000	1.70000	6.664	0.69	6.660	0.004
14	0.50000	1.70000	6.307	0.57	6.305	0.002
15	0.50000	1.70000	4.326	2.40	4.326	0.001
16	0.50000	1.70000	3.132	1.06	3.132	0.000
17	0.50000	1.70000	5.646	1.41	5.646	0.000
18	0.50000	1.70000	4.625	1.50	4.625	0.000
19	0.50000	1.70000	4.710	2.34	4.709	0.000
20	0.50000	1.70000	4.905	-0.02	4.905	0.000
21	0.50000	1.70000	6.430	0.17	6.430	0.000
22	0.50000	1.70000	3.181	-0.46	3.181	0.000
23	0.50000	1.70000	1.880	-2.27	1.880	0.000
24	0.50000	1.70000	0.158	-1.49	0.158	0.000
25	0.50000	1.70000	-3.780	-1.35	-3.780	0.000
26	0.50000	1.70000	-4.423	-0.22	-4.423	0.000
27	0.50000	1.70000	-4.507	-0.28	-4.507	0.000
28	0.50000	1.70000	-2.627	-0.80	-2.627	0.000
29	0.50000	1.70000	-1.790	0.51	-1.790	0.000
30	0.50000	1.70000	-2.255	-0.19	-2.255	0.000
31	0.50000	1.70000	-0.260	-2.30	-0.260	0.000
32	0.50000	1.70000	-0.453	-0.41	-0.453	0.000
33	0.50000	1.70000	-4.137	-1.66	-4.137	0.000
34	0.50000	1.70000	-2.765	-1.57	-2.765	0.000
35	0.50000	1.70000	-4.205	0.05	-4.205	0.000
36	0.50000	1.70000	-4.771	-0.62	-4.771	0.000
37	0.50000	1.70000	-2.301	-2.00	-2.301	0.000
38	0.50000	1.70000	-2.204	-0.84	-2.204	0.000
39	0.50000	1.70000	-4.502	0.99	-4.502	0.000
40	0.50000	1.70000	-3.679	-0.06	-3.679	0.000

FIGURE 26-9   (continued)

(b) Period	ϕ_1	$W_t = Y_t - Y_{t-1}$	e_t	Total Forecast	$\hat{W}_t + e_t$	e'_t
5	0.50000	6.412	0.904	6.412	7.316	0.000
6	0.50000	6.640	0.452	6.640	7.092	0.000
7	0.50000	5.751	0.226	5.751	5.977	0.000
8	0.50000	3.760	0.113	3.760	3.873	0.000
9	0.50000	0.809	0.057	0.809	0.865	0.000
10	0.50000	3.668	0.028	3.668	3.697	0.000
11	0.50000	3.840	0.014	3.840	3.854	0.000
12	0.50000	6.731	0.007	6.731	6.738	0.000
13	0.50000	6.664	0.004	6.664	6.667	0.000
14	0.50000	6.307	0.002	6.307	6.309	0.000
15	0.50000	4.326	0.001	4.326	4.327	0.000
16	0.50000	3.132	0.000	3.132	3.133	0.000
17	0.50000	5.646	0.000	5.646	5.646	0.000
18	0.50000	4.625	0.000	4.625	4.625	0.000
19	0.50000	4.710	0.000	4.710	4.710	0.000
20	0.50000	4.905	0.000	4.905	4.905	0.000
21	0.50000	6.430	0.000	6.430	6.430	0.000
22	0.50000	3.181	0.000	3.181	3.181	0.000
23	0.50000	1.880	0.000	1.880	1.880	0.000
24	0.50000	0.158	0.000	0.158	0.158	0.000
25	0.50000	-3.780	0.000	-3.780	-3.780	0.000
26	0.50000	-4.423	0.000	-4.423	-4.423	0.000
27	0.50000	-4.507	0.000	-4.507	-4.507	0.000
28	0.50000	-2.627	0.000	-2.627	-2.627	0.000
29	0.50000	-1.790	0.000	-1.790	-1.790	0.000
30	0.50000	-2.255	0.000	-2.255	-2.255	0.000
31	0.50000	-0.260	0.000	-0.260	-0.260	0.000
32	0.50000	-0.453	0.000	-0.453	-0.453	0.000
33	0.50000	-4.137	0.000	-4.137	-4.137	0.000
34	0.50000	-2.765	0.000	-2.765	-2.765	0.000
35	0.50000	-4.205	0.000	-4.205	-4.205	0.000
36	0.50000	-4.771	0.000	-4.771	-4.771	0.000
37	0.50000	-2.301	0.000	-2.301	-2.301	0.000
38	0.50000	-2.204	0.000	-2.204	-2.204	0.000
39	0.50000	-4.502	0.000	-4.502	-4.502	0.000
40	0.50000	-3.679	0.000	-3.679	-3.679	0.000

This completes (26-12) in two stages: the first estimating $\hat{W}_t = \delta_1 W_{t-1}$ $+\omega_0 Z_{t-2}$ and the second estimating $e_t = \phi_1 e_{t-1}$. As can be seen from Figure 26.9, the final errors, e'_t, are all zero. These errors are used in forming the D_{ij} matrix as well as all other quantities needed to update the parameters of the model using the non-linear estimation procedure of steepest descent.

The inputs and outputs of the NLEST program are shown below:

```
NLEST

       ***** NON-LINEAR ESTIMATION *****
NEED HELP?N
NO..OF OBSERVATIONS?40
DO YOU WANT TO ESTIMATE FOR UNIVARIATE ARMA(BOX-JENKINS)?N
NAME OF FILE OF INDEPENDENT VARIABLE?X
NAME OF FILE OF DEPENDENT VARIABLE?Y
ENTER R,S AND B ?1,0,2
SHORT,LONG DIFFERENCES, LENGTH OF SEASONALITY?1,0,0
CONCERNING THE NOISE MODEL SPECIFY THE FOLLOWING :
ENTER P IN AR(P), Q IN MA(Q), 1 IF SEAS. PARAM. IN AR (0 IF NOT), 1 IF
SEAS. PARAMETER IN MA (0 IF NOT ) ?1,0,0,0
LENGTH OF SEASONALITY OF NOISE MODEL ?0
ENTER INITIAL ESTIMATES OF 3    PARAMETERS IN ORDER SPECIFIED
?.5186,1.7177,.4657
NUMBER OF ITERATIONS?11
NO. OF ITERATION = 1     MSE = 1.05678E-02
    0.519        1.718        0.466
NO. OF ITERATION = 2     MSE = 8.31729E-05
    0.502        1.699        0.495
NO. OF ITERATION = 3     MSE = 3.52425E-08
    0.500        1.700        0.500
FINAL MSE = 3.52425E-08
CORRELATION MATRIX OF THE PARAMETERS
    1.00     -0.65     -0.05
    1.00     -0.26
    1.00
ESTIMATE          95PERCENT CONFIDENCE LIMITS
                  LOWER              UPPER
    δ1 = 0.500        0.500              0.500
    ω1 = 1.700        1.700              1.700
    φ1 = 0.500        0.500              0.500

DO YOU WANT THE AUTOCORRELATIONS OF THE RESIDUALS?N
```

The final model is

$$\hat{W}_t = .5\hat{W}_{t-1} + 1.7Z_{t-2} + .5e_{t-1} \tag{26-20}$$

The correlations between the parameters—δ_1, ω_0 and ϕ_1 — are found to be $-.65$ for δ_1 and ω_0, $-.05$ for δ_1 and ϕ_1 and $-.26$ for ω_0 and ϕ_1. (These are found through the J_{ij} matrix used in the last iteration.) Confidence intervals are also computed. However, in this example the confidence intervals are constant since the MARMA model of (26-2) was generated without any randomness.

13.   *Checking the Model's Adequacy*

There are two tests used to determine whether (26-20) is an adequate model. First, the cross autos between the prewhitened X_t series, X_{et} (equation [26-15]), and e_t' are calculated. If these indicate white noise, it suggests that the MARMA model has been correctly specified. If they do not indicate white noise, one must return to step 9. Second, the autocorrelations of e_t' (26-20) are found. If these autos are random, it indicates that the noise model has been correctly specified. If they are not random, one must return to step 11. Finally, if both the cross autos and the autos indicate significant correlation, it suggests that while the MARMA model may be correctly specified, the noise model may not be. The residuals of (26-20), e_t', are nearly zero (the MSE of the final iteration is .000000035) making it clear that the tentative model is adequate. (In most cases, these residuals would not be nearly as small, simply due to randomness in the original data.)

14.   *Forecasting*

Equation (26-20) refers to the differenced values of X_t and Y_t. Thus, after the forecasts have been calculated they must be dedifferenced back to the original data. The forecasts for periods 41 through 45 are (see Figure 26-9 for values of W_t, Z_t and e_t):

$$\hat{W}_{41} = .5\hat{W}_{40} + 1.7Z_{39} + .5e_{40} = .5(-3.679) + 1.7(.99) + .5(0)$$

$$= -.1565$$

$$\hat{W}_{42} = .5W_{41} = 1.7Z_{40} = .5(-.1565) + 1.7(-.06) = -.180$$

$$\hat{W}_{43} = .5W_{42} + 1.7(0) = -.09 \text{ or } \hat{W}_{43} = .5W_{42} + 1.7(\hat{Z}_{42})$$

$$= -.144$$

$$\hat{W}_{44} = .5(-.09) = -.045 \text{ or } \hat{W}_{44} = .5W_{43} + 1.7(\hat{Z}_{42}) = -.094$$

$$\hat{W}_{45} = -.02 \text{ or } \hat{W}_{45} = .5W_{44} + 1.7(\hat{Z}_{43}) = -.061$$

In the second set of forecasts the value of the independent variable, Z_t, is forecast beyond period 4 since its generating process $Z_t = .9Z_{t-d} + e_t - .4e_{t-1}$ is known. For example:

$$\hat{Z}_{41} = .9Z_{40} - .4e_{40} = -.054$$

The actual forecasts, Y_t, are found by adding $Y_{t-1} + W_t$.

$$Y_{41} = Y_{40} + \hat{W}_{41} = 62.227 + (-.1565) = 62.07$$

$$Y_{42} = Y_{41} + \hat{W}_{42} = 62.07 - .180 = 61.89$$

$$Y_{43} = Y_{42} + \hat{W}_{43} = 61.89 - .09 = 61.8 \text{ or } 61.656$$

$$Y_{44} = Y_{43} + \hat{W}_{44} = 61.8 - .045 = 61.76 \text{ or } 61.666$$

and

$$Y_{45} - Y_{44} + W_{45} = 61.76 - .02 = 61.74 \text{ or } 61.679$$

The forecast for period 43, $\hat{W}_{43}$, however, is based only on extrapolations of the dependent variable W_t while the independent variable, Z, takes the value 0 since there are no actual values beyond period 40. This can be partially corrected by substituting the forecast of the independent variable. Since $Z_t - .794Z_{t-1} - .286e_{t-1}$ (where $Z_t = X_t - X_{t-1}$) Z_t can be predicted beyond period 40. For example:

$$Z_{41} = .794(18.15 - 18.21) - .286(-.3485)$$

$$(e_{40} = -.3484)$$

Thus $Z_{41} = .052$

$$\hat{Z}_{42} = .794\hat{Z}_{41} - .286e_{41} = .794(.052)0 - .286(0) - .0413$$

Similarly $\hat{Z}_{43} = .0328$, $\hat{Z}_{44} = .026$, and $\hat{Z}_{45} = .0207$

The estimated values – $\hat{Z}_{41}, \hat{Z}_{42}, \ldots, \hat{Z}_{45}$ – could therefore be substituted into (26-20). Thus

$$\hat{W}_{43} = .5\hat{W}_{42} + 1.7\hat{Z}_{41} = .5(-.18) + 1.7(.052) = -.0016$$

$$\hat{W}_{44} = .5\hat{W}_{43} + 1.7\hat{Z}_{42} = .5(-.0016) + 1.7(.0338) = .069$$

$$\hat{W}_{45} = .5\hat{W}_{44} + 1.7\hat{Z}_{43} = .5(.069) + 1.7(.026) = .090$$

(Prior to period 42, the values of W are as before.) Thus

$$Y_{43} = Y_{42} + \hat{W}_{43} = 61.89 - .0016 = 61.89$$

$$Y_{44} = Y_{43} + \hat{W}_{44} = 61.89 + .069 = 61.96$$

and

$$Y_{45} = Y_{44} + \hat{W}_{45} = 61.96 + .09 = 62.05$$

It is also possible to determine confidence intervals, as shown below, although in this case their range is zero because there is no randomness present.

| | | 95% BOUNDS | | FORECAST OF |
PERIOD	FORECAST	LOWER	UPPER	INDEPENDENT VARIABLE
41	62.07	62.07	62.07	18.20
42	61.89	61.89	61.89	18.24
43	61.89	61.89	61.89	18.28
44	61.96	61.96	61.96	18.30
45	62.05	62.05	62.05	18.32

COMPARISON OF PARAMETER VALUES

The generating process (26-1), for the original univariate model was

$$Z_t = .9Z_{t-1} + e_t - .4e_{t-1} \tag{26-5}$$

where

$$Z_t = X_t - X_{t-1}$$

and e_t was normally distributed with mean zero and variance one. (This is minimal randomness for most ARMA models encountered in practice.)

The initial estimates of ϕ_1 and θ_1 using the Z_t data are:

$$\phi_1 = .794 \text{ and } \theta_1 = .286$$

The final estimates are:

$$\phi_1 = .845 \text{ and } \theta_2 = .423$$

The final estimates are much closer to the generating parameters, $\phi_1 = .9$ and $\theta_1 = .4$, even though not exact. This is because the randomness, e_t, was introduced.

The generated MARMA model is

$$Y_t = .5Y_{t-1} + 1.7X_{t-2} \tag{26-6}$$

The initial estimates of the MARMA parameters are:

$$\delta_1 = .5186$$

$$\omega_0 = 1.718$$

These are quite close to the actual ones since (26-6) includes no randomness.

The final MARMA estimates are

$$\delta_1 = .5$$

$$\omega_0 = 1.7$$

These are the exact generating parameters in (26-6) since, again, there is no randomness. The noise parameter, $\phi_1 = .5$, is also exact (.9 − .4 = .5, see (26-5)) since there is no randomness in (26-6).

CHAPTER 27

Multivariate Generalized Adaptive Filtering (MGAF)

Multivariate Generalized Adaptive Filtering (MGAF) is very similar to the multivariate MARMA models examined in chapter 26. The main difference between the two methods is the way in which the optimal parameter values are obtained. In MGAF the parameters vary with each observation because they are updated at each time period using the equivalent of (17-13).

The MGAF model, for the bivariate case, is:

$$Y_t = \delta_{1t} Y_{t-1} + \delta_{2t} Y_{t-2} + \ldots + \delta_{rt} Y_{t-r} + \omega_{ot} X_{t-b}$$

$$- \omega_{2t} X_{t-b-1} - \ldots - \omega_{st} X_{t-b-s} + e_t \qquad (27\text{-}1)$$

where e_t can be

$$e_t = \phi_{1t} e_{t-1} + \phi_{2t} e_{t-2} + \ldots + \phi_{pt} e_{t-p} + e_t'$$

$$- \theta_{1t} e_{t-1} \ldots \theta_{qt} e_{t-q} \qquad (27\text{-}2)$$

and where e_t' is white noise.

To update the parameters of (27-1), the following equation is used. (Note that the subscript t has been omitted from the equation.)

$$V_i' = V_i + 2Ke_t^{*'} R_{t-i}^* \qquad (27\text{-}3)$$

where V_i includes the δ_i, ω_i and ϕ_i parameters and R_t includes the Y_t, X_t and e_t variables. Also,

$$\theta_i' = \theta_i - 2Ke_t^{*'} e_{t-1}^* \tag{27-4}$$

where e_t^*, R_{t-i}^*, $e^{*'}$, and e_{t-i}^* are all standardized values obtained by dividing the actual values by the largest of all of them.

References for further study:

Makridakis, S. and S. Wheelwright (1978, chapters 9 and 11)

INPUT/OUTPUT FOR MGAF PROGRAM

Inputs

1. *Short term differences.*

The level of short term differences at which the data are stationary must be specified. (No long term differencing is required in GAF.)

2. *r, s and b*

(See chapter 26.)

3. *Order of p and q.*

Usually p, corresponding to AR(p), is set equal to one, two, or three if the data is not seasonal and set equal to the length of seasonality with seasonal data. The order of q, corresponding to MA(q), is set equal to one or at most 2. If the user feels that some moving average terms should be included, q should be set equal to one. This will usually suffice, but if it does not, a value of q = z can be tried.

The program automatically establishes the initial values of the parameters as

$$\delta_1 = .1, \, \delta_2 = .2, \, \ldots, \, \omega_0 = (r + t) \, (.1), \, \omega_1 = (r + 2) \, (.1), \, \ldots,$$

$$\omega_s = (r + s + 1)\ (.1),\ \phi_1 = (r + s + 2)\ (.1) + \dots,$$

unless the user specifies values directly.

Output

The final parameter values and the resulting forecasts are provided.

EXERCISE

The 40 data values of (26-1) and (26-2) were generated using

$$Y_t = .5Y_{t-1} + 1.7X_{t-2} + e_t \tag{27-5}$$

where e_t is normally distributed.

Apply MGAF to this data series to obtain parameter estimates for the appropriate MARMA model.

Solution

The values of r, s, b and p are set as in chapter 26, i.e., r = 1, s = 0, b = 2 and p = 1. No initial values are given. Figure 27-1 shows the relevant values of the parameters, how they are updated and the forecasts and errors for the first training iteration.

It can be seen in Figure 27-1 that the parameter values are not updated until period 8. This is to allow the errors to settle before updating the parameters.

The MSE of the first training iteration is 7.63 and the parameter values are

$$\delta_1 = .748$$

$$\omega_0 = 1.32$$

$$\phi_1 = .219$$

FIGURE 27-1   Parameters, Forecasts and Errors Using MGAF (1 iteration, k = .05)

Period	δ_1	ω_0	ϕ_1	Y_t-Y_{t-1}	X_t-Y_{t-1}	$\hat{W}_t$	Standardization Constant	Overall Forecast	e'_t
4	0.100	0.200	0.300	3.87	2.02	0.37	3.49	0.37	3.49
5	0.100	0.200	0.300	5.52	1.43	0.50	3.97	1.55	3.97
6	0.100	0.200	0.300	6.86	0.52	0.45	5.02	1.96	4.90
7	0.100	0.200	0.300	6.22	-0.63	0.33	6.40	2.25	3.97
8	0.102	0.203	0.333	3.88	1.92	0.14	5.89	1.90	1.97
9	0.101	0.206	0.318	0.54	1.18	-0.11	3.74	1.13	-0.59
10	0.098	0.254	0.334	4.53	2.83	0.38	3.94	0.59	3.94
11	0.103	0.268	0.382	3.69	1.94	0.34	4.15	1.72	1.97
12	0.110	0.325	0.454	6.71	1.75	0.79	4.64	2.07	4.64
13	0.117	0.346	0.506	6.47	0.69	0.73	5.92	3.41	3.06
14	0.118	0.348	0.514	4.08	0.57	0.69	5.75	3.60	0.49
15	0.138	0.368	0.613	5.42	2.40	0.32	3.39	2.07	3.35
16	0.138	0.367	0.604	2.94	1.06	0.25	5.09	3.38	-0.44
17	0.143	0.423	0.667	6.85	1.41	0.92	4.31	2.54	4.31
18	0.142	0.422	0.658	4.02	1.50	0.58	5.93	4.53	-0.51
19	0.157	0.457	0.743	5.88	2.34	0.68	3.44	2.94	2.94
20	0.155	0.453	0.732	4.04	-0.02	0.79	5.21	4.66	-0.62
21	0.177	0.518	0.821	6.47	0.17	1.18	3.25	3.56	2.91
22	0.168	0.518	0.784	2.58	-0.46	0.20	5.29	4.54	-1.96
23	0.166	0.515	0.752	1.23	-2.27	0.12	2.38	1.99	-0.75
24	0.175	0.481	0.836	1.95	-1.49	-0.22	1.33	0.62	1.33
25	0.191	0.651	0.635	-4.09	-1.35	-1.13	2.27	0.68	-4.77
26	0.228	0.735	0.722	-5.70	-0.22	-1.25	2.96	-3.13	-2.57
27	0.221	0.727	0.697	-3.38	-0.28	-1.28	4.45	-4.49	1.11
28	0.276	0.737	0.787	-3.80	-0.80	-0.44	2.10	-1.91	-1.89
29	0.265	0.730	0.702	0.98	0.51	-0.33	3.95	-2.97	3.95
30	0.340	0.914	0.400	-3.71	-0.19	-0.67	1.31	0.25	-3.96
31	0.348	0.908	0.434	-2.02	-2.30	0.24	3.04	-0.98	-1.04
32	0.353	0.904	0.382	0.10	-0.41	-0.09	2.26	-1.07	1.17
33	0.357	0.988	0.375	-3.98	-1.66	-2.11	2.30	-2.04	-1.94
34	0.338	0.985	0.359	-1.46	-1.57	-1.16	2.11	-1.86	0.40
35	0.396	1.068	0.374	-3.52	0.05	-2.03	1.66	-2.13	-1.39
36	0.555	1.191	0.491	-6.25	-0.62	-2.48	2.03	-3.04	-3.21
37	0.505	1.192	0.416	-0.33	-2.00	-1.32	3.77	-3.16	2.84
38	0.773	1.318	0.214	-4.52	-0.84	-1.40	1.32	-0.99	-3.52
39	0.779	1.327	0.228	-4.79	0.99	-3.72	3.11	-4.39	-0.41
40	0.748	1.320	0.219	-3.13	-0.06	-4.01	3.72	-4.26	1.13

After 10 training iterations the MSE is reduced to 1.33 (very close to the generated variance of 1) and the final parameters are

$$\delta_1 = .59$$

$$\omega_0 = 1.67$$

$$\phi_1 = -.33$$

These are close to the generated values of $\delta_1 = .5$ and $\omega_0 = 1.7$. The results of the final (10th) training iteration are shown in Figure 27.2. The values at period 10, for example, are calculated as follows:

$$\hat{W}_{10} = \delta_1\hat{W}_2 + \omega_0(X_8 - X_7) = .55(1.64) + 1.672(1.92) = 4.11$$

FIGURE 27-2 Parameters, Forecasts and Errors Using MGAF (10th Iteration, K = .05)

Period	δ_1	ω_0	ϕ_1	Y_t-Y_{t-1}	X_t-X_{t-1}	$\hat{W}_t$	Standardization Constant	Overall Forecast	$e_t{}'$
4	0.592	1.669	-.327	3.87	2.02	3.10	1.86	3.10	0.76
5	0.592	1.669	-.327	5.52	1.43	5.69	3.10	5.45	0.08
6	0.592	1.669	-.327	6.86	0.52	6.74	5.69	6.80	0.05
7	0.592	1.669	-.327	6.22	-0.63	6.38	6.74	6.35	-0.12
8	0.579	1.668	-.327	3.88	1.92	4.65	6.38	4.70	-0.82
9	0.550	1.672	-.322	0.54	1.18	1.64	4.65	1.89	-1.35
10	0.553	1.675	-.324	4.53	2.83	4.11	1.92	4.47	0.06
11	0.543	1.672	-.325	3.69	1.94	4.25	4.11	4.12	-0.42
12	0.531	1.664	-.323	6.71	1.75	7.04	4.25	7.22	-0.51
13	0.522	1.662	-.323	6.47	0.69	6.97	7.04	7.07	-0.60
14	0.485	1.652	-.320	4.08	0.57	6.55	6.97	6.71	-2.62
15	0.490	1.653	-.322	5.42	2.40	4.31	6.55	5.10	0.31
16	0.495	1.654	-.321	2.94	1.06	3.05	4.31	2.70	0.24
17	0.539	1.688	-.322	6.85	1.41	5.48	3.05	5.52	1.33
18	0.533	1.687	-.324	4.02	1.50	4.74	5.48	4.30	-0.28
19	0.549	1.692	-.326	5.88	2.34	4.91	4.74	5.14	0.74
20	0.531	1.686	-.329	4.04	-0.02	5.23	4.91	4.91	-0.87
21	0.519	1.681	-.327	6.47	0.17	6.73	5.23	7.12	-0.65
22	0.505	1.681	-.326	2.58	-0.46	3.46	6.73	3.54	-0.96
23	0.473	1.679	-.318	1.23	-2.27	2.03	3.46	2.32	-1.08
24	0.548	1.662	-.347	1.95	-1.49	0.19	2.03	0.44	1.51
25	0.548	1.654	-.341	-4.09	-1.35	-3.67	2.27	-4.28	0.19
26	0.586	1.669	-.336	-5.70	-0.22	-4.48	3.67	-4.33	-1.37
27	0.562	1.662	-.343	-3.38	-0.28	-4.88	4.48	-4.46	1.08
28	0.565	1.662	-.344	-3.80	-0.80	-3.10	4.88	-3.62	-0.18
29	0.470	1.653	-.366	0.98	0.51	-2.22	3.10	-1.98	2.96
30	0.474	1.655	-.371	-3.71	-0.19	-2.37	3.20	-3.54	-0.17
31	0.568	1.634	-.317	-2.02	-2.30	-0.28	2.37	0.22	-2.24
32	0.568	1.634	-.318	0.10	-0.41	-0.47	1.75	0.09	0.01
33	0.566	1.624	-.315	-3.98	-1.66	-4.02	2.30	-4.20	0.23
34	0.529	1.620	-.315	-1.46	-1.57	-2.94	4.02	-2.96	1.50
35	0.488	1.598	-.295	-3.52	0.05	-4.25	2.94	-4.72	1.19
36	0.523	1.610	-.300	-6.25	-0.62	-4.58	4.25	-4.80	-1.45
37	0.490	1.611	-.312	-0.33	-2.00	-2.31	4.58	-1.81	1.49
38	0.566	1.631	-.377	-4.52	-0.84	-2.13	2.31	-2.75	-1.76
39	0.612	1.674	-.326	-4.79	0.99	-4.47	2.38	-3.57	-1.22
40	0.592	1.670	-.328	-3.13	-0.06	-4.14	4.47	-4.04	0.91

where $\hat{W}_{10}$ is the MARMA forecast.

$$e_{10} = (Y_{10} - Y_9) - \hat{W}_{10} = 4.53 - 4.11 = .418$$

Similarly,

$$\hat{W}_{11} = \delta_1 \hat{W}_{10} + \omega_0(X_9 - X_8) = .553(4.11) + 1.675(1.18) = 4.25$$

and

$$e_{11} = (Y_{11} - Y_{10}) - \hat{W}_{11} = 3.69 - 4.25 = -.56$$

and so on. The errors - $e_4, e_5, \ldots, e_{40}$ - are then forecast.

For period 11, for example, the forecast of the error is

$$e_{11} = \phi_1 e_{10} = -.324(.418) = -.135$$

The total forecast then becomes

$$F_{11} = 4.25 - .135 = 4.12$$

and the total error is

$$e'_{11} = (Y_{11} - Y_{10}) - F_{11} = 3.69 - 4.12 = -.42$$

Once the errors are found, the parameter values can be updated using (27-3) and (27-4). The updated values for period 11 are:

$$\delta'_1 = \delta_1 + 2Ke_t^{*'}\hat{W}_{t-1}^* = \delta_1 + 2Ke_{11}^{*'}\hat{W}_{10}^* - .553$$

$$+ 2(.05)\,(-.42/4.11)\,(4.11/4.11) = .543$$

where

$$e_{11}^{*'} = e'_{11}/4.11, \text{ and } W_{11}^* = W_{11}/4.11$$

and 4.11, the largest of the values used in updating the parameters, is used as the standardization constant. (In MGAF the largest value is used because it is easier computationally than the standardization procedure applied in Chapter 17.)

$$\omega'_o = \omega_o + 2Ke_t^{*'}\,(X_{t-b-1} - X_{t-b-2})^* = \omega_o + 2Ke_{11}^{*'}\,(X_8 - X_9)^*$$

$$= 1.675 + 2(.05)\,(-.42/4.11)\,(1.18/4.11) = 1.672$$

Finally,

$$\phi'_1 = \phi_1 + 2Ke_t^{*'}\,e_{t-1}^* = \phi_1 + 2Ke_{11}^{*'}\,e_{10}^*$$

$$= -.324 + 2(.01)\,(-.42/4.11)\,(.418/4.11) = -.325$$

The parameters for other periods are updated in a similar manner.

Cases

LYON PLATE GLASS

In late 1972, Jean-Pierre Laurent, a recent graduate of a well-known European business school, inherited a sizeable block of stock in the Lyon Plate Glass Company (LPG). LPG was a closely held French corporation in the glass production business. The firm sold most of its products to other industrial companies, particularly auto manufacturers and construction operations, although it did have a growing retail business amounting to about one-eighth of its total operation in 1971.

Because the firm's stock was closely held by two families, neither of whose name happened to be Laurent, they were anxious to buy Jean-Pierre's stock, rather than take a chance on his wanting a say in the management of the company. Early in 1973, negotiations were completed and Jean Pierre was faced with choosing between two alternative buy-out plans. (He had definitely decided to sell the stock which his stepmother had left him.)

The first plan was a lump sum agreement that would pay him 8,200,000 Francs for this stock. The alternative plan was based on 1973 sales of LPG. (He had negotiated this plan because an advisor from a well-known U.S. business school had predicted that 1973 would be another boom year for the French glass industry.) The amount Jean-Pierre would receive under this alternative would be a lump sum of 7,750,000 Francs if the sales of LPG were 775 million Francs or *less* during 1973. If 1973 sales exceeded this level, he would receive an amount equal to 1% of 1973 sales. Since under both alternatives Jean-Pierre would receive his money in cash on January 30, 1974, he did not feel it was necessary to discount the amounts involved in order to make his decision.

To help him in making his decision, Jean-Pierre had tried to apply much of what he had learned in business school about decision-making under uncertainty. Graphically, he viewed his decision as follows:

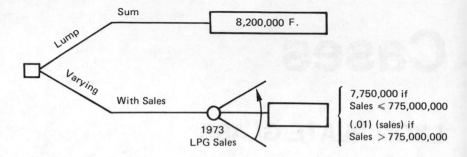

Jean-Pierre saw as his next task determining the possible sales levels for 1973. Because LPG sold mainly to the auto and construction industries, and since the French government prepared forecasts for these two industries, regression analysis looked like a useful forecasting tool. After gathering the data in Table 1. Jean-Pierre ran several regressions, the results of which are attached. He must now decide how to complete the analysis of his problem.

TABLE 1   Relevant Data for Lyon Plate Glass Company

Year	Net Sales LPG (millions of francs)	Automobile Production (in millions)	Building Contracts (in millions)
1956	280.0	3.909	9.43
1957	281.5	5.119	10.36
1958	337.4	6.666	14.50
1959	404.2	5.338	15.75
1960	402.1	4.321	16.78
1961	452.0	6.117	17.44
1962	431.7	5.559	19.77
1963	582.3	7.920	23.76
1964	596.6	5.816	31.61
1965	620.8	6.113	32.17
1966	513.6	4.258	35.09
1967	606.9	5.591	36.42
1968	629.0	6.675	36.58
1969	602.7	5.543	37.14
1970	656.7	6.933	41.30
1971	778.5	7.638	45.62
1972	877.6	7.752	47.38
1973*		6.400	48.51
1974*		7.900	51.23

*Estimates (forecasts) prepared by the French government.

RESULTS OF REGRESSION ANALYSIS

Run 1 $(Y = A + B_1 X_1)$

Y = LPG Sales X_1 = Automobile Production

$A = -33.98$

$B_1 = 95.11$

SEE = 125.1

Adjusted R^2 = .473

Using these results to forecast Y for 1973 gives the following fractiles:

.01	.10	.25	.50	.75	.90	.99
282.3	415.0	490.4	575.0	659.5	735.2	867.4

Run 2 $(Y = A + B_2 X_2)$

Y = LPG Sales X_2 = Building Contracts

$A = 180.93$

$B_2 = 12.69$

SEE = 54.1

Adjusted R^2 = .899

Using these results to forecast Y for 1973 gives the following fractiles:

.01	.10	.25	.50	.75	.90	.99
670.5	727.4	760.0	796.5	833.1	865.6	922.4

Run 3 $(Y = A + B_1X_1 + B_2X_2)$

Y = LPG Sales X_1 = Automobile Production

X_2 = Building Contracts

A = 19.13

B_1 = 35.67

B_2 = 10.86

SEE = 42.5

Adjusted R^2 = .947

Using these results to forecast Y for 1973 gives the following fractiles:

.01	.10	.25	.50	.75	.90	.99
672.3	716.8	742.7	771.4	800.1	825.9	880.5

ALPHA CONCRETE PRODUCTS

Early in the Spring of 1974, Mr. Rick Woodbury, Secretary-Treasurer of ALPHA Concrete Products was anticipating how he might evaluate the long range forecasts of company sales and profits that had recently been prepared by a local consulting firm. He realized that the other members of top management at ALPHA were expecting him to make a formal presentation of his projections within the next week so that as a group they could complete their five-year plan. Thus, he wanted to determine the limitations and strengths of the consulting firm's approach, perform any additional analysis that might be required, and determine just what his presentation should cover.

ALPHA was in the business of manufacturing concrete pipe and reinforced concrete boxes used in highway, utility and farm construction. The company maintained three production facilities, one each in Utah, Colorado and Idaho. Sales were handled through the company's own distributors located throughout the intermountain area. Approximately two-thirds of the company's sales had gone to the major intermountain public utilities in recent years with the balance going mostly to agricultural construction in the Idaho region and to about an equal mix of highway and agricultural construction in the other regions. In 1973 total annual sales were 60% from Utah, 22% from Idaho and 18% from Colorado.

While the company generally submitted bids on all of the major projects in its marketing area, it was only awarded about 30% of those contracts. In aggregate, such major contracts accounted for over 80% of its volume. The company had noticed in the past that the total volume of bid requests was growing at a modest rate each year. However, the fraction of bid requests from each of the major market segments — utilities, highways and agricultural — varied substantially from year to year.

Rick Woodbury had asked a local consulting firm to prepare a set of sales and profit forecasts for ALPHA through 1980. The consultant who actually conducted the project had spent several hours with Rick at the outset, asking questions about the firm and its skills, products, customers and competition. The consultant also obtained from Rick a set of ALPHA's annual reports covering the past 15 years. The attached report had been given to Rick following a summary presentation made by the consultant.

PROFORMA SALES AND PROFITS 1974-1980 ALPHA CONCRETE PRODUCTS, INC.

CONTENTS

A. EXECUTIVE SUMMARY

The major impetus for the preparation of this study was the need that ALPHA's management felt to prepare more accurate forecasts of both sales and income statements. Thus, rather than simply relying on linear extensions of past growth rates, the company decided to determine some of the basic relationships between economic factors and sales and then base prediction on these fundamental relationships.

The approach taken in this study involved the application of regression analysis to the sales forecasting problem at ALPHA. There were three main reasons for selecting this approach to forecasting. First, it is a systematic approach that has been tested in a wide range of situations and has been found to give significant results when the proper requirements are met. Second, it provides not only a best estimate forecast for each of several years into the future but also allows the computation of confidence intervals so that management can determine the uncertainty that surrounds such estimates. Finally, this approach uses the historical data available to determine causal relationships between the dependent variable (sales) and the several independent variables that are felt to be the most important determinants of sales.

The results of applying this statistical approach to sales forecasting were indeed satisfactory. The model actually used explained 95% of the historical variation in ALPHA's sales. Thus, if this model had been used in the past 15 years in forecasting sales, on average the actual sales value would have deviated from the forecast value by less than $500,000.00. This would appear to be a fairly accurate model, at least on an historical basis.

This model, which bases the sales forecast on population growth, new dwelling unit permits, and construction employment, was then used to prepare forecasts of sales for each of the next seven years (through 1980). Of course, the basic assumption in such usage of the model is that those relationships that have determined the company's sales in the past will continue to do so in the future. In addition to giving a single sales forecast for each of the next seven years, this model was also used to determine a 90% confidence interval around each of those forecasts. This interval delineates the range of values within which it is 90% certain that the actual outcome will fall. Thus, this provides management with a measure of the uncertainty in these forecasts.

The final step was to take the sales forecasts and use these in preparation of *pro forma* income statements for the past six years of company operations. Thus, the assumption is that the cost structure will remain the same in the future as it has been in the past. (In fact, the cost structure has been very stable over the past six years.)

Since this is a first application of statistical forecasting at ALPHA, it will be necessary for management to explain the basis of this forecast to those who will actually use it. It will undoubtedly be the case that as the forecast is used other variables and sources of data which might be used in the future will be recognized and thus the application of this technique can be continually improved.

B.   Development of a Sales Forecasting Model

While there are several alternative sales forecasting methods available, the one that was selected for use in forecasting ALPHA's sales was regression analysis. This technique develops a basic model using historical data and then applies this model to forecast future sales values. The model itself is a causal one in that the dependent variable (ALPHA's sales) is stated as a function of several independent variables on which data can be collected. Thus, in very elementary terms, the essence of this approach is to identify those factors (variables) that are felt to have the greatest impact on ALPHA's sales and then determine the relationship between these factors and company sales based on historical data. This relationship is then used to predict future values and sales.

Through discussions with ALPHA management, several factors were identified that might have a possible impact on ALPHA's sales. These included:

1. Construction activity
2. Expansion by major utilities
3. Highway construction
4. Agricultural expansion
5. Interest rates
6. Population

Based on these factors a number of variables were identified for which historical data could be collected and which could also be projected into the future. Figure B1 indicates the nine variables that were identified for possible use in this study. All of these variables are defined in Figure B1 and should be clear with the possible exception of variables 8 and 9. Variable 8 is the number of new dwelling unit permits from the previous year. That is, this variable takes *1970* dwelling unit permits and uses them as an indicator of *1971* sales. The reason for this is that dwelling unit permits generally precede the actual construction activity. Thus, the issuance of such permits would be a leading factor in predicting ALPHA's sales. Variable number 9 is what is referred to as a *lag variable* and is based on construction employment. Here the assumption is that actual construction employment will affect ALPHA's sales in a prior year. For example, ALPHA's sales in 1970 would be affected by 1971 construction employment.

The actual data collected on each of the variables to be considered for use in this model are shown in Figure B2. This data covers the 15-year period from 1959 through 1973. One of the attractive features of the technique of regression analysis is that using a computer, several different models (that is, models that include different combinations of the variables) can be tested without much difficulty and then the most appropriate model (the one that makes the most sense and gives the best results) can be actually selected for use in forecasting. This approach of trying several models is the one that was followed in this study.

The results of this statistical analysis can first be seen by examining the simple correlation matrix shown in Figure B3. This shows the extent of the relationship (correlation) between each pair of variables. The value of any correlation is between −1.00 and +1.00. A correlation of 0 incicates no relationship between the two variables, a correlation of +1.00 indicates perfect positive correlation (an increase in one variable is always accompanied by an increase in the other), and a correlation of −1.00 indicates a perfect negative correlation (an increase in one variable is always accompanied by a decrease in the other variable).

FIGURE B1 Definition of Variables

	Variable	Definition (and source of data)	Units*
1.	Time	The year with 1959 = 1	1's
2.	Population	Total Utah population (Utah Statistical Abstract)	100,000's
3.	Residential Units	Total residential units in Utah (Mountain Bell)	100,000's
4.	New Dwelling-Unit Permits	Total of new housing permits for Utah (Utah Real Estate Facts)	100,000's
5.	Construction Employment	Total employment in Utah construction industry (Utah Statistical Abstract)	100,000's
6.	Mountain Bell Telephone Mains	Total Mountain Bell telephone mains in Utah (Mountain Bell)	100,000's
7.	ALPHA's Sales	Total annual sales of ALPHA (Accounting statements)	100,000's
8.	Lead-Dwelling Unit Permits (8)	Same as variable 4	100,000's
9.	Lag-Construction Employment (5)	Same as variable 5	100,000's

*Units used in regression model.

FIGURE B2   Historical Data Variables

Year	Time	Population	Res. Units Mtn. Bell	New Dwelling Unit Permits	Const. Empl.	Mtn. Bell Tel. Mains	ALPHA's Sales $	Lead New Dwelling Permits	Lag Constr. Employment
1959	1	868,000	227,000*	8,272	15,715	218,500*	2,904,000	8,000*	14,851
1960	2	890,627	232,627	6,700	14,851	226,600	2,868,000	8,272	15,569
1961	3	936,000	237,700*	8,030	15,560	234,400*	3,303,000	6,700	17,790
1962	4	958,000	243,200*	8,420	17,790	242,100*	4,888,476	8,030	17,549
1963	5	974,000	248,600*	8,744	17,549	249,900*	5,879,591	8,420	17,035
1964	6	978,000	253,800*	6,494	17,035	257,700*	5,947,587	8,744	15,971
1965	7	991,000	259,100	5,657	15,971	265,400	5,905,301	6,494	15,507
1966	8	1,009,000	264,800*	3,982	15,507	275,000*	5,442,447	5,657	13,420
1967	9	1,019,000	270,500*	5,116	13,420	285,000*	4,327,223	3,982	13,689
1968	10	1,029,000	276,200*	5,490	13,689	294,700*	6,237,503	5,116	13,964
1969	11	1,047,000	281,900*	5,964	13,964	304,300*	5,921,922	5,490	14,583
1970	12	1,059,000	287,700	9,070	14,583	314,000	7,619,577	5,964	16,951
1971	13	1,099,000	276,100	12,777	16,951	330,500	7,863,210	9,070	20,669
1972	14	1,128,000	313,800	17,320	20,669	349,100	9,853,870	12,777	22,100
1973	15	1,157,000	323,800	13,450	22,100	369,100	11,979,262	17,320	20,000*
Mean	8.0	1,009,660	266,347	8,364	16,358	281,033	6,082,720	7,947	16,763
Std. Deviation		81,242	28,169	3,623	2,459	45,542	2,506,079	3,345	2,821

*Approximate

FIGURE B3   Simple Correlation Matrix

	Var. No.	Time	Popu- lation	Res. Units	New Const.	Constr. Emp.	Bell Tel. Mains	ALPHA's Sales	Lead (4)	Lag (5)
Time	1	1.000								
Population	2	.983	1.000							
Res. Units	3	.963	.965	1.000						
New Constr. (Permits)	4	.509	.581	.566	1.000					
Constr. Emp.	5	.349	.483	.481	.794	1.000				
Bell Tel. Mains	6	.987	.986	.978	.607	.465	1.000			
ALPHA's Sales	7	.878	.922	.917	.689	.709	.917	1.000		
Lead (4)	8	.396	.502	.532	.790	.943	.526	.727	1.000	
Lag (5)	9	.491	.593	.538	.947	.845	.591	.715	.843	1.000

From Figure B3 it can be seen that all of the variables show a high positive correlation with ALPHA's sales with the strongest correlation being with population (.922) and the weakest correlation being with new construction permits (.689). It will also be noted that previous year's building permits (variable 8) correlates more highly with ALPHA's sales (.727) than does current year's building permits (.689) and next year's construction employment (variable 9) correlates more highly with ALPHA's sales (.715) than does current year's construction employment (.709).

The approach actually used in determining which set of variables gives the best model for forecasting ALPHA's sales is to try a variety of combinations (using a computer-based analysis) and then to select the best of those. In practice the computer uses a systematic approach to formulate the most promising combinations of variables. For purposes of this written analysis, it is only necessary to state that this has been done and that from the many alternative models available, the one shown in Figure B4 has been identified as being best in this situation.

The model shown in Figure B4 bases the prediction of ALPHA's sales on three variables:

x_2 = Population

x_5 = Construction Employment

x_8 = Lead Construction Permits (previous year's permits)

This model was chosen as the best based on its ability to explain ALPHA's sales over the past 15 years. Some of the measures used in selecting this model are shown in Figure B4.

a. $R^2 = .9457$

This value indicates that the model emplains 95% of the variation in ALPHA's sales.

b. Std Error of Estimate = $658,820

Approximately two thirds of the actual values are within $658,820 of the value predicted by this model.

c. Positive Coefficients

All three of the coefficients (B_2, B_5, and B_8) are positive indicating positive correlation with sales. This is what one would expect.

d. Residual Values

These values are the difference between the sales value predicted by the model and the actual sales value. On the average the error over the past 15 years is less than $450,000.00.

FIGURE B4   Results of 4-Variable Regression Model

Model

$$Y = B_0 + B_2 X_2 + B_5 X_5 + B_8 X_8$$

where

Y = ALPHA's Sales
X_2 = Population
X_5 = Construction Employment
X_8 = Lead (1 year) Building Permits (i.e., building permits in previous year)
B_0 = Constant Factor
B_2, B_5 and B_8 — Coefficients for independent variables

Measures of Goodness of Fit

Multiple R = .9725
R^2 (% variance explained) = .9457
Standard Error of Estimate = $658,820

Coefficient Values (Based on 15 years historical data)

B_0 (Constant)   = −208.69
B_2 (Population)   = 22.91
B_5 (Construction Employment) = 157.69
B_8 (Lead Building Permits)   = 155.78

Residual Values (Using model to predict past 15 years sales)

	Time	Residual*		Time	Residual*
(1959)	1	$252,687		9	$−954,523
	2	−284,433		10	573,518
	3	−754,892		11	−256,115
	4	−293,394		12	988,885
	5	352,426		13	−534,986
	6	359,352		14	−372,551
	7	972,812	(1973)	15	155,020
	8	203,823			

*Residual — Actual-Forecast

In summary this model fits very well with the historical data of the past 15 years and assuming the same basic relationship in the future, provides a good model for forecasting ALPHA's sales for the next several years.

C.   Forecasts of Alpha's Sales

Having selected a regression model to be used in sales forecasting, the next step is to obtain projected values for each of the independent variables. Since the three variables we are using are ones of major interest to the government and several businesses, these projections are readily available. Figure C1 indicates the values used for 1974-1980 for the variables-population, construction employment, and new dwelling unit permits. Based on these independent variables the ALPHA's sales projections shown in Figure C1 can be made simply by substituting the appropriate values for each year into the forecasting equation.

As indicated in Figure C1, ALPHA's sales are expected to reach $16,854,000 by 1980. Again, however, it must be remembered that these projections assume that the basic relationships that have affected sales in the past will continue into the future.

Additional information on future ALPHA sales can be obtained by computing a 90% confidence interval for each of these annual sales forecasts. This is done in Figure C2. The basis for this computation is the standard error of estimate that was determined when computing the coefficients for this regression model. (See Figure B4). The notion is a statistical one which determines the range around any forecast such that the actual value will fall

FIGURE C1   ALPHA Sales Forecasts

$$Y = B_0 + B_2 X_2 + B_5 X_5 + B_8 X_8$$
$$Y = -208.69 + 22.91 X_2 + 157.69 X_5 + 155.78 X_8$$

Year	X_2[1]	X_5[2]	X_8[3]	Y
1974	1,168,000	23,800	13,450	11,738,000
1975	1,194,000	25,490	15,800	12,966,000
1976	1,221,000	26,250	15,000	13,580,000
1977	1,248,000	27,000	15,300	14,364,000
1978	1,276,000	27,750	16,000	15,232,000
1979	1,304,000	28,500	16,000	15,993,000
1980	1,333,000	29,260	16,500	16,854,000

[1] Based on Mtn. Bell Forecasts.
[2] & [3] Based on Bureau of Economic and Business Research, University of Utah estimates.

FIGURE C2   ALPHA Sales Forecasts-Confidence Intervals

Year	Sales Forecast	Confidence Interval (90%)
1974	$11,738,000	$10,538,000-$12,938,000
1975	12,966,000	11,746,000- 14,187,000
1976	13,580,000	12,200,000- 14,860,000
1977	14,364,000	13,084,000- 15,644,000
1978	15,232,000	13,900,000- 16,530,000
1979	15,993,000	14,687,000- 17,300,000
1980	16,854,000	15,537,000- 18,171,600

within that range (that confidence interval) 90% of the time. For example, the 90% confidence interval for ALPHA's sales in 1980 is $15,537,000 to $18,171,600. This means, assuming that the same relationships that have held in the past continue to hold in the future, there is a 90% chance that actual ALPHA sales in 1980 will fall within this range. Of course one must also remember that it means there is a 10% chance that the actual sales will fall outside of this range.

D.   Alpha Pro Forma Income Statements

With the sales forecasts in hand, *pro forma* income statements can be prepared for 1974-1980. The one additional prerequisite to doing this is determining ALPHA's cost structure. Figure D1 indicates what this cost structure has been for each of the past six years. This figure also includes a composite set of figures based on the historical data *and* ALPHA management's judgment as to what can be expected in the next few years. It is this composite set of figures that has been used in preparing annual *pro forma* income statements through 1980.

Figure D2 indicates the income statements that ALPHA can expect based on the sales forecasts of Figure C1 and the composite cost structure of Figure D1. The short-term downturn in profits is due to the fact that the 1973 cost structure gave considerably higher profits (as a per cent of sales) than does the composite structure used in these projections. This simply underscores the fact that the cost structure is a major determinant of profitability, as are sales.

The relative effects of cost structure and sales can be seen from Figure D3. This figure shows the projected income statement based on sales being at either the upper or lower value of the 90% confidence interval. From these figures it should be clear that if ALPHA is to accomplish its long-term goals, it must manage both its sales and costs effectively.

FIGURE D1   ALPHA's Historical Cost Structure*

	1968	*1969*	*1970*	*1971*	*1972*	*1973*	*Composite*
Net Sales	100.0	100.0	100.0	100.0	100.0	100.0	100.0
Cost of Sales	66.1	66.1	65.5	66.2	62.7	65.5	66.0
Gross Profit	33.9	33.9	34.5	33.8	37.3	34.5	34.0
Selling, G & A	24.8	28.0	22.3	23.1	20.9	20.8	23.0
Operating Profit	9.1	5.9	12.2	10.7	16.5	11.7	11.0
Interest Expense	2.3	2.8	2.3	2.4	1.8	2.2	3.0
Pretax Income	7.1	4.6	9.2	7.9	15.3	13.0	8.0
Income Taxes	3.4	2.3	4.6	3.7	7.3	5.8	3.8
Net Income	3.7	2.3	4.6	4.2	8.0	7.2	4.2

*Costs as a percent of sales.

FIGURE D2   ALPHA Pro Forma Income Statements 1974-1980

	Composite*	1974	1975	1976	1977	1978	1979	1980
Net Sales	100.0	$11,738,000	$12,966,000	$13,530,000	$14,364,000	$15,232,000	$15,993,000	$16,854,000
Cost of Sales	66.0	7,747,000	8,558,000	8,953,000	9,480,000	10,053,000	10,555,000	11,124,000
Gross Profit	34.0	3,991,000	4,408,000	4,617,000	4,884,000	5,179,000	5,438,000	5,730,000
Selling G & A	23.0	2,701,000	2,982,000	3,123,000	3,304,000	3,503,000	3,678,000	3,876,000
Operating Profit	11.0	1,291,000	1,426,000	1,494,000	1,580,000	1,676,000	1,760,000	1,854,000
Interest Expense	3.0	352,000	389,000	407,000	431,000	457,000	480,000	506,000
Pretax Income	8.0	939,000	1,037,000	1,087,000	1,149,000	1,219,000	1,280,000	1,348,000
Income Taxes	3.8	446,000	493,000	516,000	546,000	579,000	608,000	640,000
Net Income	4.2	493,000	544,000	571,000	603,000	640,000	672,000	708,000

*Costs as a per cent of sales (see Figure D1).

FIGURE D3   ALPHA 1980 Pro Forma Income Statements

	Low* Sales Value	Sales Forecast	High* Sales Value
Net Sales	$15,537,000	$16,854,000	$18,172,000
Cost of Sales	10,254,000	11,124,000	11,993,000
Gross Profit	5,283,000	5,730,000	6,179,000
Selling, G & A	3,574,000	3,876,000	4,179,000
Operating Profit	1,709,000	1,854,000	2,000,000
Interest Expense	466,000	506,000	545,000
Pretax Income	1,243,000	1,348,000	1,555,000
Income Taxes	590,000	640,000	690,000
Net Income	653,000	708,000	865,000

*Based on 90% confidence interval for the sales forecast (see Figure C2).

HARMON FOODS

INTRODUCTION

"Mac" MacIntyre, general sales manager of the Breakfast Foods Division of Harmon Foods, Inc. was having difficulty in forecasting the sales of Treat. Treat is a ready-to-eat breakfast cereal with an important share of the market. Treat was also the major component of production in those company plants where it was manufactured. Mac was responsible for sales forecasts from which production schedules were prepared. In recent months, Treat sales had varied from 50% to 200% of his forecast. Mac's difficulty with sales forecasts can be easily understood from Exhibit 1, historical monthly sales of Treat.

The manager of manufacturing expressed his unhappiness over these forecasts and the associated production schedule changes in very positive terms. The accounting system was such that sales were debited on day of shipment so that Exhibit 1 represents both unit sales and unit shipments. The jump in sales from August to September 1969 represents sales nearly tripling in one month. It was this sort of volatility of sales and shipments that was the cause of Mac's errors in forecasts and production schedules.

MANUFACTURING PROBLEMS

Accuracy in production forecasts was essential for the health of the entire business. The individual plant managers received these schedules and certified

EXHIBIT 1   HARMON FOODS, INC., Shipments of Treat
(Standard Cases)

	YEAR			
Month	1966	1967	1968	1969
January	425,075	629,404	655,748	455,136
February	315, 305	263,467	270,483	247,570
March	432,101	468,612	429,480	732,005
April	357,191	313,221	260,458	357,107
May	347,874	444,404	528,210	453,156
June	435,529	386,986	379,856	320,103
July	299,403	414,314	472,058	451,779
August	296,505	253,493	254,516	249,482
September	426,701	484,365	551,354	744,583
October	329,722	305,989	335,826	421,186
November	281,783	315,407	320,408	397,367
December	166,391	182,784	276,901	269,096

their ability to meet them. Acceptance of a schedule by a plant manager represented a "promise" to deliver. Crews and machines were assigned, materials ordered and storage space allocated.

Changes in schedule were expensive. The lead time on raw material orders was several weeks. Shortages were expensive in lost production time and disappointed customers. After schedule reductions, this same raw material could not be used as fast as it arrived and storage space was short. The resultant tie-up of rail cars, barges and trucks was extremely expensive in demurrage charges.

Demurrage charges are arbitrary assessments by a shipper against a consignee for delays in unloading (or the initiation of unloading) or the transport carrier. Typically, there is an allowance of one free hour for trucks before unloading begins, three free days for rail cars, and one free day for barges or vessels, including unloading time. Typical charges might be as low as $5 an hour for a truck and $8 per day for a rail car to $1000 a day for a large motor vessel.

More important than the storage problem is the problem of efficient man-power utilization. Production schedules were always tight to ensure optimal costs. Overtime was expensive and interfered with weekend maintenance. The labor force was highly skilled and difficult to increase over the short run. Lay-offs were avoided to preserve the skills of the crew. This job security resulted in a high level of employee morale. It was the policy of the company to maintain this job security.

BUDGETS AND CONTROLS

The error in forecasts was also the subject of complaints by the Breakfast Foods Division controller. Each brand prepared a budget based on forecasted shipments. This budget "promised" a contribution to division overhead and profits. Long-term dividend policy and corporate expansion plans depended upon the accuracy of these forecasts. The fact that high months balanced low months in the long run was not sufficient for at least two reasons.

First, regular quarterly increases in earnings over prior years had resulted in a high P/E capitalization of these earnings. Public confidence in the constant future growth of both earnings and dividends should be maintained if this market valuation were to be maintained. Since the market value of the common stock is a chief interest of the owners, profit planning was an important part of the management control system.

Secondly, quarterly earnings were difficult to predict in the presence of erratic advertising expenditures. It was the policy at Harmon Foods, as at many companies, to fix advertising expenditures at a fixed amount per unit sold. Whenever shipments ran high, the brand managers tended to spend more on network advertising. Most of the advertising dollars for Treat were spent on Saturday morning network shows for children. These shows were relatively expensive, over $20,000 per one-minute commercial. Time on these shows was purchased up to a year or more in advance.

It was the opinion of all of the brand managers in the Breakfast Foods Division that these network programs delivered the best value for each advertising dollar. This opinion was based upon $/M messages delivered, viewer recall scores and measures of audience composition. Brand managers tended to contract for time on these programs to the limit of the budget allowance which was based on sales forecasts. Whenever sales ran ahead of forecast, brand managers would seek contracts for time from other brand managers at Harmon Foods, Inc., who were shipping behind budget. Failing this, they would seek network time through the agencies as available. Failing this, they would seek spot advertising as close to prime program time as possible.

This discretionary "overspending" on advertising by brand managers did not have budgetary approval. Until a new forecast for the fiscal year on a new "budget base" (i.e., sales forecast) was approved at all levels, such overspending was merely borrowing ahead on the current fiscal year. The controller's office debited the budgeted advertising each quarter and carried the excess over, since it was unauthorized for the present quarter. This brought a surge in earnings in the current quarter since the higher sales brought more contribution from which only a fixed amount of advertising dollars was subtracted. Since advertising dollars are like an investment, with a delayed return income stream, this accounting practice had worked best for the company over the years.

In the past fiscal year, Treat, along with several other brands, had overspent in early quarters due to shipments fluctuating on the high side. By the time a new budget base proposal had been processed for authorization, it was clear that the high level of shipments had been temporary. It was necessary to cut advertising expenditures in the final quarter in order to match sales levels, which came out to be not much more than the original budget base forecast over a year before.

Final quarter earnings were badly distorted as a result. Shipments were low and current advertising expenditures were matched. But deferred advertising debits made total advertising for the quarter higher than current shipments allowed and reported profits were proportionately lower. Last fiscal year this had occurred in several brands in the Breakfast Foods Division. Divisional earnings were lower than budgeted by more than one million dollars as a result of these deferred advertising debits. The division manager, as well as his sales manager, advertising manager and controller, had felt very uncomfortable in the meetings and conferences which had been held as a result of this shortage of reported profits. The extra profits reported in earlier quarters had offset other divisional shortages. In the final quarter there had been no one to offset theirs. The fact that they had reported extra profits in earlier quarters was only slight consolation for the disappointment over their fourth-quarter figures.

THE BRAND MANAGER

Don Carswell, the advertising brand manager for Treat, was a key man in the resolution of such difficulties for his brand. The advertising brand manager prepared the budget base upon which network time was purchased. This budget base meant monthly, quarterly and annual forecasts. It was Don's budget base which was used by Mac for sales forecasts and product schedules. All of the advertising brand managers worked closely with Mac since they were dependent on the field sales force to realize their budget forecasts. Mac regularly circulated all of the weekly (district managers) and monthly (unit mangers) letters from the field which mentioned a given brand to the appropriate advertising brand manager.

This use of the advertising brand manager's forecast as the basis for the official sales manager's forecast required mutual confidence and understanding. From the sales manager came information of capabilities and commitments of the sales force and information of both own and competitive activity and pricing at the store level. The Treat advertising brand manager was responsible for knowledge of market trends for Treat and its competitors, as well as for that subdivision of breakfast foods for which Treat was a part. He kept records of all available market research reports on his and similar brands. He also knew of package design and product formulations under development for Treat or being tested by competitors. These and other inputs were the basis for the forecasts made by Mac in consultation with Don Carswell.

Don knew that the problem was his to solve. It was his forecast that was the basis of problems. It was his decision to change level of advertising media expenditures if sales level changed. Furthermore, he held back part of his advertising dollars for price-competitive moves. These tactical moves could produce wide variations in sales and shipments. Don talked to analysts in the Market Research, Systems Analysis, and Operations Research Departments and concluded that a better forecast was possible. In particular, Bob Haas of the Operations Research Department offered to work with him on the project. Don recieved enthusiastic support in his efforts from the sales and advertising managers and from the controller. Although such projects were outside the normal scope of an advertising brand manager's duties, he recognized his opportunity to convert routine failure to a successful solution that would have company-wide application.

Don and Bob discussed at great length the factors that influenced sales. A 12-month moving average of the data in Exhibit 1 was constructed which confirmed the long-term rising trend in sales. This trend confirmed the A. C. Nielsen store audit which reported a small but steady rise in market share for Treat, plus a steady rise for the commodity group to which Treat belonged.

In addition to trend, Don felt that seasonal factors might be important. In November and December, sales slowed down as inventory levels among stores and jobbers were drawn down for year-end inventories. Summer sales were often low due to plant shutdowns and sales vacations. There were fewer selling days in February. Salesmen often started new fiscal years with a burst of energy in order to get in a good quota position for the rest of the year. Don and Bob agreed to add monthly seasonal variables to basic trend.

CONSUMER PACKS

Those advertising dollars used for consumer and store promotions represented about 25% of the advertising budget. These dollars were spent in two ways. The first was price-off to the consumer or equivalent value in coupon or premium. These were called "consumer deals" and are reported for each brand by A. C. Nielsen from their store audit. The usual offer was 5¢ off per package. These offers were usually made twice a year during two different sales canvass periods. (A sales canvass period is the time it takes a salesman to make a complete round of all customers in his assigned area. Harmon Foods scheduled 10 five-week canvass periods each year. The remaining two weeks were split between one at midsummer and one at year end to allow for holidays and vacations.) The offer was sometimes a coupon, an enclosed premium, or a mail-in offer. These offers were always submitted to a test panel before acceptance. Only those offers or premiums which outsold 5¢ off were acceptable. Don decided that since all consumer deals were at least equivalent to 5¢ off, he would group all consumer deals together as "5¢ off or equivalent."

Securing numerical measures of consumer deals in the past was not easy. The "special packs" were produced ahead of the assigned canvass period by the plants and then sold throughout the five weeks. Special cartons and containers, as well as supporting advertising material had to be ordered in advance. This meant that the quantity of each special consumer pack had to be set before the start of the canvass period. The canvass period overlapped into two and sometimes three different months. Bob Haas suggested to Don that they average weekly shipments of consumer deals for each of the five weeks of every canvass period that contained a special pack. They were able to find three years of history covering six special packs. They found that 35% of a pack moved out the first week, 25% the second week, 15% the third and fourth weeks and 10% during the last week of the canvass.

Don and Bob decided to divide every canvass period among the appropriate months in fractional weeks. They then weighted these weeks by the appropriate percentage times the total scheduled cases of pack to develop an allocation of special pack cases to calendar months. They would use these weights to predict the calendar months for expected shipments of future special packs. The actual pattern of consumer pack shipments during the past four years is shown in Exhibit 2.

No attempt would be made to predict the small movement of special pack cases between the two or three promotional canvasses each year. Such movement was usually only a substitute for regular stock or delayed shipment of an order. Of great concern to Don and Bob was the "loading" of both the stores and the consumers by such promotions. Since the consumer ate Treat at a more or less constant rate over time, the wild fluctuations in monthly sales could only represent overbuying by stores, jobbers and consumers to take advantage of the "deals." This inventory would be depleted in subsequent months with a resultant decline in sales. Mac said that his sales force felt this reaction throughout the next canvass period. This implied that any variable for pack in one month should show as a pack reaction variable one month later and probably as a third column variable two months later.

EXHIBIT 2 HARMON FOODS, INC., Consumer Packs
(Standard Shipping Cases)

Month	1965	1966	1967	1968	1969
			YEAR		
January	−15	75,253	548,704	544,807	299,781
February	−47	15.036	52,819	43,704	21,218
March	−7	134,440	2,793	5,740	157
April	−1	119,740	27,749	9,614	12,961
May	15,012	135,590	21,887	1,507	333,529
June	62,337	189,636	1,110	13,620	178,105
July	4,022	9,308	436	101,179	315,564
August	3,130	41,099	1,407	80,309	80,206
September	−422	9,391	376,650	335,768	5,940
October	−8	942	122,906	91,710	36,819
November	5	1,818	15,138	9,856	234,562
December	220	672	5,532	107,172	71,881

DEALER ALLOWANCES

Sales were even more sensitive to dealer allowances for cooperative promotional efforts. These dealer allowances represented $1 to $2 a case discount to those dealers purchasing during the canvass period of the promotion allowance. (This represents about the same cost to the brand as 5¢ off on a case of 24 individual cartons.) These dealer allowances or dealer payments were to reimburse the dealer for participation in joint promotional effort. This could take the form of "giant spectacular aisle-end" displays, newspaper ads, coupons, flyers, etc. In actual practice, when one company offered a dealer allowance, other companies might respond with minimum requirements of proof of expenditure by the store management.

Dealer allowances were the largest single contributor to shipment and sales variation. A giant spectacular aisle-end display near a cash register would average five weeks' business in a single weekend. Dealers receiving merchandise at what amounted to a 10% or higher discount would load their warehouses. (A two-months' extra supply represented 60% return on capital with no risk.) Bob and Don decided that three columns would be needed for dealer allowances as for consumer packs. One for the original boost in sales and two for subsequent reaction.

The data were obtained for past payments to dealers over the last 48 months. These data are reported in Exhibit 3.

EXHIBIT 3   HARMON FOODS, INC., Cooperative Merchandising
Allowances ($)

Month	YEAR				
	1965	*1966*	*1967*	*1968*	*1969*
January	99,194	114,433	0	166,178	8,133
February	38,074	63,599	78,799	134,206	5,867
March	39,410	64,988	176,906	137,890	1,125,864
April	61,516	66,842	49,616	37,520	125,226
May	83,929	39,626	119,720	145,200	0
June	81,578	107,503	114,293	108,770	0
July	65,821	97,129	-177,370	90,286	11,526
August	122,169	56,404	11,345	24,461	23,063
September	8,482	260,576	7,020	7,593	1,217,488
October	56,007	243,523	27,880	37,581	94,139
November	76,001	75,473	66,800	73,261	94,139
December	88,218	19,037	88,576	40,697	138,134

CONCLUSION

Don and Bob felt that the most important factors had been isolated and
measured to the best of their ability. They knew competitive advertising and
price moves were important but unpredictable. They wished to restrict their
model to those variables which could be measured or predicted in advance.

Bob Haas agreed to write the model, construct the data matrix, and write
an explanation of how the solution of the model could be used to evaluate
promotion strategies, as well as to forecast sales and shipments. Don and Bob
would then join in planning a presentation to divisional managers.

ASSIGNMENT

Formulate the model and write the first and last five lines of the data matrix.
Prepare yourself to do the presentation that Don and Bob propose. In
particular, address yourself to the problem of trend extrapolation into the
next year or two for profit planning and television budgeting purposes. Those
who can find the time are encouraged to do the full computer solution.

VALLEY LITHO

In the early summer of 1972, Mr. Larry Griffin, president and owner of Valley Litho, met with Kent Bowen, Head of Valley Litho's yearbook operations, to determine what changes should be made in preparing high school yearbook cost estimates for the coming academic year. Generally, the company received a set of specifications from each of several high schools during June-August. Kent Bowen, or his assistant, would then develop a cost estimate for each school's contract based on the detailed components of that book (number of pages, material cost, binding cost, cover complexity, extra colors, number of books, etc.). In all, about 24 separate costs were estimated and then these were added together to get a total cost estimate. Once this was done (which took several hours for each contract), Kent would sit down with Larry Griffin and together they would decide what bid price to submit for that school's contract. (Although most schools requested bids from three or more printers, less than half of such contracts were awarded strictly on the basis of price.)

It had been three years (1969) since Valley Litho had revised and updated its yearbook cost estimating procedures. During that period the company's costs had changed significantly, and in fact the previous summer (1971) Kent had started with the basic 1969 cost rates, but then found it necessary to add an overall adjustment factor of 14.6% in order to reflect the higher 1971 cost levels. In light of this and also because of the company's recent labor problems, both Kent and Larry felt that either an entirely new set of cost rates should be developed *or* that a new procedure for estimating should be developed. In line with the latter approach, Kent had suggested that regression analysis might provide the basis for an entirely new estimating procedure that would be more accurate, less costly and yet simpler to use than just an updated version of the existing procedure. With Larry's approval, he decided to pursue this possibility.

As a first step in developing a regression analysis approach, Kent collected data on the 20 high school contracts that Valley Litho had just completed. The four variables he examined initially were:

- Actual cost
- Number of books
- Number of pages/book
- Number of color pages/book

The actual data he ended up with is shown in Exhibit 1. Using a local time sharing service, Kent obtained the results of three regression runs one of which is shown in Exhibit 2. Kent's problem now was to determine which, if any, of these models he should use.

After studying the results of Exhibit 2 and considering several important questions, Kent presented his analysis and recommendations to Mr. Griffin. Much to Kent's surprise, Mr. Griffin was not enthusiastic about immediately adopting the regression analysis model. He raised a number of questions about how the regression results compared with the results of the cost estimating procedure actually used for the 71/72 contracts and how these regression results might have compared with regression performed on some prior year's data. As a first step in answering Mr. Griffin's questions Kent obtained the data on Valley Litho's 69/70 high school contracts shown in Exhibit 3. He then obtained the regression results shown in Exhibit 4. These results made Kent even more uncertain as to what model could and should be used in estimating the costs associated with a specific yearbook contract.

EXHIBIT 1   Valley Litho — 71/72 High School Yearbook Data

School	Actual Cost	Number of Books Delivered	Number of Pages/Book	Number of Color Pages/Book
1	$ 9,539	1,456	212	4
2	6,039	1,110	186	1
3	8,807	1,785	216	4
4	9,625	1,360	240	8
5	10,974	1,710	224	12
6	9,763	1,560	210	8
7	11,293	2,460	208	4
8	8,498	1,410	210	4
9	3,076	220	112	8
10	6,161	1,235	152	4
11	9,946	1,760	200	4
12	7,230	975	164	4
13	11,011	1,890	224	4
14	9,492	1,660	200	4
15	6,789	1,170	168	4
16	9,374	1,710	200	4
17	8,932	1,310	192	8
18	10,195	1,710	224	4
19	8,176	1,360	192	4
20	7,656	1,020	192	4

EXHIBIT 2   Initial Regression Results — 71/72 High School Yearbook Data

```
REGRESSION NUMBER 1                      :DEPENDENT VARIABLE 13 1

VARIABLE MEAN          STANDARD DEVIATION
2        1443.55          451.978
         136.3            29.446
4        5.05             2.45967
1        9628.8           2014.85

CORRELATION COEFFICIENTS
VAR     3     3     4      1
 2   1.000 0.743 -.103 0.895
 3   0.743 1.000 0.093 0.985
 4   -.102 0.093 1.000 0.148
 1   0.895 0.885 0.149 1.000

VARIABLE B            STD. ERROR  T-TEST   ACT. VARIATION
                                           EXPLAINED

CONSTANT -1643.59     927.952     -1.77121

 2       3.65211      .436252     6.07939   .532331
 3       29.2823      6.68978     4.36874   .378073
 4       139.728      53.9004     2.59234   2.54598E-D2

R-SQUARED=0.936  R=0.967
F-TEST= 77.82,  STD. OF REGR.=556.05
DEGREES OF FREEDOM FOR NUMER=3   FOR DENUMER.=16
DO YOU WANT THE RESIDUALS TO BE PRINTED?Y
ACTUAL    PREDICTED   RESIDUALS    % ERROR
 3539     3974.05      564.946     5.93249
 6033     6877.3      -833.197    -13.8747
 3307     9963.53    -1156.53     -13.1919
 9885    10096.8      -471.867     -4.90253
10974    11116.3      -142.303     -1.29673
 3763     3750.32      12.6794      .129872
11293    11519.8      -226.947     -3.00974
 3498     8793.59     -295.591     -3.47836
 3076     3331.72     -255.718     -9.31331
 6161     6634.       -472.995     -7.67735
 3946     9439.51      516.491     5.19295
 7230     6295.23      934.766     12.928
11011    10475.9       535.14      4.96005
 3433     9164.3       327.703     3.45241
 6789     6929.33     -140.326     -2.06697
 9374     9296.9        77.0975     .822451
 3332     8561.11      370.883     4.15237
10195     3998.48      196.521     1.92762
 3176     8134.8        41.1962     .503867
 7656     7233.08      422.915     5.53397

DURBIN-WATSON STAT.= .94846

MEAN SQUARED ERROR (MSE) = 247350.
MEAN ABSOLUTE PERCENTAGE ERROR (MAPE) = 5.10863
MEAN PERCENTAGE ERROR (MPE) OR BIAS = -.566923
```

EXHIBIT 3   Valley Litho — 69/70 High School Yearbook Data

School	Actual Cost	Number of Books Delivered	Number of Pages/Book	Number of Color Pages/Book
1	$ 7,397	910	192	8
2	8,123	1,112	176	8
3	6,960	1,320	180	8
4	10,352	1,845	204	10
5	6,986	1,320	168	8
6	12,085	2,168	220	8
7	12,179	2,083	240	22
8	9,980	1,272	185	8
9	6,255	900	418	8
10	2,937	512	116	0
11	9,795	1,941	184	8
12	6,528	1,162	144	8
13	2,030	297	112	0
14	9,256	1,560	208	8
15	13,810	2,831	208	8
16	8,900	1,960	192	8
17	1,427	180	84	0
18	7,462	1,220	178	8
19	10,970	1,887	228	12

EXHIBIT 4   Regression Results — 69/70 High School Yearbook Data

```
REGRESSION NUMBER 2                :DEPENDENT VARIABLE IS 1

VARIABLE B              STD. ERROR  T-TEST   PCT. VARIATION
                                             EXPLAINED

CONSTANT  1318.73       533.293     2.49647

2         3.9817        .444693     8.34133   .752126
4         163.843       63.6758     8.57312   .173547
```

```
R-SQUARED=0.926   P=0.962
F-TEST= 39.63     STD. OF REGP. = 973.34
DEGREES OF FREEDOM FOR NUMER.=8   FOR DENUMER.= 16
```

```
DO YOU WANT THE RESIDUALS TO BE PRINTED?Y
ACTUAL    PREDICTED    RESIDUALS    % ERROR
7397      6207.35      1199.66      16.0831
3123      7001.54      1121.46      13.806
6360      7819.33      -959.331     -12.3467
10352     10211.2      140.838      1.36049
6986      7819.33      -333.331     -11.9385
12035     11153.4      931.592      7.70366
13179     13113.1      -934.051     -7.66936
9930      7630.61      2349.39      83.541
6255      6168.02      36.9917      1.33059
2937      3347.48      -410.434     -13.9767
3795      10360.3      -465.914     -4.75665
6523      7198.12      -670.123     -10.8654
2050      2486.44      -456.442     -52.4848
9256      9762.84      433.063      5.32695
13810     13760.1      49.8772      .361167
3900      10335.4      -1435.62     -18.1305
1437      3026.43      -599.434     -43.0066
7462      7426.16      35.8391      .430288
10970     10704.      366.016       8.42494
```

```
DURBIN-WATSON STAT.= 2.18606

MEAN SQUARED ERROR (MSE) = 799840.
MEAN ABSOLUTE PERCENTAGE ERROR (MAPE) = 11.8657
MEAN PERCENTAGE ERROR (MPE) OR BIAS = -3.63539
```

FORECASTING IBM STOCK PRICES-PARIS EXCHANGE

Mr. Puzo, an INSEAD participant taking a statistics course, wanted to determine whether he could apply his statistical knowledge to make a profit on the Paris stock exchange. For this purpose he collected the data in Exhibit 1, the closing stock prices of IBM common stock traded on the New York Stock Exchange for the period April 30 to October 28, 1976. (These prices have been converted to French francs using the exchange rates prevailing in the New York money market each day.) He also collected the data in Exhibit 2, corresponding stock prices for IBM common as traded on the Paris Stock Exchange.

Mr. Puzo sought to determine whether the New York prices affected the Paris prices of IBM stock. If this proved to be the case, he felt he could profit from such a relationship by using it to forecast IBM stock prices on the Paris exchange. From some preliminary analysis, he found that transaction costs and taxes ran about 1.3% of the total investment. Since the IBM price was about 1300 FF at the time, he realized that he had to make a profit of at least 33 FF per share to cover broker's fees and taxes. Mr. Puzo decided to use only 124 of the available data points and to test his results on the remaining 6 observations.

A. REGRESSION APPROACH

As a first step Mr. Puzo decided to utilize regression analysis. He lagged the IBM-NY prices several periods but soon realized that little information was being obtained by introducing additional time lags. [Exhibit 3 shows the correlation matrix for IBM-Paris (dependent variable) and IBM-NY lagged for up to four periods (independent variables).]

Mr. Puzo's first reaction was that the high intercorrelations between pairs of independent variables was causing multicollinearity problems.

The authors would like to thank Mr. Puzo, a 1976-77 INSEAD participant, for providing the data in this case and for his efforts at forecasting the prices of IBM stock traded on the Paris exchange.

EXHIBIT 1 IBM-New York Stock Prices (closing) for Period of 31 April
to 28 October 1976
    (converted to French Francs at prevailing daily exchange rates)

PERIOD	OBSERVATION	PERIOD	OBSERVATION	PERIOD	OBSERVATION
1	1160.247	44	1301.542	87	1351.502
2	1178.102	45	1304.462	88	1365.187
3	1161.314	46	1307.179	89	1361.009
4	1166.310	47	1310.655	90	1365.956
5	1170.085	48	1320.639	91	1372.100
6	1206.924	49	1314.149	92	1379.000
7	1210.950	50	1325.852	93	1366.164
8	1204.045	51	1334.408	94	1368.594
9	1197.337	52	1327.047	95	1377.096
10	1184.629	53	1334.015	96	1368.344
11	1194.089	54	1325.102	97	1369.150
12	1197.512	55	1341.844	98	1360.990
13	1200.277	56	1353.285	99	1388.050
14	1215.610	57	1332.120	100	1390.819
15	1207.425	58	1327.343	101	1406.405
16	1181.317	59	1331.075	102	1409.975
17	1185.787	60	1335.014	103	1385.139
18	1183.500	61	1352.252	104	1397.799
19	1156.353	62	1339.202	105	1393.212
20	1209.280	63	1336.676	106	1407.584
21	1213.936	64	1342.068	107	1393.987
22	1215.810	65	1340.291	108	1384.135
23	1219.262	66	1337.968	109	1388.375
24	1210.800	67	1336.792	110	1389.077
25	1158.588	68	1350.403	111	1385.720
26	1155.730	69	1359.477	112	1387.086
27	1201.252	70	1370.745	113	1385.946
28	1196.379	71	1363.922	114	1391.906
29	1209.281	72	1385.885	115	1379.077
30	1211.072	73	1381.882	116	1382.077
31	1215.931	74	1387.500	117	1356.372
32	1224.940	75	1400.280	118	1366.233
33	1223.028	76	1394.081	119	1321.707
34	1273.177	77	1392.125	120	1325.405
35	1260.973	78	1393.657	121	1322.079
36	1284.537	79	1370.951	122	1314.338
37	1275.219	80	1362.611	123	1308.278
38	1250.164	81	1362.782	124	1280.333
39	1291.904	82	1339.145	125	1277.306
40	1303.600	83	1354.220	126	1293.607
41	1298.829	84	1336.432	127	1317.829
42	1308.005	85	1334.993	128	1330.266
43	1310.748	86	1336.067	129	1332.300
				130	1358.886

EXHIBIT 2   IBM-Paris Stock Prices (closing) for Period of 31 April to 28 October 1976

PERIOD	OBSERVATION	PERIOD	OBSERVATION	PERIOD	OBSERVATION
1	1186.000	44	1305.000	87	1340.000
2	1187.000	45	1317.000	88	1340.000
3	1169.000	46	1305.000	89	1352.000
4	1190.000	47	1309.000	90	1377.000
5	1172.000	48	1317.000	91	1378.000
6	1181.000	49	1312.000	92	1377.000
7	1185.000	50	1320.000	93	1377.000
8	1223.000	51	1326.000	94	1383.000
9	1212.000	52	1342.000	95	1371.000
10	1217.000	53	1350.000	96	1361.000
11	1211.000	54	1348.000	97	1370.000
12	1190.000	55	1345.000	98	1375.000
13	1200.000	56	1353.000	99	1355.000
14	1213.000	57	1358.000	100	1355.000
15	1200.000	58	1336.000	101	1386.000
16	1227.000	59	1322.000	102	1396.000
17	1208.000	60	1329.000	103	1392.000
18	1195.000	61	1356.000	104	1408.000
19	1187.000	62	1353.000	105	1401.000
20	1196.000	63	1360.000	106	1383.000
21	1205.000	64	1343.000	107	1391.000
22	1229.000	65	1351.000	108	1413.000
23	1219.000	66	1352.000	109	1400.000
24	1218.000	67	1344.000	110	1383.000
25	1220.000	68	1350.000	111	1412.000
26	1220.000	69	1350.000	112	1397.000
27	1212.000	70	1361.000	113	1392.000
28	1214.000	71	1372.000	114	1368.000
29	1200.000	72	1378.000	115	1370.000
30	1200.000	73	1380.000	116	1398.000
31	1213.000	74	1391.000	117	1370.000
32	1221.000	75	1399.000	118	1373.000
33	1233.000	76	1402.000	119	1354.000
34	1223.000	77	1404.000	120	1358.000
35	1241.000	78	1406.000	121	1308.000
36	1263.000	79	1407.000	122	1317.000
37	1260.000	80	1403.000	123	1323.000
38	1275.000	81	1378.000	124	1318.000
39	1278.000	82	1360.000	125	1302.000
40	1293.000	83	1379.000	126	1270.000
41	1301.000	84	1362.000	127	1262.000
42	1304.000	85	1350.000	128	1300.000
43	1299.000	86	1334.000	129	1315.000
				130	1337.000

Question: What are your feelings about multicollinearity in this case?

Exhibit 4 shows a regression run with five independent variables. The dependent variable in this, as in all subsequent runs, is the IBM-Paris price. The independent variables are current and lagged values of IBM-NY prices. These change with each run and are identified by number. (See Exhibit 3 for identification.)

The only t-tests that possibly could be significant in Exhibit 4 are those of variables 3 and 4. A new regression model with these two variables alone was therefore run. (See Exhibit 5.)

EXHIBIT 3 Correlation Coefficients of IBM-Paris and IBM-NY Lagged up to Four Periods

Variable	2	3	4	5	6	1	
2	1.000	0.984	0.975	0.960	0.950	0.973	IBM-NY
3	0.984	1.000	0.986	0.977	0.963	0.983	$IBM\text{-}NY_{t-1}$ ⎞ Independent
4	0.975	0.986	1.000	0.986	0.978	0.991	$IBM\text{-}NY_{t-2}$ ⎟ Variables
5	0.960	0.977	0.986	1.000	0.986	0.979	$IBM\text{-}NY_{t-3}$ ⎟
6	0.950	0.963	0.968	0.986	1.000	0.968	$IBM\text{-}NY_{t-4}$ ⎠
1	0.973	0.983	0.991	0.979	0.968	1.000	IBM-Paris — Dependent Variable

EXHIBIT 4 Regression Run on All Variables

Variable	B	Std. Error	T-Test	Pct. Variation Explained
Constant	30.2216	15.718	1.02270	
2	4.34367E−02	6.99842E−02	.620665	4.15841E−02
3	.163372	9.25408E−02	1.76712	.160767
4	.766717	.093954	8.16055	.774932
5	5.69823E−02	9.8855E−02	.626969	5.76406E−02
6	−4.97418E−02	6.96846E−02	−.713813	−5.06258E−02

R-Squared = 0.984   R = 0.992
F-Tests = 1432.45   Std. of Regr. = 9.17
Degrees of Freedom for Numerator = 5 For Denominator = 114

Durbin-Watson Stat. = 1.32777

EXHIBIT 5   Regression Run of IBM-NY$_{t-1}$, IBM—NY$_{t-2}$

**REGRESSION NUMBER 2   :DEPENDENT VARIABLE IS 1

Variable	B	Std. Error	T-Test	Pct. Variation Explained
Constant	32.268	15.2693	2.11325	
3	.209924	6.85575E−02	3.06202	.206578
4	.769356	6.73101E−02	11.43	.777599

R-Squared = 0.984   R = 0.992
F-Test = 3647.53   Std. of Regr. = 9.09
Degrees of Freedom for Numerator = 2 For Denominator = 117

Durbin-Watson Stat. = 1.33226

The model in Exhibit 5 provides a large F-test value, an unusually large R^2 value, and t-tests that are all significant. However, multicollinearity may be a problem since the intercorrelation between dependent variables 3 and 4 is .986 (r_{34} = .986 and R = .992 indicates that the large intercorrelation may result in a multicollinear relationship). Other than multicollinearity, however, the relationship estimated in Exhibit 5 has no apparent faults. The Durbin-Watson statistic is rather low, but it is known that the D-W statistic does not apply to equations with lagged variables.

Mr. Puzo was unclear on what should be done next. He decided to make another regression run with only one independent variable, NY$_{t-2}$ (this variable explains the highest percentage of variation as shown in Exhibit 5), and obtained the results shown in Exhibit 6.

Exhibit 6 indicates a correct statistical relationship except that the assumption of uncorrelated residuals is violated. The Durbin-Watson statistic, valid in this case, is only 1.417. A value greater than 1.72 would be needed to show the absence of autocorrelation.

In an effort to correct the violation of the assumption of random residuals, Mr. Puzo took the first difference of all variables. The simple correlation matrix shown in Exhibit 7 was the result.

Question: How do you explain such large differences in the correlations of Exhibits 3 and 7? How do you think this will affect the regression results? (Notice the correlations of variables 1 and 3, 1 and 4, and 1 and 5 for the two exhibits.) Exhibit 8 shows the regression results for the same two variables as shown in Exhibit 5, but is based on the first differenced data.

Question: A comparison of Exhibits 5 and 8 indicates that the regression coefficients are very similar, except for the constant, which is expected to be different since the data of Exhibit 8 have been differenced. However, the R^2 values of the two exhibits are very dissimilar. To what can such large

EXHIBIT 6   Regression Run of IBM-Paris and IBM-NY$_{t-2}$

REGRESSION NUMBER 3   :DEPENDENT VARIABLE IS 1

Variable	B	Std. Error	T-Test	Pct. Variation Explained
Constant	41.4319	15.4902	2.67471	
4	.972492	.011787	82.5051	.982912

R-Squared = 0.983   R = 0.991
F-Test = 6807.07   Std. of Regr. = 9.40
Degrees of Freedom for Numerator = 1 For Denominator = 118

Durbin-Watson Stat. = 1.41734

EXHIBIT 7   Correlation Coefficients of IBM-Paris and IBM-NY Lagged Up to Four Periods (First Differenced Data)

CORRELATION COEFFICIENTS

Variable	2	3	4	5	6	1
2	1.000	−.181	0.228	−0.82	−.014	0.142
3	−.181	1.000	−.196	0.215	−.188	−.038
4	0.228	−.196	1.000	−.202	0.216	0.637
5	−.082	0.215	−.202	1.000	−.217	−.047
6	0.014	−.088	0.216	−.217	1.000	0.066
1	0.142	0.038	0.637	.047	0.066	1.000

EXHIBIT 8   Regression Run of IBM-Paris and IBM-NY$_{t-1}$, IBM-NY$_{t-2}$, for First Differenced Data

REGRESSION NUMBER 2   :DEPENDENT VARIABLE IS 1

Variable	B	Std. Error	T-Test	Pct. Variation Explained
Constant	4.67517E−02	.976941	4.78552E−02	
3	.187683	7.89749E−02	2.37548	6.43793E−03
4	.74362	7.90587E−02	9.40591	.427065

R-Squared = 0.434   R = 0.658
F-Test = 44.38   Std. of Regr. = 10.53
Degrees of Freedom for Numerator = 2 For Denominator = 116

Durbin-Watson Stat. = 2.64199

differences be attributed? Why is the Durbin-Watson statistic of Exhibit 8 different from that of Exhibit 5?

Finally, Mr. Puzo made a regression run of Exhibit 9 with only one variable, IBM-NY$_{t-2}$. The D-W value is still outside the limits used to accept the hypothesis that the residuals are random. He recognized, however, that the assumption of random residuals is not as necessary in simple regression (as in Exhibit 9) because there is only one independent variable and thus no covariance is involved.

In an effort to eliminate possible patterns from the residuals another regression run was made. The same data were used except that 0.4 of the first difference was taken instead of the whole first difference as was the case in Exhibits 7, 8 and 9. Exhibit 10 shows the matrix of simple correlation coefficients for these data.

Question: Compare the correlation coefficients of Exhibits 3, 7 and 10. How can you account for the differences in simple correlations?

Exhibit 11 shows the regression coefficients of variables 3 and 4, IBM-NY$_{t-1}$ and IBM-NY$_{t-2}$, when the dependent variable is IBM-Paris$_t$. The F-test and t-tests (except for the constant) are statistically significant. R^2 is .96, suggesting an extremely good fit and the value of the D-W statistic is 2.11 which is within the allowable limits denoting lack of autocorrelated residuals.

Question: Could you use the regression equation of Exhibit 11 to forecast IBM-Paris prices for periods 125 to 130? If you were to do so, how confident would you be of the predictions obtained?

Finally, Exhibit 12 shows the regression results with one independent variable, IBM-NY$_{t-2}$.

Question: Are the models of Exhibits 11 and/or 12 statistically correct? Are there any assumptions of regression that are violated?

Question: Compare the regression coefficients of Exhibits 5, 8 and 11 as well as those of 6, 9 and 12. What can you say about their magnitude? Do the same for the t-tests, F-tests, R^2 and D-W statistics. How and why do they differ?

B. MULTIVARIATE TIME SERIES APPROACH

Mr. Puzo recently had been introduced to several time series forecasting concepts. One of them, cross autocorrelations, seemed to be appropriate for what he wanted to do so he ran his data through the appropriate computer program obtaining the cross autos for time lags of 1 through ±15 shown in Exhibit 13.

EXHIBIT 9   Regression Run of IBM-Paris and IBM-NY$_{t-2}$, for First Differenced Data

**REGRESSION NUMBER 3   *:DEPENDENT VARIABLE IS 1*

Variable	B	Std. Error	T-Test	Pct. Variation Explained
Constant	.317999	.989334	.321427	
4	.706805	7.90509E−02	8.94113	.405922

R-Squared = 0.406   R = 0.637
F-Test = 79.94   Std. of Regr. = 10.74
Degrees of Freedom for Numerator = 1 For Denominator = 117

Durbin-Watson Stat. = 2.59003

EXHIBIT 10   Correlation Coefficients of IBM-Paris and IBM-NY Lagged up to Four Periods (0.4 of First Difference)

CORRELATION COEFFICIENTS

Variable	2	3	4	5	6	1
2	1.000	0.941	0.947	0.921	0.913	0.938
3	0.941	1.000	0.944	0.951	0.929	0.946
4	0.947	0.944	1.000	0.946	0.952	0.977
5	0.921	0.951	0.946	1.000	0.947	0.940
6	0.913	0.926	0.952	0.947	1.000	0.933
1	0.938	0.946	0.977	0.940	0.933	1.000

EXHIBIT 11   Regression Run of IBM-Paris and IBM-NY$_{t-1}$, IBM-NY$_{t-2}$, for 0.4 of First Differenced Data

**REGRESSION NUMBER 2   *:DEPENDENT VARIABLE IS 1*

Variable	B	Std. Error	T-Test	Pct. Variation Explained
Constant	19.2391	14.8149	1.29863	
3	.214865	5.66656E−02	3.79181	.20257
4	.764548	5.56966E−02	13.727	.757492

R-Squared = 0.960   R = 0.980
F-Test = 1394.04   Std. of Regr. = 8.64
Degrees of Freedom for Numerator = 2 For Denominator = 116

Durbin-Watson Stat. = 2.10825

EXHIBIT 12   Regression Run of IBM-Paris and IBM-NY$_{t-2}$ for 0.4 of First Differenced Data

**REGRESSION NUMBER 3   :DEPENDENT VARIABLE IS 1*

Variable	B	Std. Error	T-Test	Pct. Variation Explained
Constant	31.5328	15.2592	2.06648	
4	.964009	1.93214E—02	49.8932	.955111

R-Squared = 0.955   R = 0.977
F-Test = 2489.33   Std. of Fegr. = 9.12
Degrees of Freedom for Numerator = 1 For Denominator = 117

Durbin-Watson Stat. = 2.20771

EXHIBIT 13   Cross Autocorrelations Between IBM-Paris and IBM New York

(original series)

```
TIME                                                    CROSS AUTO-
LAG     -1        -.5        0        .5        1       CORRELATION
        I----I----I----I----I----I----I----I----I
-15                        .  I  .              *          0.599
-14                        .  I  .              *          0.625
-13                        .  I  .               *         0.647
-12                        .  I  .               *         0.674
-11                        .  I  .                *        0.696
-10                        .  I  .                *        0.720
 -9                        .  I  .                 *       0.745
 -8                        .  I  .                 *       0.766
 -7                        .  I  .                  *      0.791
 -6                        .  I  .                  *      0.810
 -5                        .  I  .                   *     0.834
 -4                        .  I  .                   *     0.863
 -3                        .  I  .                    *    0.893
 -2                        .  I  .                    *    0.921
 -1                        .  I  .                     *   0.946
  0     I----I----I----I.---I----.----I----I----I----*    0.975
  1                        .  I  .                     *   0.971
  2                        .  I  .                     *   0.967
  3                        .  I  .                     *   0.939
  4                        .  I  .                    *    0.916
  5                        .  I  .                    *    0.888
  6                        .  I  .                   *     0.862
  7                        .  I  .                   *     0.834
  8                        .  I  .                  *      0.812
  9                        .  I  .                  *      0.790
 10                        .  I  .                 *       0.770
 11                        .  I  .                 *       0.745
 12                        .  I  .                *        0.717
 13                        .  I  .                *        0.692
 14                        .  I  .               *         0.674
 15                        .  I  .               *         0.647
        I----I----I----I----I----I----I----I----I
        -1        -.5        0        .5        1
```

Mr. Puzo was most surprised by the fact that not only did IBM-New York prices for preceding days influence IBM-Paris prices, but at the same time IBM-Paris prices exerted similar influences on the IBM-New York prices. That did not seem to make much sense. He noted that the cross autocorrelation coefficients of 0, 1, 2, 3 and 4 time lags in Exhibit 13 were about the same as the simple correlations between the dependent variable and each of the independent variables as shown in Exhibit 3. It was on these correlations that the regression runs of Exhibits 4, 5 and 6 were based. However, if Exhibit 13 was correct, then causality was in both directions and the regression results of Exhibits 5 and 6 had little meaning.

Using a Multivariate Autoregressive Moving Average (MARMA) methodology, the original data, X_t (IBM-New York) and Y_t (IBM-Paris), were reduced to white noise (prewhitened). This was accomplished quite easily by taking the first difference of the data which gave a series whose residuals were random (see Exhibits 14 and 15).

From Exhibits 14 and 15, Mr. Puzo concluded that the stock prices of IBM-New York and IBM-Paris are random around their first differences. That is, the changes from one day to the next cannot be predicted.

Exhibit 16 shows the cross autos of the first differenced (prewhitened) data.

Question: How do you explain the differences in cross autos of Exhibits 13 and 16? Why do such large differences exist?

Question: How do the positive lag cross autos compare to the simple correlation coefficients between the dependent and independent variables shown in Exhibit 17?

Question: After reviewing Chapter 25, * examine Exhibit 16 and specify the type of relationship between Y_t and X_{t-1}.

Assuming a tentative MARMA model of the form:

$$Y_{et} = W_o X_{e_{t-2}} + e_t, \tag{1}$$

Mr. Puzo found an initial estimate for W_o. Its value was .72 (not much different than the value obtained in Exhibit 9). Exhibit 17 shows the autocorrelations and partial autocorrelations of the residuals of model (1).

Question: What univariate ARMA model would you identify from the autos and partials of Exhibit 17?

Interactive Forecasting: Univariate and Multivariate Methods, 2nd Edition by Spyros Makridakis and Steven C. Wheelwright, San Francisco, Holden-Day, 1978.

EXHIBIT 14 Autocorrelation Coefficients of IBM-New York First Differenced Data

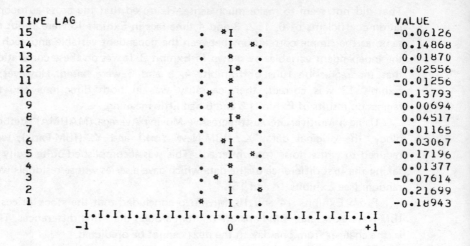

TIME LAG	VALUE
15	-0.06126
14	0.14868
13	0.01870
12	0.02556
11	-0.01256
10	-0.13793
9	0.00694
8	0.04517
7	0.01165
6	-0.03067
5	0.17196
4	0.01377
3	-0.07614
2	0.21699
1	-0.18943

EXHIBIT 15 Autocorrelation Coefficients of IBM-Paris First Differenced Data

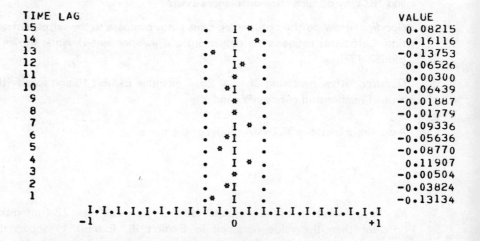

TIME LAG	VALUE
15	0.08215
14	0.16116
13	-0.13753
12	0.06526
11	0.00300
10	-0.06439
9	-0.01887
8	-0.01779
7	0.09336
6	-0.05636
5	-0.08770
4	0.11907
3	-0.00504
2	-0.03824
1	-0.13134

EXHIBIT 16   Cross Autocorrelations Between IBM-Paris and IBM-New York
(Prewhitened Series

```
TIME                                                        CROSS AUTO-
 LAG    -1        -.5        0        .5        1           CORRELATION
        I----I----I----I----I----I----I----I----I
 -15                        .    *    .                        -0.015
 -14                        .    *    .                         0.008
 -13                        .    *    .                        -0.012
 -12                        .    I  *.                          0.147
 -11                        .   *I    .                        -0.032
 -10                        .    *    .                         0.017
  -9                        .    *    .                         0.019
  -8                        .  * I    .                        -0.121
  -7                        .    I *  .                          0.109
  -6                        .  * I    .                        -0.118
  -5                        .    I *  .                          0.085
  -4                        .    *    .                         0.017
  -3                        .    I *  .                          0.083
  -2                        .    *    .                        -0.021
  -1                        .    *    .                        -0.012
   0    I----I----I·---I.---I--*.I----I----I----I              0.153
   1                        .    *    .                         0.001
   2                        .    I    .        *                0.629
   3                        .   *I    .                        -0.059
   4                        .    I*   .                          0.064
   5                        .    *    .                        -0.019
   6                        .    I*   .                          0.061
   7                        .    I*   .                          0.065
   8                      . "  I    .                          -0.117
   9                        .    *    .                          0.016
  10                      .    I*  .                            0.068
  11                        .   *I    .                        -0.055
  12                        .   *I    .                        -0.057
  13                      . "  I    .                          -0.096
  14                        .    I *. .                          0.163
  15                        .   *I    .                        -0.054
        I----I----I----I----I----I----I----I----I
        -1        -.5        0        .5        1
```

EXHIBIT 17   Autocorrelations and Partial Autocorrelations of Residuals of MARMA Mode (1)

(They suggest an MA(1) model)

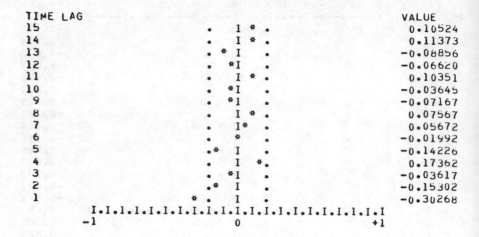

```
TIME LAG                                            VALUE
15                        .   I  *  .               0.10524
14                        .   I  *  .               0.11373
13                        .  *  I   .              -0.08856
12                        .   *I    .              -0.06620
11                        .   I  *  .               0.10351
10                        .   *I    .              -0.03645
 9                        .   *I    .              -0.07167
 8                        .   I  *  .               0.07567
 7                        .   I*    .               0.05672
 6                        .    *    .              -0.01992
 5                        .*   I    .              -0.14226
 4                        .   I   *. .              0.17362
 3                        .   *I    .              -0.03617
 2                        .*   I    .              -0.15302
 1                     *  .   I    .               -0.30268
        I.I.I.I.I.I.I.I.I.I.I.I.I.I.I.I.I.I.I.I.I
       -1                 0                 +1
```

PARTIAL AUTOCORRELATIONS

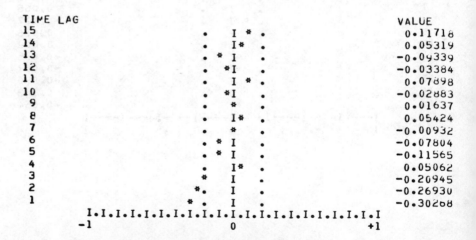

```
TIME LAG                                            VALUE
15                        .   I  *  .               0.11718
14                        .   I*     .              0.05319
13                        .  *  I    .             -0.09339
12                        .   *I     .             -0.03384
11                        .   I  *   .              0.07898
10                        .   *I     .             -0.02883
 9                        .    *     .              0.01637
 8                        .   I*     .              0.05424
 7                        .    *     .             -0.00932
 6                        .  *  I    .             -0.07804
 5                        .  *  I    .             -0.11565
 4                        .   I*     .              0.05062
 3                        .*   I     .             -0.20945
 2                       *.   I      .             -0.26930
 1                      *  .   I     .             -0.30268
        I.I.I.I.I.I.I.I.I.I.I.I.I.I.I.I.I.I.I.I.I
       -1                 0                 +1
```

Assuming a MA (1) univariate ARMA model, the tentative MARMA and noise models are:

$$Y_{et} = W_o X_{e_{t-2}} - \theta_1 e_t + e_t \tag{2}$$

Estimating final parameters for (2) yields the values

$$(Y_t - Y_{t-1}) = .952(X_t - X_{t-1}) - .82e_t + e'_t \tag{3}$$

The residuals of (3), e'_t, are random noise as can be seen in Exhibit 18.

EXHIBIT 18   Autocorrelations of Residuals of MARMA Model
$$Y_{et} = .952X_{e_{t-2}} - .82e_{t-1} = e'_t$$

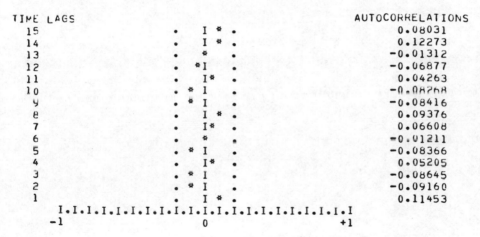

TIME LAGS		AUTOCORRELATIONS
15	. I * .	0.08031
14	. I * .	0.12273
13	. * .	-0.01312
12	. *I .	-0.06877
11	. I* .	0.04263
10	. * I .	-0.08268
9	. * I .	-0.08416
8	. I * .	0.09376
7	. I* .	0.06608
6	. * .	-0.01211
5	. * I .	-0.08366
4	. I* .	0.05205
3	. * I .	-0.08645
2	. * I .	-0.09160
1	. I * .	0.11453

```
      I.I.I.I.I.I.I.I.I.I.I.I.I.I.I.I.I.I.I.I.I
      -1                  0                  +1
```

BOX-PIERCE Q-STATISTIC AND CORRESPONDING VALUES OF CHI-SQUARE

LAGS	1-12	1-15
Q-STATISTIC	8.695	11.297
CHI-SQUARE	18.300	22.400
D.F	10.000	13.000

Questions: a) How does expression (3) vary from the regression runs of Exhibits 6, 9, and 12? (b) Why does the regression coefficient of Exhibit 12 have about the same value as that of W_o in expression (3)? (c) The R^2 of Exhibit 12 is .955 while the corresponding measure for (3) is only .452. Why does such a large difference exist between the two R^2 values? (d) Could you have specified the equation of Exhibit 12 from the simple correlation matrix of Exhibit 10 or the regression run shown in Exhibit 11?

Assignment: (A) Discuss and contrast the two different approaches used to forecast IBM-Paris prices. State the procedure, advantages and drawbacks of the two approaches.

(B) Forecast the six periods (125 through 130), which were not used in developing the model. What types of errors do you observe? Could Mr. Puzo make any money given the 1.3% charge he has to pay for transaction costs? (For expression (3), the error at period 124 was 5.986 FF.)

Preparing the Executive Forecast

It was illustrated in the earlier chapters that time series methods perform as expected in periods of "normal" economic conditions, while their predictions can be outside expected limits in times of rapid changes in the economic environment. This is the main reason one cannot apply the predictions of quantitative methods directly but must first adjust them for the effects of the cycle. Another reason that adjustments are often necessary is the occurrence of special events that are known to be forthcoming but that cannot be predicted by mechanistic methods of forecasting. A strike or a war, for example, will definitely affect production and sales. Changes in governmental policies, new legislation, or ecological considerations can have similar effects, but their influence may be slow to take effect and can, therefore, be detected by time series methods. On the other hand, it is step changes that are hard to anticipate and that, when they occur, require adjustments in the quantitative forecasts. Another consideration when preparing final forecasts is that one may use more than one method and, if the results differ, have to decide which predictions to adopt as the final ones. This last consideration will necessitate deciding what the final forecast should be, even when economic conditions are "normal" and there are no special events anticipated.

When and to what extent one should adjust quantitative predictions depends upon the specific situation (see Figure 28-1). However, as a general rule, the larger the extent of the forthcoming step change, the greater the adjustment should be. If the changes are "normal," the main concern should be to choose the most appropriate of the quantitative methods and, with possible minor modifications, use its predictions as the executive forecasts for the coming periods. The program RUNNER provides comparative information for determining the most appropriate method(s) and allows one to combine forecasts if desired (see Figure 28-1).

RUNNER is a control program through which one can run any of the quantitative methods available. RUNNER can be used after SIBYL, or it can be run by itself. Figure 28-2 shows the beginning of RUNNER. The user has the choice of several options of data — the program data, any available data file, or data generated to one's own specification. This data can then be used with any method available. Figure 28-2 illustrates running the data stored in the file INSEL (see Exercise 2 for a listing of this data) with Winters' Exponential Smoothing Model. Running any individual program under the program RUNNER is exactly the same as running the individual program independently. Figure 28-3 shows the usage of the Classical Decomposition Program through RUNNER, while Figure 28-4 shows the application of Harrison's Harmonic Smoothing through RUNNER.

Once all desired methods have been run, the user inputs EXECUTIVE (final) forecasts used during the last 12 periods, and the program then provides summary statistics on how well each method did during the last 12 periods as well as information about the mean percentage error (MPE) or bias, the mean absolute percentage error (MAPE), and the mean squared error (MSE). These measures of accuracy are for both the last 12 periods and all the data. Finally, the forecasts of each method are summarized (see Figure 28-5).

In addition to the quantitative methods, Figure 28-5 includes four additional "methods." These are Naive Method Number 1 (NAIVE1), Naive Method Number 2 (NAIVE2), Executive Forecasts (EXECUTIVE), and the Optimal Forecast (OPTIMAL). NAIVE1 is a method that assumes the current value as the forecast for next period. The value for period 133 is 116, thus the forecast for period 134 is also 116, and similarly for the next few periods. NAIVE2 is similar to NAIVE1, except that the value is seasonally adjusted. EXECUTIVE forecasts are those used in practice in the last 12 periods, while OPTIMAL contains values as if the seasonality and trend-cycle were known beforehand. That is, the OPTIMAL predictions are those excluding the random component (see chapters 15 and 16 on how time series components are decomposed). Therefore, they indicate the minimum forecasting error one can expect.

The last 12 periods are from February 1974 to January 1975, not the easiest periods to forecast. The errors of the methods are well above the minimum possible. The EXECUTIVE forecasts as adjusted by the person in charge of forecasting do better than the quantitative methods. This means that the adjustment works in the right direction. Had the opposite occurred, it would indicate that the quantitative predictions should have been weighted

FIGURE 28–1 Features Contained in RUNNER

Type of Forecast	*Economic or General Conditions*	*"Normal Changes in in Economic Conditions*	*Rapid Changes in Economic Conditions*	*Special Events (Strikes, Wars, etc.)*
Aggregate Forecasting or Forecasting for Important Products		Accept the Predictions of Quantitative Methods (e.g. Box-Jenkins, GAF, CENSUS II) with minor modifications	Be extremely careful in applying quantitative predictions. Attempt to predict the Cycle by using ORACLE or a decomposition method.	Assess influence of special events. Adjust quantitative predictions accordingly.
Forecasting for Individual Items and Less Important Products		Accept the predictions of quantitative methods (e.g. Exponential or Harmonic Smoothing) with minor or no adjustments.	Adjust the quantitative predictions across the board, preferably on a percentage basis.	Adjust the quantitative predictions across the board.

FIGURE 28-2   The Beginning of a Run of RUNNER

```
EXE RUNNER
 8K


       ***** RUNNER ** INTERACTIVE FORECASTING *****

DO YOU WANT TO USE..
(1) PROGRAM DATA
(2) YOUR OWN DATA, IN A FILE
(3) YOUR OWN DATA, TO BE TYPED IN AT THE TERMINAL
(4) DATA GENERATED BY THE PROGRAM ,WHICH (1,2,3 OR 4)?2
DATA FILENAME?FIIP
HOW MANY OBSERVATIONS DO YOU WANT TO USE?133
HOW MANY SEASONAL PERIODS PER YEAR(0=NONE)?12
SEASONAL INDICES BASED ON MEDIAL AVERAGE(Y OR N)?Y


          *** SUMMARY STATISTICS ***
COMPOSITION OF     AVERAGE PC     APPROXIMATE
PER-TO-PER CHANGE    CHANGE       % EXPLAINED

ORIGINAL DATA       11.93            100.
TREND-CYCLE          .54               4.
 TREND(LINEAR)       .47
 CYCLE               .36
SEASONALITY         9.39              76.
RANDOMNESS(NOISE)*  2.45              20.

*OPTIMAL..LOWEST ERROR POSSIBLE IN PERFECT FORC. MODEL



SEASON MEDIAL AVERAGE NORMALIZED
  2       105.99         105.76
  3       105.59         105.36
  4       105.93         105.71
  5       104.14         103.91
  6       105.26         105.03
  7        92.45          92.25
  8        60.09          59.97
  9       102.36         102.14
 10       105.64         105.41
 11       106.43         106.20
 12       105.86         105.63
  1       102.85         102.63

TOTAL     1202.57        1200.00
DO YOU WANT TO OVERRIDE THESE SEAS. FACTORS(Y OR N)?N

DO YOU WANT TO APPEND TO THE FORECAST METHOD RESULTS
ALREADY ON FILE (Y OR N)?N
RUNNER**WHICH FORECASTING METHOD DO YOU WANT TO RUN

TYPE METHOD NAME (H=HELP, N=NO MORE)?EXPOW

DO YOU WANT A GRAPH OF YOUR DATA?N
DO YOU WANT TO DESEASONALIZE YOUR DATA WITH THE
SEASONAL INDICES FROM CLASSICAL DECOMPOSITION (Y OR N)?N

*** LINEAR AND SEASONAL (WINTERS) EXPONENTIAL SMOOTHING ***

NEED HELP (Y OR N)?N

ENTER VALUES FOR A,B AND C (0,0,0=HELP)?0.4 0.3 0.2
DO YOU WANT A TABLE OF ACTUAL AND PREDICTED (Y OR N)?N
MEAN PC ERROR (MPE) OR BIAS     =        -2.19%
MEAN SQUARED ERROR (MSE)        =        247.4
MEAN ABSOLUTE PC ERROR   (MAPE) =         12.1%
```

FIGURE 28-2 (continued)

FORECAST FOR HOW MANY PERIODS AHEAD (0=NONE,24=MAX)?12

PERIOD	FORECAST
134	119.52
135	113.34
136	109.39
137	101.74
138	101.12
139	87.81
140	57.50
141	94.16
142	96.80
143	96.04
144	94.11
145	90.65

FIGURE 28-3 Using the Classical Decomposition Program Through RUNNER

***** RUNNER ** INTERACTIVE FORECASTING *****

TYPE METHOD NAME (H=HELP, N=NO MORE)?DCOMP

DO YOU WANT A GRAPH OF YOUR DATA?N

*** CLASSICAL DECOMPOSITION ***

DO YOU WANT A TABLE OF MOVING AVERAGES AND RATIOS?N
SEASONAL INDICES BASED ON MEDIAL AVERAGE(Y OR N)?Y

SEASON	MEDIAL AVERAGE	NORMALIZED
2	105.99	105.76
3	105.59	105.36
4	105.93	105.71
5	104.14	103.91
6	105.26	105.03
7	92.45	92.25
8	60.09	59.97
9	102.36	102.14
10	105.64	105.41
11	106.43	106.20
12	105.86	105.63
1	102.85	102.63

TOTAL   1202.57   1200.00
DO YOU WANT TO OVERRIDE THESE SEAS. FACTORS(Y OR N)?N

**** TREND ****
INTERCEPT (A) =   62.95
SLOPE (B) =   .42

DO YOU WANT A LIST OF CYCLICAL FACTORS AND FORECASTS ?N
MEAN PC ERROR (MPE) OR BIAS   =   -.23%
MEAN SQUARED ERROR (MSE)   =   24.6
MEAN ABSOLUTE PC ERROR   (MAPE) =   3.7%

FIGURE 28-3   (continued)

```
FORECAST FOR HOW MANY PERIODS AHEAD (0=NONE,24=MAX)?12

DO YOU WANT TO ESTIMATE THE   12 CYCLICAL FACTORS FOR YOUR
FORECASTS (IF NOT THE PROGRAM WILL ASSIGN A VALUE OF 100 TO EACH)?N


    **** FORECAST ****

PERIOD      FORECAST
134         125.46
135         125.67
136         126.37
137         124.11
138         126.03
139         110.89
140          72.65
141         124.26
142         128.33
143         129.83
144         130.06
145         127.11

WOULD YOU LIKE AUTOCORRELAIONS ON THE RESIDUALS ?N
```

FIGURE 28-4   Using Harrison's Harmonic Smoothing Through RUNNER

```
    ***** RUNNER ** INTERACTIVE FORECASTING *****

TYPE METHOD NAME (H=HELP, N=NO MORE)?HARM

DO YOU WANT A GRAPH OF YOUR DATA?N
DO YOU WANT TO DESEASONALIZE YOUR DATA WITH THE
SEASONAL INDICES FROM CLASSICAL DECOMPOSITION (Y OR N)?N

    *** HARRISONS HARMONIC SMOOTHING ***

         SIGNIFICANT HARMONICS
HARMONIC    COSINE        SINE        AMPLITUDE
   1        -.099         .015         .0999
   2        -.083         .024         .0863
   3        -.073         .001         .0730
   4        -.063         .016         .0652
   5        -.058         .012         .0594
   6        -.037         .000         .0367

      R E P L A C E D   O U T L I E R S
PERIOD          OUTLIER        REPLACEMENT
                              (NON-SEAS ADJUSTED)
   65           61.000          83.083

         SIGNIFICANT HARMONICS
HARMONIC    COSINE        SINE        AMPLITUDE
   1        -.098         .010         .0989
   2        -.088         .023         .0909
   3        -.073         .006         .0731
   4        -.058         .016         .0604
   5        -.058         .007         .0584
   6        -.042         .000         .0415
```

FIGURE 28-4   (continued)

```
PERIOD      SEASONAL INDICES
  2                 104.18
  3                 108.11
  4                 104.08
  5                 106.01
  6                 101.76
  7                  94.22
  8                  58.34
  9                 104.11
 10                 103.47
 11                 108.22
 12                 103.27
  1                 104.24
```

```
        A N A L Y S I S   O F   V A R I A N C E

        MEAN SQUARED ERRORS
SEAS. =        .1660    RES. =          .0012
F-TEST =   135.7222
DO YOU WANT A TABLE OF ACTUAL AND PREDICTED (Y OR N)?N
MEAN PC ERROR (MPE) OR BIAS        =         .04%
MEAN SQUARED ERROR (MSE)           =       12.2
MEAN ABSOLUTE PC ERROR    (MAPE) =         2.7%
```

```
FORECAST FOR HOW MANY PERIODS AHEAD (0=NONE,24=MAX)?12
```

```
PERIOD           FORECAST
 134               133.93
 135               139.30
 136               135.36
 137               138.66
 138               133.86
 139               124.65
 140                77.92
 141               139.28
 142               139.20
 143               146.40
 144               140.47
 145               142.57
```

```
WOULD YOU LIKE AUTOCORRELATION ON THE RESIDUALS?N
```

FIGURE 28-5   RUNNER Results for Quantitative and Naive Methods

```
***** RUNNER ** INTERACTIVE FORECASTING *****

TYPE METHOD NAME (H=HELP, N=NO MORE)?N


DO YOU WANT TO ENTER THE EXECUTIVE FORECASTS USED BY YOUR
ORGANIZATION FOR THE LAST 12 MONTHS(Y OR N)?Y
ENTER THE 12 EXECUTIVE FORECASTS
?135 134 134 133 133 132 132 132 132 131 131 130
```

FIGURE 28-5 (continued)

```
************** COMPARISON OF RESULTS ***************
** FORECASTING ACCURACY (%ERRORS) LAST 12 PERIODS **
        *** FORECASTING  M E T H O D S ***
PER   EXPOW   DCOMP    HARM   NAIVE1   NAIVE2  EXECUTV OPTIMAL
122    4.65    8.64    4.77    1.52     1.19    -2.27    .75
123    -.35    7.07   -1.04   -1.54      .18    -3.08   -.63
124     .48    6.54    2.19       0    -1.08    -3.08   -.69
125    -.24    8.20    -.09       0    12.17    -2.31   1.00
126    1.61    6.76    3.72       0    34.99    -2.31   -.81
127  -17.68    5.61   -2.42  -15.04   -95.95   -16.81  -2.26
128  -47.41   12.64   11.27  -41.25   -45.77   -65.00   5.78
129   35.11    2.80   -5.66   34.96    34.47    -7.32  -2.61
130   -3.26    3.52   -1.52    3.91     4.42    -3.13    .14
131   -6.47     .84   -8.50   -1.59     1.30    -3.97    .43
132   -7.66   -2.61   -7.56   -3.28    -3.28    -7.38   -.93
133   -8.82   -5.48  -14.86   -5.17    -4.78   -12.07  -3.82
MPE OR
BIAS  -4.17    4.54   -1.64   -2.29    -5.18   -10.73   -.30
MAPE  11.14    5.89    5.30    9.02    19.97    10.73   1.65


************* FORECASTS **********************************
134  119.52  125.46  133.93  116.00   119.54       0       0
135  113.34  125.67  139.80  116.00   119.09       0       0
136  109.39  126.37  135.36  116.00   119.48       0       0
137  101.74  124.11  138.66       0   117.45       0       0
138  101.12  126.03  133.86       0   118.72       0       0
139   87.81  110.89  124.65       0   104.27       0       0
140   57.50   72.65   77.62       0    67.78       0       0
141   94.16  124.26  139.28       0   115.45       0       0
142   96.80  128.33  139.20       0   119.15       0       0
143   96.04  129.83  146.40       0   120.04       0       0
144   94.11  130.06  140.47       0   119.40       0       0
145   90.65  127.11  142.57       0   116.00       0       0

*********** ACCURACY ALL DATA ***********************
MPE OR
BIAS  -2.19    -.23     .04   -1.67    -4.48   -10.73   -.30
MSE  247.43   24.64   12.18  222.49   843.74   293.58   5.39
MAPE  12.08    3.69    2.67   11.12    21.68    10.73   1.37


PERIOD      BEST FORECAST      CORRESPONDING METHOD
122            130.43              NAIVE2
123            129.76              NAIVE2
124            130.00              NAIVE1
125            130.00              NAIVE1
126            130.00              NAIVE1
127            115.74              HARM
128             70.98              HARM
129            119.56              DCOMP
130            129.94              HARM
131            124.94              DCOMP
132            125.19              DCOMP
133            121.54              NAIVE2

DO YOU WANT SOME EXPLAINATION ON HOW TO INTERPRET
THE ABOVE TABLES(Y OR N)?N
```

more than his or her own estimates. The fact that the MPE of the EXECU-
TIVE forecast is positive suggests that the person in charge has a tendency to
overestimate the Index of Industrial Production. The person should be aware
of that. With regards to the quantitative methods, HARM performs better in
this instance than the others, but the accuracies of both DECOMP and
EXPOW are close to those of HARM. A disadvantage of HARM is that it does

not do well in the last three periods, while DCOMP does much better and EXPOW is in between. This is the type of information that is supplied by Figure 28-5 and that should be used to determine the EXECUTIVE forecast.

Figure 28-6 provides additional information as to the best method during the last 12 periods. This is followed by the improvement of each method over the NAIVE1 and NAIVE2 methods and the maximum improvement possible. Thus, method (2) — DECOMP — did 3.13% better than NAIVE1, 14.07% better than NAIVE2 and 4.24% worse than the optimal forecast. This means its possible future improvement, at most, is on the average, 4.24%.

The last part of Figure 28-6 shows the possibility of combining the forecasts of the quantitative methods. In this particular run, the predictions of methods two and three were combined, and the results are shown at the bottom of Figure 28-6. The combining is proportional to the inverse of the MSE of each method.

Once the quantitative forecasts have been combined, one can stop and use these as the EXECUTIVE (final) forecasts. Alternatively, one can stop with ORACLE and combine the quantitative forecasts with personal judg- mental predictions. This is allowed at the end of ORACLE and should be

FIGURE 28-6   Additional Information Provided by RUNNER on Choice of Method

```
IMPROVEMENT OF THE QUANTITATIVE METHODS OVER THE NAIVE
***METHODS AND MAXIMUM POSSIBLE IMPROVEMENT IN MAPE***

METHOD     I M P R O V E M E N T   OVER      MAXIMUM
              NAIVE1            NAIVE2       POSSIBLE IMPROVEMENT
EXPOW        -2.12              8.82            9.49
DCOMP         3.13             14.07            4.24
HARM          3.72             14.67            3.65
NAIVE1          0              10.95            7.37
NAIVE2       -10.95              0             18.31
EXECUTV       -1.71             9.24            9.07
OPTIMAL        7.37            18.31             0




DO YOU WANT TO COMBINE THE ABOVE FORECASTS(Y OR N)?Y
HOW MANY FORECAST METHODS DO YOU WANT TO COMBINE?2
ENTER THE NAMES OF THE METHODS TO BE COMBINED, ONE PER LINE?HARM
?DCOMP




**COMBINED QUANTITATIVE FORECAST**
PERIOD      FORECAST
  134        131.13
  135        135.13
  136        132.38
  137        133.85
  138        131.27
  139        120.10
  140         75.97
  141        134.31
  142        135.61
  143        140.92
  144        137.03
  145        137.46
ARE YOU INTERESTED IN THE CYCLICAL BEHAVIOR
OF YOUR DATA(Y OR N)?Y
```

done after the user has analyzed the cyclical behavior of the data and has made allowances for special events. The steps involved in combining judgmental assessments are shown in Figure 28-7.

FIGURE 28-7   Preparing the Executive Forecast in RUNNER

```
***** CYCLICAL ANALYSIS OF TIME SERIES *****

DO YOU WANT TO LIMIT THE OUTPUT TO A GRAPH
OF TREND-CYCLE AND ITS CHANGE(Y OR N)?N

    *** T I M E   S E R I E S   C O M P O N E N T S ***
PERIOD  ACTUAL  SEAS. ADJ.  TREND-CYCLE  TREND  CYCLE   RANDOM
122     132.0   124.81      123.9        125.8  98.5    131.0
123     130.0   123.38      124.2        125.0  99.3    130.8
124     130.0   122.98      123.8        124.1  99.7    130.9
125     130.0   125.10      123.9        123.3  100.4   128.7
126     130.0   123.77      124.8        122.5  101.9   131.0
127     113.0   122.50      125.3        121.6  103.0   115.6
128     80.0    133.41      125.7        120.8  104.1   75.4
129     123.0   120.43      123.6        119.9  103.0   126.2
130     128.0   121.43      121.3        119.1  101.8   127.8
131     126.0   118.64      118.1        118.3  99.9    125.5
132     122.0   115.49      116.6        117.4  98.9    123.1
133     116.0   113.03      117.4        116.6  99.4    120.4

    *** PERCENTAGE CHANGE IN THE TIME SERIES COMPONETS **
PERIOD  ACTUAL  SEAS. ADJ.  TREND-CYCLE  TREND  CYCLE   RANDOM
122     1.54    -1.47       1.02         -.66   1.70    4.11
123     -1.52   -1.14       .23          -.67   .91     -.14
124     0       -.33        -.27         -.67   .40     .06
125     0       1.73        .02          -.68   .70     -1.68
126     0       -1.07       .74          -.68   1.43    1.82
127     -13.08  -1.03       .40          -.69   1.09    -11.82
128     -29.20  8.91        .35          -.69   1.04    -34.77
129     53.75   -9.73       -1.70        -.69   -1.01   67.44
130     4.07    .83         -1.87        -.70   -1.18   1.28
131     -1.56   -2.30       -2.57        -.70   -1.88   -1.84
132     -3.17   -2.65       -1.32        -.71   -.96    -1.85
133     -4.92   -2.13       .67          -.71   .48     -2.20

    *** PERCENTAGE ERROR OF EACH COMPONENT ***
PERIOD  SEAS. ADJ  TREND-CYCLE  TREND   CYCLE   RANDOM
122     5.44       6.15         4.68     8.66    .75
123     5.09       4.49         3.86     6.41    -.63
124     5.40       4.74         4.50     6.03    -.69
125     3.77       4.73         5.15     5.38    1.00
126     4.79       4.03         5.79     4.03    -.81
127     -8.40      -10.85       -7.63    -11.62  -2.26
128     -66.76     -57.12       -50.99   -59.30  5.78
129     2.09       -.46         2.48     -2.57   -2.61
130     5.13       5.27         6.95     2.60    .14
131     5.84       6.24         6.13     2.92    .43
132     5.33       4.45         3.75     .69     -.93
133     2.56       -1.17        -.51     -4.95   -3.82
```

FIGURE 28-7 (continued)

```
*** ABSOLUTE PERCENTAGE ERROR CONTRIBUTION OF EACH COMPONENT ***

PERIOD   SEAS. ADJ   TREND-CYCLE    TREND     CYCLE    RANDOM
122        21.20       23.96        18.22     33.71      2.91
123        24.85       21.91        18.84     31.30     -3.10
124        25.26       22.20        21.07     28.23     -3.24
125        18.81       23.62        25.72     26.86      4.99
126        24.65       20.71        29.80     20.71     -4.14
127       -20.62      -26.62       -18.73    -28.50     -5.54
128       -27.82      -23.80       -21.25    -24.71      2.41
129        20.50       -4.50        24.30    -25.16    -25.54
130        25.55       26.23        34.58     12.93       .71
131        27.10       28.95        28.46     13.52      1.97
132        35.19       29.35        24.71      4.58     -6.17
133        19.67       -8.98        -3.92    -38.03    -29.40
```

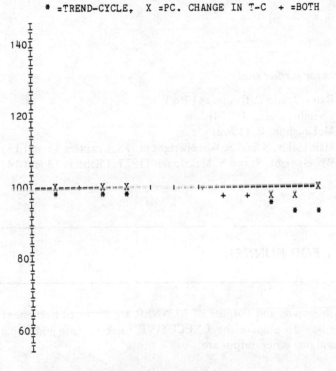

```
--- TREND-CYCLE AND PERCENTAGE CHANGE IN TREND-CYCLE ---
   * =TREND-CYCLE,  X =PC. CHANGE IN T-C  + =BOTH
```

```
DO YOU WANT TO COMBINE THE QUANTITATIVE FORECASTS WITH
YOUR OWN PERSONAL ONES(Y OR N)?Y
HOW MANY PREDICTIONS DO YOU WANT TO MAKE?6
ENTER YOUR  6 PREDICTIONS?6 120 116 114 115 115 115
WHAT WEIGHT ON A SCALE FROM ZERO TO ONE DO YOU GIVE
FOR YOUR PREDICTIONS.
?.7
```

FIGURE 28-7   (continued)

```
COMBINED FORECASTS (QUANTITATIVE AND JUDGEMENTAL)
PERIOD     FORECAST
  134      123.34
  135      121.74
  136      119.52
  137      120.65
  138      119.88
  139      116.53
FORECAST RESULTS ARE STORED IN FILE NAMED  FRENCH
DO YOU WANT TO SAVE THE RESULTS FOR USE LATER ?Y

CRUS:     260.07
CAT

YOUR FILES ARE
FRENCH  (B)   246
RUNBXJN (X)  6254
METHOD  (X)   178
SIBYL   (X)   178
RUNNER  (X)   178
DECOMPX      1481
AIRLIN        101
HARM1         204
STOP
```

References for further study:

> Bates, J. and C. Granger (1969)
> Chambers, et al. (1974)
> McLaughlin, R. (1962)
> Makridakis, S. and S. Wheelwright (1978, Chapters 15 and 18)
> Wheelwright, S. and S. Makridakis (1977, Chapters 13 and 14)

INPUT/OUTPUT FOR RUNNER

Many of the inputs and outputs of RUNNER are discussed in the text above describing how to prepare the EXECUTIVE forecast. One additional input question and one other output are:

1. *Do you want to append to the forecast method results already on file (Y or N)?*   This question appears when a file of forecasting results has been found in your directory. It allows results from previous runs to be compared and combined with results of the present run.

2. *Forecast results are stored in a file named _____ . Do you want to save the results for use later?*   A no answer permanently erases the results of the present run.

CHAPTER 29

Program Files and Data Handling

PROGRAM FILES

The interactive forecasting system consists of a number of different computer programs and data files. Each of these represents either a specific forecasting technique, a control program, or a data manipulation program. A short description of every program file in the system is contained in $INFOS. This file is accessed by SIBYL when the user requests information on a particular program. Additionally, the file $INFOS can simply be listed at the terminal and then used as a reference document when information is wanted on a specific program file, or the program METHOD can be used to obtain a summary listing of available methods. The user has three main options in using SIBYL/RUNNER. First, a program can be run by itself. This is very simple even though it varies from one computer system to another. In H-P BASIC, for example, this is done by typing: GET—$EXPO and then RUN. These two commands are sufficient to run the method of single exponential smoothing. In FORTRAN, each method can be run individually using METHOD. The second possibility is to run a program through SIBYL/ RUNNER. The final alternative is to use a program facilitating the handling of data.

To aid the user in understanding the relationship among the various program files and the ways that they can be accessed, the METHOD program has been created. Application of this program is shown in Figure 29-1. As can be seen from METHOD, there are several program files that represent individual forecasting techniques. These can be used individually (with or without METHOD) or they can be accessed through either SIBYL or RUNNER. These latter two programs are control programs that aid in running one or more of the basic technique programs. To use either SIBYL or RUNNER, it is necessary to have an output file called RESULT and a data file. (The creation of data files is described in the next section of this chapter.)

The third type of program file referenced in Figure 29-1 is for data management. These programs are used in creating and maintaining data files. Since every forecasting situation for which the interactive forecasting system might be used involves a set of historical values, these data manipulation programs have been developed to simplify the task of creating and maintaining data files. The user always has the option of inputting the data at the terminal each time a program is run, but in most instances it is likely that multiple runs will be made, and these are greatly facilitated through the use of a data file.

DATA HANDLING

There are five principal programs included in the system that are designed to do data handling. The remainder of this chapter describes each of these and provides examples of their use.

FINPUT

This program enters data into a file for later use with other programs in the system. The user is asked to furnish a file name, the number of observations, and the value of each observation. Once the data file is created, the program allows the user to list the contents of the new data file or to edit those contents. (See Figure 29-2.)

FLIST

This program lists all or part of the data in a data file. The user must specify the name of the data file and the number of observations to be listed. This program can be run alone or used in conjunction with FINPUT, FEDIT, or UPDATE. (See Figure 29-3.)

FIGURE 29-1   Overview of Program Structure Given by METHOD

```
EXE METHOD
  8K
WHICH PROGRAM WOULD YOU LIKE TO RUN?
DO YOU WANT A LIST OF AVAILABLE METHODS?Y
AVAILABLE FORECASTING METHODS
              BOXJEN
              CENSUS
              DCOMP
              GAF
              HARM
              MREG
              EXPO
              EXPOQ
              EXPO2
              EXPOTL
              EXPOW
              EXPOH
              MAVE
              MEAN
              MAVE2
              SCURVE
              SREG
              EXGROW


AVAILABLE ANALYTICAL METHODS
              ACORN
              ORACLE
              MUVING


AVAILABLE DATA MANAGEMENT ROUTINES
              FPLAY
              DGEN
              FPLAY
              FINPUT
              UPDATE
              FEDIT
              FLIST
WHICH PROGRAM WOULD YOU LIKE TO RUN?FEDIT


* INPUTS,EDITS,UPDATES AND LISTS DATA FILE *

NEED HELP (Y OR N)?N
INPUT - THIS PROGRAM INPUTS DATA ON A FILE
EDIT - THIS PROGRAM MODIFIES DATA ON A FILE
LIST - THIS PROGRAM LISTS THE CONTENTS OF A DATA FILE
UPDATE - THIS PROGRAM ADDS DATA TO AN EXISTING FILE OR DELETES
THE BEGINNING OF A FILE

WHAT PROGRAM DO YOU WANT TO RUN?
```

FIGURE 29-2   Application of FINPUT Program

```
STOP

CRUS:      22.08
EXE METHOD
   8K
WHICH PROGRAM WOULD YOU LIKE TO RUN?FINPUT

* INPUTS,EDITS,UPDATES AND LISTS DATA FILE *

NEED HELP (Y OR N)?N
INPUT - THIS PROGRAM INPUTS DATA ON A FILE
EDIT - THIS PROGRAM MODIFIES DATA ON A FILE
LIST - THIS PROGRAM LISTS THE CONTENTS OF A DATA FILE
UPDATE - THIS PROGRAM ADDS DATA TO AN EXISTING FILE OR DELETES
THE BEGINNING OF A FILE

WHAT PROGRAM DO YOU WANT TO RUN?INPUT
DATA FILENAME?ABCDATA
FILE ABCDATAALREADY EXISTS
DATA FILENAME?ABDATA
NAME OF VARIABLE?VARAB


***** ENTERS DATA IN A FILE *****
HOW MANY OBSERVATIONS DO YOU WANT TO ENTER?20

TYPE AN OBSERVATION EVERY TIME A ? APPEARS

PERIOD      OBSERVATION
   1        ?20
   2        ?22
   3        ?24
   4        ?26
   5        ?28
   6        ?30
   7        ?32
   8        ?34
   9        ?36
  10        ?38
  11        ?40
  12        ?42
  13        ?44
  14        ?46
  15        ?48
  16        ?50
  17        ?52.
  18        ?54.0
  19        ?56.000
  20        ?58

DO YOU WANT TO RUN ANOTHER PROGRAM (Y OR N)?Y

WHAT PROGRAM DO YOU WANT TO RUN?LIST


***** LISTS THE DATA OF A FILE *****
PERIOD   OBSERVATION  PERIOD   OBSERVATION  PERIOD   OBSERVATION
   1        20.00        7        32.00       13        44.00
   2        22.00        8        34.00       14        46.00
   3        24.00        9        36.00       15        48.00
   4        26.00       10        38.00       16        50.00
   5        28.00       11        40.00       17        52.00
   6        30.00       12        42.00       18        54.00
                                              19        56.00
                                              20        58.00

DO YOU WANT TO RUN ANOTHER PROGRAM (Y OR N)?Y
```

FIGURE 29-3   Application of FLIST Program

```
WHAT PROGRAM DO YOU WANT TO RUN?EDIT

***** MODIFIES DATA FILES *****
HAVE YOU DATA ON THE FILE ABDATA ?Y

IF YOU WANT TO CHANGE AN OBSERVATION TYPE ITS
PERIOD AND ITS NEW VALUE ?1 2000

MORE OBSERVATION TO CHANGE?N

DO YOU WANT TO RUN ANOTHER PROGRAM (Y OR N)?Y

WHAT PROGRAM DO YOU WANT TO RUN?LIST


***** LISTS THE DATA OF A FILE *****
PERIOD   OBSERVATION  PERIOD    OBSERVATION  PERIOD    OBSERVATION
  1        2000.00      7          32.00       13         44.00
  2          22.00      8          34.00       14         46.00
  3          24.00      9          36.00       15         48.00
  4          26.00     10          38.00       16         50.00
  5          28.00     11          40.00       17         52.00
  6          30.00     12          42.00       18         54.00
                                                19         56.00
                                                20         58.00
```

FEDIT

This program edits an existing data file by accessing the observation (s) specified by the user and replacing their previous values by those given by the user. Any number of observations can be altered and the final data file can be listed. (See Figure 29-4.)

UPDATE

This program updates the contents of a data file by adding new data at the end of the file and deleting data from the beginning of the file. The user can specify the number of data values to be added or deleted, if any. (See Figure 29-5.)

FPLAY

This program allows for almost any type of file manipulation (including transformations). The user has several options as he runs the program. (See Figure 29-6.)

FIGURE 29-4   Application of FEDIT Program

```
DO YOU WANT TO RUN ANOTHER PROGRAM (Y OR N)?Y

WHAT PROGRAM DO YOU WANT TO RUN?EDIT


***** MODIFIES DATA FILES *****
HAVE YOU DATA ON THE FILE ABDATA ?Y

IF YOU WANT TO CHANGE AN OBSERVATION TYPE ITS
PERIOD AND ITS NEW VALUE ?1 20

MORE OBSERVATION TO CHANGE?Y
PERIOD AND ITS NEW VALUE ?2 22.09

MORE OBSERVATION TO CHANGE?N

DO YOU WANT TO RUN ANOTHER PROGRAM (Y OR N)?Y

WHAT PROGRAM DO YOU WANT TO RUN?LIST


***** LISTS THE DATA OF A FILE *****
PERIOD    OBSERVATION  PERIOD    OBSERVATION  PERIOD    OBSERVATION
   1          20.00       7         32.00       13         44.00
   2          22.09       8         34.00       14         46.00
   3          24.00       9         36.00       15         48.00
   4          26.00      10         38.00       16         50.00
   5          28.00      11         40.00       17         52.00
   6          30.00      12         42.00       18         54.00
                                                19         56.00
                                                20         58.00

DO YOU WANT TO RUN ANOTHER PROGRAM (Y OR N)?Y
```

FIGURE 29-5   Application of UPDATE Program

```
WHAT PROGRAM DO YOU WANT TO RUN?UPDATE

***** UPDATES DATA FILES *****

HAVE YOU DATA ON THE FILE ABDATA ?Y

HOW MANY OBSERVATIONS DO YOU WANT TO ADD?4
TYPE AN OBSERVATION EVERYTIME AN ? APPEARS


PERIOD     OBSERVATION
  21       ?60
  22       ?62
  23       ?64
  24       ?66

DO YOU WANT TO DELETE ANY OBSERVATIONS AT THE BEGINNING OF YOUR DATA?Y

HOW MANY DO YOU WANT DELETED?3

YOU HAVE   21  OBSERVATIONS NOW

DO YOU WANT TO RUN ANOTHER PROGRAM (Y OR N)?Y

WHAT PROGRAM DO YOU WANT TO RUN?LIST


***** LISTS THE DATA OF A FILE *****
PERIOD    OBSERVATION  PERIOD   OBSERVATION  PERIOD   OBSERVATION
  1          26.00        8        40.00       15        54.00
  2          28.00        9        42.00       16        56.00
  3          30.00       10        44.00       17        58.00
  4          32.00       11        46.00       18        60.00
  5          34.00       12        48.00       19        62.00
  6          36.00       13        50.00       20        64.00
  7          38.00       14        52.00       21        66.00

DO YOU WANT TO RUN ANOTHER PROGRAM (Y OR N)?N
```

FIGURE 29-6 Application of FPLAY Program

```
***** MANIPULATES THE CONTENT OF FILE (S) ****

NEED HELP (Y OR N)?N
DO YOU WANT TO
1 - DELETE ELEMENTS FROM A FILE
2 - DO TRANSFORMATIONS ON DATA IN A FILE
3 - ADD ADDITIONAL ELEMENTS INTO A FILE
4 - DO ARITHMETIC OPERATIONS ON TWO FILES
5 - STOP

ENTER MANIPULATION CODE 1,2,3 OR 4?2

WHAT IS THE NAME OF THE FILE YOU WANT TO WORK ON?ABDATA
HOW MANY OBSERVATIONS DO YOU WANT TO USE?21
WHICH OF THE FOLLOWING TRANSFORMATIONS DO YOU WANT
1 - LOGARITHMIC
2 - EXPONENTIAL
3 - POWER
4 - RECIPROCAL
5 - S-CURVE
6 - MULTIPLY OR DIVIDE BY A CONSTANT
7 - ADD OR SUBTRACT A CONSTANT
ENTER TRANSFORMATION CODE 1,2,3,4,5,6 OR 7?3

INPUT THE POWER TO WHICH YOU WANT TO TRANSFORM YOUR DATA?0.5

DO YOU WANT THE RESULTS
1  IN YOUR ORIGINAL FILE
2  IN A NEW FILE
TYPE 1 OR 2?2

NEW FILENAME?ABTFM

DO YOU WANT THE CONTENTS OF ABTFM  LISTED?Y

PERIOD    OBSERVATION  PERIOD   OBSERVATION  PERIOD   OBSERVATION
   1          5.10        8        6.32        15        7.35
   2          5.29        9        6.48        16        7.48
   3          5.48       10        6.63        17        7.62
   4          5.66       11        6.78        18        7.75
   5          5.83       12        6.93        19        7.87
   6          6.00       13        7.07        20        8.00
   7          6.16       14        7.21        21        8.12

ENTER MANIPULATION CODE 1,2,3 OR 4?
```

DGEN

This program allows the user to generate an artifical data set (with certain specified properties) that can be used as any regular data file. A data set containing up to 150 observations along with seasonal, cyclical, trend, and randomness components can be created and graphed using this program. (See Figure 29-7.)

SUMMARY

As an overview of the steps that might be followed in getting started with an application of SIBYL/RUNNER, Figure 29-8 presents an annotated protocol

FIGURE 29-7   Application of DGEN Program

```
EXE METHOD
  8K
WHICH PROGRAM WOULD YOU LIKE TO RUN?DGEN

********** DATA GENERATION PROGRAM **********

HOW MANY OBSERVATIONS DO YOU WANT TO GENERATE?120
WHAT LENGTH SEASONAL PATTERN DO YOU WANT (0=NONE)?12
WHAT LENGTH CYCLICAL PATTERN DO YOU WANT (0=NONE)?66
ON A SCALE OF 0 TO 1 , ENTER THE WEIGHT (IMPORTANCE DESIRED FOR
EACH ONE OF THE FOLLOWING COMPONENTS).
 ** SEASONAL  **   ?0.8

WEIGHT FOR ** CYCLICAL **   ?0.6

WEIGHT FOR ** RANDOMNESS **   ?0.4

WEIGHT FOR ** TREND **?0.5

DO YOU WANT A GRAPH OF YOUR DATA?Y
HOW MANY OF YOUR 120 OBSERVATIONS DO YOU WANT PLOTTED?48
```

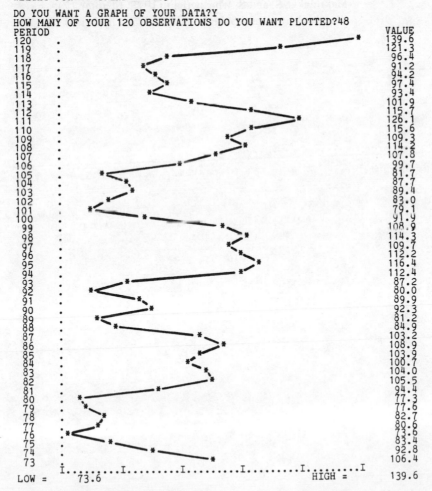

```
PERIOD                                                   VALUE
120   .                                              .   139.6
119   .                                   *              121.3
118   .                        *    *                    96.4
117   .                      *                           91.2
116   .                       *                          94.2
115   .                       *                          97.4
114   .                    *                             93.4
113   .                  *                               101.9
112   .                                   *              115.7
111   .                                    *             126.1
110   .                             *                    115.6
109   .                          *                       109.3
108   .                           *                      114.2
107   .                 *                                107.8
106   .             *                                    99.7
105   .       *                                          81.7
104   .         *                                        87.7
103   .          *                                       89.4
102   .      *                                           83.0
101   .    *                                             79.1
100   .            *                                     91.9
 99   .                    *                             108.9
 98   .                       *                          114.3
 97   .                   *                              109.7
 96   .                    *                             112.2
 95   .                      *                           116.4
 94   .                  *                               112.4
 93   .       *                                          87.2
 92   .      *                                           80.0
 91   .           *                                      89.9
 90   .            *                                     92.3
 89   .      *                                           81.2
 88   .        *                                         84.9
 87   .                *                                 103.2
 86   .                 *                                108.9
 85   .              *                                   103.9
 84   .             *                                    100.7
 83   .               *                                  104.0
 82   .            *                                     105.5
 81   .        *                                         94.4
 80   .    *                                             77.3
 79   .      *                                           77.6
 78   .       *                                          82.7
 77   .      *                                           80.6
 76   . *                                                73.6
 75   .         *                                        83.4
 74   .              *                                   92.8
 73   .                *                                 106.4
      I.........I.........I.........I.........I........I
LOW =    73.6                                   HIGH =    139.6

DO YOU WANT TO RE-DO THIS EXERCISE?N
DO YOU WANT TO SAVE THE DATA ?Y
WHAT IS THE NAME OF YOUR FILE ?DGDATA
```

based on the FORTRAN version of the system. The user first accesses the system and then calls the program, METHOD. The data are then input at the terminal using INPUT, but a couple of typographical errors are made in so doing. These are corrected using EDIT.

Next, the program UPDATE is used to add twelve more data values to the file printed. Finally, the program FPLAY is used to transorm the data, store the results in the original file and print them. This data file could then be used with any of the forecasting technique programs, SIBYL, or stored for later use. (See Figure 29-8.)

References for furthers study:

Makridakis, S. and S. Wheelwright (1978, Chapter 17)
Wheelwright, S. and S. Makridakis (1977, Chapter 9)

FIGURE 29-8   Results of Solved Exercise

```
USER NO.?Z1661
CODE?
████
  ID - DBS                                        ⎫
                                                  ⎬  Signon Procedure
TERMINAL 102   PORT  26                           ⎭
TIME: 14:14 EDT, DATE: 07/16/76. OFF AT 00:30 EDT.

USE*3243 METHOD                                   - Access METHOD
WHICH PROGRAM WOULD YOU LIKE TO RUN?FINPUT        - Run FINPUT

* INPUTS,EDITS,UPDATES AND LISTS DATA FILE *      - FINPUT is incorrect name
                                                    so program lists what's
NEED HELP (Y OR N)?N                                available.
INPUT - THIS PROGRAM INPUTS DATA ON A FILE
EDIT - THIS PROGRAM MODIFIES DATA ON A FILE
LIST - THIS PROGRAM LISTS THE CONTENTS OF A DATA FILE
UPDATE - THIS PROGRAM ADDS DATA TO AN EXISTING FILE OR DELETES
THE BEGINNING OF A FILE

WHAT PROGRAM DO YOU WANT TO RUN?INPUT             - Run INPUT
DATA FILENAME?TESTOUT                             - File name will be TESTOUT
NAME OF VARIABLE?TEST1                            - Variable name will be TEST1


***** ENTERS DATA IN A FILE *****
HOW MANY OBSERVATIONS DO YOU WANT TO ENTER?48     - 48 observations to be entered

TYPE AN OBSERVATION EVERY TIME A ? APPEARS

PERIOD      OBSERVATION
  1         ?112                                  - Input data values, 1 through 48
  2         ?118
  3         ?132
  4         ?129
  5         ?121
  6         ?135
  7         ?148
  8         ?148
  9         ?136
 10         ?119
 11         ?104
 12         ?118
 13         ?115
 14         ?126
 15         ?141
 16         ?135.00
 17         ?125
 18         ?149
 19         ?170
 20         ?170
 21         ?158
 22         ?133
 23         ?114
 24         ?140
 25         ?145
 26         ?150
 27         ?178
 28         ?163
 29         ?172
 30         ?178
 31         ?199
 32         ?199
 33         ?184.56
 34         ?162
 35         ?146
 36         ?166
 37         ?171
 38         ?180
 39         ?193
 40         ?181
 41         ?183
 42         ?218
 43         ?230
 44         ?242
 45         ?209
 46         ?191
 47         ?172
 48         ?196
```

FIGURE 29-8 (continued)

DO YOU WANT TO RUN ANOTHER PROGRAM (Y OR N)?Y

WHAT PROGRAM DO YOU WANT TO RUN?EDIT   - Running EDIT to correct input
                                              errors

***** MODIFIES DATA FILES *****

HAVE YOU DATA ON THE FILE TESTOUT?Y

IF YOU WANT TO CHANGE AN OBSERVATION TYPE ITS
PERIOD AND ITS NEW VALUE ?33 184   - Input value number

MORE OBSERVATION TO CHANGE?Y   and corrected
PERIOD AND ITS NEW VALUE ?48 194

MORE OBSERVATION TO CHANGE?N   value

DO YOU WANT TO RUN ANOTHER PROGRAM (Y OR N)?Y

WHAT PROGRAM DO YOU WANT TO RUN?UPDATE   - Run UPDATE

***** UPDATES DATA FILES *****

HAVE YOU DATA ON THE FILE TESTOUT?Y

HOW MANY OBSERVATIONS DO YOU WANT TO ADD?12
TYPE AN OBSERVATION EVERYTIME AN ? APPEARS

PERIOD	OBSERVATION
49	?196
50	?196
51	?236
52	?235
53	?229
54	?243
55	?264
56	?272
57	?237
58	?211
59	?180
60	?201

Add 12 additional observations to the TESTOUT file

DO YOU WANT TO DELETE ANY OBSERVATIONS AT THE BEGINNING OF YOUR DATA?N

YOU HAVE   60   OBSERVATIONS NOW

DO YOU WANT TO RUN ANOTHER PROGRAM (Y OR N)?Y

WHAT PROGRAM DO YOU WANT TO RUN?LIST   Run LIST to list file contents

***** LISTS THE DATA OF A FILE *****

PERIOD	OBSERVATION	PERIOD	OBSERVATION	PERIOD	OBSERVATION
1	112.00	21	158.00	41	183.00
2	118.00	22	133.00	42	218.00
3	132.00	23	114.00	43	230.00
4	129.00	24	140.00	44	242.00
5	121.00	25	145.00	45	209.00
6	135.00	26	150.00	46	191.00
7	148.00	27	178.00	47	172.00
8	148.00	28	163.00	48	194.00
9	136.00	29	172.00	49	196.00
10	119.00	30	178.00	50	196.00
11	104.00	31	199.00	51	236.00
12	118.00	32	199.00	52	235.00
13	115.00	33	184.00	53	229.00
14	126.00	34	162.00	54	243.00
15	141.00	35	146.00	55	264.00
16	135.00	36	166.00	56	272.00
17	125.00	37	171.00	57	237.00
18	149.00	38	180.00	58	211.00
19	170.00	39	193.00	59	180.00
20	170.00	40	181.00	60	201.00

DO YOU WANT TO RUN ANOTHER PROGRAM (Y OR N)?N

FIGURE 29-8   (continued)


```
USE*3243 METHOD
WHICH PROGRAM WOULD YOU LIKE TO RUN?FPLAY         - Run FPLAY to transform data


***** MANIPULATES THE CONTENT OF FILE (S) ****

NEED HELP (Y OR N)?N
DO YOU WANT TO
1 - DELETE ELEMENTS FROM A FILE
2 - DO TRANSFORMATIONS ON DATA IN A FILE
3 - ADD ADDITIONAL ELEMENTS INTO A FILE
4 - DO ARITHMETIC OPERATIONS ON TWO FILES
5 - STOP

ENTER MANIPULATION CODE 1,2,3 OR 4?2            - Indicates that a transformation

WHAT IS THE NAME OF THE FILE YOU WANT TO WORK ON?TESTOUT      is to be made
HOW MANY OBSERVATIONS DO YOU WANT TO USE?60
WHICH OF THE FOLLOWING TRANSFORMATIONS DO YOU WANT
1 - LOGARITHMIC
2 - EXPONENTIAL
3 - POWER
4 - RECIPROCAL
5 - S-CURVE
6 - MULTIPLY OR DIVIDE BY A CONSTANT
7 - ADD OR SUBTRACT A CONSTANT
ENTER TRANSFORMATION CODE 1,2,3,4,5,6 OR 7?6    - Transformation type is specified

ENTER THE CONSTANT BY WHICH YOU WANT TO

MULTIPLY OR DIVIDE YOUR DATA?.1

DO YOU WANT THE RESULTS
1  IN YOUR ORIGINAL FILE
2  IN A NEW FILE
TYPE 1 OR 2?1                                               Transformation

DO YOU WANT THE CONTENTS OF TESTOUTLISTED?Y                 and its results
```

PERIOD	OBSERVATION	PERIOD	OBSERVATION	PERIOD	OBSERVATION
1	11.20	21	15.80	41	18.30
2	11.80	22	13.30	42	21.80
3	13.20	23	11.40	43	23.00
4	12.90	24	14.00	44	24.20
5	12.10	25	14.50	45	20.90
6	13.50	26	15.00	46	19.10
7	14.80	27	17.80	47	17.20
8	14.80	28	16.30	48	19.40
9	13.60	29	17.20	49	19.60
10	11.90	30	17.80	50	19.60
11	10.40	31	19.90	51	23.60
12	11.80	32	19.90	52	23.50
13	11.50	33	18.40	53	22.90
14	12.60	34	16.20	54	24.30
15	14.10	35	14.60	55	26.40
16	13.50	36	16.60	56	27.20
17	12.50	37	17.10	57	23.70
18	14.90	38	18.00	58	21.10
19	17.00	39	19.30	59	18.00
20	17.00	40	18.10	60	20.10

```
ENTER MANIPULATION CODE 1,2,3 OR 4?5           - Stop
 05430,STOP

CRUS:     312.36
```

Cases

GRUNER AND JAHR AG & CO.

In early 1973, Mr. Gustav Becker, head of planning and corporate development at Gruner & Jahr, was concerned with improving the company's profit performance through better matching production quantities with sales demand. Although the problem was certainly not new to either the company or the industry, Gustav felt that some of the things he had recently learned about forecasting methods and computer models might enable him to make substantial improvements in the existing procedures. As a starting point, Gustav wanted to look at the problem in the context of the company's two most important magazines: ELTERN, a monthly publication comparable to PARENTS, and DER STERN, a weekly magazine comparable to LIFE.

The basic difficulty the company faced in planning production for these two magazines was that only about 35% of the copies produced for each issue were sold through subscription. The remaining 65% reached the final consumer through thousands of retailers who were served by a few hundred independently-owned wholesalers. Since both retailers and wholesalers could return unsold copies of any issue to the publisher for full credit and since there was clearly a cost associated with being out of stock when the final customer desired a specific issue, the problem Gustav faced was setting up a system that would forecast demand as accurately as possible and then trade off the costs of overstocking against those of understocking in coming up with the most desirable production and shipping schedule for each issue.

GERMAN MAGAZINE INDUSTRY

The demand for magazines in the Federal Republic of Germany (West Germany) in the early seventies was in a period of general stagnation. While there was always some shifting of market positions among the major magazines, the total demand for magazines in the country did not seem to be changing very rapidly. There were about 50 magazines in Germany that had sales in excess of 100,000 copies per issue with the largest having sales of 3.5 million copies per issue.

The most important categories of magazines in Germany were

    a. T.V. magazines,
    b. topical illustrated magazines for the general public,
    c. yellow press, and
    d. women's magazines.

Four publishing houses whose aggregate sales accounted for 65% of the magazines sold (through 30 major titles) were the dominant factors in the German market. Gruner & Jahr ranked third in this group, with 12% of the country's total sold circulation. (Total sold circulation included both subscription sales and retail newsstand sales.)

In most instances, publishers obtained only 10%-20% of their sold circulation through subscriptions, with the remainder coming from sales at retail outlets. The thousands of magazine retailers were served by a relatively small number of wholesalers. Both the wholesalers and the retailers were independent of the publishers and it was the publisher who assumed all of the risks associated with unsold copies. Thus, the motivation for accurate forecasting and ordering for individual wholesalers and retailers lay mainly with the publisher.

THE ECONOMICS OF MAGAZINE PUBLISHING

Like most German magazines, the titles published by Gruner & Jahr depended heavily on advertising for a major part of their revenues. In fact, the individual costs of production and of distribution for each issue clearly exceeded the revenue the publisher received from sale of that issue to the final customer. Thus it was necessary for the publisher to sell substantial amounts of advertising space in each issue, in order to make the magazine profitable. The sale of this advertising was generally handled through the company's direct sales force of four people who contacted the major advertising agencies in the country on a regular basis. This sales force also arranged individual contracts with a few of the larger advertisers. The price that could be charged for an advertising page depended on many factors, including the percentage of advertising in the magazine, the sold circulation (both retail and subscription sales), the quality of the magazine, the audience being reached, and the number of colors used in the ad.

Due to the high fixed costs associated with the editorial work and the production setup required for each issue, publishers were continually seeking to increase their sold circulation. One way of doing this was to increase the level of service by supplying retailers with additional copies and thus reducing the number of stockouts. This, in turn, resulted in a substantial number of unsold copies that were returned to the publisher. Gruner & Jahr was typical of the industry, having experienced about a 15%-18% return rate over the past year for their weekly, DER STERN, and a somewhat higher return rate for their monthly, ELTERN. This meant that for every five copies sold, on the average, one copy would remain unsold. (Gustav mentioned that top management of the company considered these returns "over-production," whereas he considered them a mixture of both over-production and improper allocation to individual sales outlets.) In absolute figures per year, this meant that Gruner & Jahr was printing and distributing (under six different titles) 122 million copies per year and that about 20 million of these were being returned unsold. The direct costs associated with these returns (variable production and distribution costs) amounted to DM 22 million annually.[1]

DISTRIBUTION THROUGH WHOLESALERS AND RETAILERS

While the cost of these returns made unsold copies a major management concern, the fact that circulation of both DER STERN and ELTERN was through 180 wholesalers, each with a delivery volume of between 1,000 and 50,000 copies per issue, made it an extremely difficult problem to deal with. It was further compounded in that these major distributors supplied magazines to a total of 60,000 retailers. The wholesalers held exclusive regional

[1] After the dollar devaluation in early 1973, the Deutsch Mark had a value of about $.35 in U.S. currency (i.e., 3 Marks to the dollar).

monopolies and thus distributed magazines, newspapers and newsprint novels of all publishers to retail outlets in a particular geographical region. The typical wholesaler would have a line that would include about 800 titles, although the 30 magazines published by the four leading companies in the German publishing industry would account for just over half of the wholesaler's volume.

It was the job of the wholesaler to deliver the quantity that had been arranged between him and the publisher at the right time (for each issue there was a certain "first-sales" day), in the right quantity and to the right place. For his services, Gruner & Jahr paid the wholesaler 16% of the retail selling price for each copy actually sold. Although there were certain norms in the industry concerning the terms of agreement between the wholesaler and the publisher, there was still considerable flexibility on the part of the publisher to set up individualized agreements wherever advantageous, as long as the wholesaler continued to receive his basic commission of 16%. In the past, Gruner & Jahr had not done much in the way of special agreements, but was aware that other publishers had done so with some success.

In general, German publishers were well satisfied with the physical distribution achieved by the industry wholesalers. Thanks to the transport and distribution organization of the wholesale trade, a publisher could deliver magazines to distributors in the early morning and have them placed in 60,000 retail outlets by the end of the day. However, German publishers were generally disatisfied with the allocation of copies to individual retail outlets. Frequently, a substantial number of retailers would be sold out on a particular issue, while an equally large number of retailers would be overstocked.

At the retail level, there were actually 80,000 individual sales outlets but only three-fourths of those handled ELTERN and DER STERN. This meant that there was roughly one retail outlet handling these two magazines for every 1,000 inhabitants in Germany. Of these outlets, 60% were independent sole proprietors. The remaining 40% belonged to chains, such as food stores. For 90% of the retail outlets, magazine sales represented less than DM 2000 per month. Since the retailer was paid 20% of the retail price for each copy sold, the gross earnings from magazine sales for all but 10% of these outlets was less than DM 400 per month.

In 50% of these retail outlets, food items represented the major category of merchandise in terms of sales volume. In another 16% of the outlets, tobacco was the major sales item and in only 7% of these retail outlets were published products the major sales category. In terms of space utilized by the magazine line, in 75% of the outlets, magazines accounted for less than 10% of the floor space on which merchandise was displayed.

The typical sales outlet handling Gruner & Jahr publications would sell 15 copies of each issue of DER STERN (weekly) while returning 3 copies unsold and would sell 8 copies of each issue of ELTERN (monthly) while returning 2 copies unsold. Because retailers as well as wholesalers had the right to return unsold copies, most retailers did not avail themselves of the possibility of continuously changing and adjusting the number of copies delivered to them. Rather, determination of the quantities to be delivered to each retailer was left in the hands of the wholesaler. (This was a common practice throughout the industry.)

UNSOLD RETURNS

In early 1973, Gustav Becker was particularly concerned with the number of unsold copies being returned. After some preliminary analysis, he had determined that returns were due to three causes:

1. Individual customers (readers) did not buy a copy of every issue.
2. Individual customers did not always buy from the same retail outlet.
3. Technical returns (damaged or soiled copies).

Gustav had determined that only 30% or 40% of the nonsubscription customers of Gruner & Jahr would buy a copy of every issue of a magazine. However, because of the large number of individual customers, each buying different issues, the fluctuations in total demand for each issue were not as great as would otherwise be the case. For the magazines EI TERN and DER STERN, Gustav had gathered the data in Exhibits 1 and 2, which gave an indication of the range of fluctuations in total demand over the past two years. Gustav attributed these fluctuations in total demand to (a) trends in general demand and seasonal factors (magazine sales typically declined from January through May, increased from June through October and then dropped off again in December), (b) differences in individual issues such as cover photo, special features and total pages, (c) market factors (weather, political situation, holidays, etc.), and (d) advertising and promotion.

The fact that customers did not always buy the same magazine from a specific retail outlet was a particularly aggravating reason for unsold returns. Because of the high density of retail outlets in a given area, the copies sold through individual retail outlets could vary considerably from week to week, even though the national demand and the copies sold by individual wholesalers remained relatively constant. (Exhibit 3 indicates the magnitude of these demand fluctuations at individual retail outlets.) To allow for these fluctuations, most wholesalers planned deliveries to each of their retailers on the basis of average sales by that retailer for recent issues plus an additional complement of copies. The size of the complement of copies depended on the magazine sales volume handled by that retailer.

A final cause of unsold returns was due to technical reasons. Included here were copies used by the owner of the retail outlet, copies damaged in production or shipment, and soiled display copies. This category was generally only a small portion of total returns.

PLANNING PRODUCTION AND DELIVERY QUANTITIES

After gathering historical data on production, shipment and sales quantities, Gustav turned to an examination of the current procedure used at Gruner & Jahr to determine production and delivery plans. Although a number of staff people in the company ascribed considerable sophistication to decision making in this area, most of top management and the line people saw it as a rather straightforward matter. In the latter group the consensus was that a four-step procedure was involved in planning each issue.

1. *Preparation of a forecast of total demand.* This gave a single number corresponding to the expected number of copies that would be demanded.
2. *Setting production at the forecast value of demand plus 15%-20%.* This reflected management's past experience that around 16% unsold returns provided a reasonable level of service at the retail level.
3. *Allocating the total production by first meeting subscription requirements and then dividing the remaining copies among the wholesalers.* The allocation to each wholesaler was based largely on the *percentage* of wholesale distributed copies that had gone to that wholesaler for each of the last several issues. Occasionally, this was modified slightly if for the recent series of issues the number of unsold returns for that wholesaler had been unusually small or unusually large.
4. *The wholesaler would then distribute the copies he received to those retailers in his service area, as he thought most appropriate.* Management at Gruner & Jahr believed that most wholesalers made their allocation to each retailer based on the percentage of each issue generally handled by that retailer. However, Gustav's examination of recent data indicated that a wholesaler would generally react very quickly to a significant number of unsold returns from a given retail outlet. As a result, the number of

copies of each issue going to a single retailer was likely to fluctuate widely over a period of a few months.

One of the things that Gustav had become aware of during his preliminary analysis of the unsold returns problem was that the 16% figure used as a norm in decision-making was not based on any kind of economic analysis. Rather it had simply become an accepted standard for unsold returns based on habit and tradition of the past few years.

Pursuing his study of the existing procedure for determining total production and its allocation, Gustav had learned from the corporate marketing staff that the forecast of total demand for each issue of DER STERN and ELTERN was based largely on subjective estimates. One week prior to the start of each production run, the managing editor, the marketing director and a couple of key staff people would meet and review the production quantities, wholesale deliveries and unsold returns for the past several issues. After discussing the reasons for possible unexpected results on recent issues, they would consider those factors that they felt would be important determinants of demand for the coming issue. These factors would include such things as characteristics of the next issue, possible seasonal effects, planned advertising expenditures and holiday effects. Finally, after much discussion they would arrive at a single estimate of total demand for the coming issue. To this figure, 15%-20% would be added to represent planned returns, and this total quantity would be communicated to the production manager as the amount to be produced.

It was not until two days before the production run that detailed plans were made on the quantities to be shipped to each wholesaler. This task was handled by one of the staff people in distribution and shipping. This person would start with a set of basic percentage figures which indicated what fraction of total wholesaler shipments had been sent to each of the 180 wholesalers on the previous issue. He would then use data on the four most recent issues to determine which wholesalers had "excessive" or "insufficient" unsold returns. Although the actual criteria used to identify "excessive" and "insufficient" had never been stated explicitly, it was clear that "excessive" unsold returns was a much more frequent occurrence than "insufficient" returns. Typically only three or four wholesalers were identified as falling into one of these two categories during the planning cycle for a single issue. For these three or four, an adjustment would be made in the percentage of wholesale distributed copies each was to receive. The amount of the adjustment was based on the staff person's experience in doing that kind of thing.

Although Gustav had hoped to identify the procedure(s) most commonly used by individual wholesalers in allocating copies to their retailers, contact with half a dozen of these wholesalers had led him to believe that each used a different method. He had also concluded that over all these procedures tended to be very unsystematic and often over-reacted to variations in unsold returns from individual retailers.

A BETTER SYSTEM

Having become familiar with the procedures for planning allocating production, Gustav was more convinced than ever that substantial improvements could be made. Some of the areas that he thought might be particularly fruitful to investigate included:

1. Determination of the most appropriate level of planned returns. (Was it really 16%?)
2. Application of systematic techniques for forecasting total demand, allocation to wholesalers, and allocation to retailers.
3. Automation of a complete procedure for not only preparing forecasts but determining production quantity and delivery schedules in detail.

Although he thought it best to proceed by working on one of the company's most important magazines — either the monthly ELTERN, or the weekly DER STERN — he hoped that as he progressed he would be able to apply his findings to all six of the firm's magazines.

EXHIBIT 1 Monthly Production and Sales of ELTERN (1971-1972)
(All Figures in 000's)

| | | | | | REPRESENTATIVE WHOLESALERS | | | |
| | | | | | Wholesaler A | | Wholesaler B | |
Month	Production	Sold Circulation	Unsold Returns	Delivery Wholesalers	Delivery	Sales	Delivery	Sales
1/71	1,184.	1,033.	151	814.	12.6	11.4	18.0	13.7
2/71	1,180.	977.	203	815.	12.6	10.5	17.9	11.1
3/71	1,184.	902.	282	804.	12.6	9.7	17.5	9.5
4/71	1,190.	1,015.	175	803.	12.6	10.8	16.0	12.3
5/71	1,154.	965.	189	775.	12.3	10.4	13.1	10.7
6/71	1,149.	897.	252	760.	12.0	8.9	11.3	8.1
7/71	1,365.	1,201.	164	915.	14.0	12.8	15.7	14.3
8/71	1,318.	1,087.	231	868.	13.4	11.0	12.9	10.4
9/71	1,347.	1,030.	317	885.	14.0	10.8	14.2	10.2
10/71	1,327.	1,040.	287	881.	14.0	11.4	15.0	10.3
11/71	1,280.	972.	318	851.	14.0	11.0	16.0	11.1
12/71	1,254.	904.	350	836.	13.3	9.6	16.0	9.1
1/72	1,235.	960.	275	826.	13.0	11.0	16.0	9.6
2/72	1,219.	892.	327	812.	12.8	9.2	13.3	7.6
3/72	1,202.	830.	372	801.	12.8	8.2	13.5	7.4
4/72	1,150.	810.	340	748.	12.3	8.0	12.0	6.4
5/72	1,046.	726.	320	657.	11.0	7.1	10.0	4.9
6/72	999.	750.	249	630.	11.3	7.5	8.1	5.5
7/72	976.	760.	216	608.	10.0	7.4	8.0	5.2
8/72	1,036.	875.	161	870.	10.0	8.0	8.4	7.3
9/72	1,046.	860.	186	683.	10.0	8.2	12.0	8.6
10/72	1,038.	880.	158	687.	10.3	8.7	12.9	10.5
11/72	1,056.	880.	176	686.	10.3	8.9	13.4	9.7
12/72	1,024.	830.	194	662.	10.0	7.8	12.3	5.5

EXHIBIT 2 Weekly Production and Sales of DER STERN (1971-1972)
(All Figures in 000's)

			REPRESENTATIVE WHOLESALERS			
			WHOLESALER A		WHOLESALER B	
Week	Delivery to Wholesalers	Sales of Wholesalers	Delivery	Sales	Delivery	Sales
1/71	1,247	1,025	11.600	9.318	76.850	65.480
2/71	1,239	1,031	11.600	9.867	76.850	66.875
3/71	1,230	1,062	11.680	10.161	76.850	68.090
4/71	1,226	1,040	11.705	9.896	76.852	67.377
5/71	1,242	1,044	11.700	9.883	77.140	67.235
6/71	1,239	1,065	12.000	10.431	77.350	68.250
7/71	1,240	1,055	12.000	10.406	77.350	66.885
8/71	1,239	976	12.000	9.246	77.350	62.965
9/71	1,246	975	12.000	9.218	77.550	64.194
10/71	1,244	974	12.000	9.111	77.350	63.335
11/71	1,232	967	11.500	9.101	77.350	62.705
12/71	1,231	969	11.500	9.184	77.350	61.449
13/71	1,220	947	11.000	8.609	76.350	59.711
14/71	1,203	971	10.700	8.975	75.350	62.685
15/71	1,199	888	10.700	8.078	74.050	57.780
16/71	1,184	976	10.700	9.162	72.560	62.462
17/71	1,171	928	10.700	8.782	72.335	61.795
18/71	1,163	903	10.700	8.399	71.000	59.900
19/71	1,153	920	10.595	8.734	71.001	59.541
20/71	1,142	884	10.900	8.347	71.001	58.041
21/71	1,144	943	10.706	9.036	70.750	60.870
22/71	1,139	950	10.400	8.566	69.750	63.810
23/71	1,139	949	10.400	8.793	69.750	62.225
24/71	1,125	911	10.100	8.452	68.750	58.485
25/71	1,126	900	10.010	7.984	68.500	59.000
26/71	1,127	945	10.000	8.697	68.380	59.545
27/71	1,130	924	10.102	8.704	68.542	59.618
28/71	1,135	973	10.000	9.044	68.750	61.230
29/71	1,131	975	10.200	8.822	68.000	60.988
30/71	1,113	970	10.000	8.654	67.000	58.725
31/71	1,123	970	10.200	8.651	67.150	58.342
32/71	1,127	988	10.400	9.078	67.420	60.390
33/71	1,154	1,093	10.400	9.494	67.990	65.108
34/71	1,152	1,011	10.400	9.233	68.730	61.234
35/71	1,169	1,020	10.800	9.375	70.410	62.358
36/71	1,202	1,006	11.300	9.190	72.700	63.199
37/71	1,212	995	11.600	9.015	74.500	60.787
38/71	1,225	995	11.600	9.276	75.500	63.499
39/71	1,229	990	11.600	9.336	75.500	63.048
40/71	1,226	995	11.600	9.661	76.000	63.518
41/71	1,208	1,005	11.600	9.739	75.000	64.657
42/71	1,201	1,051	11.200	10.172	75.200	68.126
43/71	1,200	1,106	11.575	10.982	75.070	72.254
44/71	1,218	1,040	12.200	10.546	75.915	68.734
45/71	1,233	1,057	12.200	10.587	76.215	69.733

EXHIBIT 2   (continued)

| | | | REPRESENTATIVE WHOLESALERS | | | |
| | | | WHOLESALER A | | WHOLESALER B | |
Week	Delivery to Wholesalers	Sales of Wholesalers	Delivery	Sales	Delivery	Sales
46/71	1,246	1,089	12.600	11.333	77.455	71.854
47/71	1,248	1,059	12.675	11.161	78.595	69.673
48/71	1,252	1,076	12.800	11.327	79.000	71.305
49/71	1,268	1,103	13.000	11.530	79.640	71.876
50/71	1,274	1,118	13.300	11.805	79.500	73.432
51/71	1,275	1,141	13.600	12.249	79.675	73.954
52/71	1,278	1,186	13.660	12.891	79.811	77.696
2/72	1,289	1,196	13.000	12.486	82.020	77.455
3/72	1,298	1,194	13.770	12.374	81.650	75.190
4/72	1,319	1,201	14.630	12.010	82.620	77.265
5/72	1,353	1,142	14.800	11.948	84.650	72.260
6/72	1,375	1,200	14.800	12.576	86.000	77.430
7/72	1,134	1,079	14.550	11.321	85.700	69.619
8/72	1,355	1,143	14.550	12.008	85.705	72.199
9/72	1,346	1,113	14.400	11.424	85.200	70.104
10/72	1,347	1,178	14.400	12.307	85.378	74.728
11/72	1,328	1,089	14.000	11.281	85.200	68.842
12/72	1,315	1,122	13.600	11.742	82.440	72.560
13/72	1,309	1,022	13.800	10.277	82.200	64.872
14/72	1,304	1,097	13.400	11.111	79.000	70.475
15/72	1,299	1,090	13.800	11.219	79.000	67.670
16/72	1,295	1,086	12.000	11.007	78.000	67.850
17/72	1,284	1,041	13.000	10.858	76.000	66.588
18/72	1,299	1,045	13.400	10.909	76.000	64.049
19/72	1,304	955	13.400	9.143	75.003	62.683
20/72	1,299	985	13.400	9.842	75.000	61.040
21/72	1,293	1,025	13.000	10.288	74.500	62.908
22/72	1,288	1,000	13.000	10.056	73.700	61.495
23/72	1,278	928	13.000	8.425	73.500	60.690
24/72	1,245	985	12.400	9.620	72.400	61.745
25/72	1,229	960	12.000	9.185	72.200	62.775
26/72	1,224	977	11.600	9.495	71.250	60.000
27/72	1,222	1,018	11.400	9.421	71.000	62.390
28/72	1,207	1,012	11.000	9.717	71.000	61.670
29/72	1,212	1,037	11.000	9.925	70.500	61.847
30/72	1,200	1,019	11.000	9.628	69.500	58.652
31/72	1,181	1,011	11.000	9.354	69.500	58.655
32/72	1,182	991	11.000	9.072	69.800	58.750
33/72	1,187	1,003	11.000	9.083	69.890	58.545
34/72	1,195	1,013	11.000	9.240	70.300	60.275
35/72	1,196	1,023	11.500	9.244	71.200	62.150
36/72	1,208	1,005	11.500	9.281	71.695	61.720
37/72	1,214	1,019	11.500	9.548	72.500	61.560
38/72	1,234	1,075	11.500	10.267	73.500	61.595
39/72	1,218	1,027	11.500	9.858	73.500	62.540

EXHIBIT 2   (continued)

			REPRESENTATIVE WHOLESALERS			
			WHOLESALER A		WHOLESALER B	
Week	Delivery to Wholesalers	Sales of Wholesalers	Delivery	Sales	Delivery	Sales
40/72	1,220	1,030	11.500	9.778	73.008	63.777
41/72	1,224	1,009	11.800	9.765	73.002	62.872
42/72	1,224	1,026	11.800	10.282	72.504	64.344
43/72	1,225	1,025	11.800	10.309	72.504	65.020
44/72	1,223	1,031	11.950	10.266	72.510	64.385
45/72	1,230	1,029	12.000	10.433	72.750	64.370
46/72	1,219	1,037	12.000	10.381	73.001	65.392
47/72	1,191	976	11.700	9.883	71.501	64.067
48/72	1,191	956	11.700	9.539	71.180	60.974
49/72	1,210	924	12.200	9.573	73.000	59.040
50/72	1,197	1,006	12.000	10.608	72.000	63.816
51/72	1,193	987	12.000	10.320	72.015	62.174
52/72	1,156	918	11.500	9.516	70.000	57.239
53/72	1,160	968	11.500	9.715	70.000	57.230

EXHIBIT 3   Delivery and Sales for Typical Retailers

	ISSUE 30		ISSUE 31		ISSUE 32	
	Delivery	Sales	Delivery	Sales	Delivery	Sales
Retailer A	7	4	7	5	7	5
Retailer B	10	7	10	9	10	4
Retailer C	10	10	10	3	10	7
Retailer D	9	7	6	2	6	6
Retailer E	8	5	8	7	8	3
Retailer F	19	13	19	19	19	19
Retailer G	3	2	3	3	33	2
Retailer H	18	14	18	18	18	17
Totals For Wholesaler	20,300	16,395	19,300	16,496	19,300	16,290

ATWOOD APPLIANCE (A)

The Atwood Appliance Company manufactures small appliances which it markets through manufacturing representives. Both of the company's products, electric knives and electric can openers, are produced in its plant in Hammond, Illinois. In spite of its modest size, Atwood has been profitable in each of the past several years and they have realized an average growth in sales of 10% per year. (A budgeted income statement for the year ending June 30, 1972, is shown in Exhibit 1.)

Part of the company's success can be attributed directly to the efficiency of its production operations which are run by experienced personnel and to the company's ability to keep costs under control. As evident in Exhibit 2, Atwood's cost structure involves mostly variable costs, which reflects the relatively low capital investment required in the small appliance industry. The variable marketing costs are primarily sales commissions to manufacturing representatives and freight expenses. The variable manufacturing costs are approximately 40% labor expense and 60% material and component costs.

Although each of the products is made on a different production line, experience has shown that variable manufacturing costs increase by about 25% when total annual production volume exceeds 425,000 units for either product. For example, if the volume of knives exceeds 425,000 units, then the manufacturing cost/unit increases from $4.00 to $5.00 for each unit in excess of 425,000.

Another factor that has contributed to Atwood's performance has been its marketing efforts, headed by Bob Sanderson. During recent years, Bob has maintained an annual advertising budget of $100,000. This expenditure has usually been allocated about equally between the two products, electric knives and electric can openers. Recently Bob has been questioning the effect of this advertising on profits, and he has been wondering whether this amount should be increased or decreased and whether it should be divided differently between the two products.

Bob has also been considering the possibility of initiating some price changes on the two products, in addition to changing the advertising. These decisions on advertising the price for each product seem to be the major variables under his control that affect the number of units sold. Since it is now late spring, 1972, he would like to determine the most appropriate set of marketing decisions so that he can implement them for the coming fiscal year (July 1, 1972 – June 30, 1973).

To help in doing this Bob has been going over some of the statistics available on the small appliance industry. From these, he has concluded that demand for both Atwood's knives and can openers is such that volume decreases as prices increase and vice versa. These demand relationships are roughly illustrated in Exhibit 3.

In order to obtain more precise data that he can use in making these marketing decisions, Bob has recently conducted a market test for both knives and can openers using two price levels and two advertising levels for each product. From the results of this test and his other experience, Bob has prepared the generalized results shown in Exhibit 4.

EXHIBIT 1   ATWOOD APPLIANCE COMPANY 1972 Budgeted Financial Statement

INCOME STATEMENT (thousands of dollars)	
Sales (400,000 knives and 400,000 can openers)	$10,000
Manufacturing costs	5,400
GROSS MARGIN	4,600
Marketing costs	2,700
General and Administrative	500
Advertising costs	100
PROFIT BEFORE TAXES	1,300
Taxes (50 percent of PBT)	650
PROFIT AFTER TAXES	650

EXHIBIT 2 ATWOOD APPLIANCE COMPANY Analysis of Costs (Based on 1971-72 Results)

	PRODUCT LINE ($/unit)	
	Electric	*Electric*
VARIABLE COSTS	*Knives*	*Can Openers*
Selling price	$10.00	$15.00
Variable manufacturing costs		
Volume < 425,000	4.00	7.00
Volume > 425,000	5.00	8.75
Variable marketing costs	1.00	2.00

FIXED COST (per year)

Manufacturing costs — $1,000,000
Marketing costs — $1,500,000
General and Administrative — $300,000 plus 2% of total dollar sales

BUDGETED COSTS (per year)

Advertising costs — $100,000 (historically, $50,000 each product)

EXHIBIT 3 ATWOOD APPLIANCE COMPANY Product Demand Curves
(under existing advertising budget)

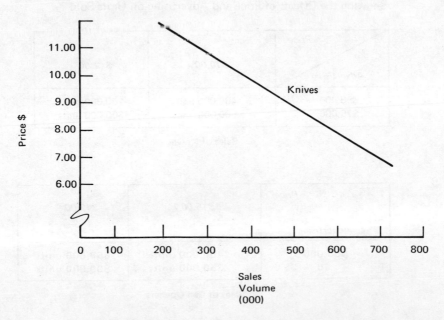

EXHIBIT 3   (continued)

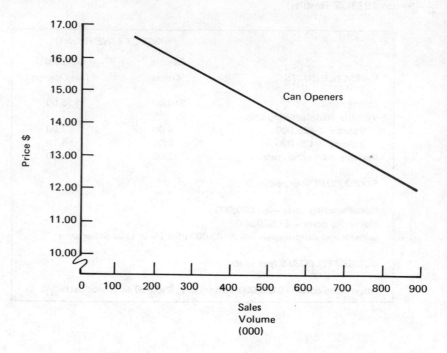

EXHIBIT 4   Atwood Appliance Company Results From Market Tests
Showing the Effects of Price and Advertising on Units Sold

Advertising \ Price	$10.00	$12.00
$50,000	400,000 units	200,000 units
$75,000	500,000 units	300,000 units

Sales of Knives

Advertising \ Price	$15.00	$12.00
$50,000	400,000 units	850,000 units
$0	350,000 units	800,000 units

Sales of Can Openers

SHASTA TIMBER

For the past two years, Bob Cohen and J.B. Sullivan have been serving as timber estimator apprentices under Al Beers, one of the two master estimators presently retained by Shasta Timber, a moderately large wood products firm in northern California. The timber estimator is a crucial link in maintaining a steady flow of logs from the forests to the mill, for based on his estimates, logging decisions are made and the flow thus established. If the flow is too low, the milling operations are disrupted and, if too high, excessive waste will result. Consequently, a good estimator is a highly valued asset to any company. Shasta will soon begin developing several new areas and a third master estimator will be required to cover these tracts. Mr. Cohen and Mr. Sullivan are prime candidates for this assignment.

A timber estimator surveys the tracts by foot and by helicopter estimating the forest density, the average height of the trees, and the average diameter. From this data and a certain feel that is developed over many years of close association with logging operations, an estimate of the number of cubic feet of usable timber is made. This estimate is given to the manager of the Forest Operations Division where it is locked and stored in his safe. The division manager is the only person in the company who has access to the estimate as well as to the actual yield from each of the surveyed tracts. They yield is roughly measured when the cut logs are transported from the tract to the mill.

The estimator works in an environment of little feedback since Shasta does not inquire as to how he made his prediction nor do they tell him the actual yield. In addition, two master estimators are never assigned to survey the same tracts. Thus the estimator does not have a standard against which he can compare his performance. In fact, the only indication that he had been performing well is that he has retained his job. When an estimator goes bad, he is transferred to another area.

The training program operates in precisely the same way except that the master estimator assists in the apprentice's estimates. For the past six months the two apprentices have been operating without Mr. Beer's assistance and the following performance has resulted:

Based on these records, the division manager must decide which apprentice is to be retained.

COHEN		SULLIVAN	
Estimate*	Actual*	Estimate*	Actual*
1500	1410	1425	1570
1625	1940	1625	2000
1225	1660	1400	1330
1375	1140	1100	1250
1850	1200	1500	1780
1450	1550		

*Cubic feet per acre

TEMPO, INC.

Louis De Rosa, the president of Tempo, Inc., had just acquired a new computer for his firm's data processing needs. Tempo was a medium-sized temporary help company operating throughout the New York-New Jersey metropolitan area. The new computer was intended primarily for payroll work (a particularly complex task for Tempo, which used a large number of part-time and occasional workers) but De Rosa had read enough about the uses of computers to realize that he should also consider the potential for having the machine contribute to the business in other ways. In the past, he had personally taken charge of forecasting the demand for workers in the 25 different categories supplied by Tempo: he prepared such forecasts several days in advance. He hoped that the new computer system could simplify this forecasting task and, perhaps, take over the bulk of the forecasting work.

THE TEMPORARY HELP INDUSTRY

In 1974 the U.S. temporary help industry was only about 25 years old. It had grown from a volume of $100 million in 1960 to one of a billion dollars in 1973. Initially, most of the workers provided by the temporary help agencies were clerical workers, but as the industry grew it had begun to provide a much wider range of employees, including some with specialized technical skills. The industry was dominated by Manpower, Inc., a 25-year-old company which accounted for a quarter of the industry's business in 1973. There were other large firms (Kelly Services, Inc. and Olsten Temporary Services) and also several thousand smaller (and more localized) firms in the field. Altogether, temporary workers comprised approximately 3% of the

U.S. labor force in 1973; the fraction of temporary workers was expected to rise markedly.

Customers of these temporary help firms hired temporary workers for varied reasons. In some cases, the temporaries were used to staff once or twice a year efforts such as the mail work associated with special promotions. Other firms used temporaries to fill in for regular workers who were on vacation. Sometimes employees who had reached the mandatory retirement age of their employer company were transferred to a temporary help agency and then hired through the agency by their original employers so that they could continue to work while technically not violating the retirement rules.

The conservative business climate and the inflationary pressure in 1974 had provided a particularly good opportunity for a more widespread and more sophisticated use of temporary help. Instead of maintaining personnel at the levels needed for times of maximum or near-maximum work load, companies whose work loads varied over the year instead were beginning to maintain their permanent staffs at lower levels, nearer the minimum required levels, and to fill in during peak times with temporary help. Such temporary workers generally cost the customer companies more than would their own permanent workers on a per hour basis, but the temporaries were nevertheless useful because they could be hired (and hence paid) for only the hours which they worked. The customer's own permanent help, on the other hand, had to be paid both wages and fringes during both the times when they were badly needed and the times when they were largely superfluous. The *total* cost of the temporaries turned out to be lower because the temporaries were not paid when they were not needed.

Another factor aiding the growth of the temporary work industry was the fact that work weeks in many industries were being shortened at the same time that the hours of stores and banks were being extended. Part-time and sometimes temporary help was used to provide staffing for the stores and banks over their extended working days.

Competition among temporary help firms was based primarily on the quality of the workers and the ability of the agencies to respond to customer needs, most of the agencies charged comparable rates and so price competition was not important.

In 1974 the industry had been growing at more than 20% a year and its firms were finding it more difficult to find high quality employees. Manpower's V.P. for marketing was quoted in *Business Week* (August 3, 1974) as saying that past marketing efforts had been concentrated on generating job orders; the current problem was to build up Manpower's pool of workers. The industry was spending more effort on recruiting and training its employees and was also working to reduce its extremely high turnover rates (75% each year for Manpower, with many employees working only a few months) and to establish closer ties with its customer firms.

TEMPO, INC.

Tempo operated twelve different offices scattered throughout the New York-New Jersey metropolitan area. The company prided itself on (and advertised strongly) its high quality personnel and its ability to fill many customer requests for help on extremely short notice — often within a matter of hours. Each of the twelve offices maintained a call list of workers in each of the twenty-five diverse fields in which the company claimed competence. These areas included secretarial work, typing, outdooor maintenance, janitorial work, electrical wiring, bookkeeping, drafting, driving and moving. Normally each office would receive most of the requests for workers at least one day early, although some of the requests came in at the last minute. The company was not always able to meet last minute requests for workers but, especially if the requests came from regular customers, the managers did try to provide the requested help. For orders for help placed at least a day early, the managers of the individual offices had orders to meet the demands, even if doing so involved unusual effort; for example, if there were an unusually high demand for outdoor maintenance workers, one office might exhaust its own call list in this area and might have to try to call up workers from other offices' lists.

De Rosa had established a system under which the work orders were reported from the individual offices to his office so that he could examine them as they came in. He also kept the demand figures for each office for each category of worker for the past several weeks to use in his forecasting. With this information, he tried to anticipate the demand figures a full week in advance so that the individual offices could have early warning of abnormally high or low demand levels. In addition, he knew that the early demand estimates were important to his managers in maintaining the particularly high quality work force which was extremely important for Tempo's business. Many of the workers on Tempo's lists were particularly good workers who, for one reason or another, chose to work part time and, often, required advance warning so that they could make arrangements to work on any given day. Many of the other temporary help firms in the metropolitan area refused to hire such people because managing them required extra effort, but De Rosa's policy was to encourage such arrangements if the workers were particularly good. His theory was that they would be likely to remain with Tempo for much longer than was common in the industry, simply because the company was willing to arrange for their needs. He saw the policy as a way of making Tempo's work force significantly better in quality than were those of some of the competitors.

Tempo had a policy for all of its workers that they would be paid in full for every day for which they were called to work, even if a job did not materialize for them on some days. Thus, the company could not afford to call up enough workers to be certain to meet all of the last minute requests which came in. On the other hand, De Rosa did ask the office managers to call in a few workers for filling such requests.

De Rosa felt that the business benefited in several ways from the demand forecasts which he prepared. The forecasts gave the branch managers advance warning of their labor needs so that they could call up appropriate numbers of workers; often a manager used the forecasts to call up workers before firm orders for those individuals had been received. In some cases the early warning allowed managers to call on workers who could work only on advance notice. In other cases, a manager was warned by the forecasts of an impending particularly heavy demand in time to be able to ask for names from the lists of other branch offices to fill the demand. De Rosa believed that without the forecasts his managers would have been reluctant to call up large numbers of workers to meet any anticipated heavy load of last minute requests for workers; he feared that without forecasts the managers would not have wanted to go out on a limb before substantial numbers of firm orders were received. With the forecasts, on the other hand, De Rosa had himself taken responsibility for such orders and his managers were far less reluctant to plan in anticipation of demand.

THE FORECASTING PROBLEM

As a first step in setting up a forecasting system, De Rosa sat down with a consultant in the field to discuss the important characteristics of his business which any worthwhile system would have to consider. He explained:

I do my planning in half-day time periods. Full-day periods just aren't fine enough and I've never had the time to consider more than two periods per day. Actually, though, I really don't think more than two slots per day would make sense.

It's essential to keep track of the figures for the separate offices rather than to try to guess at total figures for the business. The whole pattern of demand at the office near Wall Street is fundamentally different from the one in Long Island City and also different from the various Jersey offices. The offices have different busy days, different patterns of demand for the various categories of workers — I guess I'd just call them basically different.

I have to keep the figures for the 25 different worker categories separate, too. In fact, if I had the opportunity, I'd really like to break the categories down more finely than I have in the past. I think it would help to match workers to jobs more effectively. Anyhow, the categories behave differently, in terms of demand. Some are higher at one time of the year, others higher at other times. Also, the pattern of use of the various categories over the year isn't the same at all 12 of our offices.

There are real differences in the demand pattern over the course of an individual week, too. Demand is particularly high on Mondays, for example, and there is often a rush of last minute orders at the end of the week as our customers try to finish one job or another. I don't think that the pattern within a week depends on the time of the year for any one of our offices, but I'm pretty sure that the patterns at the different offices are different.

There's another set of problems which complicate the forecasting process. They're what I call special events — a bad snowstorm which raises the demand for outdoor maintenance help, for example. I can predict some of these special events (conventions, for example) but not others. And after one of those things, it's never quite clear to me how to use the demand figures which were affected by a special event.

There are, as I see it, two main sources for information for my forecasting. One is the demand information from the past few weeks. Unfortunately, I haven't kept good demand information over the years, but I do always have reasonable information from the past month or so and I use that information, together with any hunches I have based on the time of year, or special events or something, to come up with a preliminary set of demand figures a full week in advance. I use those figures to give the managers a really early indication of what to expect a week later.

There's a second set of numbers which I've decided should be used separately from the demand numbers. Those are the actual work orders as they come in. I know, for example, that work orders for a given day straggle in over the week or maybe two weeks preceding. If I see an unusually high number of orders for 5 days from now, I can do some work to try to find out whether something unusual is happening and has to be prepared for or whether there's just a fluke in the timing. I'm pretty certain that I want the forecasts prepared from the past demand kept separate from the forecasts prepared from the early orders because I wouldn't want to have to look at a single high figure, say, and try to decide which of the two factors (past demand or early orders) made it high. Also, I'd like the computer system to generate the figures based on past demand automatically. Maybe it should consider the early orders and also check whether those orders look unusual. It could send out a warning only if something strange is happening which needs my attention.

I've been assured that there is computer time available for doing some forecasting. We'll probably need to buy some extra storage space of one type or another for the data. I'm anxious to keep the costs of the computer installation down, but I do think some kind of automated forecasting would be worthwhile.

ASSIGNMENT

Design, in some detail, a forecasting system using the past demand data. In particular, how would you consider the differences among offices and work categories? What would you do about special events? How might you set up a separate procedure for using the early order information? (For each of the two procedures, consider not only how the procedure would operate when fully implemented, but also what information would have to be collected to make such full implementation possible and how the procedures could be modified if necessary for use during the implementation stage.)

PERKIN ELMER INSTRUMENT DIVISION[1]

OPERATING PLANS AND FORECASTS

In October 1974, Gaynor Kelley, Vice President and Manager of Perkin Elmer's Instrument Division, perceived disgruntling differences among three plans central to his division's operations.

The first plan was the division's annual business plan. This plan, dated July 1974, was the result of the division's annual planning activities. It was based on a projected division sales level of $93.4 million. The second plan had just reached Kelley. It was an updated forecast of orders received made by the product managers in early October 1974. It forecast 1974/75 fiscal year orders at $103.1 million. The third plan, also recently received by Kelley, was a manufacturing plan, which forecast the sales value of the product which was to be manufactured during the 1974/75 year. This "build plan" indicated that manufacturing planned to build $111.8 million worth of product for sale in the year. Inventories had already increased by over a million dollars since August in anticipation of the increase in sales projected by the product managers.

Kelley was more troubled by the high volumes indicated by the later two plans than by the discrepancies between them. Division sales had almost doubled in the previous three years, but signs that the economy was weakening had long been expected and were already appearing in other major industries. Moreover, the corporate financial staff had indicated that they had not planned to provide funds to meet the unexpected surge in business predicted by the marketing and build plans, and that they would have difficulty in obtaining them economically on such short notice.

[1] To protect the confidentiality of company data, certain figures in this case have been discussed.

Company Background

Perkin Elmer was an international developer and producer of high technology scientific instrumentation. The company was founded in 1938 to design and produce ultra-precise optical equipment. Growth in this area was rapid as these optical instruments found wide application in defense and space programs. In the 1970s under the leadership of Chairman Chester W. Nimitz, Jr., and President Robert H. Sorensen, the company began to emphasize the development of other types of laboratory analytical instrumentation. Products were marketed to researchers in college and hospital clinical laboratories, and to research and process control groups in a variety of industries. The rapid growth of the Instrument Group, which was responsible for manufacturing and selling these non-optical instruments resulted in a decrease in the percentage of the company's business that went to the U.S. government agencies. This percentage fell to only 27% of total sales in 1974. By this time, Instrument Group's sales accounted for more than one-half of the sales of the total company.

The company emphasized research and development to maintain its position as a company on the leading edge of rapidly expanding optical and electrical technologies. In 1974, the company spent $16.7 million on research and development, not counting a much larger amount spent on government contract research (see Exhibit 1 for financial data). Almost 15% of the company's 8600 employees were graduate scientists and engineers.

The Instrument Group, under the guidance of Senior Vice President Horace McDonell, had sales of $150 million in 1974, a 17% increase over 1973 sales levels. This group also emphasized research and development as an integral part of their competitive strategy. Historically, research and development expenses had averaged 8-9% of instrument sales. Over 70% of the orders taken by the group in 1974 were for products that had not existed in 1969.

The Instrument Division

The Instrument Division, headed by Gaynor Kelley was the largest organizational entity in the Instrument Group. Kelly had profit and asset responsibility for the marketing and manufacture of eight product lines. Reporting to him were eight product managers, the director of manufacturing, and the technical director who was responsible for engineering and development. Kelley also worked closely with the vice president of the Instrument Marketing Division, which was responsible for field sales and service.

Instrument Division products were sold through the field sales and service representatives of the Instrument Marketing Division. These people called on and advised potential customers, attended trade shows, and provided spare parts and technical service advice to users of Perkin Elmer equipment.

Marketing strategy, product line selection, and promotional strategy were the responsibility of the eight product managers in the Instrument Division. Each product manager had profit responsibility for an entire line of instruments and associated accessory items.

The price of an instrument varied from $3,000 − $35,000, while the price range for the accessory items used to operate the instrument was from $200 − $5,000. Price was not the primary basis for competition, however. Perkin Elmer's product managers claimed that the most important selling points were the features of their products. Perkin Elmer customers tended to be highly sophisticated from a technological standpoint. Thus, new instruments which could perform new tests or measurements with a high or greater degree of accuracy and reliability were preferred by them. Most major competitors tended to compete on the same basis.

A secondary selling point was delivery time and service. In the words of Jack Kerber, product manager for the atomic absorption instrument line, "A customer may take up to a year deciding whether he needs an instrument or in obtaining funds to purchase it, but once they decide and have the money, they want immediate delivery. And a lot of times, if they can't get it, they'll turn to our competitors if they have a comparable instrument."

Because the mechanical or electrical failure of an instrument could delay the completion of a research project, or shut down the production of industrial customers, rapid customer service was also felt to be important. To respond rapidly to these customer service needs, over 60 domestic field sales and service offices were maintained by the U.S. Sales Division.

The Instrument Division dealt with delivery lead times in two ways. First, they maintained an informal standard delivery lead time of 4-0 weeks. In other words, they attempted to maintain enough components in parts inventory or in-process, to be able to assemble, inspect, package, and ship an instrument within 4-6 weeks after it was ordered. During 1974, the average delivery lead time had slipped to about 12 weeks, however. According to Bill Chorske, General Manager, U.S. Sales Division, "We just haven't been able to catch up with our orders. The purchase lead times on some of the critical parts and materials we use have gone out to 10-12 months from an average of three months. Add to that the manufacturing time required to assemble parts into an instrument and you've got an impossible planning horizon. Couple that with an unexpected increase in demand and the result is inevitable."

The second way that the Instrument Division coped with delivery lead times was to keep the field sales force constantly informed of changes in them. This was partially accomplished with a monthly "Export Instrument Shipment Schedule" for export items and sales. In the U.S., delivery lead times for an item could be supplied to the field sales force almost instantaneously if they called the marketing service department at Perkin Elmer's home office in Norwalk, Connecticut. Marketing service representatives queried computerized files from the centralized production control system on video display units to determine if enough parts were on hand to assemble the item immediately, or to determine the length of time required to obtain unavailable parts. This information was relayed to the field representative. If a firm order was made, parts were reserved for that order and an assembly order issued.

The Planning Process

The total planning process at Perkin Elmer's Instrument Division involved the coordination of the plans of the four major functional entities: engineering, marketing, manufacturing, and finance. The degree of coordination in these individual efforts was most apparent in the annual business and financial planning cycle.

The Business Plan

The annual business plan started off with a detailed instrument by instrument forecast of order receipts. This forecast was made from a "bottom up" sales forecast from the field sales and service personnel in the Instrument Marketing Division, and a top down forecast by the product managers. Differences between these two forecasts were resolved by further investigation and negotiation between the two groups. These forecasts specified orders by month for the two immediately following quarters, and projected orders by quarter for the last half of the fiscal year (Exhibit 2).

In addition to the forecast of orders for instruments, the sales of spare parts for technical service and accessory items were also forecasted. Since these items were generally stocked in inventory for immediate delivery, sales rather than orders were forecasted. A computer program incorporating some simple smoothing methodologies was used to forecast most of these items. All in all, accessories and spare parts sales accounted for about 20% of the division's annual sales. Instruments accounted for only a small percentage of the total number of saleable items, but made up 80% of sales.

Yet a third source of forecast information was Gaynor Kelley himself. He reserved the right to make "management stock orders." These were essentially hedges on major products with long lead times. For example, one new product which was still partially in the preproduction design stage was expected to be completely ready for manufacture by February 1975. Since the procurement and manufacturing lead times for this item were in excess of a year, and the company wished to be in a position to promise six weeks delivery when it was introduced, Kelley, after weighing both the market and engineering risks, might wish to forecast sales for this item in April. This would serve as an authorization to purchase or start manufacturing the long lead time parts that went into the instrument. Similarly, if a major sale to a single customer (such as the government) of a particular high value, long lead time instrument was expected (over and above regular sales), a management stock order might be placed to ensure that a reasonable delivery lead time could be offered to the prospective customer, or to make sure that the sale could be made within the current fiscal year.

Management stock orders were collected together in a special "Z" plan. In general, the Z plan served to ensure that the major risks of the division were handled and monitored by the highest levels of management. In October of 1974, the sales value of the instruments in the Z plan for fiscal year 1975 was $4.3 million, while the product manager's forecast plus the forecast for spare parts totaled $72.2 million.

The three forecasts—the product manager/marketing forecast of major instrument orders, the computer generated forecast of accessory and spare parts sales, and the "Z" plan for management stock orders — were given to manufacturing. The production planning group then proceeded to project the manufacturing function, and the business as a whole. This was accomplished with the help of a computerized material and capacity requirements planning system.

First, the forecasts were used to construct a series of master plans. At least a part of the forecast data was forecasted order receipts. To consturct the "exploded" master plan, which showed forecasted shipments, the standard delivery lead times were added to the forecasted order receipt dates. For example, instrument number 874ZX had sales orders for 51 systems projected for October. The "exploded" master plan broke this monthly forecast down into a weekly forecast for 13 units in each of the four weeks of the month. Then, since the Instrument Division tried to maintain a fourweek position on this part, that is, be able to deliver four weeks after an order, the forecast orders for 13 units in each week in October were translated into forecast shipments of 13 in each week of November.

Parts and accessories which were stocked for immediate shipment were transposed directly into a "sales" master plan without adding lead times. The Z plan, which already reflected delivery time hedging, was the third type of master plan used by the Instrument Division.

The business plan was formulated in the manner described above at the beginning of each fiscal year. The process was repeated at mid-year in what was called the phase II plan. Thus, the company maintained a rolling six-month plan.

Short-term Planning

Between the business planning cycles, the process of forecasting and replanning was carried out on a continual basis. Sales and orders forecasts for major items were reviewed monthly by product managers. Sales and orders forecasts for less important items were reviewed bi-monthly or every quarter. These changes in forecasts meant that changes in the master plans were made monthly.

Preparing for a Business Turndown

Gaynor Kelley felt that the differences between the forecasts in the annual business plan, the product manager forecast, and the build plan were largely due to differences in the perceptions of the people involved in making these forecasts. The product managers, after having been caught short during the '74 boom, were bullish. Many manufacturing people, and especially the product planners, after having lived through a period of shortages, extended lead times, and vendor unreliability, were cautious. They were attempting to gain back the four to six-week delivery position lost earlier in the year, and were still projecting protracted lead times for use in timing orders in the material requirements planning system.

Kelley felt that fiscal year 1975 sales would actually turn out to be very close to those projected in the original business or financial plan. The Instrument Division had generally followed a cyclical pattern close to that of the national economy. Some lead times were falling and purchase commitments for the division were rising. Order cancellations threatened in other industries. But, he was at this point undecided as to whether the Instrument Division should position themselves to handle the upside or downside risks.

The upside risks of losing sales and market position would be felt if they cut back, and a recession did not materialize. The downside risks of large inventories and a high fixed cost position would be felt if they increased production and a recession did materialize.

EXHIBIT 1   Perkin Elmer Instrument Division (R)

FINANCIAL DATA

(Dollar amounts in thousands except per share amounts)

Financial Operations — years ended July 31

	1974	1973	1972	1971	1970	1969	1968	1967	1966	1965
Net sales	$272,042	$233,323	$216,813	$202,550	$232,948	$227,102	$171,962	$128,542	$103,511	$ 78,383
Cost of sales	161,524	141,286	136,864	129,818	159,049	157,814	114,428	80,199	62,466	46,380
Research and development	16,727	13,120	10,772	10,076	10,814	10,394	8,869	8,135	6,357	4,747
Marketing	43,194	35,178	30,240	28,053	27,052	23,751	20,409	18,025	14,878	11,699
General and administrative	21,681	17,455	16,365	14,892	16,197	15,244	13,009	10,546	8,992	7,018
Interest expense	1,385	919	1,575	1,443	1,563	1,628	1,564	1,146	732	764
Other incomes net	(3,082)	(1,867)	(1,912)	(1,729)	(1,843)	(570)	(541)	(440)	(576)	(613)
Income before provision for income taxes	30,616	27,232	22,909	19,997	20,116	18,841	14,224	10,931	10,662	8,388
As a percent of sales	11.3%	11.7%	10.6%	9.9%	8.6%	8.3%	8.3%	8.5%	10.3%	10.7%
Net income	$ 17,159	$ 15,698	$ 13,133	$ 10,122	$ 9,647	$ 9,297	$ 6,601	$ 5,193	$ 5,486	$ 4,201
As a percent of sales	6.3%	6.7%	6.1%	5.0%	4.1%	4.1%	3.8%	4.0%	5.3%	5.4%
Net income per share	$ 0.98	$ 0.90	$ 0.77	$ 0.60	$ 0.58	$ 0.57	$ 0.42	$ 0.34	$ 0.38	$ 0.33
Dividends per share	0.225	0.214	0.206	0.15	—	—	—	—	—	—
Return on shareholders equity at year-end	11.9%	12.4%	11.9%	10.6%	11.1%	12.3%	10.7%	9.9%	12.4%	11.1%

EXHIBIT 1 (continued)

(Dollar amounts in thousands except per share amounts)

	1974	1973	1972	1971	1970	1969	1968	1967	1966	1965
Financial Condition — at July 31										
Working capital	$109,085	$ 91,233	$ 86,179	$ 79,548	$ 71,540	$ 60,002	$ 47,368	$ 37,778	$ 29,860	$ 27,051
Fixed assets at cost	71,206	65,577	58,503	57,249	55,284	52,169	47,929	39,181	31,186	24,256
Long term debt	4,295	5,704	7,072	16,015	18,213	19,055	19,718	13,681	8,274	8,605
Shareholders equity	144,390	126,493	110,814	95,750	86,996	75,858	61,876	52,331	44,318	37,725
General										
Average number of common shares outstanding including common stock equivalents (in thousands)	17,667	17,514	17,305	16,740	16,621	16,445	15,890	15,081	14,441	12,899
Employees	8,627	7,933	7,334	7,509	7,827	8,204	7,623	6,978	5,769	4,417
Shareholders	10,668	10,559	10,087	10,176	10,925	9,562	9,510	8,118	6,427	5,588

The above data include Interdata, Incorporated on a pooling of interests basis. Interdata results included for the fiscal years 1971 and 1973 are for the twelve months ended July 31 and for the total years 1967 (the first year of operations) through 1972 are for the twelve months ended December 31. In addition certain other amounts have been reclassified to conform with the 1974 presentation.

EXHIBIT 1 (continued)

Assets

Assets	at July 31 1974	1973
Current Assets		
Cash, including time deposits	$ 7,554,375	$ 9,727,595
Marketable securities, at cost (approximate market)	17,535,029	16,608,661
Accounts receivable, less allowance for doubtful accounts of $608,630 ($583,809 in 1973)	62,787,044	55,732,439
Inventories, at lower of cost or market	75,287,832	54,857,102
Prepaid expenses and other current assets	6,612,049	4,426,662
	169,776,329	141,352,459
Marketable securities maturing beyond one year, at cost (approximate market)	5,753,005	6,598,684
Property, Plant and Equipment, at cost		
Band	4,285,966	4,499,384
Buildings	36,207,139	34,540,674
Machinery and equipment	30,712,593	26,536,833
	71,205,698	65,576,891
Accumulated depreciation and amortization	(34,625,214)	(31,290,702)
	36,580,454	34,286,189
Other Assets		
Excess of purchase price over net assets of companies acquired	4,207,142	4,194,168
Other investments, patents, deferred charges, etc.	3,534,311	3,873,891
	7,741,453	8,068,059
	$219,851,241	$190,305,391

Liabilities

Liabilities	at July 31 1974	1973
Current Liabilities		
Loans payable, United States	$ 7,500,000	$ 3,385,994
Loans payable, foreign	7,220,911	3,828,395
Accounts payable, including advances from customers	17,671,796	16,023,352
Accrued salaries and wages	11,347,676	10,041,667
Accrued taxes on income	7,086,132	7,745,084
Other accrued expenses	9,865,276	9,095,211
	60,691,791	50,119,703
Long-Term Debt		
United States	—	5,704,420
Foreign	4,294,631	6,677,774
Other Long-Term Liabilities	8,957,309	
Minority Interest	1,517,760	1,310,986
Shareholders' Equity		
Capital stock		
Preferred stock, $1 par value: Shares authorized 1,000,000	—	—
Common stock, $1 par value: Shares authorized 20,000,000 Shares issued 17,435,881 (17,264,034–1973)	17,435,881	17,264,034
Capital contributed in excess of par value	28,386,472	24,284,280
Retained income	98,567,397	84,944,194
	144,389,750	126,492,508
	$219,851,241	$190,305,391

EXHIBIT 2 Perkin Elmer Instrument Division

INSTRUMENT ORDERS FORECASTS (FISCAL YEAR ENDS JULY 31)

Instrument	Fiscal Year Actual/(Forecasts)*			Fiscal Year 1975 (Forecasts)								
	72	73	74	A	S	O	N	D	J	3rd Q	4th Q	12 month Total*
874X	—	—	78/(101)	19	33	51	27	29	48	133	129	469
90475Z	174/(169)	235/(214)	153/(160)	8	8	18	15	8	14	18	20	109
9903751	177/(185)	122/(111)	110/(85)	2	3	9	—	2	3	11	—	30
4753ZY	22/(15)	14/(21)	15/(15)	—	—	2	1	—	1	—	1	5
976427	3/(5)	14/(10)	18/(20)	—	—	7	1	—	2	1	3	14
274678	—	4/(0)	33/(45)	4	7	12	2	6	11	29	26	97

*Annual Forecasts are simply the sum of the forecasts for **6** single months plus two quarters prepared as a part of the business plan just prior to the start of each fiscal year.

References For Further Study

Aaker, D. A., ed., *Multivariate Analysis in Marketing: Theory and Application,* Wadsworth Publishing Company, Belmont, California, 1971.

Anderson, B. D., "A Qualitative Introduction to Wiener and Kalman-Bucy Filters," *Proceedings,* The Institution of Radio and Electronics Engineers, Australia, March 1971, pp. 93-103.

Bates, J. M., and C. W. J. Granger, "Combination of Forecast," *Operation Research Quarterly,* Vol. 20, no. 4, 1969, pp. 451-468.

Bode, H. W., and C. E. Shannon, "A Simplified Derivation of Linear Least Square Smoothing and Prediction Theory," *Proceedings,* I.R.E., Vol. 38, 1950, pp. 417-425.

Booton, R. C., "An Optimization Theory for Time-Varying Linear Systems With Non-Stationary Statistical Inputs," *Proceedings,* I.R.E., Vol. 40, 1952, pp. 977-981.

Box, G. E. P., and G. M. Jenkins, "Some Recent Advances in Forecasting and Control," *Applied Statistics,* Vol. 17, 1968, p. 91.

Box, G. E. P., and G. M. Jenkins, *Time Series Analysis, Forecasting and Control,* Holden-Day, San Francisco, 1976 (revised edition).

Box, G. E. P., and D. A. Pierce, "Distribution of Residual Autocorrelations in Autoregressive-Integrated Moving-Average Time-Series Models," *Journal of the American Statistical Association,* Vol. 65, 1970, pp. 1509-1526.

Brown, R. G., *Statistical Forecasting for Inventory Control,* McGraw-Hill, New York, 1959.

Brown, R. G., *Smoothing, Forecasting and Prediction of Discrete Time Series,* Prentice-Hall, New Jersey, 1963.

Brown, R. G., and R. F. Meyer, "The Fundamental Theorem of Exponential Smoothing," *Operations Research,* Vol. 9, 1961, pp. 673-685.

Burman, J. P., "Moving Seasonal Adjustments of Economic Time Series," *Journal of the Royal Statistical Society,* Series A, Vol. 128, 1965, pp. 534-558.

Chambers, J. C., et al., "How to Choose the Right Forecasting Technique," *Harvard Business Review,* July-August, 1971, pp. 45-74.

Chambers, J.C., S. K. Mullick and D. D. Smith, *An Executive's Guide to Forecasting,* John Wiley and Sons, New York, 1974.

Chatfield, C, *The Analysis of Time Series,* Chapman and Hall, Ltd., London, 1975.

Chatfield, C., and D. L. Prothero, "Box-Jenkins Seasonal Forecasting: Problems in a Case-Study," *Journal of the Royal Statistical Society,* Vol. 136, Series A, Part 3, pp. 295-336.

Chisholm, Roger K. and Gilbert R. Whitaker, Jr., *Forecasting Methods,* Richard D. Irwin, Inc., Homewood, Illinois, 1971.

Daniell, P. J., "Discussion on 'Symposium on Autocorrelation in Time Series'," Supp. *Journal of the Royal Statistical Society,* Series B, Vol. 8, 1946, p. 88.

Dauten, Carl A. and Lloyd M. Valentine, *Business Cycles and Forecasting*, South-Western Publishing, Cincinnati, Ohio, 1974.

Dixon, W. J., "Further Contributions to the Problem of Serial Correlations," *Annals of Mathematical Statistics*, Vol. 15, 1944, pp. 119-144.

Draper, N. and H. Smith, *Applied Regression Analysis*, John Wiley & Sons, N.Y., 1966.

Durbin, J., "Efficient Estimation of Parameters in Moving-Average Models," *Biometrica*, Vol. 46, 1959, pp. 306-316.

Durbin, J., "Estimation of Parameters in Time-Series Regression Models," *Journal of the Royal Statistical Society*, Vol. 22, 1960, pp. 139-153.

Durbin, J., "Testing for Serial Correlation in Least-Squares Regression When Some of the Regressors are Lagged Dependent Variables," *Econometrica*, Vol. 38, 1970, pp. 410-421.

Durbin, J., "Trend Elimination for the Purpose of Estimating Seasonal and Periodic Components," in M. Rosenblatt, (ed.), *Time-Series Analysis*, Wiley, New York.

Farley, J. U., and M. J. Hinich, "Detecting 'Small' Mean Shifts in Time Series," *Management Science*, Vol. 17, no. 3, November 1970, pp. 189-199.

Granger, C. W. J., "Investigating Causal Relations by Econometric Models and Cross Spectral Methods," *Econometrica*, 37, pp. 424-438.

Granger, C. W. J., and P. Newbold, "Economic Forecasting: the Atheist's Viewpoint," *Nottingham University Forecasting Project*, Note 11, 1972.

Granger, C. W. J. and P. Newbold, "Spurious Regressions in Econometrics," *Journal of Econometrics*, 2, 1974, pp. 111-120.

Groff, G. K., "Empirical comparison of Models for Short-Range Forecasting," *Management Science*, Vol. 20, no. 1, September 1973, pp. 22-31.

Gross, Charles W. and Robin T. Peterson, *Business Forecasting*, Houghton Mifflin Company, Boston, Massachusetts, 1976.

Handbook of Forecasting Techniques, A Report Submitted to the U.S. Army Engineer Institute for Water Resources, IWR Contract Report 75-7, December, 1975.

Hannan, E. J., "The Estimation of Seasonal Variation in Economic Time Series," *Journal of the American Statistical Association*, Vol. 58, 1963, pp. 31-44.

Harrison, P. J., "Short-Term Forecasting," *Applied Statistics*, Vol. 14, 1965, pp. 102-139.

Holt, C. C., "Forecasting Seasonal and Trends by Exponentially Weighted Moving Averages," Carnegie Institute of Technology, Pittsburgh, Pennsylvania, 1957.

Jacoby, S. L. S., J. S. Kowalik, and J. T. Pizzo, *Iterative Methods for Non-Linear Optimization Problems*, Prentice-Hall, Englewood Cliffs, New Jersey, 1972.

Jenkins, G. M., "A Survey of Spectral Analysis," *Applied Statistics*, Vol. 14, no. 1, 1965, pp. 2-32.

Jenkins, G. M., and M. B. Priestley, "The Spectral Analysis of Time Series,"

Journal of the Royal Statistical Society, Series B, Vol. 19, no. 1, 1957, pp. 1-12.

Jenkins, G. M., and D. G. Watts, *Spectral Analysis and Its Applications,* Holden-Day, San Francisco, 1968.

Johnston, J., *Econometrics* Second Edition, McGraw-Hill, New York, 1972.

Kalman, R. E., "A New Approach to Linear Filtering and Prediction Problems," *Journal of Basic Engineering,* D. 82, 1960, pp. 35-44.

Kalman, R. E., and R. S. Bucy, "New Results in Linear Filtering and Prediction Theory," *Journal of Basic Engineering,* D. 83, 1961, pp. 95-107.

Kendall, M. G., *Time Series,* Hafner Press, New York, 1973.

Leser, C. W. V., "A Survey of Econometrics," *Journal of the Royal Statistical Society,* Series A, Vol. 131, 1968, pp. 530-566.

Levinson, N., "The Wiener RMS (Root Mean Square) Error Criterion in Filter Design and Prediction," *Journal of Mathematical Physics,* Vol. XXV, no. 4, January 1947, pp. 261-278.

Lewis, Colin D., *Demand Analysis and Inventory Control,* Lexington Books, Lexington, Massachusetts, 1975.

Mabert, Vincent A., "An Introduction to Short-Term Forecasting Using the Box-Jenkins Methodology," AIIE, (PP & G – 75 – 1), Norcross, Georgia, 1975.

Macauley, F. R., *The Smoothing of Time Series,* National Bureau of Economic Research, 1930, pp. 121-136.

Makridakis, S., "A Survey of Time Series," *International Statistical Reveiw,* Vol. 44, No. 1, 1976.

Makridakis, S., et al., "An Interactive Forecasting System," *American Statistician,* November, 1974.

Makridakis, Spyros and Steven C. Wheelwright, *Interactive Forecasting,* Holden-Day, San Francisco, 1978.

Makridakis, S. and S. Wheelwright, *Forecasting: Methods and Applications,* John Wiley and Sons, New York, 1978.

Marquardt, D. W., "An Algorithm for Least Squares Estimation of Nonlinear Parameters," *Soc. Indst. Appl. Math.,* Vol. 11, no. 2 (June, 1963).

McLaughlin, R. L., *Time-Series Forecasting,* Marketing-Research Technique, Series No. 6, American Marketing Association, 1962.

McLaughlin, R. L., "A New Five-Phase Economic Forecasting System," *Business Economics,* Sept. 1975, pp. 49-60.

McLaughlin, R. L., and James J. Boyle, *Short-Term Forecasting,* American Marketing Association, New York, 1968.

McClain, J. O., and L. J. Thomas, "Response-Variance Tradeoffs in Adaptive Forecasting," *Operations Research,* Vol. 21, no. 2, March-April, 1973, pp. 554-568.

Mehra, R. K., "On the Identification of Variances and Adaptive Kalman Filtering," IEEE, *Transactions on Automatic Control,* Vol. 10, AC-15, no. 2, April 1970, pp. 175-184.

Milne, Thomas E., *Business Forecasting, A Managerial Approach,* Longman Group, New York, 1975.

Montgomery, Douglas C. and Lynwood A. Johnson, *Forecasting and Time Series Analysis,* McGraw-Hill Book Company, New York, 1976.

Morrison, Norman, *Introduction to Sequential Smoothing and Prediction,* McGraw-Hill Book, New York, 1969.

Naylor, T. H., and T. G. Seaks, "Box-Jenkins Methods: An Alternative to Econometric Models," *International Statistical Review,* Vol. 40, no. 2, 1972, pp. 123-137.

Nelson, Charles R., *Applied Time Series Analysis for Managerial Forecasting,* Holden-Day, Inc., San Francisco, 1973.

Nerlove, M., "Spectral Analysis of Seasonal Adjustment Procedures," *Econometrica,* Vol. 32, July 1964, pp. 241-286.

Page, E. S., "On Problems in which a Change in Parameter Occurs at an Unknown Point," *Biometica,* Vol. 4, 1957, p. 249.

Parzen, Emanuel, *Time Series Analysis Papers,* Holden-Day, San Francisco, 1967.

Pierce, D. A., "Relationships – And the Lack Thereof – Between Economic Time Series, with Special Reference to Money, Reserves, and Interest Rates," forthcoming 1977, *Journal of American Statistical Association.*

Quenouille, M. H., "The Joint Distribution of Serial Correlation Coefficients," *Annals of Mathematical Statistics,* Vol. 20, 1949, pp. 561-571.

Rao, A. G., and Shapiro, A., "Adaptive Smoothing Using Evolutionary Spectra," *Management Science,* Vol. 10, no. 3, November 1970, pp. 208-218.

Reid, D. J., "Forecasting in Action: A Comparison of Forecasting Techniques in Economic Time Series," *Joint Conference of O.R. Society's Group on Long-Range Planning and Forecasting,* 1971.

Shiskin, J., "Tests and Revisions of Bureau of the Census Methods of Seasonal Adjustments," Bureau of the Census, Technical Paper, no. 5, 1961.

Shiskin, J. et al., "The X-11 Variant of the Census II Method Seasonal Adjustment Program," Bureau of the Census, Technical Paper, no. 15, 1967.

Singer, R. A., and P. A. Frost, "On the Relative Performance of the Kalman and Wiener Filters," IEEE *Transactions on Automatic Control,* Vol. AC-14, August 1969, pp. 390-394.

Sorenson, H. W., "Kalman Filtering Techniques" in (ed.) Leondes, C.T., *Advances in Control Systems,* Academic Press, New York, 1968.

"Spectral Analysis and Parametric Methods for Seasonal Adjustment of Economic Time Series," Bureau of the Census, Technical Paper, no. 23.

Theil, H., and S. Wage, "Some Observations on Adaptive Filtering," *Management Science,* Vol. 10, no. 2, January 1964, pp. 198-224.

Trigg, D. W., "Monitoring a Forecasting System," *Operational Research Quarterly*, Vol. 15, 1964, pp. 271-274.

Trigg, D. W., and D. H. Leach, "Exponential Smoothing with an Adaptive Response Rate," *Operational Research Quarterly*, Vol. 18, 1967, pp. 53-59.

Wheelwright, Steven C. and Spyros Makridakis, *Forecasting Methods for Management, Second Edition*, John Wiley and Sons, New York, 1977.

Whittle, P., "The Analysis of Multiple Stationary Time Series," *Journal of the Royal Statistical Society*, Series B, Vol. 15, 1953, pp. 125-139.

Widrow, B., "Adaptive Filtering I: Fundamentals," *Center for Systems Research*, Technical Report no. 6764-6, 1966, Stanford University.

Wiener, N., "Generalized Harmonic Analysis," *Acta-Math.* 55, 1930, p. 117.

Weiner, N., *Time Series* MIT Press, Cambridge, Massachusetts, 1964 (first edition 1949).

Wilde, Douglass J. and Charles S. Beightler, *Foundations of Optimization*, Printice-Hall, Inc., Englewood Cliffs, New Jersey, 1967.

Winters, P. R., "Forecasting Sales by Exponentially Weighted Moving Averages," *Management Science*, April 1960, pp. 324-342.

Wold, H., *A Study in the Analysis of Stationary Time Series*, Almquist and Wiksell, Stockholm, 1954 (first edition 1938).

Wold H. O., *Bibliography on Time Series and Stochastic Processes: An International Team Project*, Oliver and Boyd, London, 1965.

Glossary of Forecasting Terms

Accuracy

Accuracy is the most frequently used criterion for selecting a forecasting method. Different forecasting methods exhibit varying degrees of accuracy. This accuracy can be measured in two ways, both of which require calculating the error — the difference between actual and forecast values.

a. Fit the optimal model of the selected method to all data, and compute the errors. As a summary of the model's accuracy, one can calculate either the *mean squared error*

$$MSE = \frac{\sum_{i-1}^{n} (Y_i - \hat{Y}_i)^2}{n}$$

or the mean absolute percentage error

$$MAPE = \frac{\sum_{i-1}^{n} |(Y_i - \hat{Y}_i)|}{n}$$

where Y = the actual value,

$\hat{Y}$ = the predicted (or computed by the model) value,

$|Y-\hat{Y}|$ = the absolute differences between Y and $\hat{Y}$ (the fore-casting "error" without regard to sign).

Both the MSE and the MAPE indicate how well the model fits the data. Thus they allow evaluation of the accuracy of the method. Of course this approach assumes that future actual values, as well as predictions, will be about the same as those of the past and that the future MSE or MAPE will be similar.

b. The second method of measuring accuracy is to fit the selected model to only a subset of the data points excluding, for example, the last (L) values. One could then estimate the *parameters* of the model and calculate the MSE or the MAPE for the n – L values. The forecasts for the L periods could then be obtained and compared with the L actual values. Since the last L values were not used to estimate the model, this is equivalent to an ex-post testing of the accuracy of the model.

SIBYL provides estimates for the expected accuracy of all methods. These estimates are based on analysis of many data series and are a measure of the accuracy of the model fitted to those series and for forecasts 1, 2, 3, 6 and 12 periods ahead.

Adaptive Filtering (see chapter 17)

Adaptive filtering is a method of forecasting that determines the optimal parameters (weights) to be applied to past data in such a way that the squared errors will be minimized. It is a time-series method belonging to the group of autoregressive/moving average techniques.

Adaptive Response Rate (see chapter 6)

In many time-series forecasting methods a tradeoff must be made between smoothing randomness and reacting quickly to changes in the basic pattern. Adaptive response rate forecasting involves using a decision rule that instructs the forecasting methodology (such as exponential smoothing) to adapt the model's parameter(s) when it appears that a permanent change in pattern has occurred and to remain stable when it appears that no such change has occurred.

Algorithm

An algorithm is a systematic set of rules for studying a particular type of problem. For example, if one desires to find the highest common denominator for a series of numbers one can use an algorithm to do so. At a more complex level, if one desires to calculate the optimal allocaton of jobs to machines, one can use an algorithm to do so. Linear programming uses the Simplex algorithm to obtain optimal solutions.

Alpha – α (see Chapters 5 through 11)

Alpha (or α) is a constant whose value is between zero and one, that is used in computing exponential smoothing values (either single, linear, quadratic or higher order).

A small value for α will smooth the past data more than a large value (see Figure B-1) and should be used for highly fluctuating or random data. A larger value for α (closer to one) should be used when the data are changing, or when there is some pattern that the forecasting method can pick up. (See Figure B-1 for an illustration of the effect of α.)

Applicability

The applicability or complexity of a forecasting method is an important factor to consider when selecting the most appropriate forecasting method for a situation. Three things about the applicability of a given method are important:

    a. How complicated and involved is it?
    b. How long does it take to obtain a forecast?
    c. How easy is it to understand and interpret the results once they are obtained?

Often the more complex the method, the greater its sophistication and its accuracy. Thus, the user may have to choose between greater complexity and greater accuracy to minimize any perceived conflict between the two.

ARMA Models or Processes

See *Autoregressive/Moving Average*

Autocorrelation (see chapter 3)

Autocorrelation is similar to correlation in that it describes the association or mutual dependence between values of the same variable but at different time periods. Autocorrelation coefficients provide important information about the structure of a data set, and indicate a great deal about its pattern. For example, the autocorrelation coefficients in Figure B-2 reveal a seasonal pattern of 12-month duration and the existence of some trend in the data from which they were computed. More specifically, the autocorrelation coefficients reveal the existence of:

    • trend in the data,
    • seasonality (and its length).

FIGURE B-1 Effect of α on "Smoothing" the Data

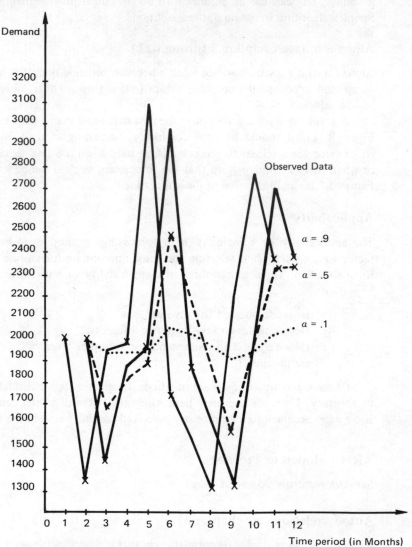

FIGURE B-2   Autocorrelation of Different Time Lags For U.K. Consumption of Gasoline Data

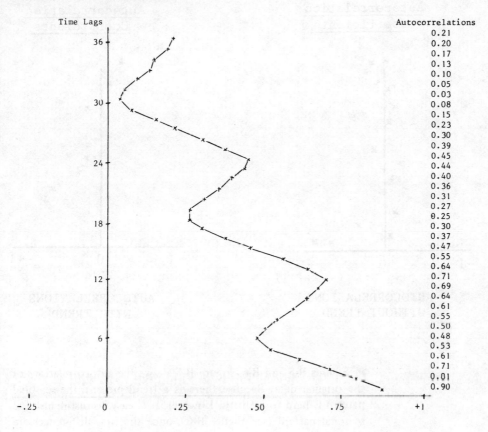

Time Lags	Autocorrelations
	0.21
36	0.20
	0.17
	0.13
	0.10
	0.05
	0.03
30	0.08
	0.15
	0.23
	0.30
	0.39
	0.45
24	0.44
	0.40
	0.36
	0.31
	0.27
	0.25
18	0.30
	0.37
	0.47
	0.55
	0.64
12	0.71
	0.69
	0.64
	0.61
	0.55
	0.50
6	0.48
	0.53
	0.61
	0.71
	0.01
	0.90

The following steps may be taken to determine whether these two patterns are present.

   a. Examine the graph of the autocorrelations. If there are a few high values and the rest are close to zero or oscillate around zero, it implies that there is no trend in the data. (See Figure B-3.)

   b. If there is a trend (see Figure B-4), then the autocorrelations of successive time lags will form a diagonal line. Clearly, the value of the first half of the autocorrelation coefficients will be quite different from those of the last half and not close to zero. When there is a trend, any seasonal pattern is often dominated by it and cannot be seen. Thus one should take the first differences which remove the trend and facilitate identification of the seasonal (plus cyclical) pattern in the remaining series. This can be seen clearly

FIGURE B-3

<u>Autocorrelation
Coefficients</u>

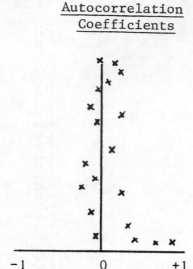

$$-1 \qquad 0 \qquad +1$$

AUTOCORRELATIONS
WITHOUT TREND

FIGURE B-4

<u>Autocorrelation
Coefficients</u>

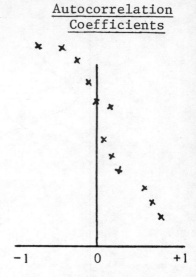

$$-1 \qquad 0 \qquad +1$$

AUTOCORRELATIONS
WITH TREND

in Figures B-5 and B-6. Figure B-5 shows the autocorrelations of the original data. Because there is a trend present, the seasonal pattern is hard to identify. However, it is easy to distinguish the seasonal pattern (see Figure B-6), once the first differences are taken.

(See Figure B-5 for the autocorrelations of the original data.) The seasonal pattern is now apparent.

The first differences can be computed as shown in Figure B-7. The effect of this transformation on the data series can be seen in Figure B-8.

The first differences of the original data are 2, 1, 2, -1, 4, -1, 2, and there is no trend in it. Thus the outcome of differencing is a horizontal or stationary series.

In order to find the second difference, one can take the first difference of the first differences — in other words, one can difference the already differenced data of Figure B-7. The result is shown in Figure B-9.

The second difference is not used very often. It is only employed if there is still some trend left after computing the first differences. Differences higher than second order are rarely used.

FIGURE B-5   Autocorrelations of Data, Including Trend and Seasonal Elements

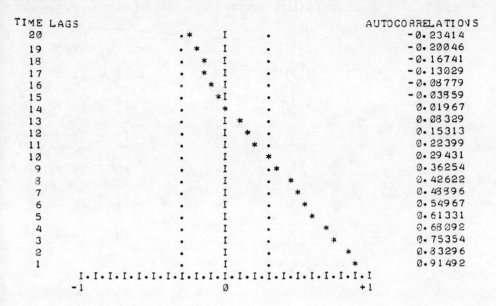

```
TIME LAGS                                              AUTOCORRELATIONS
  20             .*      I        .                       -0.23414
  19             . *     I        .                       -0.20046
  18             .  *    I        .                       -0.16741
  17             .  *    I        .                       -0.13029
  16             .   * I          .                       -0.08779
  15             .    *I          .                       -0.03859
  14             .     *          .                        0.01967
  13             .     I *        .                        0.08329
  12             .     I  *.      .                        0.15313
  11             .     I   * .    .                        0.22399
  10             .     I      *   .                        0.29431
   9             .     I      .*                           0.36254
   8             .     I      .  *                         0.42622
   7             .     I      .   *                        0.43896
   6             .     I      .    *                       0.54967
   5             .     I      .     *                      0.61331
   4             .     I      .       *                    0.68092
   3             .     I      .        *                   0.75354
   2             .     I      .         *                  0.83296
   1             .     I      .          *                 0.91492
      I.I.I.I.I.I.I.I.I.I.I.I.I.I.I.I.I.I.I.I.I.I
     -1                   0                     +1
```

FIGURE B-6   Autocorrelation After 1st Difference

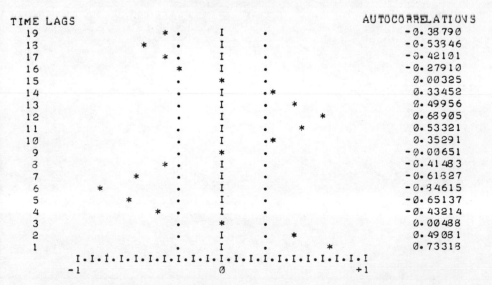

```
TIME LAGS                                              AUTOCORRELATIONS
  19            *  .      I       .                       -0.38790
  18         *     .      I       .                       -0.53346
  17          * .         I       .                       -0.42101
  16          *           I       .                       -0.27910
  15             .        *       .                        0.00325
  14             .        I       .*                       0.33452
  13             .        I       .   *                    0.49956
  12             .        I       .      *                 0.68905
  11             .        I       .    *                   0.53321
  10             .        I       .*                       0.35291
   9             .        *       .                       -0.00651
   8          * .         I       .                       -0.41483
   7        *     .       I       .                       -0.61827
   6     *        .       I       .                       -0.84615
   5       *      .       I       .                       -0.65137
   4         *  .         I       .                       -0.43214
   3             .        *       .                        0.00488
   2             .        I       .  *                     0.49081
   1             .        I       .       *                0.73318
      I.I.I.I.I.I.I.I.I.I.I.I.I.I.I.I.I.I.I.I.I.I
     -1                   0                     +1
```

FIGURE B-7 Computation of First Differences

PERIOD	ORIGINAL DATA	FIRST DIFFERENCE
1	⑩	12 – 10 = 2
2	⑫	13 – 12 = 1
3	⑬	15 – 13 = 2
4	15	14 – 15 = –1
5	14	18 – 14 = 4
6	18	17 – 18 = –1
7	17	19 – 17 = 2
8	19	

FIGURE B-8 Plot of First Differences Data and Original Data

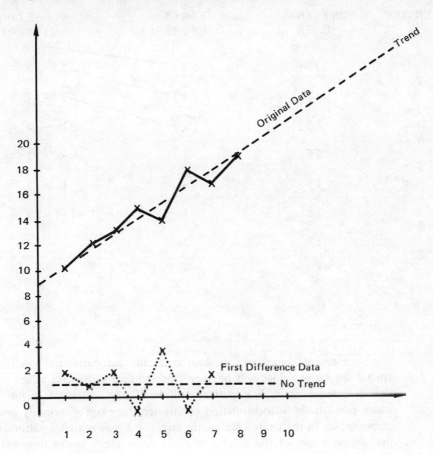

FIGURE B-9 Second Differences

PERIOD	ORIGINAL DATA	FIRST DIFFERENCE	SECOND DIFFERENCE
1			
	10	(2)	1 - 2 = -1
2			
	12	(1)	2 - 1 = 1
3			
	13	(2)	-1 - 2 = -3
4			
	15	-1	4-(-1) = 5
5			
	14	4	-1 - 4 = 5
6			
	18	-1	-2-(-1) = 3
7			
	17	2	
8			
	19		

In a set of completely random data, the autocorrelaton coefficients among successive values will be very close to zero while data values with a strong seasonal or cyclical pattern will be highly correlated. Figure B-2, which presents the autocorrelation of different time lags of monthly gasoline consumption in the United Kingdom, reveals a strong seasonal pattern since the highest values of the autocorrelations occur every twelve time periods. That figure also indicates the existence of a trend because the autocorrelations do not fluctuate around zero.

One does not need to know anything about the type of the data or their pattern to obtain the autocorrelation coefficients. These will reveal the type of data one is dealing with, and its pattern by revealing the structure of the dependence among successive values of the data. Autocorrelations, therefore, are a main form of time-series analysis available to the statistician, indicating at what level the data are stationary. If there is seasonality it indicates its length.

Autocorrelated Residuals (see chapter 3)

See *Autocorrelation and Residuals*

The importance of autocorrelated residuals is that they indicate whether or not the correct forecasting method or model is being used. If the autocorrelations of the residuals are randomly distributed around zero, as in Figure B-10, then the model is adequate. No additional pattern can be identified, in which case the residuals are said to be "white noise."

If the autocorrelations are not randomly distributed and a pattern exists as in Figure B-11, then the forecasting model has not completely identified the pattern. Whenever a pattern exists, there is a model capable of identifying it and removing it from the residuals, although in some cases it may be nontrivial to identify that pattern, thus making forecasting more accurate.

FIGURE B-10 Autocorrelations of Random Residuals

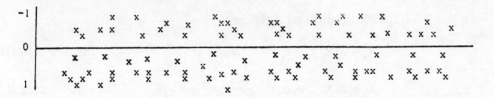

FIGURE B-11 Autocorrelations of Residuals Containing a Pattern

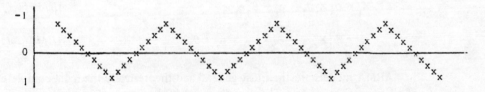

Autoregressive

This is a form of regression, but rather than the dependent variable (the item to be forecast) being related to independent variables, it is simply related to past values of itself at varying time lags. Thus an autoregressive model would be one that expressed the forecast as a function of previous values of that time series.

Autoregressive/Moving Average (ARMA)
Schemes (see chapters 17 and 18)

In time series there are two basic ways of expressing a relationship — "Autoregressive" and "Moving Average." Autoregressive (AR) processes, models or schemes assume that future values are linear conbinations of past values. Moving Average (MA) processes, models or schemes on the other hand, assume that future values are linear combinations of past errors. A combination of the two is called an "Autoregressive/Moving Average (ARMA) Scheme." These models are represented mathematically as:

$$X_t = \phi_1 X_{t-1} + \phi_2 X_{t-2} + \ldots + \phi_p X_{t-p} + e_t \qquad \text{AR}$$

$$X_t = e_t - \theta_1 e_{t-1} - \theta_2 e_{t-2} - \ldots - \theta_q e_{t-q} \qquad \text{MA}$$

$$X_t = \phi_1 X_{t-1} + \phi_2 X_{t-2} + \ldots + \phi_p X_{t-p} + e_t - \theta_1 e_{t-1}$$

$$- \theta_2 e_{t-2} - \ldots - \theta_q e_{t-q} \qquad \text{ARMA}$$

Specific examples are as follows:

$$X_t = \phi_1 X_{t-1} + e_t \qquad \text{AR(1)}$$

$$X_t = \phi_1 X_{t-1} + \phi_2 X_{t-2} + \phi_3 X_{t-3} + e_t \qquad \text{AR(3)}$$

$$X_t = e_t - \theta_1 e_{t-1} - \theta_2 e_{t-2} \qquad \text{MA(2)}$$

$$X_t = \phi_1 X_{t-1} + e_t - \theta_1 e_{t-1} \qquad \text{ARMA(1,1)}$$

$$X_t = \phi_1 X_{t-1} + \phi_2 X_{t-2} + e_t - \theta_1 e_{t-1} \qquad \text{ARMA(2,1)}$$

ARMA models are the most general and theoretically the most complete of all time-series models. They are the basis of the Box-Jenkins methodology.

Average Absolute Percentage (%) Change (see chapter 16)

As the name implies, the average absolute percentage change denotes the change that occurs on the average to either the original data or any of its components. The following are commonly used forms of the average absolute percentage change and their mathematical computation.

a. Original Data

$$\frac{\sum\limits_{t=1}^{n-1} \left| \dfrac{X_{t+1} - X_t}{X_t} \right|}{n-1}$$

where n is the number of terms in the time series, $|X_{t+1} - X_t|$

denotes absolute values (i.e., absolute errors), and X_{t+1}, X_t are the original values of the time series at periods $t + 1$ and t.

b. Trend Cycle

$$\frac{\sum\limits_{t=1}^{n-1} \left| \dfrac{TC_{t+1} - TC_t}{TC_t} \right|}{n - 1}$$

where TC_t denotes the trend cycle value at time period t.

c. Seasonal Data

$$\frac{\sum\limits_{t=1}^{n-1} \left| \dfrac{S_{t+1} - S_t}{S_t} \right|}{n - 1}$$

where S_t is the cycle component.

d. Random Component

$$\frac{\sum\limits_{t=1}^{n-1} \left| \dfrac{R_{t+1} - R_t}{R_t} \right|}{n - 1}$$

where R_t is the random component.

e. Trend

$$\frac{\sum\limits_{t=1}^{n-1} \left| \dfrac{T_{t+1} - T_t}{T_t} \right|}{n - 1}$$

where T_t is the trend component.

f. Cycle

$$\frac{\sum\limits_{t=1}^{n-1} \left| \dfrac{C_{t+1} - C_t}{C_t} \right|}{n - 1}$$

where C_t is the cycle component.

Back Forecasting

In applying quantitative forecasting techniques, particularly moving average ones, starting values for the errors are required so that recursive calculations can be made. One way to obtain these is to apply the forecasting methodology to the reversed series (the series of values obtained by simply starting at the end of the series and going to the beginning). This is referred to as back forecasting and provides a set of starting error values that can then be used as the starting point for applying that forecasting methodology to the standard (beginning to end) sequence of values. If no back forecasting is made the initial error values can be assumed to be zero. If the time series is a long one, e.g., more than fifty observations, back forecasting provides little or no benefit.

Beta — β (see chapters 8 and 10)

See *Alpha*

Beta is a smoothing constant used in Winters' and Holt's exponential smoothing models. Its value varies between zero and one and is specified by the user. A small value for β will smooth past data to a greater degree than a large value. It should be used, therefore, for highly random data.

Biased Estimators

See *Parameters*

Estimators are the values that must be estimated in order to forecast. For example, in the equation

$$\hat{Y} = a + bX$$

one must estimate the values of a and b before forecasting $\hat{Y}$. Biased estimators exist when there are distortions of the values of the estimates as a result of the sampling procedure, or the method of forecasting involved.

Box-Jenkins Methodology (see chapter 18)

G. E. Box and G. M. Jenkins popularized a methodology for dealing with Autoregressive/Moving Average Schemes, based on theory originally developed in the 1930s. Since publication of their book (1970) on how to utilize autoregressive/moving average models for time-series analysis, forecasting and control, their names have often been used synonymously with these models.

Unlike all other forecasting methods, the Box-Jenkins methodology need not assume an initial pattern. It begins with a tentative pattern, and a model is fitted in such a way that the error will be minimized. It then provides explicit information as to whether the tentative pattern employed is the correct one. If this pattern is not correct, the method provides guidelines for finding the right one.

Thus, the Box-Jenkins methodology provides enough information to tentatively identify the pattern that best describes the data, estimate the parameters involved by minimizing the sum of the squared errors, and provide for a diagnostic check of the residuals to determine whether the tentative pattern (model) is an adequate one.

Brown, R. G. (see chapters 5, 7, and 9)

Robert G. Brown wrote important books on forecasting during the 1960s that were aimed particularly at the use of exponential smoothing models for handling inventory forecasting problems. Some of the higher orders of exponential smoothing frequently carry his name, such as Brown's linear exponential smoothing and Brown's quadratic exponential smoothing. Brown is considered the most important contributor in the field of exponential smoothing models.

Box-Pierce Statistic

See *Chi-squared Test*

Business Cycle

Periods of prosperity followed by periods of depression make up business conditions over time and create a "business cycle." The various stages in such a cycle are:

1. prosperity (the peak, expansion),
2. recession,
3. depression (down swing),
4. recovery (revival).

Several explanations of cycles have been suggested, and from an analysis of their statistics different cycles have been identified (each one named after the man who developed it). The length of the cycle can vary from a few years to a few decades. Its length is not constant and is therefore hard to predict.

Causal or Explanatory Models (see chapter 19)

Causal models assume that the factor to be forecast exhibits a cause/effect relationship with a number of other factors (e.g., sales = f [income, prices, advertising, competition, etc.]). The purpose of the model is to discover that relationship so that the future value of sales can be found using the values of income, price, advertising, etc. A causal model is not as easy to develop as a time-series model, but can be particularly helpful for policy and decision-making purposes.

Suppose one develops the following causal model which indicates that sales are influenced by

- Gross National Product (GNP)
- price,
- advertising,

and that a prediction of sales for 1980 (80) is

$$\text{Sales}_{80} = 12.5 + 0.053 \, \text{GNP}_{80} - 2.5 \, \text{price}_{80} + 3.8 \, \text{advertising}_{80}$$

To determine the level of sales for 1980, one must first estimate GNP in 1980, the price in 1980, and the advertising budget. All of these factors will influence the level of sales in 1980. Estimating these values is not always simple. However, causal or explanatory models often provide material on which to base policy decisions. For example, increasing the price by 1 unit (dollar), decreases the sales by 2.5 units, while a 1 unit (dollar) increase in the advertising expenditure increases sales by 3.8 units. This type of information can be most valuable, and cannot be provided by time-series models that are mechanistic in their prediction, using what can be called a "black box" approach.

The major forms of causal forecasting models are multiple regression equations, simultaneous systems of such equations (econometric models), and multivariate time-series models.

Census II (see chapter 16)

The Census II method of forecasting is a refinement of the classical decomposition method. It attempts to decompose a series into seasonal, trend-cycle, and random components that can be analyzed separately and then it is hoped, predicted. The Census II method uses more elaborate procedures than classical decomposition in isolating the major components of a time series. Census II can be a powerful forecasting technique, particularly for short- and medium-term predictions and has a good reputation with managers as to the usefulness and accuracy of its results. Census II is intuitively appealing and provides a wide range of information so that the trend-cycle can be identified and estimated more easily. Finally, it is appropriate for most macro-data as reported and used by government agencies.

Central Limit Theorem

The central limit theorem is of critical importance in statistical theory. It states that the sampling distribution of the mean approaches a normal distribution when the sample size used is sufficiently large (greater than thirty). If the sampling is repeated many times, the mean of the resulting distribution is the real population mean, while its variance is equal to the population variance divided by the sample size. If n is the sample size, $\overline{X}$ is the sampling mean, μ is the population mean, $\sigma_{\bar{x}}^2$ is the variance of the sampling distribution, and σ^2 is the population variance, the following holds:

$$\bar{X} \to \mu, \text{ and } \sigma_{\bar{x}}^2 = \frac{\sigma^2}{n} \text{ when } n \to \infty$$

Chi-Squared Test (χ^2)

The Chi-squared Test is a statistical test that can be used to test the hypothesis that the mean value of the autocorrelations (or any other statistic of interest) is significantly different than zero. If the value of the χ^2-test is smaller than the corresponding value from the table, then the data are random, without any pattern. The opposite will imply that the data are autocorrelated.

For example, if for a set of autocorrelations

Chi-squared (computed) = 13.8

Chi-squared (table)   = 18.5

then the data are not autocorrelated, but are random. The χ^2-test is used to check whether a given method or model is appropriate. This can be done by computing the autocorrelation of the residuals and testing their χ^2-value against the corresponding value from the table. The computed Chi-squared value is also referred to as the Box-Pierce statistic.

Classical Decomposition Method

Most forecasting techniques attempt to distinguish the underlying pattern of past data from randomness, so that the pattern can be projected into the future and used in forecasting. No attempt is made to distinguish the subparts of the basic underlying pattern.

In many cases, the pattern can be broken down, allowing a more accurate forecast to be made. For example, when a cyclical or seasonal pattern exists in addition to any regular growth or decline (trend) in the data, it is important to know when higher values occur because of trend and when they are due to the weather, holidays, the business cycle, etc.

Traditionally, the decomposition method has tried to identify three separate portions of an underlying pattern. These are the trend, cyclical, and seasonal factors, with any remaining value being caused by randomness. Decomposition attempts to deal with randomness by isolating it rather than attempting to separate it directly as is done by the majority of time-series methods.

Coefficient of Autocorrelation

See *Autocorrelation*

Coefficient of Determination

See *R-Squared*

Coefficient of Multiple Correlation (see chapter 19)

The degree of association between a dependent variable and two or more independent variables is measured by the coefficient of multiple correlation. The symbol used to express this relationship is R. The square of R (R-squared or R^2) is a measure of the goodness of fit that indicates the percentage of variation in the dependent variable explained by variations in the independent variables. (Or similarly, the percentage of the total error that is explained by the regression equation.)

Coefficient of Regression (see chapter 19)

For simple regression the coefficient of regression is the slope of the regression line. It is the average number of units of either increase or decrease in the dependent variable that are associated with a one unit increase in the independent variable. A similar interpretation of the coefficients of multiple regression also applies.

Coefficient of Variation

A statistical measure of the relative dispersion of a data series is the coefficient of variation, which varies between zero and one. It is useful for time-series analysis in determining whether the data are stationary. A smaller coefficient of variation for differenced data than for the original data indicates a more stationary series. The coefficient of variation is computed by dividing the standard deviation by the mean of the original data.

$$\text{Coefficient of variation} = \frac{\text{Standard deviation}}{\text{Mean}}$$

Confidence Limits

Based on statistical theory and probability distributions, a confidence interval, or set of confidence limits, can be established for a forecast value or a series of residual values. The hypothesis is that, depending on the confidence limits established, some percentage of the actual values will fall within those confidence limits. Thus, if sample values are observed to fall within those limits, it is assumed that the original hypothesis is true. If they do not, it is assumed that the original hypothesis is not true and that the results are significantly different from that hypothesis.

Correlation Coefficient (see chapter 19)

It often occurs that two variables are related, but that it is inappropriate to say that values of one of the variables depend on, or are caused by, values of the other. Another relationship with no direction of causality — the variables are related or have some connection with each other — can then be stated.

The degree of relationship that the variables have with each other is the coefficient of correlation, denoted by r. This coefficient can vary from −1, indicating a perfect negative correlation, to +1, indicating a perfect positive correlation. When the coefficient is greater than 0, the two variables are positively correlated, and when less than 0, they are negatively correlated.

For example, there is a positive correlation between:

    a. The height of a person and his weight. (As he gets taller he becomes heavier; as one increases, so does the other.)

    b. The number of cars on the road and the number of accidents. (As traffic increases, so do the number of accidents; as one increases, so does the other.)

    c. Prosperity and spending. (As people receive higher incomes, they spend more; as one increases, so does the other.)

An example of negative correlation is the price of, say, lobsters and their demand: as lobsters become more and more expensive, the demand for them becomes smaller. (As price increases, demand decreases)

Correlation Matrix (see chapter 19)

Most computer programs that are designed to perform multiple regression analysis include computation of the simple correlaton matrix. It provides information as to the correlaton between each pair of variables used in the regression. For example:

SIMPLE CORRELATION MATRIX			
	Sales	*Production*	*Contracts*
Sales	1.00	0.56	0.99
Production	0.56	1.00	0.53
Contracts	0.99	0.53	1.00

The correlation between sales and production (.56), production and contracts (.53), and sales and contracts (.99) can be easily seen from such a matrix.

It should be noted that the correlation matrix is symmetric; that is, the upper half is the same as the lower half. This should be expected because the correlation between sales and production (.56, first row, second column) is the same as the correlation between production and sales (.56, second row, first column). The same is true with all other variables. If all values in a correlation matrix are zero, it is referred to as being singular.

Costs

When one is selecting a forecasting method, cost must be taken into account. There are three kinds of cost involved in the utilization of forecasting methods on the computer.

   a. *Development Cost:* This is the cost of writing or modifying the computer program necessary to apply a forecasting method (D_1).
   b. *Storage Cost* (S_1, S_2): In order to utilize computer programs of forecasting methods, one must have the programs, as well as the required data, stored in a memory device in the computer. This storage incurs a cost.
   c. *Running Cost:* Every time one runs the computer program to obtain forecasts, or modify the working model, a cost is incurred. This cost is mainly associated with the Central Processing Unit (CPU) time required to run the program.

A computation of all these costs will provide a true picture of the total expenses for utilizing a forecasting method. This will be a critical factor in deciding:

   a. Whether a forecasting technique is to be used at all;
   b. If one is to be used, how to select, from those available, the one that best fits the budget.

Covariance

The formula for covariance is:

$$\frac{\Sigma(X_i - \overline{X})(Y_i - \overline{Y})}{n}$$

where X is one variable,

and Y is another variable.

Covariance is the joint variation between the variables X and Y.

The term $(X_i - \overline{X})$ in the formula expresses the variations of X around its mean $(\overline{X})$, and the term $(Y_i - \overline{Y})$ expresses the variations of Y around its mean $(\overline{Y})$. Multiplying the two terms gives the combined variance of X and Y, and is called the covariance. Covariance is very similar to correlation, except that it is not standardized. Where correlations vary from -1 to $+1$, covariances can vary from $-\infty$ to $+\infty$.

Cross Autocorrelation (see chapter 24)

See *Cross Autocovariance*

The cross autocorrelation is for multivariate time-series analysis what the simple autocorrelation is for univariate time-series analysis. The cross autocorrelations (or cross autos as they are called) are standardized measures (i.e., varying between -1 and +1) of association between the present values of a given variable and past, present, and future values of another time-series variable. It is from the cross autocorrelation coefficients of the prewhitened independent and dependent variables that the form of the multivariate ARMA model can be identified.

In addition, the adequacy of the model can be determined using the cross autos of the residuals of the model fitted with the prewhitened independent variable. (This test of model adequacy is in addition to the regular one of making sure that the simple autocorrelations of the residuals are random noise.)

Cross Autocovariance (see chapter 24)

See *Cross Autocorrelation*

The cross autocovariance is a measure of association that relates present values of a given variable to past, present and future values of another time-series variable. It combines the concepts of covariance and autocovariance. The cross autocovariance forms the basis for calculating the cross autocorrelation coefficients that are standardized measures of association between two time series.

Cumulative Forecasting

It is frequently the case that, rather than forecasting values for sequential time periods of equal length, the user of forecasting would prefer to forecast the cumulative level of a variable over several periods. For example, one might forecast cumulative sales for the next twelve months, rather than forecasting an individual value for each of those twelve months.

Curve Fitting (see chapters 12, 13, and 19)

One approach to forecasting is simply to fit some form of curve, such as a polynomial, to the historical time-series data. Use of a linear trend is, in fact, a curve fitting method. Higher forms of curve fitting are also possible — such as quadratic, cubic, exponential, logarithmic, S-curve, etc.

Cycle

See *Types of Data*

Cyclical Index (see chapter 15)

A cyclical index is a number, usually based around 100, that indicates the cyclical pattern of a given set of time-series data. If, for example, the cyclical index of sales for July is 100, it signifies that there are no influences due to cyclical effects during July. If it is 110 for August, it indicates that, on average, 10% of the value of sales in August is due to cyclical factors. An index of 85 implies that the effect of cyclicality is negative, causing the observed value to be 15% below the average of all periods.

Decomposition

See *Classical Decomposition*

Degrees of Freedom (D.F.)

This is a statistical term that indicates the number of variables or data points (minus some adjustments) used for different statistical tests. It is often difficult to give a clear explanation as to how many degrees of freedom remain after adjustments are made. One usually starts with k variables and n data points or observations that are adjusted for each particular statitic. In forecasting, the following degrees of freedom (D.F.) are frequently used:

Tests	*D.F.*	*Example: if n = 40 k = 5*	*Corresponding value from statistical table (95% confidence)*
"T-test"	n−k	35	1.96
"F-test"	k−1 numerator	4	2.61
	n−k denominator	35	
"Chi-Squared test" Univariate	n−1	39	26.51
"Durbin-Watson test"	k−1	4	1.38 − 1.72
	n	40	

Delay

See *Prewhitening*

This is the time it takes until the influence of a change in a given independent variable affects the dependent variable.

Demand

In economics, the term demand expresses the quantity of goods that buyers are ready to purchase at a specific price in a particular market at a given point in time. Demand can also mean the quantity demanded at one given price.

Delphi Method

This approach to qualitative or technological forecasting seeks to systematically use the judgment of experts in arriving at a forecast of either what things are to happen, or when they are to happen.

Dependent and Independent Variables (see chapter 19)

In the equation $Y = a + bX$, Y is the value to be determined or forecast, and a + bX is the combination of information which one needs to find Y. Since Y is determined by something else — a + bX — it is dependent. The value of X, on the other hand, is not contingent on something else in the model. It is independent.

For example, for an automobile

Speed = Function (Acceleration)

Speed is dependent upon how hard the accelerator is depressed, whereas such pressure on the accelerator is not influenced by speed. Thus it is independent of speed.

The term variable is used for Y and X because their values vary with each situation.

In the general case of a regression model,

$$Y = a + b_1X_1 + b_2X_2 + b_3X_3 + \ldots + b_mX_m$$

where Y is the dependent variable,

$X_1, X_2, X_3, \ldots, X_m$ are the independent variables,

$a, b_1, b_2, b_3, \ldots, b_m$ are the parameters.

Depression

See *Business Cycle*

The term depression is used to describe a period in the business cycle when production and prices are at their lowest, unemployment is at its highest, and general morale is poor.

Deseasonalized Data

Removal of the seasonal pattern in a data series results in deseasonalized data. Deseasonalizing facilitates the comparison of month-to-month changes, such as when dealing with unemployment statistics or economic indicators.

Differences, First and Second or Method of Differencing
(see chapter 3)

See *Autocorrelations*

The method of differencing converts a nonstationary time series into a stationary one. It consists of subtracting successive values from one another and using their difference as a new series. For example, the first difference of the series Y is:

First Differences

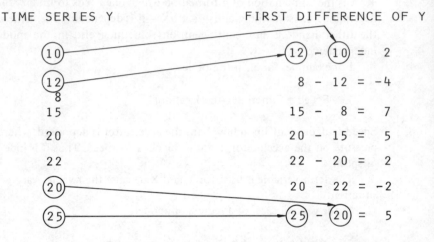

TIME SERIES Y

FIRST DIFFERENCE OF Y

⑩
⑫ – ⑩ = 2

⑫
8 – 12 = –4

8

15
15 – 8 = 7

20
20 – 15 = 5

22
22 – 20 = 2

⑳
20 – 22 = –2

㉕
㉕ – ⑳ = 5

The resulting series – 2, –4, . . . , 5 – is called the first difference. The number of observations in this series is one less than the undifferentiated series 10, 12, . . . , 25.

The second difference consists of taking the first difference of the first difference. For example:

Second Differences

FIRST DIFFERENCE

SECOND DIFFERENCE

②
–④ – ② = –6

–④
7 – (–4) = 11

7
5 – 7 = –2

5
2 – 5 = –3

2
–2 – 2 = –4

–②
⑤ – –② = 7

⑤

The number of terms in the second difference series is one less than the first, and of course, two less than the original series.

Double and Higher Order Moving Averages (see chapters 4 and 11)

See *Moving Averages*

A double moving average is a moving average of a single moving average. A triple moving average is a moving average of a double moving average, etc. The number of periods included in the moving average need not be the same for the single, double, triple, or higher order averages, Figure B-12 gives an example of the calculations for single, double, and triple moving averages.

Dummy Variables

Dummy or binary variables are special forms of variables, the values of which can be either zero or one. They are used to quantify qualitative events. Strike or nonstrike, for example, can be expressed by 1 for strike, 0 for nonstrike. The coefficients of dummy variables are like those of any other variable used in regression analysis.

For example, suppose one wants to forecast the daily sales of fish. Since fish sales are higher on Fridays than other days of the week, a dummy variable might be used to express that fact, as shown in Figure B-13.

Suppose the quarterly sales of Company XYZ are thought to be seasonal. One could use three dummy variables to indicate the seasons. (When all three

FIGURE B-12   Moving Averages

Period	Data	Single Moving Average (N=5)	Double Moving Average (5x3) N=5, N=3	Triple Moving Average (5x3x3) N=5, N=3, N=3
1	4	-		
2	6	-		
3	5	5.4	-	
4	4	5.8	5.67	-
5	8	5.8	6.00	6.09
6	6	6.4	6.60	6.60
7	6	7.6	7.20	7.16
8	8	7.6	7.67	7.53
9	10	7.8	7.73	7.71
10	8	7.8	7.73	7.60
11	7	7.6	7.33	-
12	6	6.6	-	
13	7	-		
14	5	-		

FIGURE B-13   Daily Fish Sales

Time		Sales	Dummy Variable
J. 1	M	32	0
J. 2	T	35	0
3	W	28	0
4	T	33	0
5	F	62	1
6	S	25	0
7	S	39	0
8	M	41	0
9	T	42	0
10	W	33	0
11	T	42	0
12	F	68	1
13	S	25	0
14	S	30	0
15	M	31	0
16	T	35	0
17	W	38	0
18	T	40	0
19	F	71	1
20	S	35	0
21	S	38	0

dummy variables are zero, it will denote the fourth season. This season can be thought of as the base for comparison with the other three seasons.) Figure B-14 illustrates this.

Dummy variables are treated like regular variables in regression, except that they take only two values. In the example of Figure B-14, D1 will be one for each winter quarter and zero otherwise. D2 will be one for spring data and zero otherwise. Similarly, D3 will be one and zero for summer and nonsummer data. When D1, D2, and D3 are all zero, the data will be for the autumn quarter.

Durbin-Watson Test (D-W Test) (see chapter 19)

The Durbin-Watson statistic tests the hypothesis that there is no autocorrelation of one time lag present in the residuals. As with the F-test and t-test, this test compares the computed value $(D - W_c)$ of the Durbin-Watson test with the corresponding value in the table. This requires reading two values — $D - W_L$, $D - W_u$ (where L stands for lower and u for upper) — from a D - W table corresponding to the degrees of the data. The D - W distribution is symmetrical around 2, its mean value.

FIGURE B-14   Quarterly Company Sales

Year	Quarter	Sales	D1 Winter	D2 Dummy Variable Spring	D3 Summer
1968	1	108	1	0	0
	2	112	0	1	0
	3	125	0	0	1
	4	115	0	0	0
1969	1	114	1	0	0
	2	118	0	1	0
	3	145	0	0	1
	4	125	0	0	0
1970	1	120	1	0	0
	2	130	0	1	0
	3	160	0	0	1
	4	132	0	0	0
1971	1	124	1	0	0
	2	131	0	1	0
	3	155	0	0	1
	4	129	0	0	0
1972	1	135	1	0	0
	2	146	0	1	0
	3	175	0	0	1
	4	150	0	0	0
1973	1	138	1	0	0

One can construct five different regions as shown in Figure B-15 using $D - W_L$ and $D - W_u$.

    a. less than $D - W_L$
    b. between $D - W_L$ and $D - W_u$
    c. between $D - W_u$ and $4 - (D - W_u)$
    d. between $4 - (D - W_u)$ and $4 - (D - W_L)$
    e. more than $4 - (D - W_L)$

If the computed $D - W_L$ is in either interval (a) or (e), autocorrelation exists. If $D - W_L$ is in (c), no autocorrelation exists, and if it is in either (b) or (d), the test is inconclusive, indicating that one cannot be sure whether or not there is autocorrelation.

For example, if there are four variables and thirty observations, then

$$D - W_L = 1.21$$

$$D - W_u = 1.65$$

FIGURE B-15   Durbin-Watson Statistic

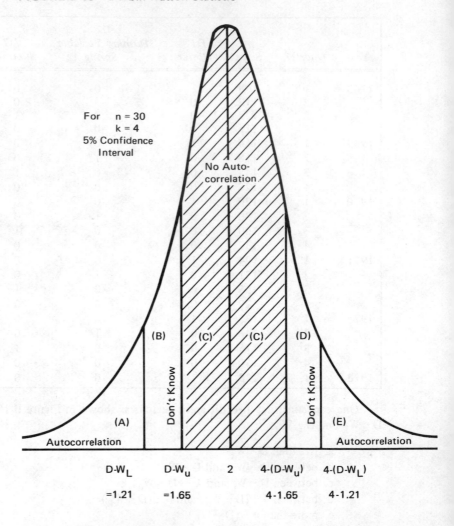

For   n = 30
       k = 4
5% Confidence
   Interval

No Auto-
correlation

(B)   (C)   (C)   (D)

Don't Know

Don't Know

(A)   (E)

Autocorrelation   Autocorrelation

D-W$_L$	D-W$_u$	2	4-(D-W$_u$)	4-(D-W$_L$)
=1.21	=1.65		4-1.65	4-1.21

If $D - W$ is less than 1.21 or more than $4 - (D - W_L) = 4 - 1.21 = 2.79$, there is autocorrelation.

If $D - W_L$ is between 1.65 and $4 - (D - W_u) = 2.35$, there is no correlation.

If $D - W_L$ is either between 1.21 and 1.65, or between 2.35 and 2.89, the test is inconclusive (see Figure B-15).

In general one can assume (with little loss of statistical rigour) that there is no autocorrelation in the data if the value of the $D - W$ test is between 1.5 and 2.5.

Econometrics

Econometrics deals with the measurement of economic phenomena. It applies statistical and mathematical techniques to either test or demonstrate economic theories. The main objective of econometrics is to give empirical content to a priori economic reasoning. For example, from economic theory one knows that price and demand are negatively related. Through econometrics one could measure such a relationship. For example, it might be

Demand = 100 − 2 (price)

Indicating that when price increases by one unit, demand will decrease by two units. Determining the parameter values, 100 and −2, involves measuring a general economic relationship. Since the general equation, Demand = a − bP, is well known in economics, its use is referred to as econometrics.

Econometric Forecasting (Econometric Models)

Econometric forecasting is an extension of regression analysis. Regression analysis assumes that each of the independent variables included in the equation is determined by outside factors. In economic or organizational relationships, however, such an assumption of independence is unrealistic. Often there is mutual independence among all variables in a forecasting equation, and regression analysis is incapable of dealing with such interdependence, Econometric forecasting enables one to deal with such a situation. It expresses a more accurate relationship by developing a system of simultaneous equations. These, by their nature, are capable of dealing with interdependent variables.

For example, suppose Sales = f(GNP, price, advertising). (This is the Multiple Regression form.) In econometric form one might have:

- Sales = f(GNP, price, advertising)
- Cost = f(Production and Inventory levels)
- Selling Expenses = f(Advertising, other selling expenses)
- Price = Cost + selling expenses.

Now, instead of one relationship, there are four. As with regression analysis, one must:

- determine the functional form of each of the equations;
- estimate in a simultaneous manner the values of their "parameters";
- test for the statistical significance of the results and the validity of the assumptions involved.

The main advantage of this method of forecasting is that it provides the values of several of the interdependent variables within the model itself, thus eliminating the need to estimate them externally. It also provides useful explanatory information as to the type of relationships involved. The estimating of the equation parameters involves problems far more complex than those encountered in regression analysis. It is these problems that make the application of econometric forecasting both difficult and expensive. However, the advantages gained by the method may well compensate for the extra costs incurred.

Economic Indicators

An economic indicator is an economic series that has a resonably stable relation (of lag, lead, or coincidence) to some other series, such that the first series is of some value in predicting the future course of the second series. For example, unemployment is considered a leading indicator of the level of economic activity. This means that when unemployment rises, there will be a slowdown in the economy. When it falls, the level of economic activity will pick up.

Elasticity of Demand and Supply

If the price of a product is increased by 1%, its elasticity indicates the percentage by which its demand will subsequently decrease. It is therefore a measurement of the relative changes between price and demand.

Endogenous Variable

An endogenous variable is one whose value is determined within the system. The price of a product can be thought of as an endogenous variable, while GNP will be an exogenous one. The term is used in econometrics and corresponds to the dependent variable used in regression.

Error

See *Residuals*

Estimating Systems of Simultaneous Equations

There are six methods of estimating simultaneous equations:

1. Ordinary least square method,
2. Full information maximum likelihood,
3. Limited information maximum likelihood,
4. Indirect least squares,
5. Two-stage least squares,
6. Three-stage least squares.

These methods vary considerably in terms of complexity, cost and statistical completeness. The most commonly used method is the two-stage least squares (2-SLS).

Estimation

Estimation consists of finding appropriate values for the parameters of an equation in such a way that some criterion will be optimized. The most commonly used criterion is that of mean squared error. Often, an iterative procedure is needed in order to determine those parameter values that minimize this criterion.

Executive Forecast

See *Final Forecast*

Exogenous Variable

An exogenous variable is one whose value is determined outside the model or system. Thus, GNP is an exogenous variable while price may be endogenous because it can be determined from factors within the system or perhaps simply specified by the decision-maker. In regression, independent variables correspond to exogenous variables.

Exponential Smoothing, Adaptive Response Rate (see chapter 6)

The exponential smoothing, adaptive response rate method is exactly the same as single exponential smoothing, except that the user does not have to specify a value for alpha. The method calculates a value for alpha depending on fluctuations in the data and the previous errors. The value of alpha is increased when the error increases and decreased when the error decreases. This provides a forecasting method that is responsive to changes in the pattern of the data.

The exponential smoothing, adaptive response rate method is based on the same equation as single exponential smoothing, (a), but α is calculated using (b).

$$F_{t+1} = \alpha_t X_t + (1 - \alpha_t) F_t \tag{a}$$

$$\alpha_t = \left| \frac{E_t}{M_t} \right| \tag{b}$$

$$E_t = \beta\, e_t + (1 - \beta) E_{t-1}$$

$$M_t = \beta |e_t| + (1 - \beta) M_{t-1}$$

(β is usually set at .1 or .2.)

As can be seen from (b), the value of α_t will vary with the errors.

Exponential Smoothing, Linear (Brown's) Method (see chapter 7)

Single exponential smoothing cannot deal with nonstationary data. Linear exponential smoothing is an attempt to deal with linear nonstationarity (i.e., trend). Its only difference from single exponential smoothing is that it introduces extra formulas that can estimate the trend and subsequently use it in forecasting. Equation (a) below is exactly the same as the formula used for single exponential smoothing. This is not so, however, with equation (b) which is introduced to estimate the trend. The basic idea behind equation (b) is that the double exponential smoothing value, S''_{t+1}, will lag the single exponential smoothing value, S'_{t+1}, by as much as the single exponential smoothing will lag the original data. By subtracting the double exponential smoothing value form the single exponential smoothing value, an estimate of the trend can be obtained. This is done in equation (c). The factor, α divided by $1 - \alpha$, is multiplied by the difference between the single and double exponential smoothing values (or the trend).

In addition to the trend, linear exponential smoothing makes an estimate of the present level of the data. This is given by equation (d) whose basic concept is like equation (c) – the single exponential smoothing value, S'_t, lags the data by as much as the difference between the single exponential smoothing and the double exponential smoothing values. Thus, if their difference is added to the single exponential smoothing value, the result would be an update that would bring the data to their current level. Finally, in order to forecast, one must use equation (e) which, starting from the current level, a_t, adds as many times the trend, b_t, as the number of periods ahead one desires to forecast. This is therefore a direct adjustment for the trend factor in the data.

$$S'_{t+1} = \alpha X_t + (1 - \alpha)S'_{t-1} \tag{a}$$

$$S''_{t+1} = \alpha S'_t + (1 - \alpha)S''_{t-1} \tag{b}$$

where S'_{t+1} is a single exponential smoothing value,

and S''_{t+1} is a double exponential smoothing value.

$$b_t = \frac{\alpha}{1 - \alpha} (S'_t - S''_t) \tag{c}$$

where b_t is an estimate of trend.

$$a_t = S'_t + (S'_t - S''_t) = 2S'_t - S''_t \tag{d}$$

$$F_{t+m} = a_t + b_t m \tag{e}$$

where m is the number of periods ahead one desires to forecast.

Exponential Smoothing, Linear (Holt's Two-Parameter Smoothing) (see chapter 8)

As with Brown's one-parameter linear exponential smoothing, Holt's two-parameter smoothing attempts to devise an estimate for the trend which is subsequently added to the data. Equation (a) below is similar to the original equation of single exponential smoothing, except that a term for the trend, T, is added. The value of this term, T, is calculated using equaiton (b). The difference between two successive exponential smoothing values is used as an estimate of the trend in Holt's method. This is because successive values have been smoothed for randomness and thus, their differences constitutes the trend. This is equivalent to subtracting the double from the single exponential smoothing value. This estimate of the trend is smoothed by multiplying it by γ, and then multiplying the old estimate of the trend by $(1 - \gamma)$. That is, equation (b) is exactly the same as equation (a) except that the smoothing is not done for the actual data but rather for the trend. The end result of (b) is to smooth the trend and thus eliminate any randomness that may be present.

Finally, in order to forecast one must multiply the trend by the number of periods ahead to be forecast, and then add these to S_t, the current level of the data.

$$S_t = \alpha X_t + (1 - \alpha)(S_{t-1} + T_{t-1}) \tag{a}$$

$$T_t = \gamma(S_t - S_{t-1}) + (1 - \gamma)T_{t-1} \tag{b}$$

$$F_{t+m} = S_t + mT_t \tag{c}$$

where m is the number of periods ahead to be forecast.

Exponential Smoothing, Quadratic (Brown's One Parameter) (see chapter 9)

Quadratic exponential smoothing is an extension of linear exponential smoothing. It aims at dealing with trend of higher order than linear. This aim is achieved by introducing a triple exponential smoothing term which, in addition to the double exponential smoothing, is used to remove nonlinear trends. Quadratic exponential smoothing is not limited to quadratic functions only but can be used for nonstationary series of higher than first degree.

Equation (a) below is the original single exponential smoothing. Equation (b) is similar to double exponential smoothing while equation (c) is triple exponential smoothing. Equation (d) determines the current value of the data. Equation (e) determines the linear parameter, and equation (f) determines the quadratic trend. Finally, equation (g) is used for predicting future periods. As can be seen in the last equation, both a linear trend and a quadratic trend are used to forecast. The last term of equation (g) is multiplied by 1/2, a factor obtained from differentiation.

$$S'_{t+1} = \alpha X_t + (1 - \alpha)S'_{t-1} \tag{a}$$

$$S''_{t+1} = \alpha S'_t + (1 - \alpha)S''_{t-1} \tag{b}$$

$$S'''_{t+1} = \alpha S''_t + (1 - \alpha)S'''_{t-1} \tag{c}$$

where S'_{t+1} is a single exponential smoothing,

S''_{t+1} is a double exponential smoothing,

S'''_{t+1} is a triple exponential smoothing.

$$a_t = 3S'_t - 3S''_t + S'''_t \tag{d}$$

$$b_t = \frac{\alpha}{2(1 - \alpha)} \left[(6 - 5\alpha)S'_t - (10 - 8\alpha)S''_t + (4 - 3\alpha)S'''_t\right] \tag{e}$$

$$c_t = \frac{\alpha^2}{(1 - \alpha)^2} (S'_t - 2S''_t + S'''_t) \tag{f}$$

$$F_{t+m} = a_t + b_t m + \frac{1}{2} C_t m^2 \tag{g}$$

Exponential Smoothing, Single (see chapter 5)

Exponential smoothing methods were developed in the beginning of the 1950s and since then have become quite popular, especially for business applications. They are used when there are many items involved, and for short- or immediate-term predictions.

The method of exponential smoothing is based on averaging (smoothing) past values of a time series in a decreasing (exponential) manner. This is achieved with formula (a) below. If expanded by substituting in previous values of F_t [see (b)], the exponentially decreasing nature of the weights given to past observations can be seen. Another way of looking at equation (a) is to express it in the form of (c), which is obtained by simple algebraic manipulation of (a). Equation (c) indicates that the forecast for the next period is equal to the forecast of the last period plus an adjustment based on the magnitude of the error. This adjustment is a multiple of the value of α. Thus, the error between the actual and the forecast value of the previous period is the basis for correcting the forecast for the next period. This is the simplest form of the control principle that is used extensively by more sophisticated forecasting methods.

$$F_{t+1} = \alpha X_t + (1 - \alpha)F_t \tag{a}$$

but

$$F_t = \alpha X_{t-1} + (1 - \alpha)F_{t-1}$$

$$F_{t-1} = \alpha X_{t-2} + (1 - \alpha)F_{t-2}$$

etc.

Thus,

$$F_{t+1} = \alpha X_t + \alpha(1 - \alpha)X_{t-1} + \alpha(1 - \alpha)^2 X_{t-2}$$

$$+ \alpha(1 - \alpha)^3 X_{t-3} \cdots \tag{b}$$

Also by expanding (a),

$$F_{t+1} = \alpha X_t + F_t - \alpha F_t,$$

or

$$F_{t+1} = F_t + \alpha(X_t - F_t)$$

$$F_{t+1} = F_t + \alpha e_{t-1} \tag{c}$$

where $X_t - F_t$ is the difference between the actual and forecasted values, the residual error, or simply error, e_t.

Exponential Smooth, Linear, and Seasonal Model (Winters' Three Parameters) (see chapter 10)

Winters' exponential smoothing extends Holt's linear exponential smoothing by including an extra equation which is used as an estimate of seasonality. The estimate of seasonality is given as an index, and is calculated with equation (a) below. The form of equation (a) is similar to those of all exponential smoothing methods. That is, a value (in this case X_t divided by S_t) is multiplied by a constant β, and then this is multiplied by $1 - \beta$ and added to its previous smoothed estimate. The factor X_t/S_t is used rather than simply X_t or S_t, so that S_t will express I_t as an index rather than in absolute terms. In addition to equation (a), Winters' smoothing uses the remaining three equations of Holt's model but introduces the seasonal index I_t into the formulas. Thus, equations (b), (c) and (d) are used to obtain estimates of the present level of the data, the trend and the forecast for some future period, $t + m$.

There is a slight difference in equation (b) and the corresponding one in Holt's model. In (b), X_t is divided by I_{t-L}. This adjusts X_t for seasonality by removing the seasonal effects from X_t. Thus, the estimate S_t excludes the seasonal effects which may exist in the data. In equation (d), a forecast is obtained that is equal to $S_t + m(T_t)I_{t-L+m}$. This is exactly the same as the corresponding formula to obtain a forecast from Holt's model. However, these estimates for the future period, $t + m$, are multiplied by I_{t-m+L}, the seasonal index. Therefore the forecast is readjusted for seasonality, introducing seasonal effects into the data. Multiplying the forecast by I_{t-L+m} has the opposite effect of dividing X_t by I_{t-L} in equation (b). Both the subscripts of I_{t-L} and I_{t-L+m} correspond to the same period.

$$I_t = \beta \frac{X_t}{S_t} + (1 - \beta)I_{t-L} \tag{a}$$

where L is the length of the seasonality.

$$S_t = \alpha \frac{X_t}{I_{t-L}} + (1 - \alpha)(S_{t-1} + T_{t-1}) \tag{b}$$

$$T_t = \gamma(S_t - S_{t-1}) + (1 - \gamma)T_{t-1} \tag{c}$$

$$F_{t+m} = (S_t + mT_t)I_{t-L+m} \tag{d}$$

Final Forecast (or Executive Forecast)

A final forecast is one that combines the quantitative predictions with those arrived at through the judgmental processes of the decision-maker.

First Difference

See *Differences*

F-Test

The F-test is a statistical test that indicates whether a regression line fits a particular set of data, that is, whether there is an overall relationship between the dependent variable and the "independent" variables that is statistically significantly different from zero.

For a situation with 40 observations and 3 variables, and a computed F-test of 5.8, one must look at a table of F-values for the intersection of:

$$3 - 1 = 2$$

$$40 - 3 = 37$$

where 3 is the number of variables,

and 40 is the number of observations.

The values 2 and 37 are called degrees of freedom for this particular F-test. Their intersection for a 95% confidence level is 3.23. Since

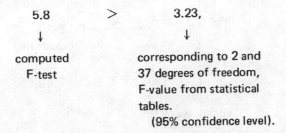

5.8   >   3.23,

↓   ↓

computed   corresponding to 2 and
F-test   37 degrees of freedom,
                F-value from statistical
                tables.
                (95% confidence level).

One would conclude (or accept the hypothesis) that the regression line is an acceptable model for describing the data and forecasting future values.

File (see chapter 29)

A file is a collection of data arranged for convenient reference. The file is stored on the computer and can be accessed at any time by referring to its name. For example, the following can be put into a file.

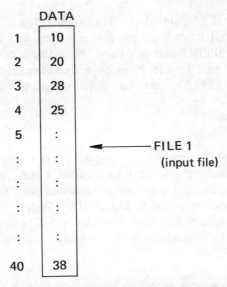

DATA

1	10
2	20
3	28
4	25
5	:
:	:
:	:
:	:
:	:
40	38

FILE 1
(input file)

The data of this "input file," which contains the past history of say, sales, can be used to forecast future values of sales.

If ten forecasts are desired, they can be put into another file, often called an "output file."

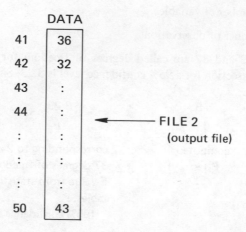

DATA

41	36
42	32
43	:
44	:
:	:
:	:
:	:
50	43

FILE 2
(output file)

Running the SIBYL/RUNNER programs requires an output file and it is suggested that the historical data be set up as an input file because it is often convenient to work with files on the computer. This is done as follows on the Hewlett-Packard time sharing system. (The commands may be different for other computer systems.)

OPEN-ANYNAM, 2 (this creates a file called ANYNAM — six characters are the maximum allowed in naming a file).

OPEN-RESULT, 6 (this opens the output file needed to run SIBYL/ RUNNER).

There are several programs that aid in manipulating files. For example, the program $FINPUT helps in getting data into a file. $FLIST aids in listing the data of a file, $FEDIT aids in editing the contents of a data file, and $UPDATE can be used to add items to a file when new data becomes available. Finally, $FPLAY can do more advanced file manipulation operations.

Foran System

The FORAN system is a method of forecasting that is based on time-series decomposition — breaking a series into seasonal, trend-cycle and random elements. Its main advantage over Census II is that the FORAN System is more business forecasting-oriented. The FORAN System is not limited to time but can deal with independent variables also. The technique is most effective when the independent variable is a leading indicator of the dependent one.

The FORAN System, in addition to decomposing a time series into seasonal, trend-cyclical, and irregular elements, provides a summary of the importance or contribution of each component. Furthermore, it provides the user with a number of forecasts together with a description of their accuracy over the last several periods and allows the user to decide which forecast or combination of forecasts to use for a final prediction. The FORAN System extends these forecasts one, two and three periods ahead, employing any selected series as the dependent variable. The user can, therefore, develop either a "causal" or "time-series" model.

Forecasting

Forecasting is the predicting of future values of a variable(s) based on historical values of the same or other variable(s). If the forecast is based simply on past values of the variable itself, it is called time-series forecasting; otherwise, it is a causal-type forecasting.

Forecasting Horizon

The forecasting horizon is the length of time into the future that the predictions cover. This length of time can be a day, a month, a week, a quarter, a year, or any other desired period. If one has monthly data and wants to forecast three months ahead, the time horizon of forecasting is three months.

The following categories are often used to summarize varying forecasting horizons.

- Immediate   Less than one month
- Short   One to three months
- Medium   Three months to two years
- Long   More than two years.

Function

The notation, $f(X)$, denotes a function of X. Thus, $Y = f(X)$ denotes that Y is a function of X, which means that as X changes, Y will change also. The purpose of forecasting is to discover the functional relationship and measure it. For example, one might write:

$$X_t = f(X_{t-1}, X_{t-2}, e_{t-1}, e_{t-2})$$

Furthermore, one might assume the relationship to be linear,

$$X_t = \phi_1 X_{t-1} + \phi_2 X_{t-2} + e_t - \theta_1 e_{t-1} - \theta_2 e_{t-2}$$

and then estimate $\phi_1, \phi_2, \theta_1, \theta_2$.

Gamma (γ)

See *Alpha*

Gamma is a smoothing constant used in Winter's linear and seasonal exponential smoothing. It varies in value between zero and one, and should be specified in such a way that it is close to one when the data are not fluctuating much, and closer to zero when there is considerable randomness.

GNP – Gross National Product

The most comprehensive measure of a nation's income, GNP, includes the total output of all goods and services. It is a key measure of the overall performance of the economy and a gauge of the health of its major sectors. It is used quite often in causal models to incorporate the extent and influence of cyclical fluctuations. It is given as a single number, such as the GNP for the United States in 1973 was $835 billion.

Generalized Adaptive Filtering

See *Adaptive Filtering*

Generalized adaptive filtering (G.A.F.) aims at extending the concept of simple adaptive filtering by applying repeated training iterations, using moving average terms, and utilizing the concept of autocorrelation.

Harmonic Smoothing (Harrison's)

A harmonic series is one that can be represented by using sine and cosine functions. Harmonic analysis is aimed at fitting some sine and consine function to a time series.

Heteroscedasticity

Heteroscedasticity is the condition where the error is not constant with respect to some variable, i.e., the variance is not the same across the entire range of available observations. When it occurs, it violates a basic assumption made by both time series and regression methods. The existence of heteroscedasticity appears graphically as below.

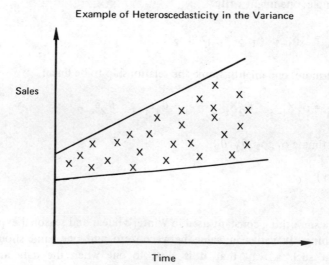

Example of Heteroscedasticity in the Variance

Heuristic

A heuristic is a method of optimization that uses a trial and error procedure. Heuristic comes from the Greek, meaning to discover or find out.

Homoscedasticity

Homoscedasticity is the condition where the error variance is the same throughout the entire range of the data. It is the opposite of heteroscedasticity. Graphically, it can be shown as follows.

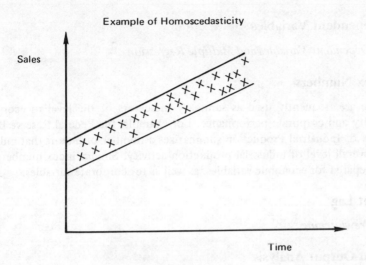

Example of Homoscedasticity

Horizontal or Stationary Data

See *Types of Data*

Hypothesis Testing

An approach commonly used in classical statistics is to formulate a hypothesis and to test the statistical significance of a set of sample data in order to determine whether that hypothesis holds true. For example, an hypothesis might be that the residuals from applying a time series method of forecasting are random. The statistical test would then be set up to determine whether those residuals behave in a pattern that makes them statistically, significantly different from zero.

Identification

Identification is a term used in econometrics to express the problems encountered in estimating the individual coefficients of systems of simultaneous equations. It is also used with ARMA models to denote the order of AR, MA or the terms of a multivariate ARMA function and delay lags.

Impulse Response Parameters

See *Multivariate Box-Jenkins Methodology*

The series of weights that appear in a multivariate ARMA model relating an independent (input) series and a dependent (output) series is referred to as the impulse response function. The parameters are simply the individual weights included in such a function.

Independent Variables

See *Dependent Variables and Multiple Regression*

Index Numbers

These are frequently used as summary indicators of the level of economic activity and corporate performance. For example, the Federal Reserve Board Index of Industrial Production summarizes a number of factors that indicate the overall level of industrial production activity. Similar index numbers can be prepared for economic variables, as well as for corporate variables.

Input Lag

See *Prewhitening*

Input-Output Analysis

Input-output analysis is an approach which deals with the problems of production and cost analysis, and is paritcularly associated with Leontief's work. His models are of a linear type and are not based on conventional methods of statistical inference but rather on accounting data.

The main purpose of input-output analysis is to show the inter-industrial structure of production and its interdependence with final consumption. Each sector produces output that may be used as input in another sector or that could be used by the ultimate consumer.

For example, a steel factory can produce utensils which are bought directly by the consumer, or it can produce a steel frame to be used by the car manufacturers.

Input Variable

See *Multivariate ARMA Model*

Interactive Forecasting

This term has been used to describe forecasting packages that are run on a time-shared computer and allow the user to interact directly with the data and with the results of alternative forecasting methods. SIBYL/RUNNER, a set of programs developed by Makridakis and Wheelwright, is one such interactive forecasting package.

Intercept A

See *Simple Regression*

In the equation $Y = a + bX$, a is the intercept, and b is the slope.

Interdependence

If two or more factors are interdependent, they can be said to be mutually dependent, or to depend on each other for their values. That is, a change in one is associated with a change in the other, and vice versa.

Intervention Analysis

Intervention analysis is a forecasting technique in the area of multivariate ARMA models that aids in determining the effects of unusual changes in the independent variable(s) on the dependent variable. The most important characteristic of this approach is that transient effects caused by such changes can be measured and their influence on the dependent variable observed.

Kalman Filters

This is an engineering approach to estimating the parameters of a function or filter. Kalman Filters are the most general approach to estimation in time series forecasting. While extremely powerful and computationally quite straightforward, they have not yet found widespread use in forecasting applications.

Lag

A lag is a length of time. As used in time series, time lag is the length between time periods. A time lag of 3 implies a difference of three time periods. An autocorrelation of 3 time lags or ρ_3, for example, indicates how values of periods 1, 2, 3, 4, ..., correlate with values of periods 4, 5, 6, 7, ..., respectively. In forecasting it is useful to relate two time series by referring to one as leading the other by one or more time periods.

Leading Indicators

The leading-indicators technique is a method of forecasting that is based on the assumption that a stable relationship exists between one economic series (the indicator) and another series (sales, GNP, prices, etc.) that is being forecast. The forecast is obtained by assuming that movements of the economic indicator will be followed by movements in the variable to be forecast.

Lead-Time

The term lead-time describes the time interval between two events. For example, if the day on which goods are ordered is the first of the month, and if the day on which they are delivered is the 24th of the same month, then the lead-time would be twenty days.

Learning Constant K

K is a parameter important in adaptive filtering. It is called the learning constant, and determines the speed of adaptation of the parameters of the ARMA process.

A high value of K (closer to one) will result in a rapid training of the parameters. However, it may overreact to changes in the data. A small K value (close to zero) will be slow to change the parameters when changes in the data occur. It will also require a large number of "training iterations" to reach the minimum squared error, resulting in additional computer costs. The behavior of the mean squared error (MSE) under different values of K is shown below. The optimal value in this example is K = 0.4, which results in

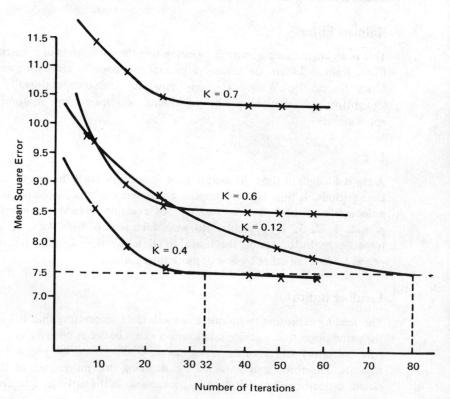

Behavior of the Mean Squared Error (MSE) Under Different Values of K

an MSE of 7.4 after 32 iterations. A smaller value of K, such as 0.12, will eventually give an MSE of 7.4, but only after a much larger number of iterations (80).

Least Squares Estimation

One approach to estimating the parameter values in an equation is to do so in a manner that will minimize the squares of the deviations that result from fitting that particular model. For example, if a trend line is being estimated to fit a data series, the method of least squares estimation could be used to minimize the mean squared error. This would give a line whose estimated values would minimize the sum of the squares of the actual deviaitons from that line for the historical data.

Linear Exponential Smoothing (see chapters 7 and 8)

Linear exponential smoothing is similar to single smoothing, except that it can be used when there is a linear trend in the data. That is, single exponential smoothing applies to stationary data only. Major forms of linear exponential smoothing models are Brown's one parameter and Holt's two parameter.

Logistic Curve

This curve has the typical S-shape form often associated with the product life cycle. It is frequently used in connection with long-term curve fitting as a technological method.

Long Difference

In order to achieve stationarity before applying the Box-Jenkins methodology to time-series forecasting, it is often necessary to take the first or second differences of the data. A long difference refers to a difference that is taken between seasonal periods that are separated by one complete season. Thus, if monthly data are used with an annual seasonal pattern, a long difference would simply compute the difference for values separated by twelve months rather than using the first difference which is for values adjacent to one another in a series.

Macro-Data

Macro-data describe the behavior of macro-economic activities such as GNP, inflation, the index of industrial producton, sales of the auto industry, etc.

Macro-Economics

Coming from the Greek work "macro" meaning large, macro-economics are economics of an aggregated level, such as nationally. Economic studies or statistics that consider large aggregates of individuals, groups of commodities, or data referring to the entire economy are referred to as macro-economics.

Marginal Analysis

Marginal Analysis is the analysis of data in an incremental fashion, similar to derivatives in calculus. Specifying changes (increases or decreases) in one variable caused by changes (increases or decreases) in some other variable(s) is a common form of marginal analysis. For example, if a unit increase in the price of an electric blender causes its demand to drop by 100 units, then 100 units is the marginal change caused in the demand by a one unit increase in price.

Marquardt's Non Linear Estimation

See *Non Linear Estimation*

Matrix

In mathematics, a matrix is a rectangular array of elements (numbers) arranged in rows or columns. For example, A is a matrix with n rows and k columns.

$$A = \begin{bmatrix} a_{11} & a_{12} & a_{13} & \cdots\cdots & a_{1k} \\ a_{21} & a_{22} & a_{23} & \cdots\cdots & a_{2k} \\ a_{31} & a_{32} & a_{33} & \cdots\cdots & a_{3k} \\ \cdot & \cdot & \cdot & \cdots\cdots & \cdot \\ \cdot & \cdot & \cdot & \cdots\cdots & \cdot \\ a_{n1} & a_{n2} & a_{n3} & \cdots\cdots & a_{nk} \end{bmatrix}$$

A vector is a special type of matrix with only one row or one column, such as V below

$$V = [a_1 \ a_2 \ a_3 \ \cdots\cdots \ a_k]$$

which is a row vector, while

$$V = \begin{bmatrix} a_1 \\ a_2 \\ a_3 \\ \cdot \\ \cdot \\ \cdot \\ a_n \end{bmatrix}$$

is a column vector.

Mean

The mean is a statistic that can be used as a method of forecasting when the pattern of the data is stationary. It is one of the simplest of all methods, and is often used without realizing that it is a form of forecasting. It involves averaging a set of historical data values and using that average as an estimate (forecast) of future values. This method is often used in market research and opinion polls. For example, before an election a survey of 1000 people might reveal that 48% of them prefer candidate A, and 52% prefer candidate B. This percentage, or mean, can be used to forecast a future event — the outcome of the election.

Mean Absolute Percentage Error (MAPE)

See *Accuracy*

The mean absolute percentage error is the average of the sum of all the percentage errors from a given data set taken without regard to sign (absolute values). An example of the calculations used in obtaining the MAPE would be:

Actual	–	Forecast	= Error	Absolute Percentage Error
10		12	−2	$\frac{2}{10} = 20\%$
100		90	10	$\frac{10}{100} = 10\%$
		TOTAL ERROR PERCENTAGE		30%
		MAPE = 30% ÷ 2 = 15% (average)		

The MAPE is a more intuitive and more easily understood measure than many alternatives such as the mean squared error.

Mean Percentage Error (MPE) (or Bias)

The mean percentage error is the average of all of the percentage errors for a given data set. This average allows positive and negative percentage errors to cancel one another. Because of this, it is sometimes used as a measure of bias in the application of a forecasting method.

Mean Squared Error (MSE)

See *Accuracy*

The mean squared error is found by squaring each individual error (Actual − Forecast) for a given data set, summing those, and finding the average. The calculations are as follows:

Actual	−	Forecast	= Error	Error^2
10		12	−2	4
110		90	10	100

TOTAL SQUARED ERROR   104

MEAN SQUARED ERROR $= \dfrac{104}{2} = 52$

Medial Average

The middle number of a data set is the median. It can be found by arranging the items in the data set in ascending order and simply identifying the middle item. The medial average includes only those items grouped around the median value. For example, the highest and lowest value may be excluded from a medial average.

Micro-Economics

Micro-Economics (from the Greek word "micro," meaning small) are economics on a small scale, often at a company level. The term refers to economic studies or statistics that consider particular individuals, single commodities, or specific organizations. For example, the demand for shoes, the sales of company XYZ, the assets of organization A, or the income of person B are micro-data.

Mixed Processes

In time-series analysis those processes or models that combine moving average forms with autoregressive forms are frequently referred to as mixed processes.

Model

A model is a symbolic representation of reality. For example, the equation Y = a + b (GNP) where Y = sales, is a symbolic representation of reality that implies that sales are influenced by (or are a function of) GNP.

Months For Cyclical Dominance (MCD) (see chapter 16)

The months for cyclical dominance is the ratio between the average percentage change in the random component over the average percentage change in the trend-cycle component. It indicates how many months it takes for the trend-cycle to dominate the random component. The months for cyclical dominance is a very important factor in understanding the magnitude of the cyclical component and its influence on the data. It is used in the Census II method of forecasting.

Moving Average (Single) (see chapter 11)

The single moving average consists of taking a set of observed values, finding their average, and then using that average either as the forecast for the coming period or as a way of detrending the data (making them stationary). The number of observations included in the average must be specified by the user. The term moving or rolling average is used because as each new observation becomes available, a new average is computed. This involves dropping the last observation in the average each time a new one is added. Thus, the number of data points in the series is always constant and always includes the most recent observations.

If a moving average is used for forecasting and if there is considerable randomness in the historical observations, a large number of observations should be used in computing the moving average. Conversely, if there is little randomness in the underlying data, a smaller number of observations can be used in computing the moving average.

An example of a 5-period moving average is shown below.

Five-Period Moving Average

Period	Data	5-Month Moving Average
1	4	-
2	6	5.4
3	5	5.8
4	4	5.8
5	8	5.8
6	6	6.4
7	6	7.6
8	8	7.6
9	10	7.8
10	8	7.8
11	7	7.6
12	6	6.6
13	7	-
14	5	-

As can be seen, two observations are lost in the beginning of the data, and two at the end because of the averaging. (In general, m - 1 observations are lost where m is the length of the moving average.)

When the aim is forecasting, instead of centering the moving average, the most recent value obtained is used to predict the next period. Thus, in the above example, at period 5 the value 5.4 would be used as a forecast for period 6.

Moving Averages (Multiple) (see chapter 4)

See *Moving Averages (Single)*

Multicollinearity

Multicollinearity exists when two or more independent variables are highly related to each other. This high correlation between independent variables creates a computational problem in multiple regression, making the results of regression of limited usefulness. Multicollinearity can be corrected by eliminating all but one of the highly correlated independent variables from the regression model.

Multiple Correlation

See *Correlation Coefficient and R-Squared*

Multiple Regression (see chapter 19)

The technique of multiple regression allows consideration of any number of independent variables in preparing a forecast. (Simple Regression can use only one independent variable.) For example, if sales are to be forecast, it may be that several factors — such as advertising, prices, competition, budget of R & D, GNP — in addition to time, influence the level of sales. In such circumstances, the technique of Simple Regression may not be appropriate and Multiple Regression may be a much better alternative.

As an illustration of multiple regression, one can consider the equation,

$$Y = a + b_1 X_1 + b_2 X_2$$

where Y = annual sales, X_1 = annual production, and X_2 = annual contracts. Y is the dependent variable to be forecast, and X_1 and X_2 are two independent variables that influence Y.

Multiple regression is a powerful and practical forecasting method and an alternative to time-series forecasting. Regression equations are often called causal or explanatory models.

Multivariate ARMA Models or Processes

See *Multivariate Box-Jenkins Methodology and Multivariate Generalized Adaptive Filtering*

Univariate ARMA models or processes express future forecasts as a linear combination of past values of a single variable or past errors. The general form for this is

$$X_t = \phi_1 X_{t-1} + \phi_2 X_{t-2} + \ldots + \phi_p X_{t-p} + e_t - \theta_1 e_{t-1}$$
$$- \theta_2 e_{t-2} - \ldots - \theta_q e_{t-q}$$

Multivariate ARMA models, on the other hand, express future forecasts as a function of both past values and past errors of the variable to be forecast (dependent) and also as a function of past values of other variables which may lead the dependent variable. For example, if Y_t is the dependent variable (often called the output variable, a name derived from control engineering) and X is the independent (input) variable (assuming there is only one independent variable), then the Multivariate ARMA model will be

$$Y_t - \delta_1 Y_{t-1} - \delta_2 Y_{t-2} - \ldots - \delta_r Y_{t-r} = \omega_0 X_{t-b} - \omega_1 X_{t-b-1} - \ldots - \omega_s X_{t-b-s}$$

where r is the number of terms of past values of the dependent variable, s is the number of past terms minus one of the independent variables, and b is the delay lag between the dependent and the independent variables, i.e., the number of periods that the dependent variable leads the independent variable.

Multivariate Box-Jenkins Methodology (see chapter 27)

See *Multivariate ARMA Models or Processes*

The multivariate Box-Jenkins methodology extends the concept of univariate ARMA models (involving just one variable) to include more than one variable. At present, most of the work has been restricted to two variables (bivariate) models, but there is no reason why the concept cannot be extended to include more than two variables. The multivariate (bivariate) Box-Jenkins methodology consists of (a) the identification of an appropriate model among a general class of available models (achieved by first prewhitening the independent and dependent variables and then calculating their cross autocorrelations and impulse response parameters from which an appropriate model can be identified), (b) estimating its parameters through a non-linear estimation procedure, (c) specifying some noise parameters over and above the regular multivariate ARMA ones, and (d) checking to make sure that the model employed is an adequate one.

The multivariate Box-Jenkins methodology refers to the multivariate models as "transfer functions," a term derived from engineering applications.

Multivariate Generalized Adaptive Filtering (see chapter 26)

See *Multivariate ARMA Models*

Multivariate generalized adaptive filtering extends the concept of univariate Adaptive Filtering (involving a single variable) to include more than one variable. The method consists of identifying an appropriate ARMA model, estimating its parameters through the estimating procedure of the method of steepest descent and then making a diagnostic check to determine whether the model selected is adequate. The method of multivariate adaptive filtering differs from the corresponding one of Box-Jenkins in that the estimated parameters vary in value at each time period rather than being fixed throughout the entire time horizon.

"n"

A small "n" is used to express the number of observations used to estimate a model. In statistical terms, "n" is the sample size, total number of observations, or number of data points.

"N"

The number of periods used in the calculation of a moving average is expressed as "N". N is always a positive number greater than zero. N is also used to denote the total size of the population.

Naive Forecasts (see chapter 2)

A naive forecast can be obtained with a minimal amount of manipulation of recent information. The kind of adjustment depends on the type of data. For example, with seasonal data one can use last year's (same month) value as a naive forecast for a particular month this year.

Naive Method 1 (see chapter 2)

Naive 1 uses the latest available information as a forecast for the future. Thus, if X_t is the value of sales for the current period, the future forecast(s) will be:

$$F_{t+1} = X_t$$

$$F_{t+2} = X_t$$

$$F_{t+3} = X_t$$

. .
. .
. .

$$F_{t+k} = X_t$$

Naive Method 2 (see chapter 2)

Naive 2 is similar to Naive 1, except that it first adjusts the data for seasonality. This is done by dividing the value, X_t, by the corresponding seasonal index, $N(J)$, to deseasonalize it and then seasonalize it to the appropriate period. The forecast(s) then become(s):

$$F_{t+1} = X_t' \cdot N(J + 1)$$

$$F_{t+2} = X_t' \cdot N(J + 2)$$

.   .
.   .
.   .

$$F_{t+k} = X_t' \cdot N(J + k)$$

where $X_t' = X_t/N(.1)$

Noise

The randomness often found to exist in data series is frequently referred to as noise. This term comes from engineering, where the purpose of a filter is to eliminate noise so that the true pattern can be identified.

Nonlinear Estimation

Nonlinear estimation procedures determine the parameter values of a given function in a nonlinear fashion. As with linear estimation, this proceedure generally involves determining those parameter values that minimize the mean squared error. A frequently used algorithm for nonlinear estimation is that developed by Marquardt (1963).

Nonstationary

A time series that contains a trend is frequently referred to as being non-stationary.

Observation

An observation is a quantified event as expressed in some measurement scale by a data point. Any number of observations referring to any number of factors or variables is possible, as illustrated below.

Observed Data Values

TIME - Variable 1	*SALES - Variable 2*	*GNP - Variable 3*
1 January 1965	10.2	400
2 February 1965	8.9	398
3 March 1965	12.8	360
4 April 1965	4.8	370
5 May 1965	6.3	375
6 June 1965	7.5	390
7 July 1965	8.6	400
8 August 1965	9.8	390
9 September 1965	10.2	410
10 October 1965	11.6	400
11 November 1965	9.5	440
12 December 1965	8.6	445
.	.	.
.	.	.
.	.	.
40 April 1968	13.4	525

This set of data includes forty observations on three variables — Time, Sales, and GNP. The totality of observations for all the variables is called a data set and can be used as the input to quantitative forecasting.

Optimal Weights (Or Parameters)

See *Adaptive Filtering*

Optimal Weights are the final weights used in Adaptive Filtering.

Outliers

An outlier is a data value which is unusually high or low. Such a value can be caused by a strike, a war, a total breakdown, etc. It is often removed from the data set because the unusual reasons which gave rise to its occurrence are not expected to be repeated in the future.

Output Lag

See *Prewhitening*

Output Variable

See *Multivariate ARMA Models*

Parameter

A parameter is a constant describing some population characteristics, or it can be the coefficient of a variable in an equation. For example, in the equation

$$Y = a + b_1 X_1 + b_2 X_2$$

a, b_1, and b_2 are parameters. The values of these parameters are estimated in regression by the classical Linear Least Square Method.

Parsimony

The concept of parsimony is to minimize the number of parameters used in fitting a model to a set of data. This concept is a basic underlying premise of the Box-Jenkins methodology for time-series analysis.

Partial Autocorrelation

This measure of correlation is used to identify the extent of the relationship between current values of a variable with earlier values of that same variable (values for various time lags), while holding the effects of all other time lags constant. Thus, it is completely analogous to partial correlation but refers to a single variable.

Partial Correlation

See *Correlation*

Partial Correlation is a measure of association between a dependent variable and one or more independent variables, when the effect of the remaining independent variables is held constant.

Pattern

The basic set of relationships or the underlying process over time is referred to as the pattern in the data.

Period

Virtually all approaches to time-series forecasting deal with the value of some variable for a defined period of time. In contrast, a technique such as spectral-analysis deals with the frequency with which an event occurs as opposed to the quantity of that event in a specific time period.

Polynomial

A polynomial is an algebraic expression containing two or more terms. For example, the equation

$$Y_t = \zeta_1 Y_{t-1} + \zeta_1^2 Y_{t-2} + \zeta_1^3 Y_{t-3} + \cdots$$

is a polynomial.

$Y = a + bX + cX^2$ is another polynomial (a quadratic expression).

Polynomial Fitting

It is possible to fit a polynomial of any number of terms to a set of data. If the number of terms (the order) equals the number of data observations, the fit can be made perfectly.

Prewhitened Series

See *Prewhitening*

Prewhitening (see chapter 24)

In multivariate ARMA models it is important to know the type of model to be fitted to the data. The process of finding an appropriate model, known as identification, is aimed at determining:

1. The number of terms of past values of the dependent variable to be included in the model. This is usually denoted by r, and is also called the number of output lags.
2. The number of terms of past values of the independent variable to be included in the model. This is usually denoted by s and is also called the number of input lags.
3. The time delay, denoted by b, or number of periods before a change in the independent variable will influence the dependent variable. This is also called the delay lag.
4. The noise parameters p and q which are similar to the order of the AR(p) and MA(q) processes of univariate ARMA models. These are whatever is left after a multivariate model has been fitted to the data.

To identify 1, 2, 3 and 4 above, both the dependent and independent variables must be prewhitened. Prewhitening is done by fitting a univariate ARMA model to the independent variable, estimating its parameters, and then calculating the residuals of the model. If the residuals are randomly distributed, this means that the chosen model is an appropriate one, and that from an estimating point of view the residuals have been reduced to white noise. (That is, they exhibit no pattern; what remains is completely random.) This reduction of the independent variable to white noise by the application of a univariate ARMA model is called prewhitening.

The same process (ARMA model) can be applied to the dependent variable using the parameters estimated in fitting the univariate ARMA model to the independent variable. This amounts to prewhitening the dependent variable, although it may not be reduced completely to white noise. It is the prewhitened residual errors of the independent and dependent variables that are used in identifying 1, 2, 3 and 4 above. Unless the series are prewhitened, the type of multivariate ARMA model to be fitted to the data cannot be identified.

Probability

The language of probability has been developed as a basis for describing uncertainty and explicitly accounting for the likelihood of various outcomes. Frequently, the probability of occurrence of a given event is developed based on the frequency with which that event has occurred in the past.

Product Life Cycle

A concept that has been found to be particularly useful in forecasting and analyzing historical data is that of the product life cycle. It presumes that demand for a product follows an S-shaped curve, where demand in the early stages grows slowly, in the middle stages it enters a period of rapid and sustained growth, and then growth slows in the mature stage.

Quadratic Exponential Smoothing

See *Exponential Smoothing, Quadratic*

Quantitiative Forecasting

Quantitative Forecasting is one of two major categories of forecasting techniques (the other being qualitative or technological). It can be applied when three conditions exist:

1. There is information about the past;
2. That information can be quantified (implying the existence of a measuring scale);
3. The pattern of past information can be assumed to continue into the future.

This last factor, the assumption of constancy, is the basis of all forms of quantitative forecasting.

Qualitative or Technological Forecasting

Qualitative forecasting and quantitative forecasting are the major categories of forecasting techniques. Qualitative methods are most appropriate when one of the following conditions exists:

1. The assumption of constancy is invalid;
2. When information about the past cannot be obtained;
3. When the forecast is about unforeseen, unlikely or unexpected events of the future.

Qualitative forecasting is unstructured and demands a great deal of personal inputs (expert judgments).

Random Sampling

Random Sampling is a sampling method used in Statistics. A sample is selected from a statistical population in such a way that every unit, or combination of units, within that population has the same probability of being selected as any other unit in the sample.

Randomness

The noise or random fluctuations in a data series are frequently described as the randomness of that data series.

Rate of Adaptation

See *K - Learning Constant*

Ratio

Ratio is the proportional relationship between two values and is obtained by dividing one value by the other. For example, 5 to 2 is a ratio written as 5 : 2 or 5/2.

Recession

See *Business Cycle, Depression*

A recession is a mild depression.

Regression Line

A regression line describes the average relationship between a dependent variable (such as sales) and one or more independent variables (such as price, GNP, advertising, etc.). The coefficients of such a line are usually estimated by the Least Squares Method. For example, Sales = 3.5 + 2.5 GNP − 1.8 Price + 4.2 Advertising is a regression equation.

Regression Coefficient

See *Parameters*

Regression Equation

The form for a regression equation is:

$$Y = a + bX_1 + b_2X_2 + b_3X_3 + \ldots + b_mX_m$$

where $Y, X_1, X_2, X_3, \ldots, X_m$ are variables, and $a, b_1, b_2, b_3, \ldots, b_m$ are parameters.

Different values of parameters result in different equations. Using the single regression equation, $Y = a + bX$, for example, one can obtain the regression equation for any of the lines shown below as a and b change their value.

A regression equation is a specific form of linear algebraic equations or functions. In order to understand the implications of linearity more fully, consider the following example. In the following data the pairs of figures under Years and Sales (each a variable) can be expressed algebraically.

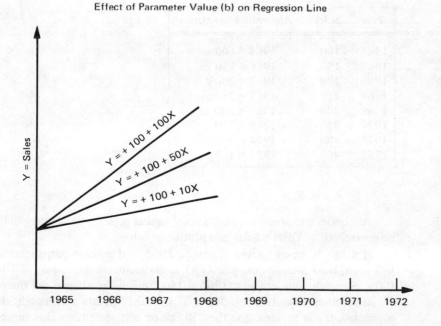

Effect of Parameter Value (b) on Regression Line

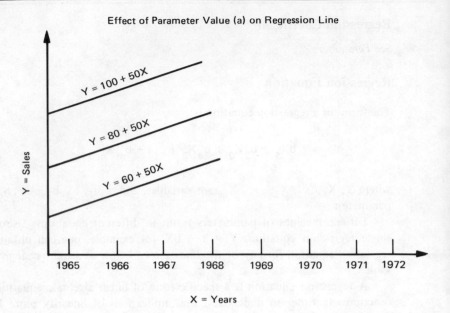

Effect of Parameter Value (a) on Regression Line

Example of Data for Simple Regression

Year	Sales	Algebraic Expression
1965	100	1965 = 100 c
1966	150	1966 = 150 c
1967	200	1967 = 200 c
1968	250	1968 = 250 c
1969	300	1969 = 300 c
1970	350	1970 = 350 c
1971	400	1971 = 400 c
1972	450	1972 = 450 c

As algebraic expressions, each pair of figures presents an identity that can be generalized as Years = Sales and plotted as below.

If a line is drawn perpendicular to 1965, and another perpendicular to 100 units, their intersection, point (A), is the geometric expression of 1965 = 100 c. Similarly, one can plot 1966 = 150 c on the graph, and call the point of intersection (B), 1967 = 200 c (C), etc. If points A through H are connected, it can be seen that they all fall on a straight line. This suggests a form of relationship between years and sales such that if the year is known, the level of sales can be predicted.

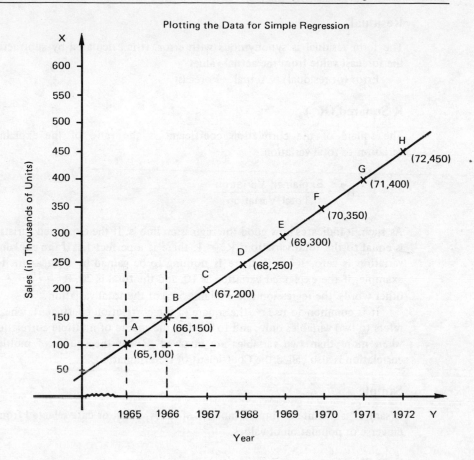

Plotting the Data for Simple Regression

In algebraic terms, this can be written as

$$\text{Sales} = f(\text{Years}),$$

indicating that sales are a function of the year (or time).

One can be more specific and say that the relationship is linear,

$$\text{Sales} = a + bX$$

where X represents years, and a and b are the parameters. (See Least Square Estimates.) If the values of a and b can be estimated, a specific relationship expressing sales as a function of time will have been determined.

Based on the above analysis, one can predict that sales will increase 50 units every year. Thus sales for 1973 will be 500 units and for 1980 they will be 850 units, assuming the relationship continues to hold true.

Residual

The term residual is synonymous with error. It is calculated by subtracting the forecast value from the actual value.

Error (or residual) = Actual – Forecast.

R-Squared (R^2)

The square of the correlation coefficient is the ratio of the explained variation to total variation.

$$R^2 = \frac{\text{Explained Variation}}{\text{Total Variation}}$$

As such, it indicates how good the regression line is. If the explained variation is equal to the total variation, $R^2 = 1$, there is a perfect fit. If the explained variation is zero, $R^2 = 0$, there is nothing to be gained by a linear fit. For example, if the explained variation is 10, and the total is 20, $R^2 = \frac{10}{20} = .5$. In other words, the regression line explains 50% of the total variation.

It is common to use r^2, the square of the correlation coefficient, when it refers to two variables only, and to use R^2, the square of multiple correlation, when more than two variables are involved. R^2 or the square of multiple correlation is also called the Coefficient of Determination.

Sample

A sample is a finite or limited number of items, units, or data selected from a universe or population of values.

Sampling Error

The sampling error is an indication of the magnitude of the difference between the real value of a population parameter and the estimation of that value through some statistic. Sampling error is a special case of what is generally called error — the difference between actual and predicted values.

S-Curve (see chapter 13)

An S-Curve represents reasonably well the life cycle of certain products. As shown below, it has a slow start, a steep growth, and a saturation level that is achieved after some time. One equation for calculating an S-curve is given by

$$X_t = e^{a - \frac{b}{t}}$$

Seasonal

See *Types of Data*

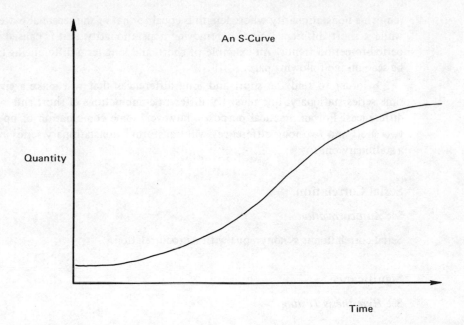

An S-Curve

Quantity

Time

Seasonal Exponential Smoothing

See *Exponential Smoothing, Seasonal*

Seasonal Index

A Seasonal Index is a number, usually with a base around 100, that indicates the seasonality for a given period (e.g., month) in relation to other periods. If, for example, the seasonal index for July is 100, it indicates that there are no influences due to seasonal factors during July. If it is 110 for August, it indicates that, on average, 10% of the value of, say, sales in August is due to seasonal factors. An index below 100, for example 90, implies that the effect of seasonality is negative. In the last case, it is 10% below the average.

Seasonal Variation

See *Percentage Change*

Seasonal Variation is the change that occurs in a series of data because of seasonal factors (e.g., months or seasons of the year, vacations, etc.).

Short Difference (see Chapters 3 and 18)

See *Differencing*

In the Box and Jenkins methodology, when using seasonal models one must distinguish between short and long differences. A long difference aims at removing nonstationarity whose length is equal to that of the seasonal pattern while a short difference aims at removing nonstationarity that is caused by period-to-period trends. An example of short- and long-term differencing can be seen on the following page.

In order to find the short and long differences that will make a given time series stationary, one must try different combinations of short and long differences. For all practical purposes, however, some combination of up to two short and two long differences will transform a nonstationary series into a stationary one.

Serial Correlation

See *Autocorrelation*

Serial correlation is synonymous with Autocorrelation.

Significance

See *Hypothesis Testing*

Simple Regression (see chapter 12)

Simple regression is a special case of regression when there is only a single independent variable. In simple regression, as in multiple, one assumes that a linear relationship exists and attempts to estimate it using the method of least squares. Simple regression is not just limited to a time-series relationship, but can estimate the relationship between any two variables and base the forecast of one variable (dependent) on the other variable (independent). Thus it can serve as a causal model.

The starting point in using simple regression is the assumption that some relationship exists between two variables, and that it is linear. Mathematically, such a relationship is written as

$$Y = f(X)$$

This says simply that the value of Y is a function of the value of X. When this is assumed to be a straight-line relationship, it can be written as:

$$Y = a + bX$$

Slope

The slope of a curve at a given point can be found by dividing the average change in the dependent variable (Y axis) by the average change in the

Example of Short and Long Differencing

Period	Data	SHORT-TERM DIFFERENCING		LONG-TERM DIFFERENCING*	
		First	Second	First	Second
1	10	12 − 10 = 2	3 − 2 = 1	20 − 10 = 10**	10 − 10 = 0
2	12	15 − 12 = 3	−1 − 3 = −4	30 − 12 = 18	−5 − 18 = 23
3	15	14 − 15 = −1	6 − (−1) = 7	25 − 15 = 10	10 − 10 = 0
4	14	20 − 14 = 6	10 − 6 = 4	40 − 14 = 26	0 − 26 = −26
5	20	30 − 20 = 10	−5 − 10 = −15	30 − 20 = 10	20 − 10 = 10
6	30	25 − 30 = −5	15 − (−5) = 20	25 − 30 = −5	30 − (−5) = 35
7	25	40 − 25 = 15	−10 − 15 = −25	35 − 25 = 10	25 − 10 = 15
8	40	30 − 40 = −10	−5 − (−10) = 5	40 − 40 = 0	15 − 0 = 15
9	30	25 − 30 = −5	10 − (−5) = 15	50 − 30 = 20	
10	25	35 − 25 = 10	5 − 10 = −5	55 − 25 = 30	
11	35	40 − 35 = 5	10 − 5 = 5	60 − 35 = 25	
12	40	50 − 40 = 10	5 − 10 = −5	55 − 50 = 5	
13	50	55 − 50 = 5	5 − 5 = 0		
14	55	60 − 55 = 5	5 − (−5) = 10		
15	60	55 − 60 = −5			
16	55				

* Assumes that the seasonal pattern is 4 periods long.

** For example, 20 is the datum corresponding to period 5 while 10 corresponds to period 1. The long difference is 4 periods or equal to the length of seasonality.

independent variable (X axis). For a straight line, the slope is the same at all points, i.e., it is constant. The slope is synonymous with the first derivative.

$$\text{Slope} = \frac{dY}{dX}$$

Slope b

See *Simple Regression*

In the equation $Y = a + bX$, b is the slope.

Smoothing

See *Exponential Smoothing*

The term to smooth means to average according to some rule in order to eliminate fluctuations in the data.

Smoothing Constant

See *Alpha, Beta and Gamma*

Spectral Analysis

This approach to data analysis is based on analyzing the frequency with which certain outcomes are observed, rather than measuring the magnitude of an outcome as summarized for a specified period of time.

Speed of Adaptation

See *K - Learning Constant*

Spencer's Weighted Moving Averages

Spencer's Weighted Moving Averages (formulas) are used in removing up to cubic trend from a set of data. There are two averages — one whose length is fifteen periods, and the other whose length is twenty-one periods. Both are widely used, particularly in CENSUS II.

Standard Deviation

The Standard Deviation is symbolized by the small Greek letter, sigma (σ). It is the square root of the variance and gives another measure of the dispersion of all values around the mean.

The formula for estimating the standard deviation is

$$S = \sqrt{\frac{\sum_{i=1}^{n} (X_i - \overline{X})^2}{n - 1}}$$

Standard Error

See *Central Limit Theorem*

The term standard error is applied to various statistical measures (such as the mean), and gives the distribution within which that measure will fall when samples are drawn from any population or distribution. The standard error is dependent on the dispersion in the population and the size of the sample. The standard error for the arithmetic means is calculated as

$$\text{Standard error} = \frac{\text{Standard deviation of the population}}{\sqrt{\quad \text{Number of units in the sample}}}$$

The standard error of an estimate can be calculated by dividing the standard deviation of the population from which the estimate comes, by the square root of the sample size, (n). In general, this can be written as:

$$\sigma_{\overline{x}} = \frac{\sigma}{\sqrt{n}} = \frac{S}{\sqrt{n}}$$

Stationary

See *Types of Data*

A *Stationary Time Series* is one that oscillates around a constant mean. It is a series that shows no growth or decline over time. Deviations around the mean are temporary, and in the longer term, oscillations are in equilibrium around the mean. A typical stationary series as well as a nonstationary one is shown below.

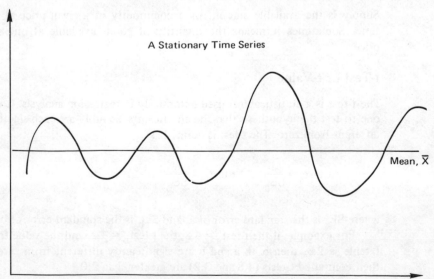

A Stationary Time Series

Mean, $\overline{X}$

Time

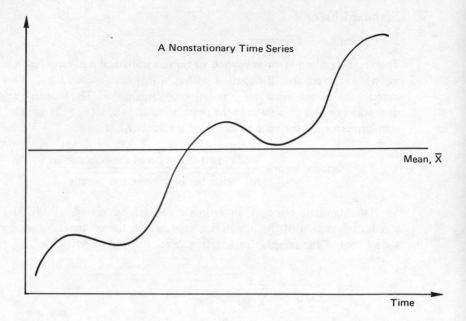

Statistic

Any summary measure that is used to describe a set of data or a population of values is referred to as a statistic. The mean, standard deviation, and variance are the most commonly used statistics.

Supply

Supply is the available amount of a commodity at a given price at a given time. Sometimes it means the quantity of goods available at one specific price.

t-Test or t-Value

The t-test is a statistical test used extensively in regression analysis. It enables one to test the hypothesis that the coefficients, a and b, are each significantly different from zero. The t-test is defined as:

$$\text{t-test}_a = \frac{a}{SE_a} \qquad \text{t-test}_b = \frac{b}{SE_b}$$

where SE_a is the standard error of a, and SE_b is the standard error of b.

For example, if the t-test$_a$ = 4.8, the t-test$_b$ = 3.2, and the value from the t table is 2.0, then both a and b are significantly different from zero, since their computed t-tests (4.8 and 3.2) are greater than 2.0.

In cases where this is not so, one can accept the hypothesis that their values will be zero. If, for example, the value of t-test$_a$ = 1.3, one can assume that a = 0. Thus, the regression equation would be:

$$Y = bX$$

Technological Forecasting

See *Qualitative Forecasting*

Time Horizon

See *Forecasting Horizon*

Time-Series Model

This is one of two important types of models available. It predicts the future by expressing it as a function of the past [e.g., Sales = Function (time)]. The objective of this method is to discover the pattern into the future.

The time-series model is easier to develop than its alternative, the causal model, but often not as useful for policy and decision-making situations as the latter.

A time-series model can be of the form:

$$Y = a + bX$$

where X will always be time. For example, simple regression might give as values of the parameters a and b, a = 30, b = 2.5. The model thus becomes:

$$Y = 30 + 2.5 \ X$$

To estimate sales for 1980 or (80), using this model,

$$Sales_{80} = 30 + 2.5 \ (80)$$

$$= 30 + 200$$

$$= 230 \ (million \ dollars)$$

Thus, estimating future sales is a rather trivial matter once it is known that a = 30 and b = 2.5. This is not true for causal models, where estimates of the values for the independent variables must be made before forecasting.

Tracking Signal

Since quantitative methods of forecasting assume the continuation of some historical pattern into the future, it is often useful to develop some measure that can be used to determine when the basic pattern has, in fact, changed. A tracking signal is the most common such measure. It is used simply to indicate to the forecaster when the basic pattern has been altered. A particularly common tracking signal is to compute the cumulative error over time and to set limits so that when the cumulative error goes outside of those limits, the forecaster is notified and a new model can be considered.

Trading Days

In many business time series the number of business days included in a month or some other specified period of time may vary. It is frequently necessary to make trading day adjustments to reflect the fact that every January may not include the same number of trading days.

Transfer Functions

See *Multivariate ARMA Models*

Transformations

It is frequently possible to transform an independent variable into its log or some exponent so that it will be related to the dependent variable in a linear fashion. This enables regression analysis to handle a much wider range of problems than would otherwise be the case.

Trend

See *Types of Data, Stationary* and *Nonstationary Times Series,* and *Differencing*

Trend Analysis

See *Simple Regression*

Trend Analysis is a special form of simple regression, where time is the independent variable. It is expressed as:

$$X_t = a + bt$$

Types of Data

See *Stationary* and *Nonstationary Time Series*

1. *Stationary Pattern*

A stationary pattern can be identified when the data are fairly evenly distributed over time. There is no growth or decline; even with a shift in the time origin, the pattern remains the same. A product whose sales do not increase or decrease over time falls into this category. An example is shown below.

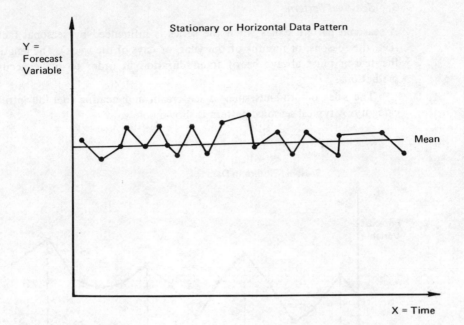

Stationary or Horizontal Data Pattern

2. *Trend Pattern*

A trend pattern exists when there is either growth or decline in the data over the relevant time span. Data with a trend pattern are referred to as nonstationary. The Gross National Product (GNP), stock prices, and other business and economic indicators follow a trend pattern in their long-run movement through time, as shown below.

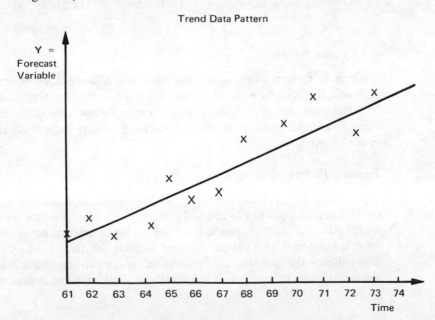

Trend Data Pattern

3. *Seasonal Pattern*

A seasonal pattern exists when the series is influenced by seasonal factors (i.e., the seasons, or months of the year, or days of the week). The length of the season must always be of fixed duration in order to make accurate predictions.

The sales of products such as ice cream and heating fuel fall into this category. A typical seasonal pattern is shown below.

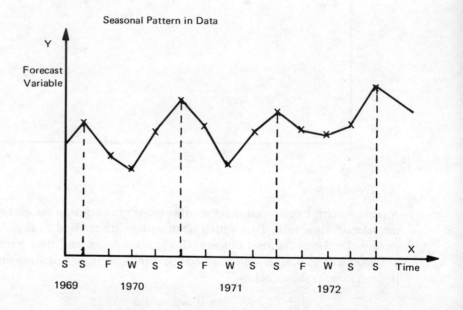

4. *Cyclical Pattern*

A Cyclical Pattern exists when the data are influenced by longer term economic fluctuations which coincide with movements of the business cycle.

Often, periods of prosperity and recession follow each other, and the sales of commodities such as cars, steel and furniture, are particularly affected by such movements.

Turning Points

Any time a data pattern changes direction it can be described as having reached a turning point. For seasonal patterns these turning points are usually predictable and can be handled by many different forecasting methods because the length of a complete season remains constant. In many cyclical data patterns the length of the cycle varies, as does its magnitude. Here the identification of turning points is a particularly difficult and important task.

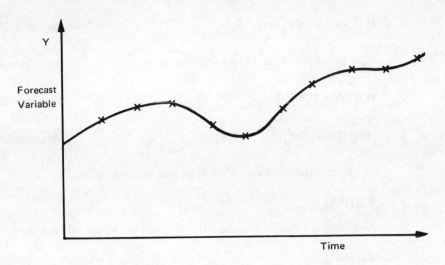

Cyclical Pattern in Data

Type of Model (see chapter 18)

In some seasonal Box-Jenkins programs, one has to specify the type of model one will like to use. The choice will generally include the following 14 models.

(1) $Y_t = (1 - \theta_1 B - \theta_2 B^2)(1 - \theta_s^* B^3)e_t$ *MA(2) seasonal in MA*

(2) $(1 - \phi_1 B)Y_t = (1 - \theta_1 B)(1 - \theta_s^* B^s)e_t$ *ARMA(1,1) seasonal in MA*

(3) $(1 - \phi_1 B - \phi_2 B^2)Y_t = (1 - \theta_s^* B^s)e_t$ *AR(2) seasonal in MA*

(4) $(1 - \phi_s^* B^s)Y_t = (1 - \theta_1 B - \theta_2 B^2)e_t$ *MA(2) seasonal in AR*

(5) $(1 - \phi_1 B)(1 - \phi_s^* B^s)Y_t = (1 - \theta_1 B)e_t$ *ARMA(1,1) seasonal in AR*

(6) $Y_t = (1 - \theta_1 B)(1 - \theta_s^* B^s)e_t$ *MA(1) seasonal in MA*

(7) $(1 - \phi_1 B)Y_t = (1 - \theta_s^* B^s)e_t$ *AR(1) seasonal in MA*

(8) $(1 - \phi_s^* B^s)Y_t = (1 - \theta_1 B)e_t$ *MA(1) seasonal in AR*

(9) $(1 - \phi_1 B)(1 - \phi_s^* B^s)Y_t = e_t$ *AR(1) seasonal in AR*

(10) $Y_t = (1 - \theta_1 B - \theta_2 B^2)e_t$ *Nonseasonal MA(2)*

(11) $(1 - \phi_1 B)Y_t = (1 - \theta_1)e_t$ *Nonseasonal ARMA(1,1)*

(12) $(1 - \phi_1 B - \phi_2 B^2)Y_t = e_t$ *Nonseasonal AR(2)*

(13) $Y_t = (1 - \theta_1 B)e_t$ *Nonseasonal MA(1)*

(14) $(1 - \phi_1 B)Y_t = e_t$ *Nonseasonal AR(1)*

Many other types of models can also be constructed.

Variable

A variable is a measurable quantity or function, whose value changes.

Variance

The variance is a measure of the distribution of all population values around the mean value. (The mean is the expected value of all values.) The formula for finding the variance, (σ^2), is:

$$\sigma^2 = \frac{\sum_{i=1}^{n}(X_i - \bar{X})^2}{n - 1}$$

where X_i is the ith data point, and $\bar{X}$ is the mean.

By knowing the mean and the variance of a distribution, one can estimate the probability that a given value, or ranges of values, will be observed.

Weight (or Parameter)

See *Parameter*

The term weight, used in a statistical sense, means the relative importance given to items of the data set in forecasting. For example, in a set of given values

$$3 \quad 4 \quad 5 \quad 7 \quad 8 \quad 6$$

one can weight the four most recent values equally:

$$1/4(5) + 1/4(7) + 1/4(8) + 1/4(6) = 6.5$$

This type of weighting scheme is employed when the moving average method is used.

One can also have a system of decreasing weights that assign decreasing importance to the older values.

$$3 \qquad 4 \qquad 5 \qquad 7 \qquad 8 \qquad 6$$

$$1/64(3) + 1/32(4) + 1/16(5) + 1/8(7) + 1/4(8) + 1/2(6) = 5.6$$

Thus, 5.6 is the forecast for the next period. Decreasing weights are employed in the exponential smoothing methods.

There are many other weighting schemes that vary considerably in complexity, sophistication, and effectiveness.

White Noise

When there is no pattern whatsoever contained in the data series, it is said to represent white noise. This is analogous to a series that is completely random.

Winters' Exponential Smoothing

See *Exponential Smoothing, Winters*

Index